INTERACTION OF RADIATION WITH SOLIDS

INTERACTION OF RADIATION WITH SOLIDS

Edited by
Adli Bishay
Chairman, Department of Physical Sciences
The American University in Cairo

Proceedings of the Cairo Solid State Conference at The American University in Cairo, held September 3-8, 1966

Springer Science+Business Media, LLC 1967

Library of Congress Catalog Card Number 67-19567

Originally published by Plenum Press in 1967.
Softcover reprint of the hardcover 1st edition 1967

ISBN 978-1-4899-5551-7 ISBN 978-1-4899-5549-4 (eBook)
DOI 10.1007/978-1-4899-5549-4

CAIRO SOLID STATE CONFERENCE

A dream which started in Brussels in the summer of 1965 was realized in Cairo in September 1966. The dedication of the New Science Building of the American University in Cairo was coupled with an International Conference at which 16 nations were represented.

"Interaction of Radiation with Solids" was the theme of the Conference. Effects of particle and electromagnetic radiation, of high or low energy, on both crystalline and non-crystalline solids were discussed.

The successful completion of this Conference would not have been possible without the initial "go ahead" and the continuous encouragement of Dr. Thomas Bartlett, President of the American University in Cairo, the full cooperation of the U.A.R. Supreme Council for Scientific Research, and the hard work and great enthusiasm of Mary Megalli and Waguih Atalla of the Department of Physical Sciences.

It was a pleasure to find that all authors submitted their manuscripts on time. However, some modifications and shortening of papers was found necessary in order to achieve some uniformity and to keep the volume to a reasonable length. In order to publish the book as early as possible it was decided not to send the final modified manuscripts to the authors before printing. My sincere apologies to those of you who may not be happy with the result.

One young lady was mainly responsible for typing the final version of this book: to my secretary, Lilian Mikhail, my congratulations and sincere appreciation. Thanks are also due to Mrs. Roda Amine who has helped with proof reading, Shafik Hanna and Kamal Ragy who typed the draft of the manuscript, Mr. Ahmed Sherif El Tony who was responsible for illustrations, and Mr. Richard Stegman for his full cooperation.

To the staff and students of the Department of Physical Sciences, this book is dedicated.

Adli Bishay

CONTENTS

PART I

MAGNETIC RESONANCE

PART II

ELECTRICAL CONDUCTIVITY

PART III

RADIATION EFFECT I (NON-CRYSTALLINE SOLIDS)

PART IV

RADIATION EFFECTS II (CRYSTALLINE SOLIDS)

PART V

METALS

PART VI

SPECTROSCOPY

PART VII

QUANTUM ELECTRONICS

A REVIEW OF MAGNETIC RESONANCE METHODS APPLICABLE TO SOLIDS

George E. Pake

Washington University

St. Louis, Missouri, U.S.A.

The spectroscopy of solids in both the radiofrequency and microwave portions of the electromagnetic spectrum has provided a wealth of fundamental information about the solid state during the 20 years since the pioneering experiments. This paper reviews magnetic resonance methods and surveys the interactions which are revealed by splittings in the resonance spectra. Relaxation phenomena and the preparation and observation of non-equilibrium spin systems are included in the discussion. The general principles are applied to both nuclear magnetic resonance and to electron spin resonance, identifying essential similarities as well as inherent differences between these two kinds of electromagnetic probes for investigating solids. Examples illustrate the influence of characteristic solid properties on the resonance spectrum: crystalline (or non-crystalline) structure, paramagnetism and ferromagnetism, and conduction properties.

INTRODUCTION

In early 1946, as a young graduate student having just arrived at Harvard University, I attended a physics colloquium which determined the direction of my future scientific interests. In that colloquium, an associate professor named Purcell, who had just returned to academic life from radar research and development during World War II, described the first detection of nuclear magnetic resonance absorption in bulk matter by himself, Henry Torrey and Robert Pound (1).

As Edward Purcell discussed the simple vector model of magnetic resonance and the possible sources of the width of the absorption line, I was fascinated by the fact that the radiofrequency field seemed to give the physicist a detailed view of the interior of a solid or liquid. The interactions were sufficiently weak and seemed adequately localized that the inferences from the experiment held promise for a penetrating probing of the internal structure of matter. I could imagine that this new technique might almost confer on the physicist the observational capabilities of a Maxwell demon clambering about in a crystal lattice.

In his talk, Purcell mentioned the earlier unsuccessful attempt of Gorter to detect lattice heating in consequence of nuclear magnetic resonance absorption. He also mentioned the 1945 experiments of Zavoisky (2) which detected barely resolved electron magnetic resonance absorption of manganous and cupric ions in crystals. Cummerow and Halliday (3) soon observed the well-resolved resonance of Fig. 1

The several pioneering experiments just mentioned began a remarkable era of investigation of solids using electromagnetic radiation in the radiofrequency and microwave frequency range. We are still very much in the midst of that period, and this conference on "The Interaction of Radiation with Solids" may be considered to mark the 21st birthday of the era of application of magnetic resonance techniques to solids. In my country, the 21st birthday is the point at which men and women become legally of age. Perhaps we can agree that magnetic resonance spectroscopy, after an exciting period of growth, now has fully come of age as

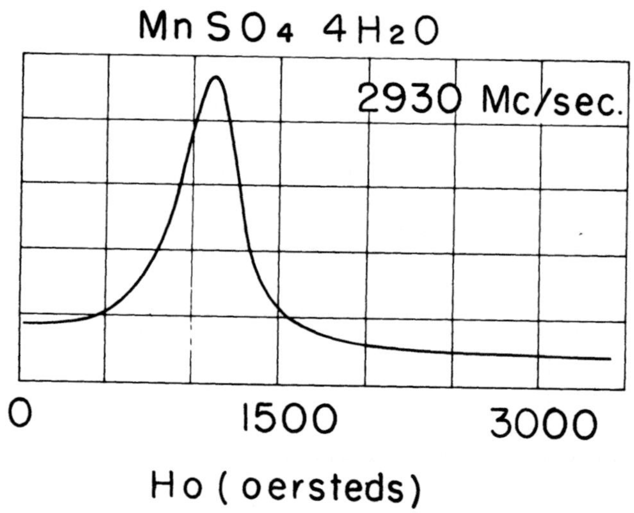

Fig. 1. Early paramagnetic resonance of Mn^{++} observed by Cummerow and Halliday

a proven tool for investigation of the structure and properties of solids.

A significant feature of the electromagnetic radiation employed in magnetic resonance is the low energy of the quanta. Most of the radiofrequency and microwave frequency range falls between 10^{-8} electron volts and 10^{-4} electron volts. This energy is appreciably less than the energy for excitation of an electron in a semi-conductor into the conduction band (about one electron volt) or the binding energy of an ion into an ionic crystal (about 10 electron volts).

Comparing these energies makes it clear that radiofrequency and microwave irradiation of a solid will leave the atomic arrangement unaffected and should, indeed, have little or no direct effect on the lattice vibrations. Any heating of the lattice depends upon the coupling of the spins to the lattice, about which we shall say more later. Thus magnetic resonance investigation of a solid in no way damages the sample.

Another feature of the magnetic interactions in solids is the weakness of the perturbing interactions so that the resonances are sharp and splittings are often well resolved. Most, if not all, electrons are in atomic and molecular closed-shell configurations which have no orbital magnetic moment and which have paired-off spin magnetic moments. The remaining occasional unpaired electron spins and the nuclear magnetic moments serve as the probes which elucidate details of the structure and properties of solids through magnetic resonance.

THE PHENOMENOLOGICAL DESCRIPTION

The weak interaction among nuclear or electron magnets is significant to our understanding, for, although we deal in reality with an assembly of many coupled nuclear or electronic magnets, we can often describe the motions using a single-particle model, with the effects of interactions introduced through simple damping constants. This is indeed the strength of the Bloch phenomenological equations of motion for the magnetization M, which is the magnetic moment per unit volume of the solid.

We shall begin with the assembly of nuclear magnets, which are readily treated independent of the electrons in the surrounding closed shells of atomic electrons. Each nucleus, in consequence of its angular momentum, has circulating currents within its structure which give rise to the magnetic moment of the ground state. The factor of proportionality between the individual nuclear magnetic moment μ, which is parallel to the total nuclear angular momentum $\hbar I$, and $\hbar I$ is denoted γ:

$$\mu = \gamma(\hbar I) \tag{1}$$

The well-known equation of motion for the magnetic moment under the influence of the torque provided by an applied magnetic field H_0 is

$$d(\hbar I)/dt = \mu \times H_0 \tag{2}$$

which becomes

$$d\,\mu/dt = -\gamma H_0 \times \mu \tag{3}$$

Eq. (3) describes a precession of μ about the H_0 direction at the Larmor frequency $\omega = -\gamma H_0$

We are able to reduce many magnetic resonance problems to just such a simple precession solution by the artifact of mathematical transformation to one or a succession of rotating coordinate frames within the last of which the motion is a simple precession about an effective magnetic field in that rotating frame.

Fig. 2 defines the rotating coordinate system associated with the appiied rf field $H_x = H_1 \cos \omega t$, $H_y = H_1 \sin \omega t$.

Fig. 3 illustrates the effective field in a frame rotating with H_1 and shows the cone of precession about the effective field H_e in the rotating frame. At resonance $H_0 = \omega/\gamma$ and $H_e = H_1$; this precession alters the z-component of μ and thus corresponds to energy absorption or emission in the H_0 field.

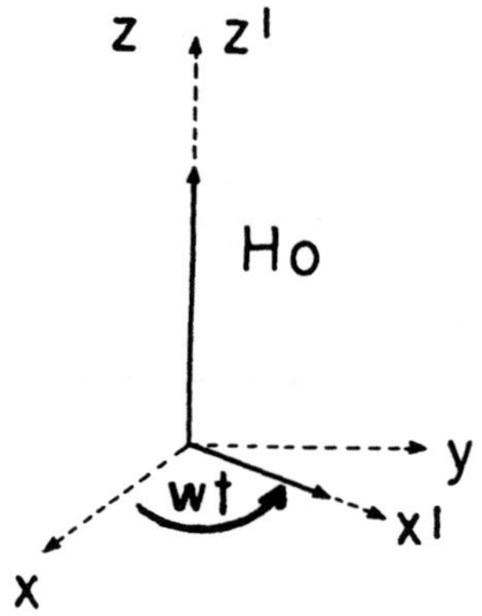

Fig. 2. Definition of rotating (primed) coordinates

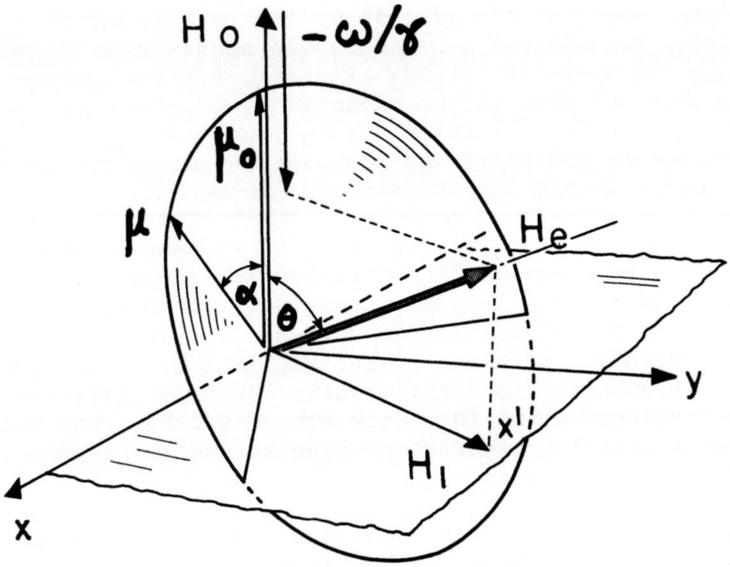

Fig. 3. Cone of precession of the magnetic moment about the effective field in the rotating frame

Fig. 4 sketches the variation of H_e as ω is swept through the resonance condition. Sweeping through resonance at the proper rate can, with many samples, carry the entire magnetic moment of the material into a non-equilibrium configuration along the final H_e direction.

The Bloch phenomenological equations (4) describe the motion of the bulk magnetization M of the sample, $M = \sum_i \mu_i$. At thermal equilibrium in zero external field we expect M = 0, since there is no energetically preferred direction in space. But when a magnetization M is established in an external static field H_0 along z accompanied by a smaller perpendicular time-dependent field H(t), the Bloch equations for the motion in the total field H superimpose upon the gyromagnetic motion simple decays of the transverse and longitudinal components of magnetization:

$$dM_z/dt = \gamma(M\times H)_z + (M_0 - M_z)/T_1$$
$$dM_x/dt = \gamma(M\times H)_x - M_x/T_2$$
$$dM_y/dt = \gamma(M\times H)_y - M_y/T_2$$

These equations, though they have well known limitations, are enormously useful in the physical interpretation of magnetic resonance experiments.

Whatever the mechanism of coupling to the lattice may be, the time constant T_1 describes the rate at which the energy-determining z component approaches the equilibrium magnetic moment $M_0 = \chi_0 H_0$ in a static field; it is in effect the rate at which the assembly of spins in the magnetic field comes into equilibrium with the reservoir of lattice vibrational energy as the magnetic moments take on the Boltzmann population distribution over their Zeeman energy levels in the external field.

The time constant T_2 for decay of transverse components must be at least as short as T_1, because transverse components will disappear as the equilibrium magnetization obtains in a static field. However, especially in solids, T_2 is often much shorter than T_1, because the magnetic interactions of each nuclear moment with its neighboring nuclear magnets create a distribution of local field values about the neighborhood of H_0, and the slightly different precession frequencies dephase the transverse components of M.

Excitation of magnetic resonance requires application of the electromagnetic radiation in the proper polarization, either linear and perpendicular to H_0 or circular about H_0 in the sense of the Larmor precession. In the more frequent linear transverse polarization, the time dependent magnetization along the x-direction

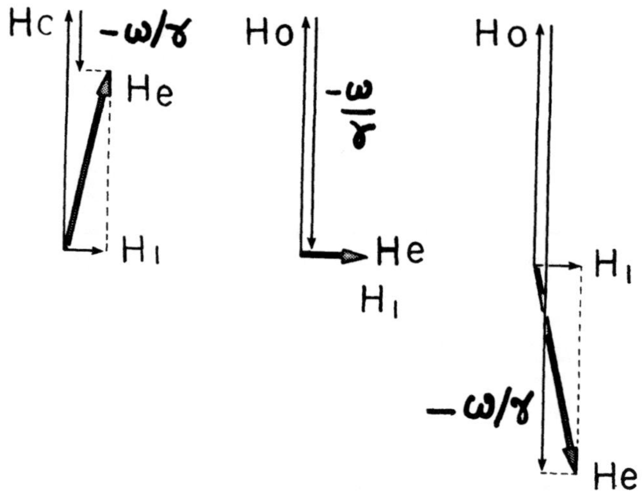

Fig. 4. Successive position of H_e as ω is swept through the resonance value

excited by the oscillating field $H_x = 2H_1 \cos \omega t$ is given by the complex susceptibility $\chi = \chi' - i\chi''$ as

$$M_x = \chi'.2H_1 \cos \omega t + \chi''.2H_1 \sin \omega t \tag{5}$$

The energy absorbed is determined by the out-of-phase component:

$$A = 2\omega\chi'' H_1{}^2 \tag{6}$$

The familiar simple non-saturation solution of the Bloch equations for χ'' is

$$\chi'' = \tfrac{1}{2}\chi_0\omega_0 T_2/(1 + T_2{}^2 (\omega - \omega_0)^2) \tag{7}$$

which represents an absorption line of frequency half-width $\Delta\omega = 1/T_2$ (see Fig. 5). This obtains provided the irradiating power level is kept low enough so that $\gamma^2 H_1{}^2 T_1 T_2 << 1$. Higher radiation intensities broaden the resonance and introduce non-linear response of the energy absorption to the irradiating energy density. There is in fact an asymptotic maximum rate of absorption at the center of the resonance line which is determined entirely by the rate $1/T_1$ at which energy can be carried off from the assembly of magnets to the lattice:

$$A_{max} = \chi_0 H_0{}^2/T_1 \tag{8}$$

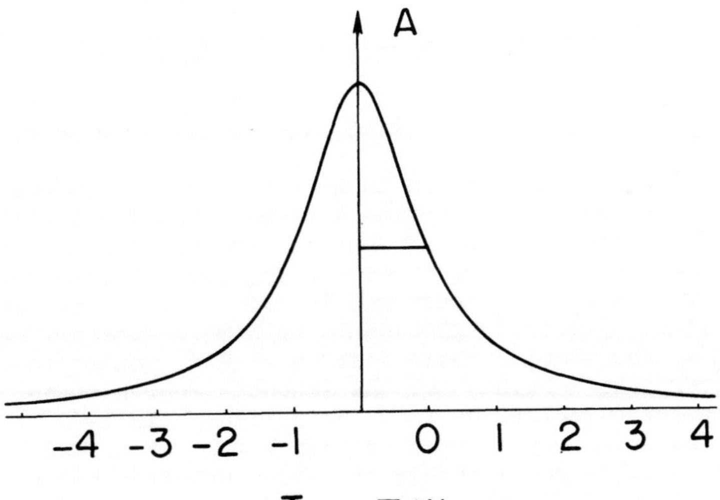

Fig. 5. Absorption line shape from solution of the Bloch equations.

THE MICROSCOPIC QUANTUM MECHANICAL FORMULATION

The Hamiltonian for an assembly of magnetic moment in external field H_0 is

$$\mathcal{H}_z = -\sum_i \mu_i . H_0 = -\gamma\hbar \sum_i I_i . H_0 \tag{9}$$

which is called the Zeeman Hamiltonian because its eigen values are the energy states, in atomic spectra of electron systems, for the Zeeman splitting. Each spin has 2I+1 energy states given by

$$(m_I | \mathcal{H}_z | m_I) = -\gamma\hbar H_0 m_I \tag{10}$$

and the magnetic dipole transitions $|\Delta m_I| = 1$ are excited by a transverse oscillating field at frequency $\hbar\omega = \gamma\hbar H_0$, the Larmor precession frequency.

Boltzmann distribution over these levels at temperature T leads to a static magnetic moment $M_0 = \chi_0 H_0$, where χ_0 is the static Curie susceptibility:

$$\chi_0 = N(\gamma\hbar)^2\ I(I+1)/3\ kT \tag{11}$$

The simple solution for isolated spins experiences considerable mathematical complication when the magnetic interactions between neighboring dipoles are introduced:

$$\mathcal{H}_{dip} = \sum_{i>j} \left[\mu_i . \mu_j - 3(\mu_i . r_{ij}/r_{ij})(\mu_j . r_{ij}/r_{ij}) \right] r_{ij}^{-3} \tag{12}$$

The resulting total Hamiltonian $\mathcal{H} = \mathcal{H}_z + \mathcal{H}_{dip}$ evidently describes a many-body problem which at first sight seems insoluble.

In seeking to explain the unexpectedly narrow resonance lines (long T_2's) observed in electron systems, Van Vleck (5) not only discovered exchange narrowing but he also used the method of moments of the line shape to provide a rigorous expression for the mean square width of the resonance in terms of simple sums over lattice sites. Precise experiments on simple nuclear moment arrays such as the simple cubic lattice of F^{19} nuclei in CaF_2 have demonstrated the accuracy of the second, fourth, and even the sixth, moment calculations (6). This triumph in 1948 represented a precise solution by Professor Van Vleck of a real many-body problem some dozen years before many-body problems became fashionable in physics.

Out of these calculations of the moments thus came the basis for interpretations of T_2 in terms of the structure of the solid. The exchange-narrowing theory also gave the basis for understanding narrowing of the nuclear resonance line (lengthening of

T_2) as a consequence of nuclear motions associated with diffusion or with rotating atomic groups within the solid. When such rotation occurs, the observed line width $\Delta\omega$ is related to the Van Vleck second moment $<(\omega - \omega_0)^2>_{av}$ through

$$\Delta\omega \approx < (\omega - \omega_0)^2 >_{av} / \omega_{rot} \tag{13}$$

where ω_{rot} is the rotation frequency of the atomic group.

Pioneering in the calculation of spin-lattice relaxation times also involves the name of Van Vleck, whose classic analysis of the early experiments on titanium and chrome alums (7) established the theoretical framework on which others built the explanation of nuclear relaxation in solids. The key to calculation of the spin-lattice relaxation time is determination, for various possible interactions which can provide energy exchange with the lattice, the transition probability for a spin flip. If W is the transition probability, the relaxation time is $T_1 \approx 1/W$. Nuclear spins are found to be coupled to the lattice either through paramagnetic impurities, which in turn have good lattice contact by virtue of spin-orbit coupling, or through interactions with nuclear moments in rotating or diffusing atomic groups within the solid. In metals, the nuclear spins are coupled to the lattice through interactions with the unpaired spins of those conduction electrons near the top of the Fermi distribution. This gives the relaxation rate $1/T_1$ a linear dependence on temperature, in direct proportionality to the fraction of electrons excited into unpaired states at the Fermi surface.

The detailed understanding of nuclear magnetic resonances stimulated interest in systems involving both nuclear moments and unpaired electron moments. In order to proceed with interactions between nuclear and electron magnets, we need to discuss paramagnetic ions in crystals.

INFLUENCE OF THE CRYSTAL FIELD ON PARAMAGNETIC IONS

A paramagnetic ion such as Cu^{++} imbedded in a crystal lattice experiences a more complicated set of interactions than does a nuclear moment, chiefly because the electric field of the crystalline environment perturbs the paramagnetic states and influences their associated magnetic moments. Time permits us only to give a qualitative summary of the effect of the crystal field interactions.

If the paramagnetic ions were isolated from the crystal, we would describe the magnetic properties of the one incomplete electron shell in terms of a total angular momentum consisting of both orbital and spin contributions. But the effect of neighboring ions or atoms in the crystal lattice is to provide an electrostatic environment of lower than spherical symmetry, in which circumstances the expectation value of the orbital angular momentum is zero. The extent of this so-called quenching of orbital angular momentum is affected by the relative strength of the low-symmetry field from external sources in comparison to the spherically symmetrical fields of the other electron shells of the ion in which the paramagnetic shell is situated. If the unpaired electrons are near the ionic surface, as in the iron group, they experience relatively greater low symmetry fields than if they lie more deeply in the ion, as in the rare earths. Therefore the quenching of orbital angular momentum is rather complete for the 3d electrons of the iron group ions, leaving spin-only paramagnetism, whereas the deeper lying 4f shells of the rare earth ions have magnetic moments arising from full orbital and spin components.

The resultant magnetic moment for a rare earth ion is then:

$$\mu = g\beta J = g\beta(L + S) \tag{14}$$

where g is the Landé factor and $\beta = e\hbar/2mc$ is the Bohr magneton. Note that the conversion between g and γ for a particular moment is

$$\gamma\hbar = g\beta \tag{15}$$

In the iron group, the quenching by the crystalline field gives a magnetic moment closer to the spin-only value, which would mean that g would be close to $g_e = 2.0023$. But here the role of spin orbit coupling, in simple cases taking the form

$$\mathcal{H}_{so} = \lambda L.S \tag{16}$$

is to restore a portion of the orbital magnetic moment. The effectiveness of this partial restoration depends upon the crystal direction along which one measures the component of the magnetic moment, but in general, the g is modified from the spin-only value by an amount

$$|g - g_e| \approx \lambda/\Delta \tag{17}$$

where λ measures the spin-orbit coupling and Δ is the splitting introduced by the crystal field as it partially lifts the orbital degeneracy of the paramagnetic shell. In the iron group, $\Delta \sim 10^4$ cm^{-1} or about 1 electron volt and $\lambda \sim 10^2$ cm^{-1} or about 0.01 electron volt.

Fig. 6 shows the splitting effect of the various interactions on a ^2D term in a tetragonally distorted cubic field.

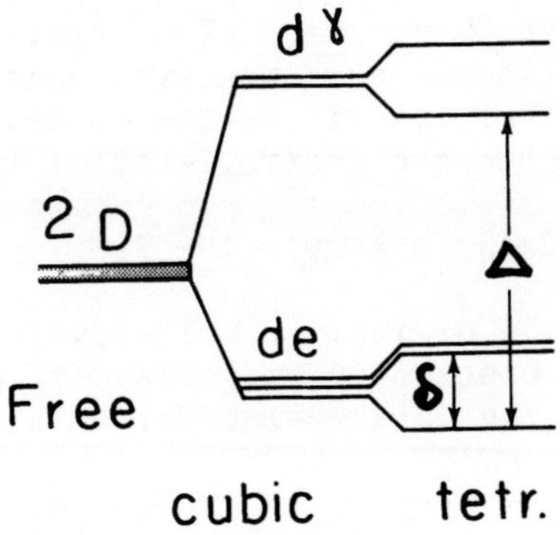

Fig. 6. Splitting of the ^2D term by a tetragonally distorted cubic field

EFFECTIVE SPIN HAMILTONIAN

Zeeman Term. We have said in the previous section that the complicated effects on an unpaired electron shell of the shielded ion core, the external crystal field, and the spin-orbit coupling within the group of unpaired electrons are reflected in an effective spectroscopic splitting factor, dependent upon external magnetic field direction with respect to the crystal, and in an effective spin vector for the shell of unpaired electrons. The interaction with the external magnetic field is usefully described in terms of an effective spin Hamiltonian:

$$\mathcal{H} = \beta\, S.G.H \tag{18}$$

where G is in general a tensor quantity. In any kind of axial symmetry, the principal values are $g_{\parallel}$, $g_{\perp}$, $g_{\perp}$ and

$$\mathcal{H} = \beta \left[g_{\perp} (S_x H_x + S_y H_y) + g_{\parallel} S_z H_z \right] \tag{19}$$

where z is the symmetry axis in the crystal.

Fine Structure. Before we turn to interactions of a specifically magnetic nature, there is yet one more term which grows out of higher order perturbation effects of the spin-orbit coupling on the crystal field interaction. This is the so-called fine structure term, measured by interaction constants of the order of λ^2/Δ. Where axial symmetry is present, this interaction constant is denoted D and we write the effective spin Hamiltonian as

$$\mathcal{H} = \beta\left[g_{\parallel}\, S_z H_z + g_{\perp}\, (S_x H_x + S_y H_y)\right] + D\left[S_z^{\,2} - 1/3\; S(S+1)\right] \tag{20}$$

For effective spin $S = \frac{1}{2}$, D has no effect. In general the D term splits the resonance into 2S lines.

Hyperfine Structure. The unpaired electron magnetic moments interact with nuclear magnets which happen to be nearby in the lattice. If we let I be the nuclear spin, then there will in general be both the isotropic Fermi contact term

$$\mathcal{H} = (16\pi/3)\ \beta\gamma\hbar\ |\psi(r_I)|^2\ I.S \tag{21}$$

as well as the ordinary dipole-dipole coupling of Eq. (12) between nuclear and electronic magnets. If we again have symmetry along the z-axis, the Hamiltonian for axial symmetry is

$$\mathcal{H}_{axial} = \beta\left[g_{\parallel}S_zH_z+g_{\perp}(S_xH_x+S_yH_y)\right] + D\left[S_z^2 - 1/3\ S(S+1)\right] + A_{\parallel}I_zS_z + A_{\perp}(I_xS_x+I_yS_y) - \gamma_I\hbar\ H.I \tag{22}$$

where we have assumed only one nuclear species and neglected dipole-dipole interactions of paramagnetic ions with one another or of nuclear magnets with one another.

The effect of the hyperfine interaction with a nucleus of spin I is to produce 2I + 1 differing local field values. Fig. 7 shows three hfs lines from $N^{14}(I=1)$ in the ion $ON(SO_3)_2^{--}$. Fig. 8 presents six hfs lines from the Mn^{55} nucleus with I = 5/2. Fig. 9 shows a portion of a more complicated pattern from interaction of the unpaired electron of an organic free radical ion with 16 H^1 nuclei and two N^{14} nuclei of the ion.

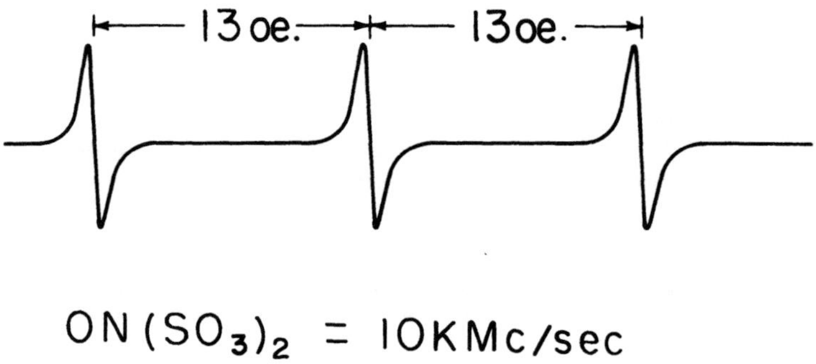

Fig. 7. Hyperfine splitting from the N^{14} nucleus.

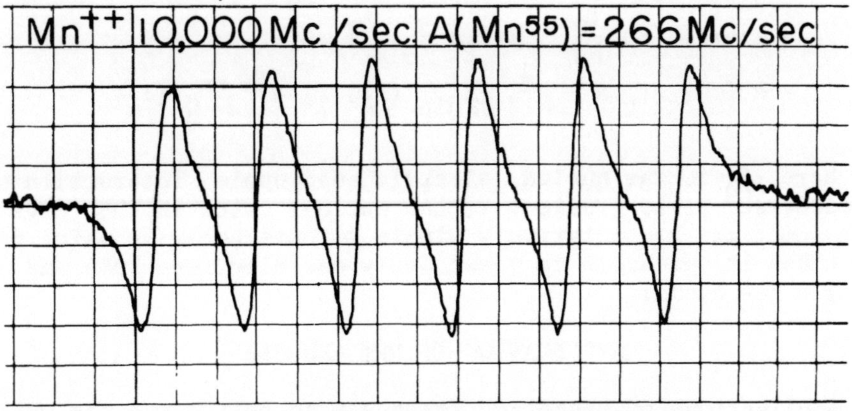

Fig. 8. Hyperfine splitting from Mn^{55}

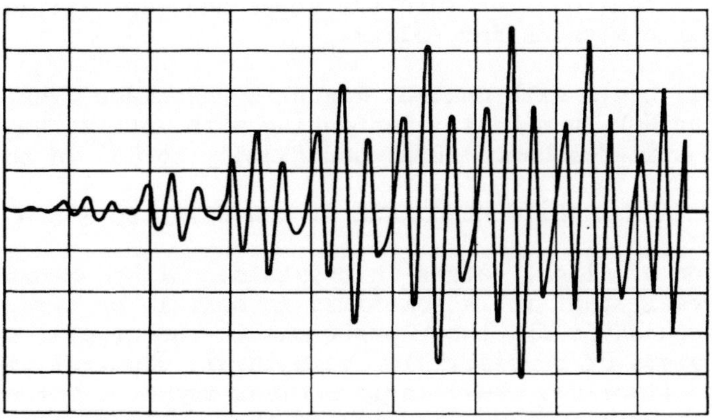

Fig. 9. Portion of the complicated hyperfine splitting arising from numerous nuclear moments in Wurster's blue ion.

Before we proceed to discussions of significant applications of radiofrequency and microwave irradiation to solids, we write down the Hamiltonian with which one usually analyzes experiments on solids containing nuclear moments or unpaired electrons or both:

$$\mathcal{H} = \beta \sum_j \left[H.G.S_j + DS_{jz}^2 + ES_{jx}^2 + FS_{jy}^2 \right]$$
$$+ \sum_{j>l} \mathcal{H}_{dip}(S_j, S_l) + \sum_{j \ast k} \mathcal{H}_{dip}(I_k, S_j) \qquad (23)$$
$$+ (16\pi/3)\, \beta\hbar \sum_{j \ast k} |\psi(r_k)|^2 \gamma_k I_k.S_j$$
$$- \hbar \sum_k \gamma_k I_k.H + \sum_{k>l} \mathcal{H}_{dip}(I_k, I_l) + \sum_k \mathcal{H}_Q(I_k)$$

Here $\mathcal{H}_Q$ is the nuclear electric quadrupole interaction with the electric field gradient at the nuclear site in the crystal. There sometimes are relatively simple circumstances in which dilution makes it possible to treat only one electron spin and only one or a few nuclei.

APPLICATION OF NMR AND ESR

Applications of magnetic resonance to solids and the uses of the Hamiltonian of Eq. (23) to interpret the experiments are so extensive that it is futile to think in terms of a complete listing. Succeeding papers of the conference will treat many of the most recent applications, and we shall mention here a series of examples selected to illustrate the scope and capabilities of magnetic resonance methods for solids.

We shall begin with nuclear magnetic resonance (NMR) in non-metals, then NMR in metals. Continuing with metals, we next turn to electron spin resonance (ESR) and finally to ESR in non-metals.

NUCLEAR MAGNETIC RESONANCE IN NON-METALLIC SOLIDS

Location of light atoms in crystals. Light atoms scatter X-rays so weakly that it is sometimes impossible by X-ray crystallographic methods to fix their positions in the crystal structure. Two light atoms of considerable structural chemical importance, hydrogen and fluorine, have large nuclear magnetic moments. In a number of crystals, NMR studies of the nuclear dipole-dipole splitting or of the second moment in comparison with the Van Vleck theory have permitted unambiguous determination of the structure (8).

Hindered rotation in solids. The rotation of an atomic complex, whether an ion such as NH_4^+ in an ionic crystal or an organic molecule such as CH_3OH in a van der Waals crystal, leads to narrowing of the NMR dipole-dipole width when the rotation frequency is not too different from the rigid-lattice line width. One can learn, from measurements of NMR line widths as functions of temperature, the identity of rotating groups in solids and the frequency of rotation as a function of temperature. From these data one obtains information on the potential barriers hindering rotation (9).

Diffusion in solids. Diffusion of the atom carrying a nucleus under NMR investigation can both narrow the resonance and influence the local-field spectrum which effects spin-lattice relaxation. Diffusion constants and their variation with temperature can be determined in favorable materials (10).

Antiferromagnetic materials. Nuclei on or near an antiferromagnetic sub-lattice experience a strong local magnetic field from the electron spins through the hyperfine coupling. This field at the nucleus is proportional to S_z and it combines with any external field to determine the total field and therefore the NMR frequency. In this way the frequency itself can provide a measure of sub-lattice magnetization as it varies with temperature. The nucleus observed may be within a magnetic ion of the lattice, as is Co^{59} in CoF_2 or it may be elsewhere in the lattice, as is F^{19} observed in MnF_2 (11).

NUCLEAR MAGNETIC RESONANCE IN METALS

Pauli paramagnetism. Any static measurement of the susceptibility of a metal gives the combined Pauli paramagnetic contribution from the conduction electrons plus the sometimes susbtantial diamagnetism of the bound electrons. The conduction electrons have a magnetic resonance (which we discuss later on) with intensity proportional to the paramagnetic or Pauli susceptibility if only one can calibrate that intensity. Schumacher and Slichter (12) used the ingenious technique of observing the Li^7 NMR and the conduction electron resonance in the same sample at the same frequency, but of course in substantially different external magnetic fields. Comparison of the relative intensities measured the Pauli susceptibility. This permits comparisons with theoretical calculations which take into account collective electron-electron effects as corrections to the simple electron gas approximation.

Diffusion in metals. Although the nuclear relaxation mechanism which usually dominates is the coupling to the conduction electrons through the hyperfine interaction, in certain temperature ranges for some metals the self-diffusion of atoms through the lattice provides intense enough components of local nuclear magnetic fields at the Larmor frequency to determine the relaxation time. This process can thus provide useful information on self-diffusion in metals (13).

Alloys and solid solutions. The static diagonal portion of the hyperfine interaction between the nuclei and the conduction electrons provides in effect a local magnetic field which shifts the nuclear resonance frequency. This Knight shift is proportional to the probability density $|\psi(0)|^2$ of the conduction electron at the nuclei, and it thus reveals significant information about the wave function of the electrons on the Fermi surface.

As the composition of an alloy is varied, the Knight shift of a given nuclear species generally varies smoothly with composition, and the NMR of any particular isotope in a multiphase region splits into multiple lines.

An interesting effect occurs when a particular impurity atom is introduced in dilute amounts into an otherwise uniform lattice. The "diffraction" of the conduction electron waves by these impurity centers leads to oscillations or ripples of the electron density as a function of position around the impurity. This in turn implies a distribution of $|\psi(0)|^2$ values for the host nuclei, and a corresponding broadening of the NMR line. Experimental results are in agreement with calculations by Friedel and coworkers of the wave function of free electrons scattered by spherical impurities (14).

Strain. The nuclear electric quadrupole moment interacts with the gradient of the crystalline electric field to split the resonance into 2I lines, where I is the nuclear spin. The splitting depends upon the angle between the principal axis of the field gradient and the external magnetic field. If the nuclear spin is half-integral, the central resonance line of the pattern is unshifted, to first order in the quadrupole interaction.

In sites with high enough symmetry (cubic or tetrahedral), electric field gradient vanishes, and there is no quadrupole splitting (nor broadening as would occur for a polycrystalline sample). But the presence of dislocations can introduce local asymmetries providing electric field gradients of random orientation and magnitude, resulting in a smearing of the quadrupole satellite resonances over a wide frequency range such that the resonance intensity at any given frequency is below the noise-level and undetectable. Only the central undisplaced quadrupole component is then observable, and the intensity will be only a fraction as large as anticipated: for $I = 3/2$, the central line contains 4/10 of the total quadrupole pattern intensity; for $I = 5/2$, it is 9/35; 4/21 for $I = 7/2$; 5/33 for $I = 9/2$.

By introducing small amounts of impurities into metals, one can observe the fall-off of intensity as strains and dislocations create electric field gradients. Effects of annealing in reducing strain can also be monitored (15).

Nuclear spin relaxation in a superconductor. A definitive

experiment by Hebel and Slichter (16) provided an important test of the BCS theory of superconductivity.

Of course the observation of NMR in a superconductor is precluded by the Meissner effect. This ingenious experiment began with aluminum at a temperature below the critical temperature but in a magnetic field *above* the critical field so that the metal is in the normal conducting state. The nuclear magnetic polarization reaches equilibrium in this field, which is then reduced rapidly compared to T_1 (of the order of 1 second) but slowly compared to T_2 so that the nuclear precessions follow the decreasing magnetic field as a thermodynamic whole. When zero external field is reached, the nuclei remain aligned with the local dipolar field within the superconductor. If, after a time interval comparable to T_1 in the superconductor, this process is reversed and the field is reestablished at its original value, the reduction in the level of the NMR signal in the normal state measures the relaxation decay which took place in the superconducting state, thus permitting measurement of T_1. The temperature dependence of the relaxation times so measured ruled out the two-fluid model and agreed with effects predicted by the BCS correlation between electrons of opposite spin and momentum.

NMR in ferromagnets. Gossard and Portis (17) first observed NMR in a ferromagnetic metal when they found the Co^{59} resonance in a sample of finely divided face-centered cubic cobalt metal at room temperature in zero external field. The frequency of 213.1 Mc / sec corresponds to 213,400 gauss at the sites of the Co^{59} nuclei. The resonance had an amazingly large amplitude, approaching 10^6 times that in zero external field.

Subsequent observation of the Fe^{57} resonance in zero external field corresponded to 330,000 gauss at the Fe^{57} site. This resonance frequency was found to *decrease* as external field was applied, indicating that the field at the nucleus is *opposite* to the magnetization. The explanation for this is core polarization (18). The *3d*-electrons should be incapable of producing isotropic hyperfine coupling at the nucleus, because the *d*-orbitals vanish at the nucleus. The core *s*-electrons experience exchange polarization with the *3d*-electrons, with the consequence that the spin magnetization of the two *s*-electrons does not add to zero at all points of the electron cloud. The resultant hyperfine field at the nucleus can point against the direction of overall magnetization of the sample.

ELECTRON SPIN RESONANCE IN METALS

Conduction Electron Resonance. We have already mentioned use of the ESR of conduction electrons to determine the Pauli susceptibility. A fraction KT/E_F of the electrons, near the Fermi energy E_F, participates in the resonance. The resonances are not observable in all metals, presumably because the spin-lattice relaxation time is very short and the resonance therefore very broad. Going to higher frequencies is complicated by the decrease in the skin depth of penetration of the microwave field into the metal particles.

Ferromagnetic resonance. The electron resonance in a ferromagnetic material occurs at a frequency dependent upon the demagnetizing factors determined by the shape of the sample. Anisotropy fields within the sample also affect the resonance frequency As one would expect, the resonance absorption is not proportional to a Curie susceptibility, and indeed the resonance is not described in terms of a χ'' component of a complex susceptibility. In comparison with magnetic resonance studies of paramagnetic and antiferromagnetic materials, ferromagnetic resonance is more influenced by macroscopic properties of the sample (19).

ELECTRON SPIN RESONANCE IN NON-METALS

Effect of high pressure on crystalline fields. We have seen that the symmetry of the g-tensor depends upon the crystal field symmetry and the magnitude of the principal g-values depends upon the crystal field splitting and the spin orbit coupling. Application of high pressure to the crystal "squeezes" the orbitals of the unpaired electron, moves external sources of the crystal field and influences core polarization and therefore hyperfine splitting A series of experiments by Walsh (20) observed high pressure effects on g and on the hyperfine splitting, from which basic information can be deduced on the sources of the crystal field and on the way in which the electron orbitals accommodate the changes in lattice parameters - certainly very fundamental data on electrons in crystals.

Defects, vacancies, and impurity centers. This is one of the most fruitful areas of application of ESR. Again, it is through hyperfine coupling of nearby nuclei to the unpaired electron that we are able to learn much about the electron wave function. Since a nuclear spin I (or a cluster of equivalently situated nuclear moments having a maximum total coupled nuclear spin I) gives rise to $2I + 1$ hyperfine components, we expect an unpaired electron to have its ESR split in a characteristic way depending upon the various nuclear spins nearby and the magnitude of their hyperfine interactions. If the resultant splitting can be observed in high enough resolution, the pattern can practically be considered to be a characteristic "finger print" for the particular crystal locale. Unfortunately, such ideal resolution is often unattainable, though we shall see that there are tricks that permit resolution of some hyperfine splittings lost in a broad ESR line.

We expect to encounter unpaired electrons associated with the paramagnetic ions of a pure crystal containing a transition element. It may also be associated with an organic free radical in a molecular crystal. But we also encounter unpaired electrons associated with defects in crystals. One of the first kinds of defects to be studied extensively are the F-centers of alkali halides, which early in their history were postulated to be electrons trapped at the point of a halide ion vacancy. In principle, we should expect to learn much about this hypothesis from the ESR of the F-center electron and the hyperfine interaction with the surrounding nuclei.

These defect centers in crystals present examples *par excellence* of the interaction of radiation with crystals, inasmuch as defects may be produced by energetic X-ray or γ-ray photons or by neutron and electron bombardment (although there are non-irradiation methods which produce some kinds of defects). The nature of the resulting defects is then determined in considerable detail by means of the very low frequency or microwave photons used in magnetic resonance.

An especially educational example is the V_k center in KCl as studied by Castner and Känzig (21). X-irradiation of the crystal at liquid nitrogen temperature produces a center with a characteristic highly resolved ESR structure. For H_0 along 100, the pattern consists of many weaker lines interspersed within a dominant pattern of seven equally spaced lines of intensity 1:2:3:4:3:2:1. The interaction of one electron equally with two identical nuclei of I = 3/2 will produce just such intensity ratios. Now Cl^{35} is 75% abundant. On this basis, 9/16 of the Cl pairs in the crystal are Cl^{35} pairs, 1/16 are Cl^{37} pairs, and the remaining pairs are mixed. Attributing the dominant pattern to Cl^{35} pairs, and making use of the fact that the Cl^{37} moment is 83% as large as the Cl^{35} moment, permits construction of a theoretical resonance pattern having intensities and spacings in agreement with the observed ESR of the V_k center.

The analysis is not as simple as this discussion has suggested, because the g-tensor and anisotropy of the hyperfine splitting must be considered (22). This adds important information, because the principal axis directions should relate to the Cl-Cl lines joining neighboring pairs of Cl ions in the crystal. Furthermore, the direction and amount of the shift of the principal g-values is helpful in interpreting whether the ESR is that of an electron or a hole. The resonance data, combined with other information, lead to the conclusion that a V_k center is a hole trapped on a pair of adjacent Cl ions.

We return to the conceptually simpler F-center, the ESR from which is however much less well resolved. If the electron trapped at an F^- vacancy in LiF, for example, interacts equally with its six Li^7 nearest neighbor nuclei, the maximum nuclear spin component of the cluster is $6 \times 3/2 = 9$, and the number of resonance lines should be 19. If the electron also interacts with the next nearest neighbor shell of F^{19} nuclei, there can be many more lines. In fact, Holton, Blum, and Slichter (23) observed 35 lines, not well resolved. They were able to clarify the origin of these 35 lines as well as many more unresolved lines by use of electron-nuclear double resonance, known as ENDOR.

In an ENDOR experiment, one observes the ESR in a condition of partial saturation while sweeping the frequency of a simultaneously applied radiofrequency field. When the radiofrequency is equal to one of the hyperfine splittings present in the unresolved ESR spectrum, (Fig. 10 is a sketch of such an unresolved pattern), nuclear spins are flipped and, as a consequence, new electron-nuclear systems of the sample may have their electron transition frequencies shifted into the portion of the ESR being saturated. This affects the absorption level for the saturated resonance and thus produces a change in the ESR signal. Holton, Blum, and Slichter were able in this way to resolve the hfs. couplings of the F-center electrons to the first *seven* shells of surrounding nuclear moments, three shells of Li^7 nuclei and four shells of F^{19} nuclei. This is rich information indeed on the spatial extent of the wave function of the trapped electron.

Donors and acceptors in semi-conductors. The model of a donor or acceptor impurity in a semi conductor as an electron or hole in an expanded hydrogen atom-like orbit about the impurity center has received striking confirmation - as well as indications of its limitations - from the ESR hyperfine structure and relaxation time. The hyperfine patterns are generally less complicated than for defects in alkali halides, and thus much simpler to interpret.

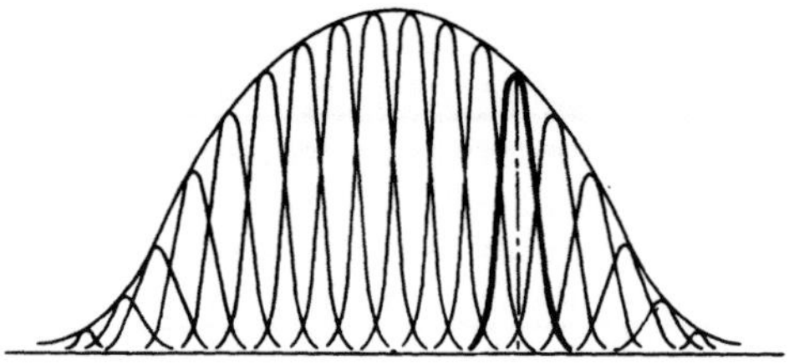

Fig. 10. Example of an unresolved splitting pattern leading to a single resonance without structure

Special magnetic arrays. The intensity of the ESR is proportional to the static paramagnetic susceptibility. In the molecular crystals of certain organic free radicals, this ESR intensity was found to increase with decreasing temperature, substantially as expected for the Curie 1/T dependence, but then to reach a smooth maximum and to decline rather rapidly at lower temperatures. Though suggestive of antiferromagnetic behavior, the susceptibility maximum was not sharp as for a typical antiferromagnetic (24).

Through both ESR studies and measurement of the NMR Knight shift in these crystals, it was verified that the magnetic properties are essentially those of a linear array of exchange-coupled spins. In order for the transition to be sharp, the lattice must be at least two-dimensional. The relatively flat free-radical molecules are stacked in the lattice in such a way that the direction of effective exchange coupling is out of the molecular plane, with very weak coupling to neighbors at the sides.

Antiferromagnetic resonance. The two interpenetrating sublattices of opposite magnetization which characterize the antiferromagnetic state may be analyzed in terms of the fundamental gyromagnetic equation

$$dM/dt = \gamma M \times H \qquad (24)$$

where, however, the motion of a particular sub-lattice magnetization is determined by a total magnetic field consisting of the external field H_0 and internal H_A and H_E associated with anisotropy forces and exchange forces. The field H_{1E} at electron spins of sub-lattice 1 is negatively proportional to the magnetization of sub-lattice 2 through the antiferromagnetic exchange integral. The resonance condition for zero external field is, for $H_A << H_E$

$$\omega(H_0 = 0) = \gamma(2H_EH_A)^{1/2} \qquad (25)$$

and Johnson and Nethercot (25) have used high microwave frequencies to measure the variation of ω with temperature as H_E declines on approaching the Neel temperature from below (Fig. 11).

The examples which have been presented here give ample illustration of the sensitive tool for investigation of solids provided by magnetic resonance. The energy of the irradiating radiofrequency and microwave quanta is weak, but the techniques are powerful.

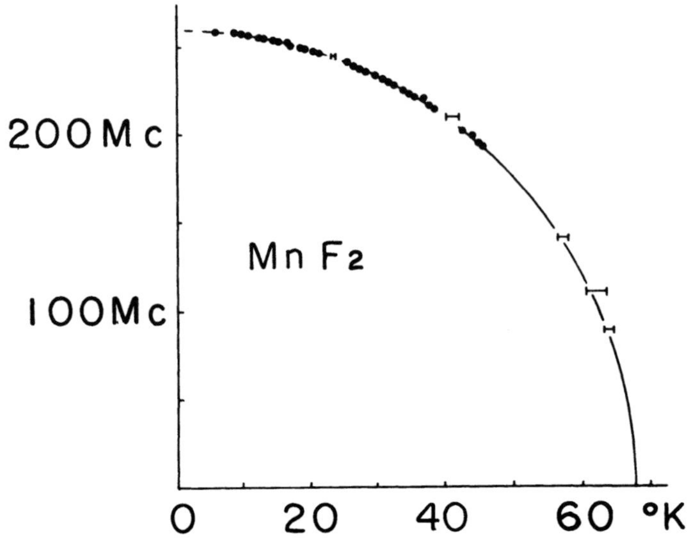

Fig. 11. Antiferromagnetic resonance frequency versus temperature for MnF_2

REFERENCES

1. E.M. Purcell, H.C. Torrey, and R.V. Pound, Phys. Rev., 69, 37 (1946).

2. E.Zavoisky, J. Phys. USSR, 9, 211, 245 (1945).

3. R.L. Cummerow and D. Halliday, Phys. Rev., 70, 433 (1946).

4. F. Bloch, Phys. Rev., 70, 460, (1946).

5. J.H. Van Vleck, Phys. Rev., 74, 1168 (1948).

6. C.R. Bruce, Phys. Rev., 107, 43 (1957).

7. J.H. Van Vleck, Phys. Rev., 57, 426 (1940).

8. H.S. Gutowsky and G.B. Kistiakowsky, G.E. Pake, and E.M. Purcell, J. Chem. Phys., 17, 972 (1949).

9. H.S. Gutowsky and G.E. Pake, J. Chem. Phys. 18, 162 (1950).

10. E.R. Andrew, "Nuclear Magnetic Resonance", 173 ff, Cambridge University Press, 1955.

11. A. Abragam, "Principles of Nuclear Magnetism", 212-213, Oxford

University Press, 1961.

12. *R.T. Schumacher and C.P. Slichter, Phys. Rev., 101, 58 (1956).*

13. *D.F. Holcomb and R.E. Norberg, Phys. Rev., 98, 1074 (1955).*

14. *T.J. Rowland, "Nuclear Magnetic Resonance in Metals", 34, Pergamon Press, 1961 (Vol. 9 of Progress in Materials Science series).*

15. *T.J. Rowland, ibid., p. 58.*

16. *L.C. Hebel and C.P. Slichter, Phys. Rev. 113, 1504 (1959).*

17. *A.M. Portis and A.C. Gossard, J. Appl. Phys. 31, 205S, (1960).*

18. *T.J. Rowland, "Nuclear Magnetic Resonance in Metals", 81, Pergamon Press, 1961 (Vol. 9 of Progress in Materials Science series).*

19. *G.E. Pake, "Paramagnetic Resonance", 157, W.A. Benjamin, Inc. 1962.*

20. *G.B. Benedek, "Magnetic Resonance at High Pressure", 41, Interscience Publishers, 1963.*

21. *T.G. Castner and W. Kanzig, I. Phys. Chem. Solids, 3, 178 (1957).*

22. *C.P. Slichter, "Principles of Magnetic Resonance",198, Harper and Row, 1963.*

23. *W.C. Holton, H. Blum, and C.P. Slichter, Phys. Rev. Letters, 5, 197 (1960).*

24. *J.H. Burgess, R.S. Rhodes, M. Mandel, and A.S. Edelstein, J. Appl. Phys. 33, 1352S (1962).*

25. *F.M. Johnson and A.H. Nethercot, Phys. Rev. 114, 705 (1959).*

DISCUSSION

A. CHARLESBY (Royal Military College of Science, England): *In many irradiated materials we see an increase in the intensity of the E.S.R. pattern, but this tends to a maximum concentration; however, the chemical or physical effects appear to increase. Thus we have a difference in the information from these two sources. Is this due to a feature of resonance theory at high concentration or of our interpretation?*

G.E. PAKE: *I am not sufficiently familiar with these experiments to know the answer. One possibility is that the chemical effects do not always, especially at higher concentrations, produce unpaired spins which remain so without picking up a stray electron to produce a magnetically alert pair. There are other possibilities such as dipole-dipole couplings of close neighbors which broaden the ESR contribution so that it is weak and undetected because it is spread over a wide frequency range.*

N.K. KREIDL (University of Missouri, USA): *In glass technology we are often concerned with the distinction of electron from hole traps. How do you judge the power and reliability of the calculation from ESR data?*

G.E. PAKE: *In favorable cases, the theory of the shift of the electron* g *from the free electron value,* g_e*, indicates the direction of the shift for an electron, and the opposite direction for a hole. If our knowledge of the source of the crystal field were complete, and if the theory were tractable, we can in principle thus distinguish an electron from a hole by observing the direction of shift. Unfortunately our knowledge of the sources is not complete, and there are often serious approximations in the calculations of this theory. We often must depend upon confirming evidence from other kinds of experiments to be confident whether the ESR is from a hole or an electron.*

MAGNETIC RESONANCE STUDIES OF BONDING, STRUCTURE, AND DIFFUSION IN CRYSTALLINE AND VITREOUS SOLIDS

P. J. Bray

Brown University

Providence, Rhode Island, U.S.A.

Nuclear magnetic resonance techniques are being employed to study glasses and crystalline compounds in various systems. Chemical shift effects and nuclear quadrupole interactions yield information concerning atomic configurations and chemical bonding. Particular attention is focused on identification in the glasses of configurations and structural groupings found in the compounds of systems. Completed studies of borate, phosphate and fluoride glasses are reported, including data obtained from the resonances of isotopes of lithium, beryllium, boron, fluorine, sodium, aluminum, phosphorous, cesium and lead. Progress reports are given on the study of NMR spectra from other nuclei in glasses. The use of NMR techniques to study diffusion in solids is illustrated with studies of alkali borates and silicates.

INTRODUCTION

Nuclear magnetic resonance (NMR) techniques have been used in recent years to study chemical bonds and atomic arrangements in a variety of glasses (1-8). The theory and experimental procedures pertinent to such studies have been presented elsewhere (1-5), but a brief summary is appropriate here.

The nuclei of most atoms possess a magnetic dipole moment $\vec{\mu}$ which can be expressed as

$$\vec{\mu} = g\,\mu_0 \vec{I} \qquad (1)$$

In Eq.1, $\vec{I}$ is the dimensionless spin vector of magnitude $\left[I(I+1)\right]^{\frac{1}{2}}$ which gives the intrinsic spin angular momentum of the nucleus in units of $h/2\pi$, the scalar I is an integer or half-integer, μ_0 is the Bohr nuclear magneton, and g is the nuclear g-factor. If the nucleus is subjected to a magnetic field $\vec{H}$, the interaction between the field and the nuclear magnetic dipole moment creates a set of possible energy levels for the system as follows:

$$E = - \vec{\mu}.\vec{H} = - g \mu_0 \vec{I}.\vec{H} = - g \mu_0 H m \qquad (2)$$

where the quantity m, called the magnetic quantum number, has 2I+1 values ranging from - I to I in integer steps.

In the absence of other interactions, the 2I+1 energy levels given by Eq. 2 are equally spaced as shown in the left-hand side of Fig. 1. Transitions between adjacent levels occur, with consequent power absorption, when the system is bathed in electromagnetic radiation whose frequency ν_0 satisfies the Bohr resonance condition:

$$h\nu_0 = \Delta E = g\mu_0 H \qquad (3)$$

Since the energy levels are equally spaced, only one resonance frequency is observed in this case. (See the left-hand side of Fig. 1)

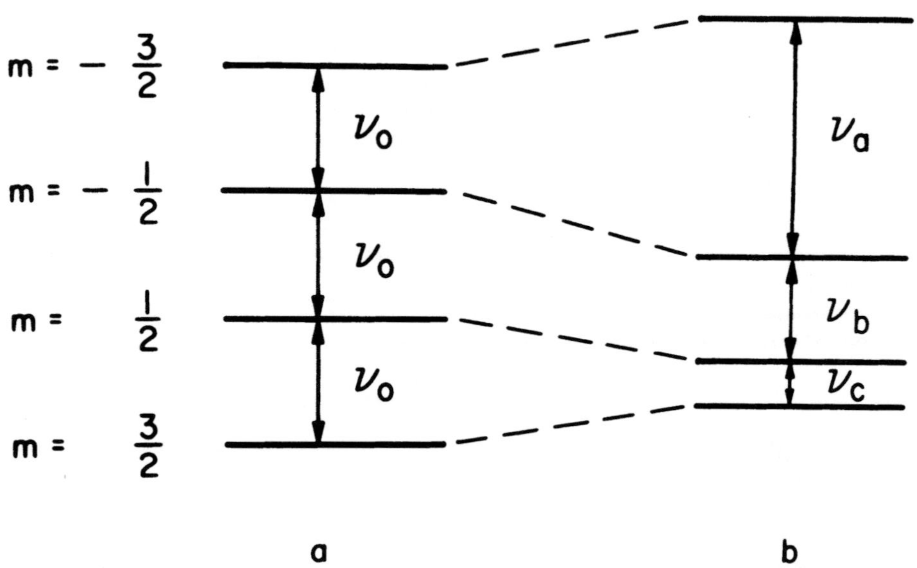

Fig. 1. Energy levels arising from the interaction of the nuclear magnetic dipole moment with a magnetic field: (a) no electrical quadrupole interaction; (b) small quadrupole interaction present. (The levels shown are appropriate for the B^{11} nucleus).

QUADRUPOLE INTERACTIONS

In addition to a magnetic dipole moment, many nuclei (all those with spin $I \geq 1$) possess an electric quadrupole moment which can interact with the gradient of any electric field that exists at the nucleus. Such electric field gradients can arise from electrons in the chemical bonds of the atom containing the nucleus and from charges on other atoms or ions. Since these field gradients depend strongly on the nature of the chemical bonds, and on the type and arrangement of ions in the environment of the atom containing the nucleus, the quadrupole interaction is a sensitive measure of bonding and structure on an atomic or microscopic scale.

When a quadrupole interaction is present, the magnetic energy levels are shifted (perturbed) as shown on the right-hand side of Fig. 1. Since the interaction is a function of the relative orientation of the magnetic field with respect to the principal axes of the electric field gradient tensor, the arrangement of levels on the right-hand side of Fig. 1 is appropriate to a particular orientation of a single crystal in the magnetic field. When considering a polycrystalline powder or a glass, the NMR response to be expected for each transition (such as the $m = -3/2 \leftrightarrow m = -1/2$ transition) will be the envelope of the range of values of the transition frequency (e.g. ν_a in Fig. 1) that would be observed for all orientations of the magnetic field.

The NMR absorption curve for a polycrystalline or vitreous sample is displayed in Fig. 2 for the case of a small quadrupole interaction, a nucleus of spin $I = 3/2$, and axial symmetry for the electric field gradient tensor. (For cases of larger half-integer values of I, the curve is similar but exhibits additional divergences such as those at $\pm A_1$, and additional shoulders such as those at $\pm 2A_1$; they arise from transitions such as $\pm 5/2 \leftrightarrow \pm 3/2$, $\pm 7/2 \leftrightarrow \pm 3/2$, etc., and are located further away from the central $+1/2 \leftrightarrow -1/2$ transition which is unaffected in first order by a quadrupole interaction). The dashed curve is the theoretical envelope noted above; the solid curve is the expected absorption curve as broadened and smoothed by the dipole-dipole interaction between the resonating nuclei and the range of small, local magnetic fields arising from the magnetic dipole moments of neighboring nuclei. It is clear that the NMR response in such cases is quadrupole interaction. (The quantity A_1 in Fig. 2 is proportional to the quadrupole coupling constant).

Fig. 3 displays the NMR of B^{11} in polycrystalline boron phosphate (BPO_4). The quadrupole coupling constant is (9) small (50.4 Kcps) because of the nearly perfect tetrahedral symmetry of the BO_4 configuration found in this compound. (The boron atom is in the center of the tetrahedron of oxygens). If the symmetry were

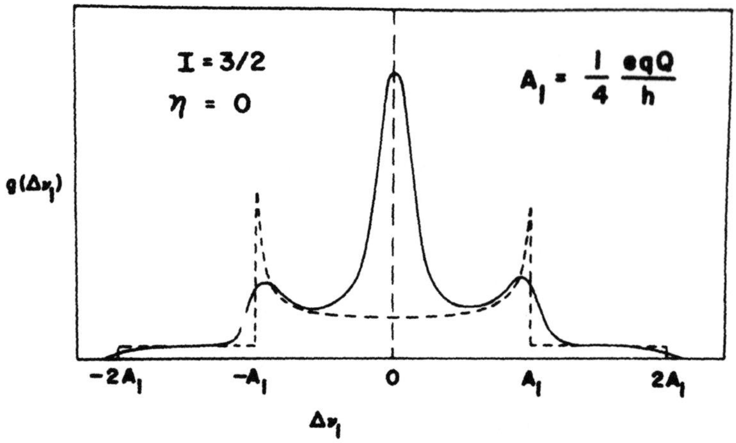

Fig. 2. Resonance line shape (solid curve) predicted for a nucleus such as B^{11} with a small quadrupole interaction in a glass or polycrystalline powder.

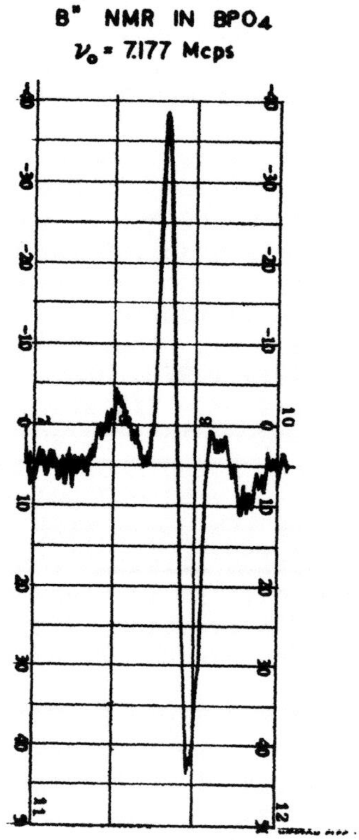

Fig. 3. The NMR of B^{11} in boron phosphate (BPO_4).

perfect, the electric field gradient would vanish and the coupling would be zero.

A larger quadrupole interaction (with axial symmetry) in a glass or polycrystalline powder produces the asymmetric absorption curve given in Fig. 4. Again, the dashed curve is the theoretical envelope without dipolar broadening while the solid includes the dipole-dipole interaction. The quantity A_2 is given by

$$A_2 = \frac{9}{64} \cdot \frac{(2I + 3)}{4I^2 (2I-1)} \cdot \frac{(e^2 qQ)^2}{\nu_0} \tag{4}$$

where $e^2 q Q$ is the quadrupole coupling constant that measures the strength of the interaction between the quadrupole moment Q of the nucleus and a component q of the electric field gradient tensor; ν_0 is the resonance frequency in the absence of a quadrupole interaction.

It should be noted that the entire curve in Fig.4 arises from the $m = 1/2 \leftrightarrow m = -1/2$ or "central" transition. A quadrupole interaction of sufficient size to produce this shape for the central transition will spread the "satellite" transitions (such as the $m = \pm 3/2 \leftrightarrow m = \pm 1/2$ transitions) over such a broad region of frequency that they are not ordinarily observed.

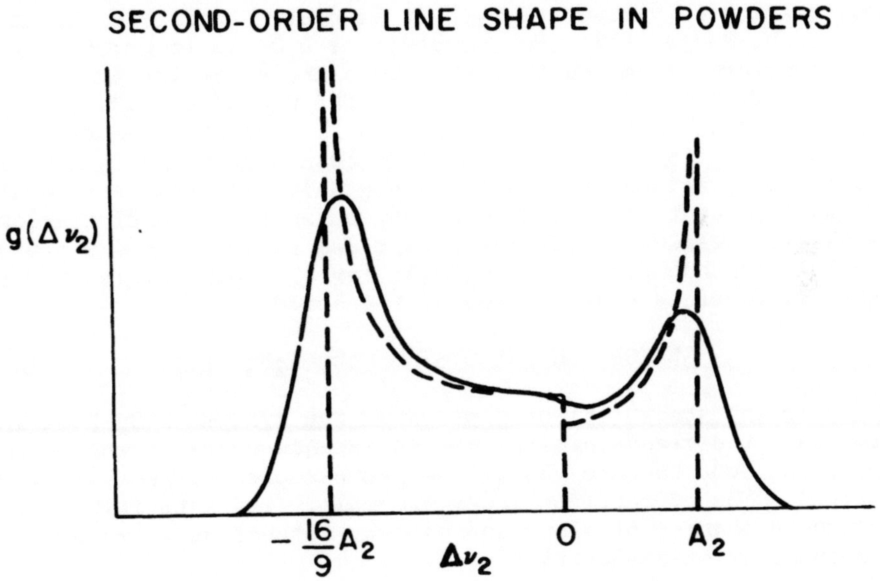

Fig. 4. Resonance line shape (solid curve) for a nucleus with a large quadrupole interaction in a glass or polycrystalline powder.

The line shape for a large quadrupole interaction shown in Fig. 4 is found in the case of B^{11} in vitreous boron oxide (B_2O_3). A dispersion mode response - which in this case produces a curve proportional to the absorption curve itself (10) rather than its first derivative - is shown (11) in Fig. 5. The usual absorption mode response - yielding the first derivative of the absorption curve - is displayed (11) in Fig. 6 along with the line shape obtained by integrating the recorder tracing. Analysis of the curves in Figs. 5 and 6 yields a quadrupole coupling constant of 2.76 Mcps, a value that lies in the range of 2.4 - 3.1 Mcps found (9) for all of the crystalline compounds known to contain the trigonal, planar BO_3 configuration. A recent dispersion mode study (11) of vitreous B_2O_3 at high levels of radiofrequency power has yielded the satellite ($\pm 3/2 \leftrightarrow \pm 1/2$) transitions. They are in accord with the 2.76 Mcps coupling constant and axially symmetric field gradient tensor deduced from the measurements of the central transition. It is clear that the boron atoms in vitreous B_2O_3 are in BO_3 triangles and not in the center of BO_4 tetrahedra.

CRYSTALLINE BORON OXIDE (12)

The B^{11} NMR response for crystalline B_2O_3 is displayed in Fig. 7 along with a trace for the vitreous material. Only one NMR response is observed for the crystalline material, and that response gives the coupling constant of 2.76 Mcps found for the BO_3 configuration in vitreous B_2O_3. It is clear that the structure proposed by Berger (13) for crystalline B_2O_3 is incorrect, since that structure requires that the boron atoms reside in two types of BO_4 configuration. The resonance for vitreous B_2O_3 in Fig. 7 exhibits a small response at ν_0 (to the right of center of the spectrum; see also Fig. 4). This response, not found in the crystalline B_2O_3, arises from boron in BO_4 units and indicates that as much as 1 percent of the borons may be in this configuration in the glass. Residual hydroxyl groups and water from the starting material (orthoboric acid, H_3BO_3) may be responsible for this degree of fourfold coordination of the boron.

GLASSES IN THE SYSTEM GeO_2-B_2O_3 (14)

A recent study (15) of glasses in the system GeO_2-B_2O_3 indicated that the glass-forming region extends over the entire range from pure GeO_2 to pure B_2O_3. No macroscopic immiscibility was observed. The refractive index and molar refractivity displayed pronounced changes of slope and minima, respectively, when plotted as a function of composition.

NMR studies (14) of B^{11} in sixteen glasses spanning the entire composition range yield only three-coordinated borons: no BO_4 tetrahedra are formed in these glasses. Measurements of the sepa-

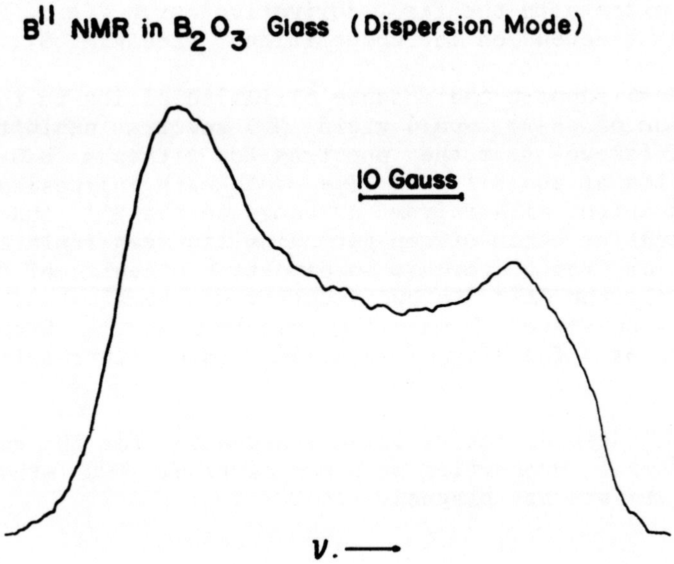

Fig. 5. The B^{11} NMR spectrum for vitreous B_2O_3 at 16 Mcps.

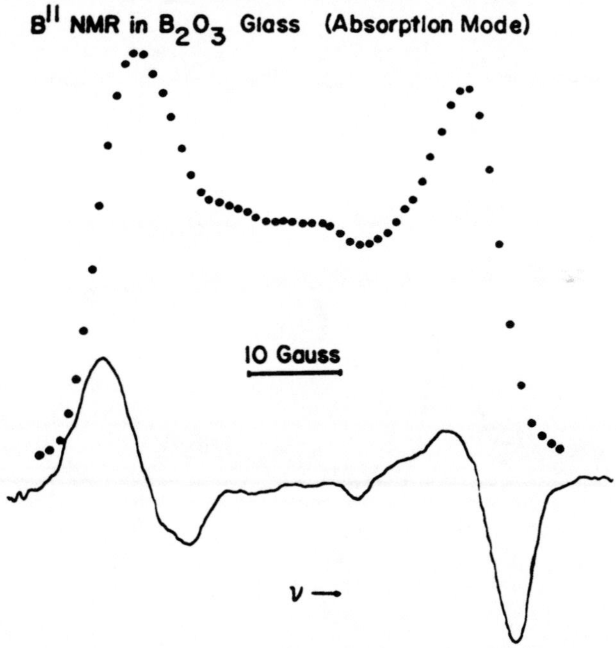

Fig. 6. Solid curve is the experimental derivative curve: open circles represent the absorption curve obtained by numerical integration of the experimental curve.

ration of the extrema in the first derivative curve (e.g. Fig. 7) yield values that depend on the composition. (See Fig. 8).

The NMR data support the absence of immiscibility in the system. Separation of B_2O_3 would yield NMR spectra exhibiting no observable difference from the spectrum for vitreous B_2O_3. The marked reduction of the B^{11} NMR line width with increasing GeO_2 content could arise either from a change in the B^{11} quadrupole coupling constant as boron-oxygen-germanium linkages replace boron oxygen-boron, or from a decrease in dipolar broadening of the B^{11} resonance. (Only the Ge^{73} isotope - having a natural abundance of 7.61 percent - possesses a magnetic dipole moment). Studies of the B^{11} spectra as a function of magnetic field will resolve this matter.

The B^{11} NMR data do not yield an explanation for the reported changes in physical properties with composition. NMR studies of Ge^{73} in these glasses are planned.

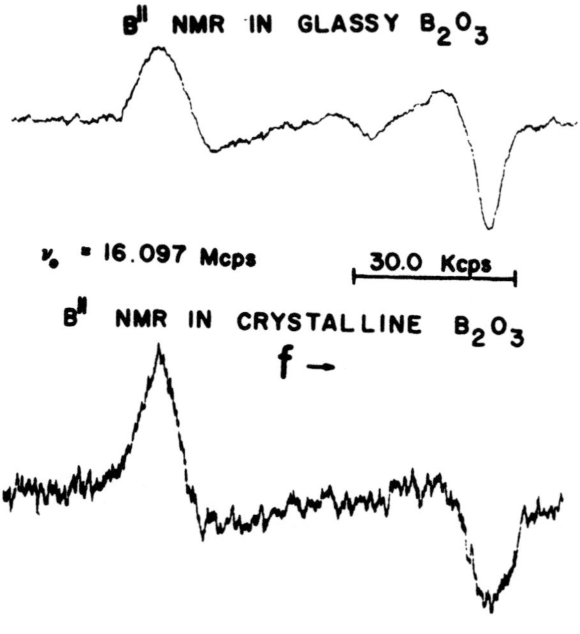

Fig. 7. The B^{11} NMR spectrum obtained for vitreous B_2O_3 (top curve) and crystalline B_2O_3 (bottom curve).

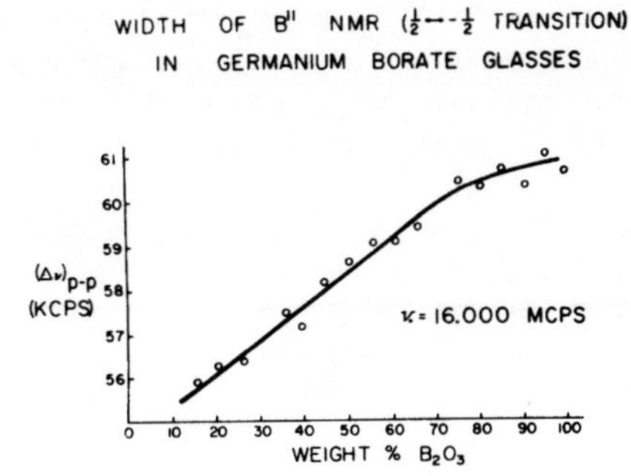

Fig. 8. Width of the B^{11} NMR for glasses in the system GeO_2-B_2O_3. (The resonance observed is the $\frac{1}{2} \leftrightarrow -\frac{1}{2}$ transition for three-coordinated borons).

ALKALI BORATE AND LEAD BORATE GLASSES

According to the X-ray studies by Warren (16), the addition of alkali oxide to boron oxide should bring about fourfold coordination of some of the boron atoms. The NMR spectra for glasses in the system Li_2O-B_2O_3 are shown (3) in Fig. 9. It is clear that addition of Li_2O does produce BO_4 units that yield the narrow resonance at ν_0. The dependence of N_4 - the fraction of boron atoms in the tetrahedral BO_4 configuration - on the molar percent of alkali oxide is shown by the data points in Fig. 10. The solid curve is the calculated value of N_4 based on the assumption that each oxygen added to the system in the form of alkali oxide converts two boron atoms from BO_3 to BO_4 configurations. This curve is of the form X/1-X for glasses of molar composition $XR_2O.(1-X)B_2O_3$ where R is an alkali. It is clear from Fig. 10 that the borate or boron anomaly (17) is not due to a maximum of N_4 at or near an alkali oxide concentration of 16 2/3 percent (18). The fraction N_4 continues to rise with increasing alkali oxide, contrary to suggestions that it might decrease rapidly to zero (19) or maintain a constant value (20).

Since N_4 follows the X/1-X relationship up to approximately 30 to 35 molar percent alkali oxide, it appeared useful to plot N_4 vs X/1-X. Figure 11 displays the result (21) for the lithium borate glasses. It is clear that the data for the entire range of compositions can be fitted by two straight-line segments. A similar plot (21) of N_4 values for lead borate glasses (Fig. 12) exhibits three straight-line segments. The corresponding plot of N_4 versus X (Fig. 13) does not reveal this division into regions of composition.

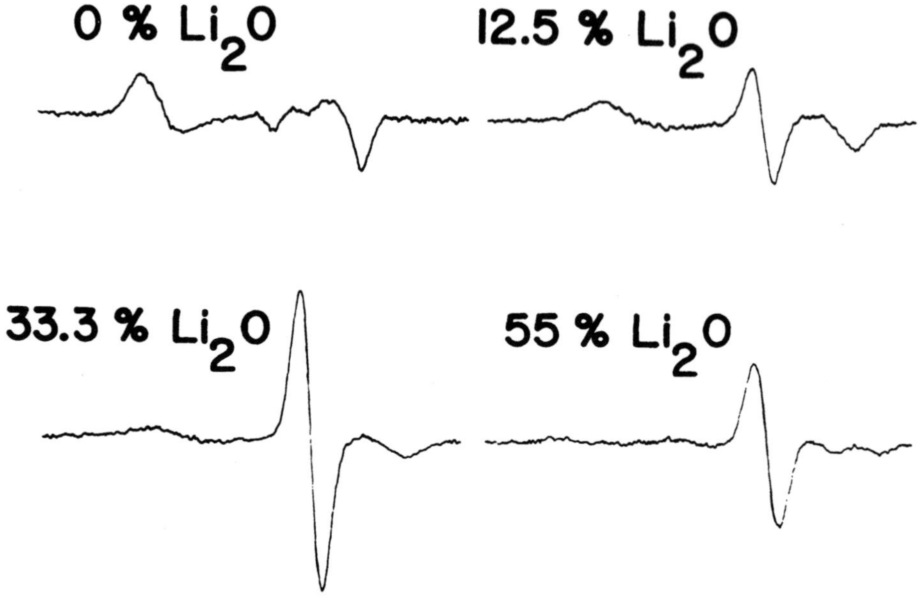

Fig. 9. The B^{11} NMR spectra for four glasses in the system $Li_2O-B_2O_3$.

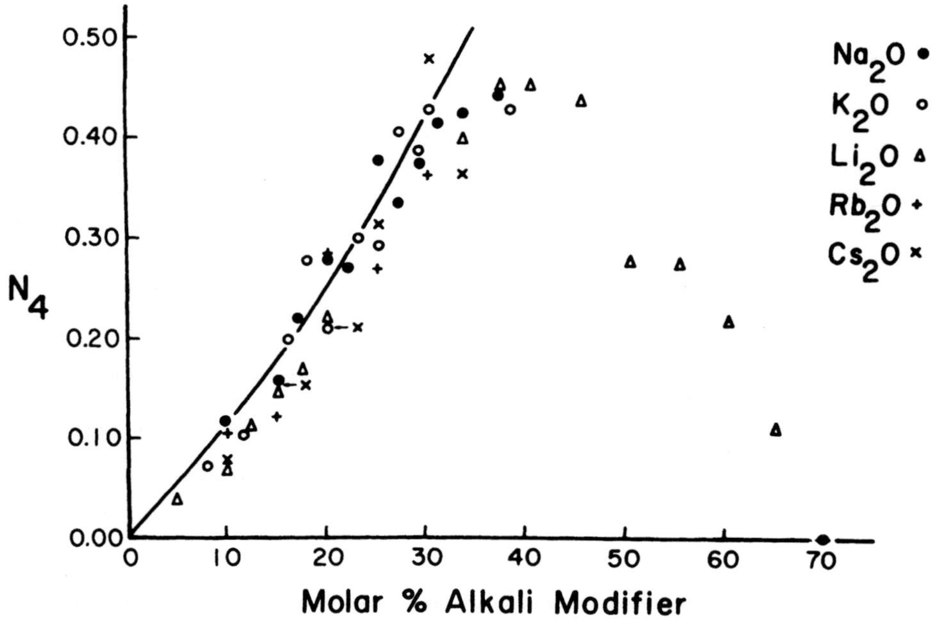

Fig. 10. The fraction N_4 of boron atoms in BO_4 configurations in alkali borate glasses plotted against the molar percent of alkali oxide.

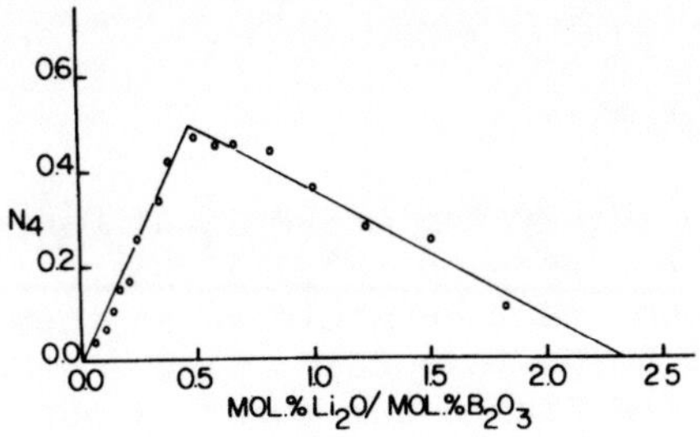

Fig. 11. The fraction N_4 of four-coordinated boron atoms in lithium borate glasses plotted against X/1-X where X is the molar fraction of Li_2O.

Fig. 12. The fraction N_4 of four-coordinated boron atoms in lead borate glasses plotted against X/1-X where X is the molar fraction of PbO.

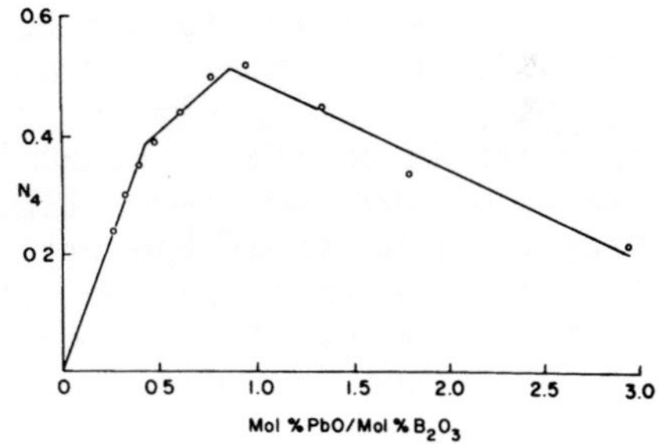

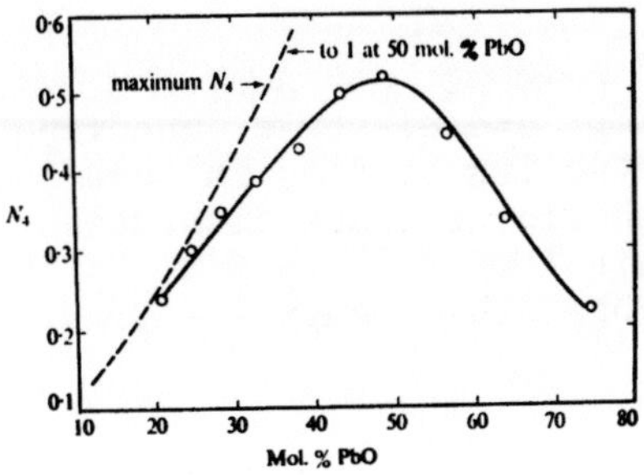

Fig. 13. The fraction N_4 of four-coordinated boron atoms in lead borate glasses plotted against the molar percent of PbO.

A simple analysis of borate glasses has been developed (21) that predicts the straight-line segments in plots of N_4 versus X/1-X. It can be used to deduce from the N_4 data the average numbers of bridging and non-bridging oxygens per boron in these glasses.

CABAL GLASSES

The glass-forming region (22) in the system $CaO-B_2O_3-Al_2O_3$ is shown in Fig. 14. It has commonly been assumed (22,23) that complete four-coordination of the aluminum must be achieved in these glasses before four-coordination of boron can take place. No glasses can be formed below the line denoting equal molar amounts of CaO and Al_2O_3; this is attributed to the presence of six-coordinated aluminum in these materials. These concepts seem to explain very well the sharp boundary of the glass-forming region along the line $CaO.Al_2O_3$.

The B^{11} NMR spectrum (7) displayed in Fig. 15 is for the cabal glass containing 15 molar percent CaO and 15 molar percent Al_2O_3. In addition to the broad line from boron in BO_3 configurations, a substantial narrow line from BO_4 configurations is present. Approximately 12 percent of the boron atoms in this glass are in fourfold coordination. NMR investigations of other glasses on the $CaO.Al_2O_3$ line yield N_4 values between 10 and 15 percent. Studies of the proton NMR spectra for these glasses show that the four-coordinated boron is not due to the presence of hydroxyl groups. It is clear that boron and aluminum must compete for the oxygen introduced as CaO; the aluminum does not get all of it.

Initial investigations (24) have been made of cabal-type glasses in which scandium oxide (Sc_2O_3) has been substituted for Al_2O_3. The NMR trace in Fig. 16 displays the Sc^{45} resonance (24) in the dispersion mode for a glass that is 0.30 CaO-0.60 B_2O_3-0.10 Sc_2O_3. (The Al^{27} and Na^{23} resonances arising from the radiofrequency probe are also visible).

Since the Sc^{45} resonance is 100 percent abundant and has a relatively large magnetic dipole moment and electric quadrupole moment, it should be quite useful in studying bonding in scandium-containing glasses. Further study should produce resolution of the broad resonance of Fig. 16; the spin I of Sc^{45} is 7/2 so that six satellite transitions should surround the central $+\frac{1}{2} \leftrightarrow -\frac{1}{2}$ transition.

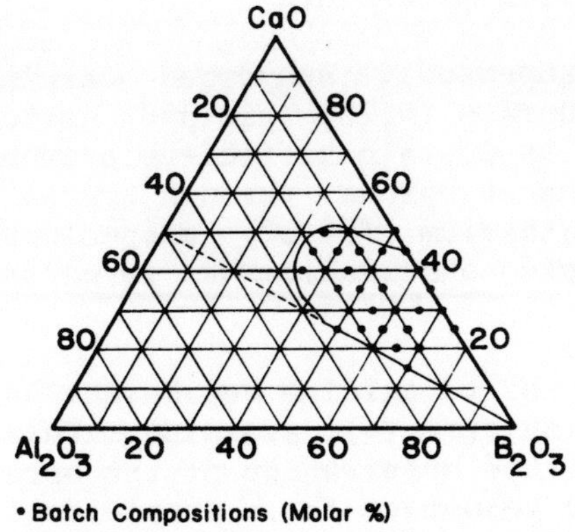

Fig. 14. Composition diagram for the system CaO-B_2O_3-Al_2O_3.

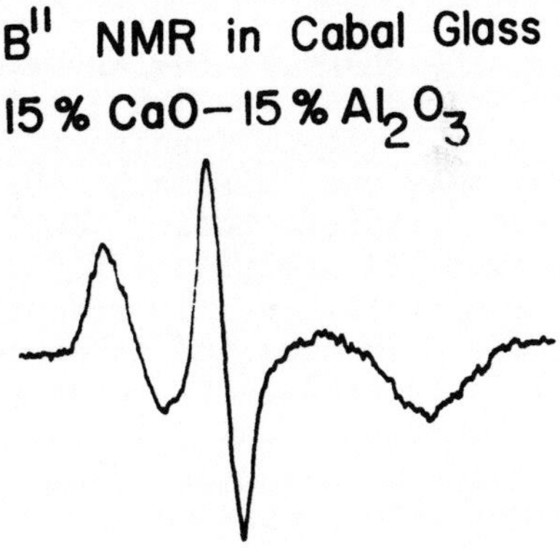

Fig. 15. The B^{11} NMR spectrum for the glass of composition 0.15 CaO-0.70B_2O_3-0.15 Al_2O_3.

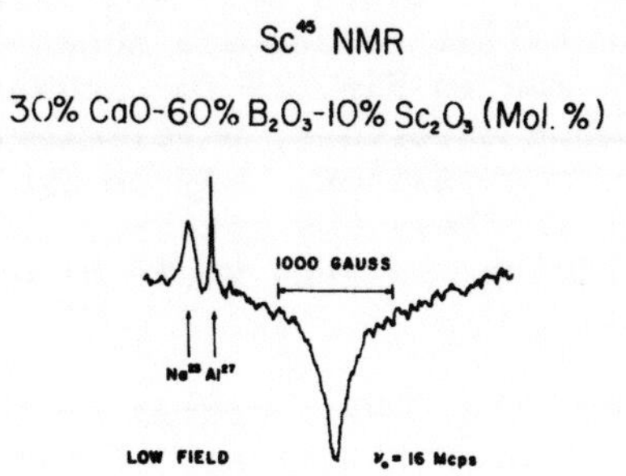

Fig. 16. The Sc^{45} NMR spectrum (dispersion mode) for the glass of composition 0.30CaO-0.60B_2O_3-0.10Sc_2O_3.

METABORATES

A. Calcium and Potassium Metaborate

Earlier NMR studies of B^{11} in supposedly anhydrous calcium metaborate (25) and potassium metaborate (9,25) yielded narrow lines. The samples were later found to contain sufficient protons to produce the observed four-coordinated borons through hydroxyl groups. Dry samples of the pure potassium (26,27) and calcium (28) metaborates have failed to yield observable NMR responses despite extensive investigations.

Addition of paramagnetic impurities to calcium metaborate has reduced the B^{11} spin-lattice relaxation time (4), avoiding saturation (4) of the resonance, and permitted observation of the boron NMR spectrum. Figure 17 displays the resonance for a sample doped with manganese chloride. (Doping had previously (4) been found useful in studying the NMR spectra of Pb^{207} in lead borate glasses).

The shape of the spectrum (Fig. 17) suggests the presence of two resonances involving different quadrupole coupling constants. However, closer analysis indicates that the spectrum may arise from only one distinct BO_3 configuration in which the electric field gradient tensor does not exhibit axial symmetry. (Full theoretical analyses (29,30) of this rather complex situation have become available only recently). A coupling constant of approximately 2.76 Mcps and an asymmetry parameter η of approximately 0.4 would produce the observed spectrum. (η measures the departure of the tensor from axial symmetry). However, careful studies as a function of magnetic field are needed in order to determine these values with precision.

The line shapes now known (29,30) to arise for cases of BO units without axial symmetry, and the NMR spectrum observed fo calcium metaborate, indicate a need to reexamine the many know cases in which splittings of the broad B^{11} lines for BO_3 unit have been reported (3,5-7,21). Representative cases are shown i Fig. 18 for an alkali borate glass (3) and in Fig. 19 for severa barium borate glasses (21). Some of these materials, particularl those in the vicinity of the metaborate composition, may be exhibiting BO_3 units without axial symmetry in their NMR spectra. (BO configurations with one or two non-bridging oxygens would be obvious possibilities in these materials).

Attempts are in progress to render the NMR spectra of th alkali metaborate compounds observable through the effect of paramagnetic centers introduced as paramagnetic ions or produced b irradiation.

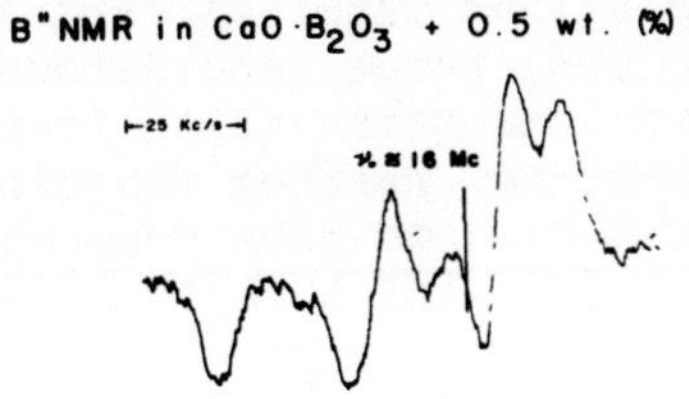

Fig. 17. The B^{11} NMR in polycrystalline calcium metaborate ($CaO.B_2O_3$) doped with manganese chloride.

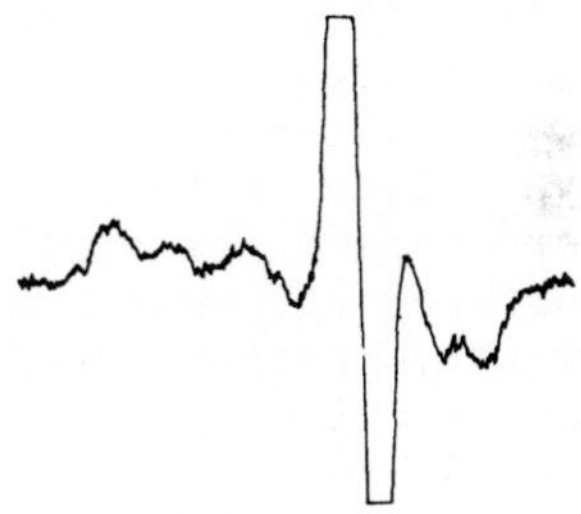

Fig. 18. The B^{11} NMR spectrum for a glass of molar composition 0.65 Li_2O-0.35B_2O_3.

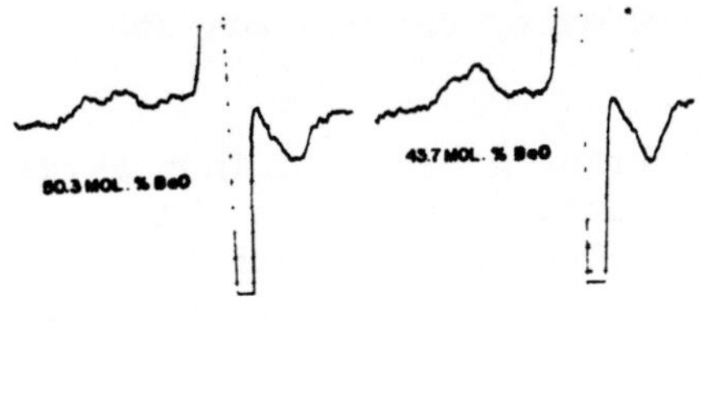

Fig. 19. The B^{11} NMR spectra for several glasses in the system BaO-B_2O_3. (Molar percents of BaO are noted for each glass).

B. Zinc Metaborate

Krogh-Moe (31) has inferred from infrared spectra that the fraction of boron in fourfold coordination is quite different for vitreous and crystalline 4 $ZnO.3B_2O_3$. This behavior is unusual in that, for many borates, the glass and crystalline material exhibit the same fractions of four-coordinated and three coordinated borons. (However, $PbO.2B_2O_3$ is a known exception (5)). Since there is disagreement in the literature as to whether the zinc metaborate is (32) 4 $ZnO.3B_2O_3$ or (33) $ZnO.B_2O_3$, we have prepared and studied vitreous and crystalline materials of both compositions.

The NMR spectra (34) for all four materials (Fig. 20) exhibit both the broad B^{11} resonance characteristic of three-coordinated borons and the narrow resonance from four-coordinated borons. (The peaks located symmetrically on both sides of the central narrow line in the crystalline material are the $m = \pm 3/2 \leftrightarrow m = \pm \frac{1}{2}$ or "satellite" transitions for the four-coordinated borons; these transitions are not observable in the glass). However, glasses of both the 1:1 and 4:3 compositions are found to have only about 15% of the boron atoms in fourfold coordination while the crystalline materials yield N_4 values of 0.60 ± 0.08. These data corroborate the earlier deduction (31) of a large difference in N_4 between vitreous and crystalline zinc metaborate, but do not support the structural model (32) in which all borons are in fourfold coordination.

$PbO-B_2O_3$ AND $PbO-SiO_2$ MATERIALS

The Pb^{207} nucleus does not have a quadrupole moment, but the resonance does exhibit a chemical shift (4,5,35); that is, with a constant value of ν_0 in Eq. 3, the resonance condition occurs for different values of the applied magnetic field H_{App}. The effect originates in diamagnetic and second-order paramagnetic changes of the atomic electron distribution caused by the applied field (36, 37). When the effect is independent of the orientation of the field with respect to the electron distribution, it can be expressed as

$$H = H_{APP} (1 - \sigma) \tag{5}$$

where H is the field actually experienced by the nucleus and σ is the chemical shift parameter.

The values of H_{APP} necessary to bring H of Eq. 5 into agreement with H of Eq. 3 and achieve the resonance condition have been measured for Pb^{207} in vitreous and crystalline materials of the systems $PbO-B_2O_3$ and $PbO-SiO_2$(4,5,14). It is convenient to deal not with H_{App} itself for each material but with the difference ΔH between H_{App} for the material in question and H_{App} for a reference material which is lead metal in the present case. The values of ΔH obtained for a fixed spectrometer frequency of 6.3208 Mcps are given in Figs. 21 and 22.

The shift of 95-100 gauss at low PbO content in the lead borate glasses (5) (Fig. 21) is characteristic of ionic lead (35). As the PbO content increases, the shift decreases rapidly and converges toward the value of 60 gauss found (4) in orthorhombic and tetragonal PbO where the bonding is strongly covalent in the PbO_4 pyramids. The shifts in the compounds agree within experimental error with the shifts in the corresponding glasses, suggesting that the lead-oxygen configurations (38) present in the compounds occur in the glasses. It is probable that the resonance for a glass of any composition is, in general, an unresolved mixture of resonances from the lead-oxygen configurations (38) present in the crystalline compounds. The substantial width of the Pb^{207} resonances (Fig. 23) probably reflects this mixture of unresolved resonances, and broadening of each resonance by anisotropic chemical shifts. (Anisotropy of the chemical shift is discussed in the next section).

Figure 22 displays the Pb^{207} chemical shifts measured (5,14) for lead silicate glasses and compounds. Again, there is agreement between the shifts for the compounds and glasses of the same composition, suggesting that the lead-oxygen configurations in the compounds occur in the glasses. At low PbO content the shifts are not the values of 95 to 100 gauss found for highly ionic lead. Instead, the value is the same as the shift for the compound PbO-SiO_2 which is the only lead silicate compound in the phase diagram between 0 and 50 molar percent PbO. It appears that the particular lead-oxygen configuration of this compound - and not Pb^{++} ions - is the single configuration present in the glasses up to 50 molar percent PbO.

CESIUM BORATE GLASSES (39)

NMR measurement of the Cs^{133} chemical shift in cesium silicate glasses have recently been reported (40). A distinct change in the shift (measured with respect to a saturated aqueous solution of Cs_2CO_3) was observed at about 25 - 30 molar percent Cs_2O; the activation energy for electrical conduction in these glasses exhibits a break at about the same composition. The measurements were confined to one spectrometer frequency and one temperature; no asymmetry of the Cs^{133} resonance was observed. Hence, effects of quadrupole interactions and anisotropic chemical shifts were not detected.

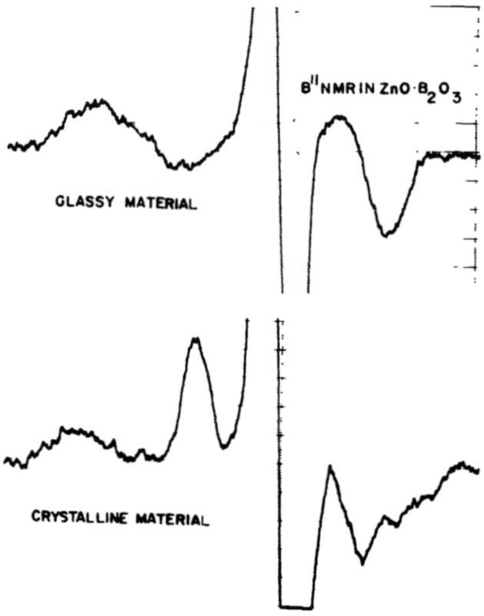

Fig. 20. The B^{11} NMR spectra for vitreous and polycrystalline zinc metaborate.

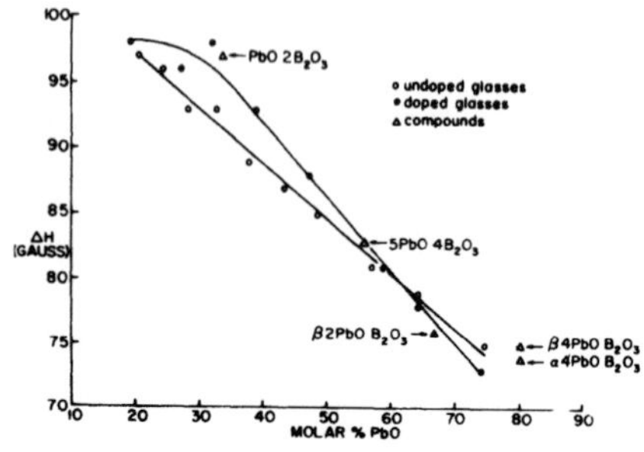

Fig. 21. The Pb^{207} chemical shifts for vitreous and crystalline lead borates.

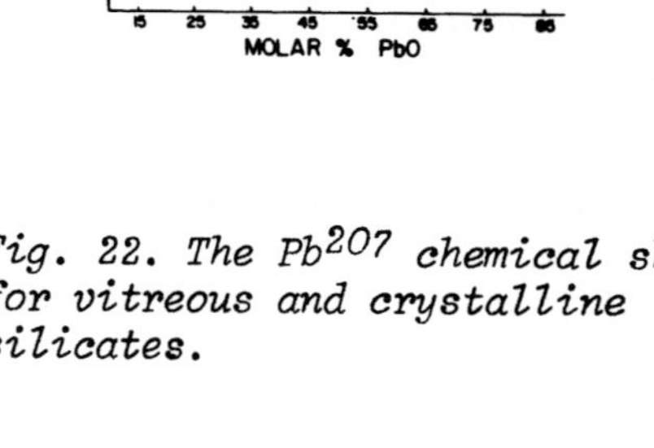

Fig. 22. The Pb^{207} chemical shifts for vitreous and crystalline lead silicates.

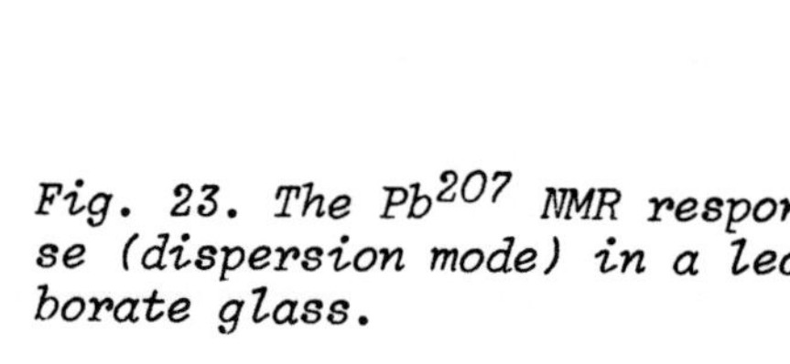

Fig. 23. The Pb^{207} NMR response (dispersion mode) in a lead borate glass.

A completed study (39) of Cs^{133} in cesium borate glasses as a function of frequency and temperature has yielded measurements of both quadrupole effects and chemical shift anisotropies that show structural changes in the glasses. Figure 24 displays the width of the Cs^{133} resonance at room temperature in various glasses as a function of spectrometer frequency. The width (measured as the separation in frequency - or magnetic field - between positive and negative extrema of the recorder tracing of the first derivative of the absorption curve) can be expressed (41,42) to a first approximation by

$$\Delta\nu = (\Delta\nu)_{DIP} + \frac{125}{9408}\,\frac{(e^2qQ)^2}{\nu_0} + (\sigma_{\parallel} - \sigma_{\perp})\,\nu_0 \qquad (6)$$

where $(\Delta\nu)_{DIP}$ is the width due to dipolar broadening, the term involving $(e^2qQ)^2/\nu_0$ is the contribution to the width due to second-order quadrupole effects on the observed $+\tfrac{1}{2} \leftrightarrow -\tfrac{1}{2}$ transition, and the final term arises from an axially symmetric chemical shifts. The quantities $\sigma_{\parallel}$ and $\sigma_{\perp}$ are the two components of the (axially symmetric) chemical shift tensor in its principal axis system.

It is clear from Fig. 24 and Eq. 6 that values for the Cs^{133} coupling constant (e^2qQ) and chemical shift anisotropy $(\sigma_{\parallel} - \sigma_{\perp})$ can be obtained for each glass. The quantities are displayed as a function of composition in Figs. 25 and 26. Both the coupling constant and the chemical shift anisotropy display distinct breaks in the vicinity of 25 molar percent Cs_2O. This abrupt change in the cesium-oxygen configuration reflects some change in the structure of the glass, but previous studies (3) have shown that the fraction N_4 of four-coordinated borons in these glasses does not exhibit any anomaly in this composition range. (See Fig. 10). The linear dependence of the measured values of e^2qQ and $(\sigma_{\parallel} - \sigma_{\perp})$ above the break could arise from a mixture of two or more cesium-oxygen configurations having different values for the two quantities and changing in concentration as a linear function of composition.

The temperature dependence of the line width $\Delta\nu$ is shown in Fig. 27 for two of the glasses studied. Narrowing of the Cs^{133} resonance by diffusion of the cesium ions is apparent; the term $(\Delta\nu)_{DIP}$ in Eq. 6 approaches zero and the frequency ν_0 of observation ($\sim$ 10 Mcps) reduces the second term on the right-hand side of Eq. 6 to something less than 250 cps. At sufficiently high temperatures the shape of the resonance (Fig. 28) is in good agreement with the shape (Fig. 29) calculated for the case of an axially symmetric chemical shift tensor with small dipolar broadening in a glass or a polycrystalline powder. (The recorder tracing of Fig. 28 is the first derivative of the resonance line shape).

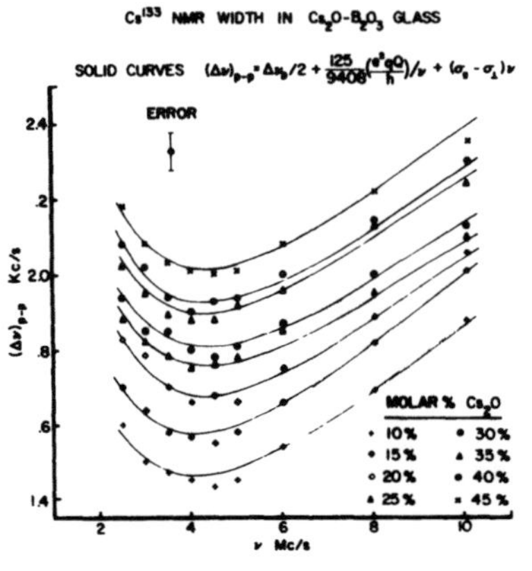

Fig. 24. Width of the Cs^{133} *NMR line for various cesium borate glasses, plotted as a function of the spectrometer frequency* ν

Fig. 25. The Cs^{133} *quadrupole coupling constant for cesium borate glasses at room temperature, plotted as a function of the molar percent of* Cs_2O*.*

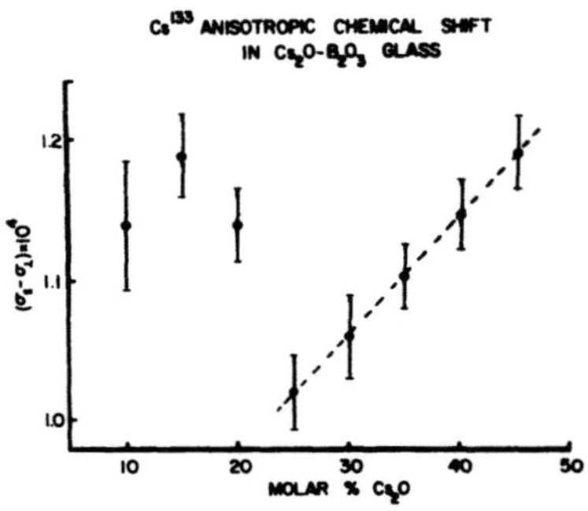

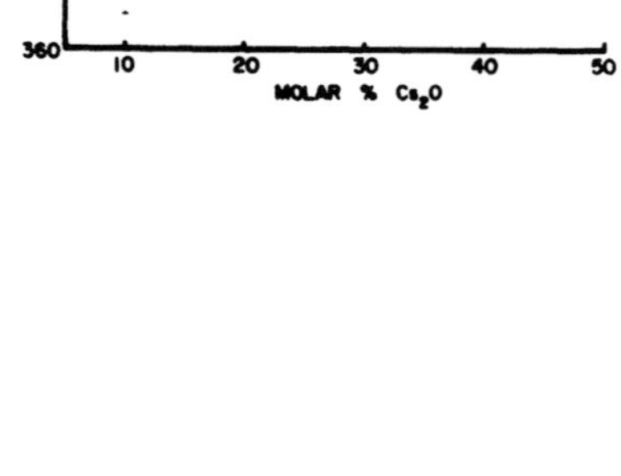

Fig. 26. The anisotropy of the (axial symmetric) chemical shift tensor for cesium borate glasses at room temperature, plotted as a function of the molar percent Cs_2O*.*

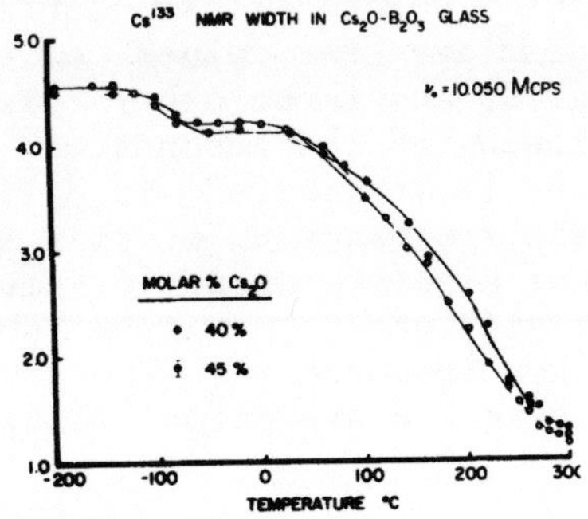

Fig. 27. Width of the Cs^{133} NMR line as a function of temperature for cesium borate glasses.

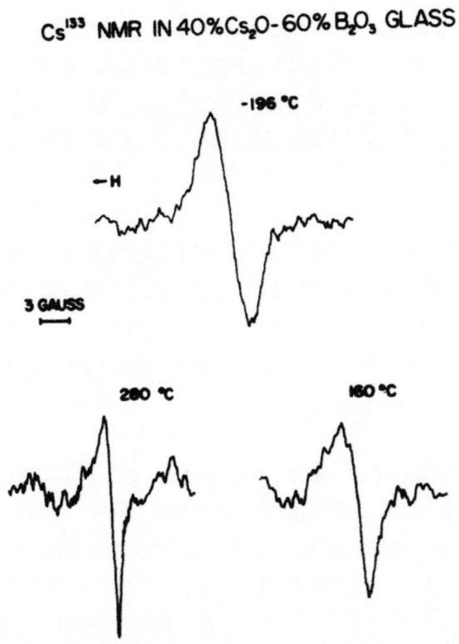

Fig. 28. The Cs^{133} NMR line at various temperatures in a glass of composition $0.40Cs_2O$-$0.60B_2O_3$.

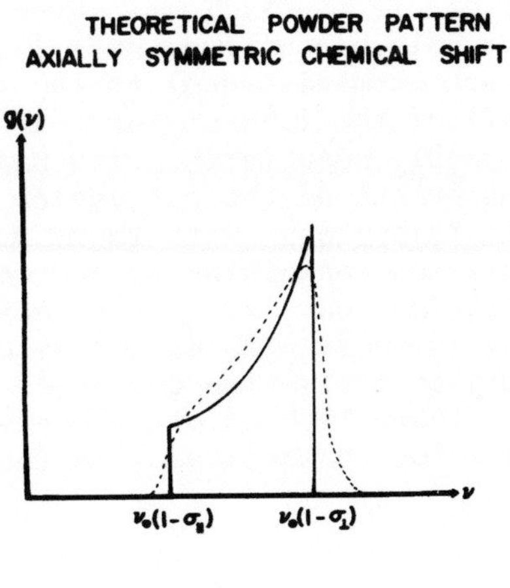

Fig. 29. Absorption line shape for a resonance exhibiting an axially symmetric chemical shift (a) solid line, no dipolar broadening; (b) dotted line, small dipolar broadening.

DIFFUSION IN LITHIUM CONTAINING GLASSES (11)

The Li^7 NMR response at three temperatures is shown in Fig.30 for a borate glass containing 60 molar percent Li_2O. Figure 31 displays the values of the width $\Delta\nu$ of the resonance for this glass and three other glasses as a function of temperature. Narrowing of the resonances with increasing temperature is evident and is attributed to diffusion of lithium ions in the glasses. The frequency ν_c of this ionic motion, which averages out the local dipolar magnetic fields experienced by the nucleus in the rigid lattice, can be obtained (43) from the measured width at each temperature. The activation energy E_a can then be evaluated using the Arrhenius relation

$$\nu_c = A \exp(- E_a/kT). \tag{7}$$

A plot of the logarithim of ν_c versus the reciprocal temperature is given in Fig. 32. The slopes of the straight lines yield activation energies of approximately 0.22 ev for the glasses of 28 and 37 molar percent Li_2O, and approximately 0.12 ev for the glasses of 50 and 60 percent Li_2O. These values are clearly less than those involved (44) in the bulk diffusion yielding ionic conductivity at higher temperatures. It is probable that, at these lower temperatures, each lithium ion is jumping from site to site within a very small region of glass.

The width of the Li^7 resonance in a lithium silicate glass containing 20 molar percent Li_2O was also measured at various temperatures. Figure 33 displays these measurements. Values of ν_c plotted versus 1/T fit a straight line whose slope yields an activation energy of 0.10 ± 0.02 ev.

In silicate glasses, the introduction of an alkali oxide produces in the silicon-oxygen network oxygen atoms that are bonded to only one silicon atom (non-bridging oxygens), and the number of such atoms is equal to the number of alkali ions (45). Therefore, all of the lithium ions in the 0.20 Li_2O-0.80 SiO_2 glass are presumably associated with non-bridging oxygens, and the excellent agreement of the activation energy for lithium diffusion measured in this glass with those energies measured in the lithium borate glasses containing 50 and 60 molar percent Li_2O suggests that the lithium ions exhibiting activation energies of approximately 0.12 ev in those glasses are associated with non-bridging oxygens. The higher activation energy of 0.22 ev found for the borate glasses of lower lithium oxide content is probably characteristic of lithium ions associated with the negatively charged BO_4 configurations (BO_4^- units).

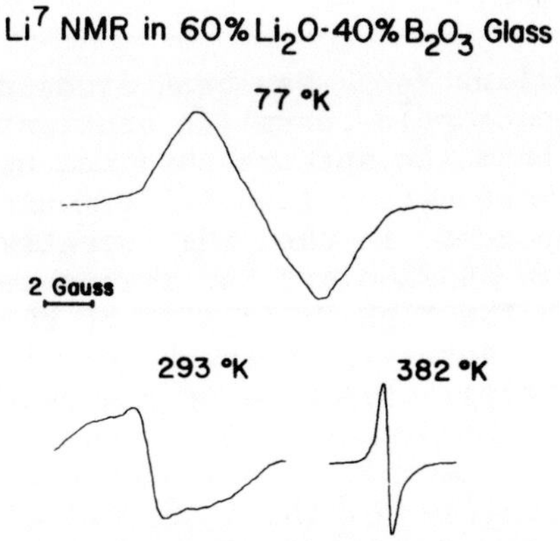

Fig. 30. The Li^7 *NMR line at three temperatures in a lithium borate glass.*

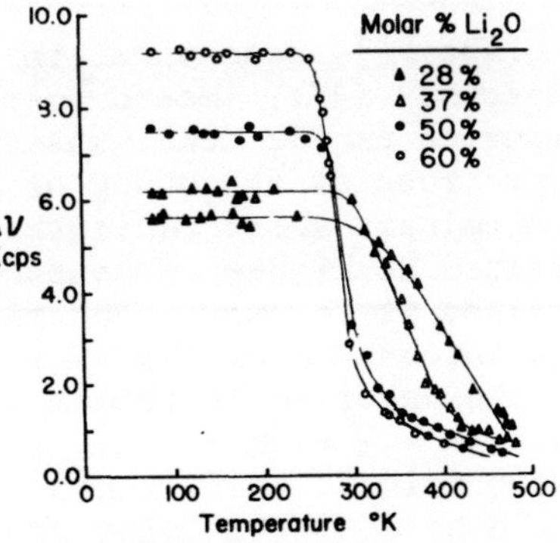

Fig. 31. Width $\Delta\nu$ *of the* Li^7 *NMR line as a function of temperature for four lithium borate glasses.*

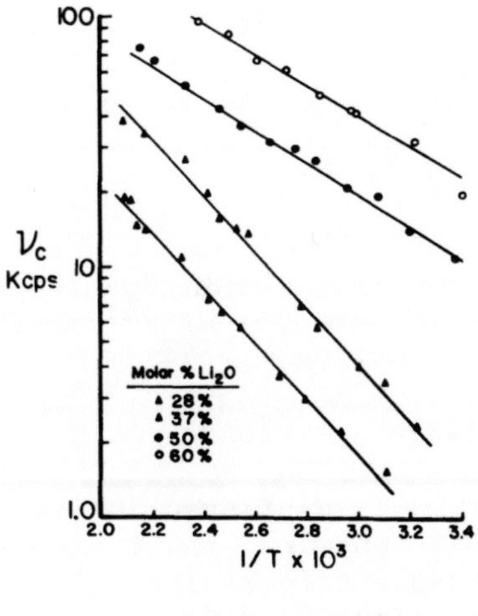

Fig. 32. Dependence of the frequency (ν_c) *of ionic motion on the reciprocal temperature for lithium ions.*

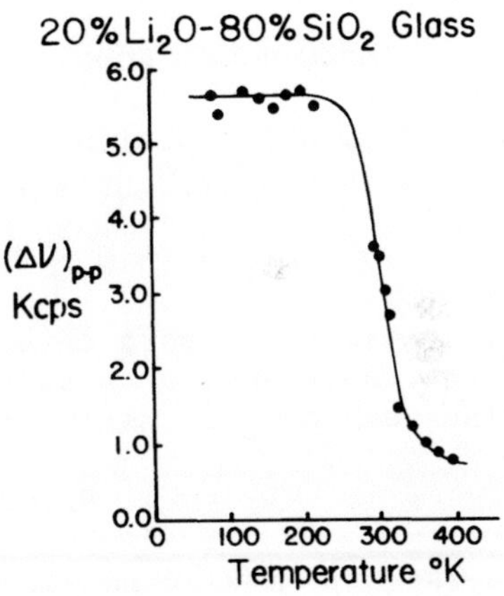

Fig. 33. Width $\Delta\nu$ *of the* Li^7 *NMR line as a function of temperature in a lithium silicate glass.*

NMR OF OTHER NUCLEI IN GLASSES

A. Vanadium (46)

The V^{51} resonance in polycrystalline V_2O_5 has been studied previously (47). Measurement of the quadrupole coupling constant (reported as 752 Kcps) was apparently based on spectra obtained at a spectrometer frequency of 10 Mcps. However, substantial effects of chemical shift anisotropy are apparent in the NMR spectra obtained at higher frequencies. Figure 34 displays the resonance obtained at 16 Mcps. The first peak on the high-field side of the most intense negative extremum is not a satellite transition; it and the negative extremum are the principal features of the central ($m = ½ \leftrightarrow m = -½$) transition in the presence of a substantial chemical shift tensor with axial symmetry. (These features should be compared with the first derivative of the theoretical line shape shown in Fig. 29). A preliminary value of $\sigma_{\parallel} - \sigma_{\perp} \simeq 1.2 \times 10^{-3}$ has been obtained from a study of this central transition as a function of the spectrometer frequency. (See Eq. 6).

The V^{51} NMR has also been studied in two vanadium-containing glasses. Figure 35 displays the resonance for a glass of composition 0.60 V_2O_5 - 0.40 PbO. The asymmetric character of the response is evident. Measurements of the line width as a function of spectrometer frequency ν_0 (Eq. 6) for the glass 0.50 V_2O_5 - 0.5C PbO have yielded the following very preliminary values

$$e^2qQ \simeq 140 \text{ Kcps}; \quad \sigma_{\parallel} - \sigma_{\perp} = (0.6 \pm 0.1) \times 10^{-3}$$

the corresponding preliminary value of $(\sigma_{\parallel} - \sigma_{\perp})$ for the glass oi composition 0.83 V_2O_5 - 0.10 P_2O_5 - 0.07 Na_2O is

$$\sigma_{\parallel} - \sigma_{\perp} = (0.9 \pm 0.3) \times 10^{-3}$$

It is clear that more than one vanadium-oxygen configuration i: present in vanadate glasses and that NMR studies of V^{51} in th glasses can provide useful structural information.

B. Phosphorous (48)

The P^{31} NMR in polycrystalline $K_4P_2O_7$ is displayed in Fig.3 for various magnetic fields. Anisotropic chemical shift effect are quite apparent. (There are no quadrupole effects for this nu cleus of spin $I = ½$). The spectra for vitreous P_2O_5, vitreou potassium metaphosphate, and the polycrystalline pyrophosphate an orthophosphate compounds are shown in Fig. 37 at one value of th field. The differences in width and structure arise from differer ent dipolar interactions and anisotropic chemical shifts. NM studies can be used to identify the various phosphorous-oxyge

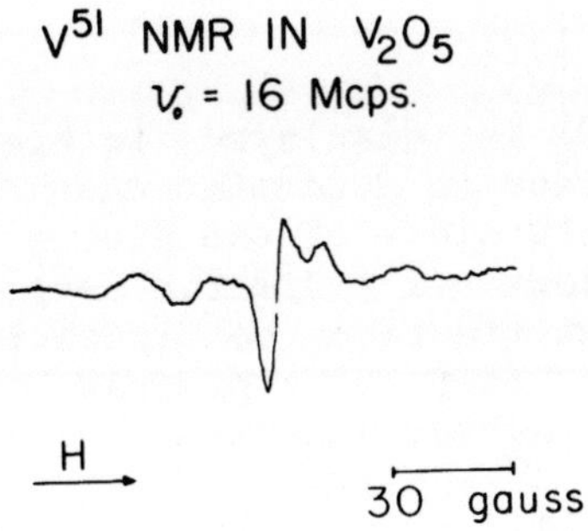

Fig. 34. The V^{51} NMR spectrum in polycrystalline V_2O_5 at room temperature.

Fig. 35. The V^{51} NMR response for a glass at room temperature

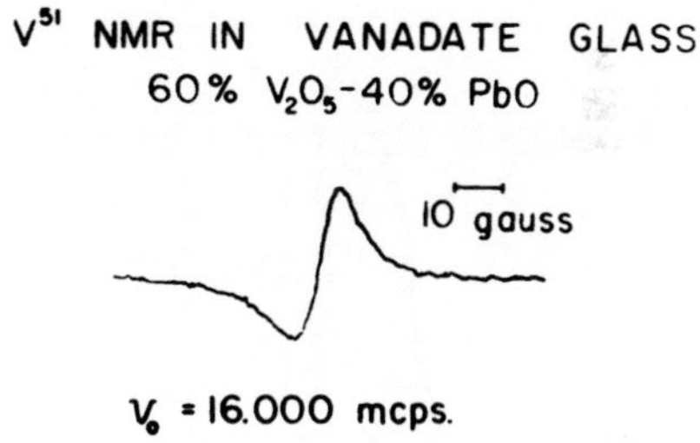

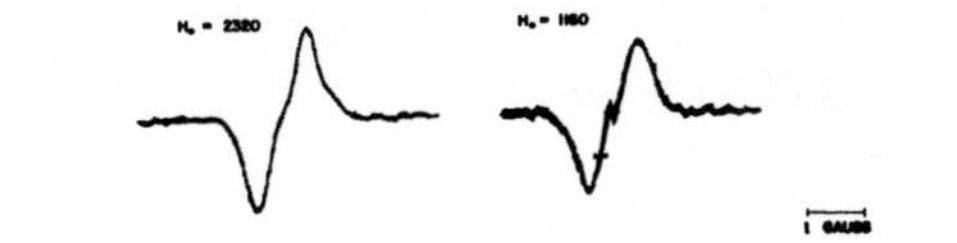

Fig. 36. The P^{31} NMR response for polycrystalline $K_4P_2O_7$ at various values of the applied magnetic field.

configurations in these materials and to determine the nature of the phosphorous-oxygen bonds.

C. Beryllium (49)

The NMR of Be^9 in vitreous BeF_2 is displayed in Fig. 38. Field dependence studies are in progress to determine whether the broad, weaker response evident on both sides of the strong resonance arises from satellite transitions ($I = 3/2$ for Be^9) or is evidence of a second type of beryllium-fluorine configuration in the glass. The study of BeF_2 is the start of a program of NMR studies of alkali fluoride-beryllium fluoride glasses.

D. Cadmium (14)

Figure 39 displays the NMR response (dispersion mode) of Cd^{113} in a glass of composition 0.40 CdO-0.60 B_2O_3. Both the Cd^{111} and Cd^{113} nuclei are being studied for chemical shift effects ($I = \frac{1}{2}$ for both isotopes) in various cadmium-containing glasses.

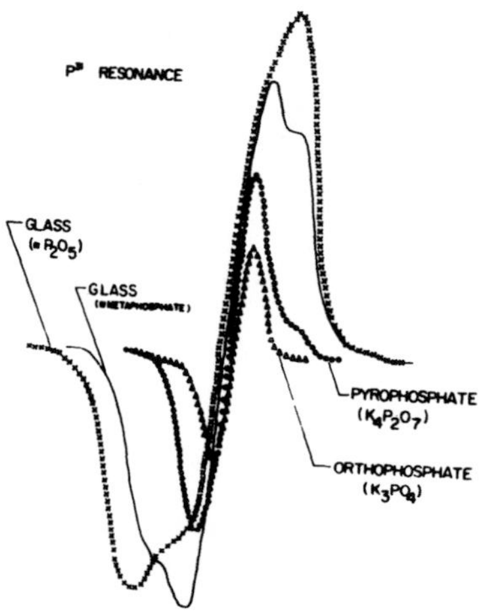

Fig. 37. The P^{31} NMR spectra for vitreous P_2O_5 and polycrystalline $K_4P_2O_7$ and K_3PO_4 at room temperature and a spectrometer frequency of 16 Mcps.

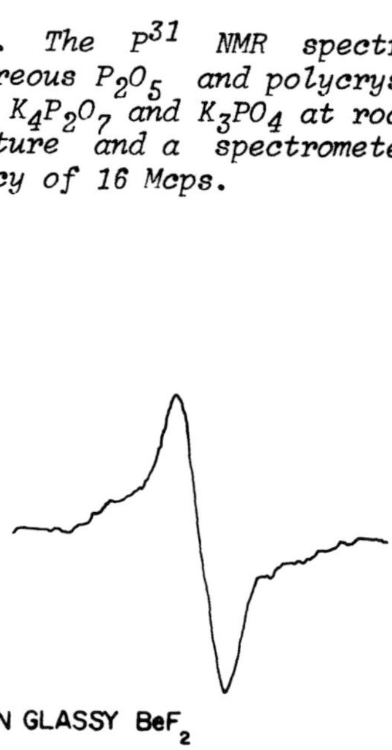

Fig. 38. The Be^9 NMR response for vitreous beryllium fluoride (BeF_2).

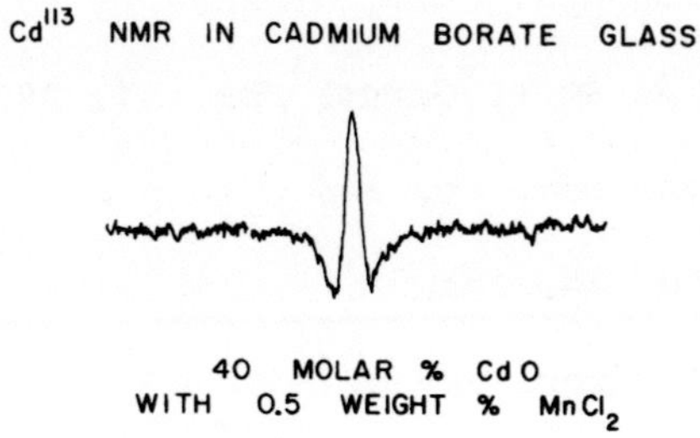

Fig. 39. The Cd^{113} NMR line (dispersion mode) for a cadmium borate glass containing 40 molar percent CdO.

REFERENCES

1. *A.H. Silver and P.J. Bray (1958). J. Chem. Phys. 29, 984.*
2. *P.J. Bray and A.H. Silver (1960). "Modern Aspects of the Vitreous State", Butterworth and Co., London, Vol. 1, Chap. 5.*
3. *P.J. Bray and J.G. O'Keefe (1963). Phys. Chem. Glasses 4, 37.*
4. *P.J. Bray, M. Leventhal and H.O. Hooper (1963). Phys. Chem. Glasses 4, 47.*
5. *M. Leventhal and P.J. Bray (1965). Phys. Chem. Glasses 6, 113.*
6. *D. Kline and P.J. Bray (1966). Phys. Chem. Glasses 7, 41.*
7. *S.G. Bishop and P.J. Bray (1966). Phys. Chem. Glasses 7, 73.*
8. *M.E. Milberg, K. Otto and .T. Kushida (1966). Phys. Chem. Glasses 7, 14.*
9. *P.J. Bray, J.O. Edwards, J.G. O'Keefe, V.F. Ross and I. Tatsuzaki (1961). J. Chem. Phys. 35, 435.*
10. *D.E. O'Reilley (1958). J. Chem. Phys. 28, 1262.*
11. *S.G. Bishop (1965). Ph.D. Thesis, Brown University.*
12. *D. Kline (1964). Ph.D. Thesis, Brown University.*
13. *S.V. Berger (1953). Acta Chem. Scand. 7, 611.*

14. J.F. Baugher (1966). Graduate student, Department of Physics Brown University.

15. M.K. Murthy and B. Scroggie (1966). Phys. Chem. Glasses 7, 68.

16. J. Biscoe and B.E. Warren (1938). J. Amer. Ceram. Soc. 21, 287

17. B.E. Warren (1941). J. Amer. Ceram. Soc. 24, 256.

18. T. Abe (1952). J. Amer. Ceram. Soc. 35, 284.

19. M.L. Huggins and T. Abe (1957). J. Amer. Ceram. Soc. 40, 287.

20. F.C. Eversteijn, J.M. Stevels and H.I. Watermann (1960). Phys. Chem. Glasses 1, 123.

21. S. Greenblatt (1965). Ph.D. Thesis, Brown University.

22. H. Moore and H.Winkelmann (1955). J. Soc. Glass Tech. 39, 254T

23. A.E. Owen (1961). Phys. Chem. Glasses 2, 87.

24. J.E. Lechus (1966). Graduate student, Department of Physics, Brown University.

25. A.H. Silver and P.J. Bray (1960). J. Chem. Phys. 32, 288.

26. J. Krogh-Moe (1962). Phys. Chem. Glasses 3, 1.

27. S.E. Svanson, E. Forslind and J. Krogh-Moe (1962). J. Phys. Chem. 66, 174.

28. S.G. Bishop and H.M. Kriz (1966). Postdoctoral research associate (S.G.B.) and graduate student (H.M.K.), Department of Physics, Brown University.

29. G.H. Stauss (1964). J. Chem. Phys. 40, 1988.

30. K. Narita, J.I. Umeda and H. Kusumoto (1966). J. Chem. Phys. 44, 2719.

31. J. Krogh-Moe (1962). Z. Krist. 117, 166.

32. P. Smith, S. Garcia-Blanco and L. Rivoir (1961). Z. Krist. 115, 460.

33. D.E. Harrison and F.A. Hummel (1956). J. Electrochem. Soc. 103, 491.

34. *P.J. Ring (1964). Ph.D. Thesis, Brown University.*

35. *L.H. Piette and H.E. Weaver (1958). J. Chem. Phys. 28, 735.*

36. *W. Lamb (1941). Phys. Rev. 60, 817.*

37. *N.F. Ramsey (1950). Phys. Rev. 78, 699.*

38. *Chemical shift effects arising from changes in geometry and bonds outside the immediate oxygen neighbors of the lead are probably smaller than the measurement errors. Hence, statements about lead-oxygen configurations are limited to the lead and its immediate neighbors.*

39. *C. Rhee (1966). Graduate student, Department of Physics, Brown University.*

40. *M.E. Milberg, K. Otto and T. Kushida (1966). Phys. Chem. Glasses 7, 14.*

41. *R.G. Barnes, W.H. Jones and T.P. Graham (1961). Phys. Rev. Letters 6, 221.*

42. *W.H. Jones, Jr., T.P. Graham and R.G. Barnes (1963). Phys. Rev 132, 1898.*

43. *N. Bloembergen, E.M. Purcell and R.V. Pound (1948). Phys. Rev. 73, 679.*

44. *B. Laurent (1953). Verres et Refr. 7, 167.*

45. *J.E. Stanworth (1950). "Physical Properties of Glass", Oxford University Press, London.*

46. *J.F. Baugher and F.R. Landsberger (1966). Graduate students, Department of Physics, Brown University.*

47. *H. Nagasawa, S.K. Takeshita and Y. Tomono (1964). J. Phys. Soc. Japan 19, 764.*

48. *J.R. Brandenberger (1964). M.Sc. Thesis, Brown University.*

49. *F.R. Landsberger (1966). Graduate student, Department of Physics, Brown University.*

DISCUSSION

C. DELBECQ (Argonne National Laboratory, U.S.A.): *What mechanism do you imagine gives rise to the low activation energy motional narrowing of the Li resonance in the Li glass?*

P.J. BRAY: *We believe that the lithium ion is jumping among identical sites within a very limited region in the material. For example, it may be jumping among a number of sites associated with non-bridging oxygens and all lying within a region of characteristic dimension 10 to 30 Angstrom units. Since the activation energies for electrical conduction are substantially higher than the energies we measure, it would appear that another distinct physical process - such as squeezing between a set of oxygen atoms - is necessary in order for the ion to move completely through the glass.*

N.J. KREIDL (The University of Missouri at Rolla, U.S.A.): *On the low diffusion activation energy of Li in borate glasses, could Dick Charles' Model be of use in describing lower potential wells?*

P.J. BRAY: *Yes. We are using his work in our efforts to understand the low activation energies.*

N.J. KREIDL: *In the excellent differentiation you name between diffusion past bridging and non-bridging oxygens, could the addition of Al_2O_3 to a lithium silicate glass be used to follow the destruction of non-bridging oxygens in a silicate (not borate) glass?*

P.J. BRAY: *Yes. This would certainly be a most interesting experiment and should be carried out.*

SOME DEFECT ELECTRON STATES OF PURE FOUR-FOLD COORDINATED OXIDES: EXPECTATIONS AND REALIZATION

R. A. Weeks

Oak Ridge National Laboratory, Oak Ridge

Tennessee, U.S.A.

The defect electron states with which this discussion will be concerned are restricted to defects which are due to either single oxygen vacancies or interstitial oxygen ions. The experimental results are primarily those obtained from electron spin resonance (ESR) measurements on SiO_2, GeO_2, P_2O_5 (both glassy and crystalline forms). Optical absorption data on the oxides are related to the ESR data.

Two models of the oxygen vacancy are proposed, based upon the structures. The electron states of these two defects are formed from a linear combination of atomic orbitals of the nearest neighbor ions. A perturbation calculation shows that for one of the oxygen vacancy models, the non-bridging oxygen vacancy, the g tensor principal values will be > 2.0023, the free electron g-value. The other model, the bridging oxygen vacancy, is more complex, and approximations in the g tensor calculation are made. The results show that for one electron the principal values are probably < 2.0023, while for three electrons they are probably > 2.0023. A comparison of these calculated values with experimental ESR data in irradiated crystalline and glassy SiO_2, GeO_2, and P_2O_5 shows reasonable agreement. Hyperfine interactions with magnetic moments of nuclei in the vicinity of the defect, in some cases, substantiate the assumed models.

Two paramagnetic forms of interstitial oxygen are considered, one being O^- and the other O_2^-. Perturbation calculations of the principal g values have been made and are utilized in the analysis of ESR spectra observed in some of the oxides listed above. The ESR spectra are compared to the spectra of known paramagnetic *forms of interstitial oxygen occurring in the super oxide, KO_2, sodium perborate monohydrate, and various plastic polymers.*

I. INTRODUCTION

The oxide glasses constitute a group of materials whose properties are, in many respects, similar to those oxides which do not form glasses and to other diamagnetic insulating crystalline compounds. Generally the glass-forming oxides are those whose low-temperature crystalline forms have four-fold coordination of the metal ion, e.g. SiO_2, GeO_2, P_2O_5, As_2O_5.

It might be expected that there would exist some correlation between the properties of the glass and crystalline forms of these materials. Indeed this is the case with respect to transparency and reflectivity to electromagnetic radiation (1). Similar absorption bands are observed in both crystalline and glassy forms of the pure material; however, many of the details present in the absorption bands of the crystal are missing in the glass (particularly with respect to the infrared vibrational modes and the fundamental absorption edge) (2). An example of this difference is shown in Fig. 1 for the fundamental absorption edge of α-quartz and silica glass. There are also large differences in the mechanical and thermal properties of glasses and their crystalline forms.

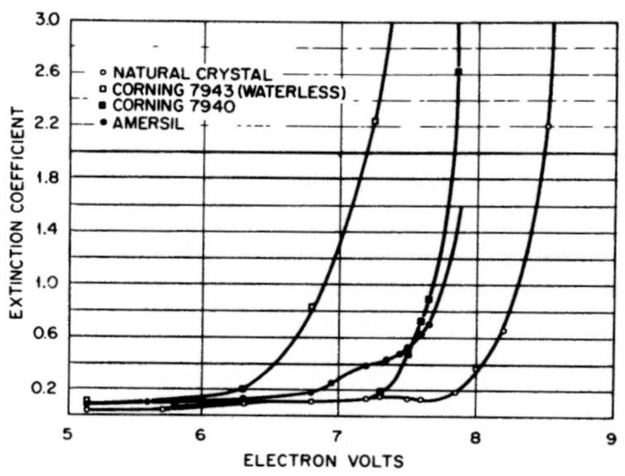

Fig. 1. Optical absorption in crystalline and glassy SiO_2. Impurity concentrations in the three glasses were < 10^{-5} ppm by weight.

There are, of course, many techniques that can be utilized in the study of the properties of glasses. This report will be concerned primarily with the information that can be gained about the structure of some glasses from a study of the electron-spin resonance (ESR) of their paramagnetic states. The discussion will consider the paramagnetism of defects in crystalline material and their correlation with the paramagnetic defects in the corresponding glass. Some consideration will be given to the effect that alkali and alkaline earth additives have upon the resonance spectra of the glasses.

II. DEFECTS IN GLASSES

In the discussion of "defect" states of a glass it will be well to start with the definition of a defect in a crystal. A general definition could be given: a defect is any characteristic of the crystal for which a set of basis vectors, defining the translational symmetry, cannot be found. Defects are vacant atom sites, atoms interstitial to proper atom sites, impurity atoms randomly distributed among lattice sites and interstitial sites, and complex arrays of all of these. There is no simple way in which these concepts of defects can be applied to glasses, since the glassy state does not have translational symmetry. In terms of the above definition of a defect, a glassy solid might be considered as one large defect.

Various proposals (3) have considered the ways in which "basic structural units" might be arranged in an orderly array over a short range in the glass. The upper limit of this short range order is usually assumed (3) to be the minimum size of crystallites required for well defined x-ray diffraction patterns (∿ 150 Å) (4). Objections to this way of considering the structure of a glass may be raised since the "basic structural unit" is not a well defined concept. The structural unit is usually assumed to be a regular MO_4 tetrahedron where M is the metal ion and O the oxygen ion. The use of the SiO_4 tetrahedron as a basic structural unit for the glass is justified by its uniform occurrence in all silicate compounds formed at normal pressures (5), although six-fold coordination has been discovered in silicate compounds found in meteor craters (6). A similar basis also exists for considering the tetrahedron as a basic structural unit in the phosphate, germanate, and arsenate glasses. These materials have stable crystal forms with four-fold coordination, while the last two are also found with stable six-fold coordinated forms (7). Various investigations (8) of the phosphate and arsenic-phosphate glasses have shown that each has several crystalline phases with four-fold coordination. The MO_4 tetrahedron may still be considered the primary "structural" unit although it is not a regular tetrahedron in many of these crystalline phases.

Therefore, it is assumed here that the basic structural units of the glasses of interest are tetrahedra and that they can be fitted together in a variety of ways. Thus one definition of a defect in a glass is: A defect is any distortion of the tetrahedral structural units found in any of the crystalline phases. In this definition the concepts of vacancies and of interstitial ions are valid and have a definite utility in a study of optical bands and paramagnetic states. Distortions of tetrahedra due to a mismatch between adjacent tetrahedra which produce local regions of strain are also defects, and a glass would be expected to have a large concentration of these distortions. The shift in the optical absorption to lower energies in SiO_2 glass as compared to α-quartz (Fig. 1) may be due, in part, to such distortions of SiO_4 tetrahedra in the glass. Similarities of optical bands and ESR spectra have been observed in silica glass (SiO_2) and α-quartz (9), crystalline and glassy P_2O_5 (10), and in germanate glass and a polycrystalline powder of the hexagonal phase of GeO_2 (11).

It is noted (12) that the thermodynamic state of a glass at 300°K can be considered to be similar to the liquid form of the material close to the temperature of solidification. It is then expected that the concentration of Schottky defects (13) (vacancies) in a glass at 300°K is approximately the same as the concentration in the liquid at the melting temperature. The energy, E, required for a formation of a Schottky defect is of the order of an electron volt (13). The concentration of Schottky vacancies is $f = \exp(-E/\kappa T)$ where κ is the Boltzman constant and T is the absolute temperature. For silica glass $T \sim 2000°K$ and, hence, $f \sim 0.01$. Other factors such as chemical reactivity of the constituent elements may determine the form of the Schottky defects. In silica glass, for example, there is some evidence that some of the defects may be in the form of SiO_2 vacancies (14). In any case a concentration of vacancies of this order in a glass must be an important factor. The melting points of the glass-forming oxides GeO_2 and SiO_2 are higher than those of P_2O_5 and As_2O_5. Assuming that the energy for forming a Schottky defect is approximately the same in all of these oxides, the number of such defects in the P_2O_5 and As_2O_5 glasses should be $\sim$ 1% of those in the SiO_2 and GeO_2 glasses.

By irradiating a crystal or glass with photons whose energy exceeds the gap between the bonding and anti-bonding orbitals of the unperturbed region, vacancy defect states may be occupied (16). Ionization of electrons at regions of local strain, permitting small changes of nuclear positions (strain relaxation), may also introduce optical bands and ESR spectra. (There is a small probability of large atomic displacement by photon irradiation) (17). Irradiation with particles, such as electrons and neutrons, of sufficient energy will produce the same effect as photon irradation, plus very large atomic displacements, thereby, introducing additional defects. (Defects produced by nuclear transformation and by decay products from transformed nuclei will not be considered). By using both kinds of irradiation (photons and particles) it is possible, in principle, to distinguish between defects which are present initially and those which are introduced by particle bombardment.

III. THEORY

A. Oxygen Vacancies

In order that the observed paramagnetic resonances be related to specific kinds of defects, it is necessary to have some information about the way in which an electron in such a state interacts with surrounding electrons. The band model of a solid (19) may be a reasonable approximation of the electron states which are of interest here. However, the use of this model is not practicable (the lack of periodicity in the structure of a glass invalidates a basic assumption of the band model). For the investigation of defect electron states the number of electrons to be considered must be limited. Karavelas (20) has assumed that a small region around the defect can be defined where there is a high probability of finding a specific number of electrons. The eigenfunctions of these electrons are considered to be those which determine the properties of the defect. The interaction of these electrons with the rest of the electrons of the solid are then treated as a perturbation on the eigenstates appropriate to the small region which has been defined. This region is usually defined as the defect and its nearest neighbor ions.

The first type of defect to be considered is an MO_4 tetrahedron from which one of the oxygen ions has been removed and in which the three remaining oxygens are each bonded to another M ion. This defect will be called a non-bridging oxygen vacancy and in this approximation the symmetry is C_{2v}. A schematic representation of this complex is shown in Fig. 2a. The M ion above the missing oxygen ion is assumed to be so far away that its effect is negligible. This type of oxygen vacancy would be expected to occur in those compounds in which non-bridging oxygen ions are characteristic of the structure, e.g., P_2O_5 and As_2O_5. It may also occur in SiO_2 and GeO_2 which contain alkali and alkaline earth oxide impurities because such additives have the effect of producing non-bridging oxygen ions.

If an oxygen ion is removed from one corner of a tetrahedron and the positions of the other ions are held fixed, all but one of the bonding and anti-bonding states will be unperturbed. This perturbed state will be simply the hybrid orbital of the M ion

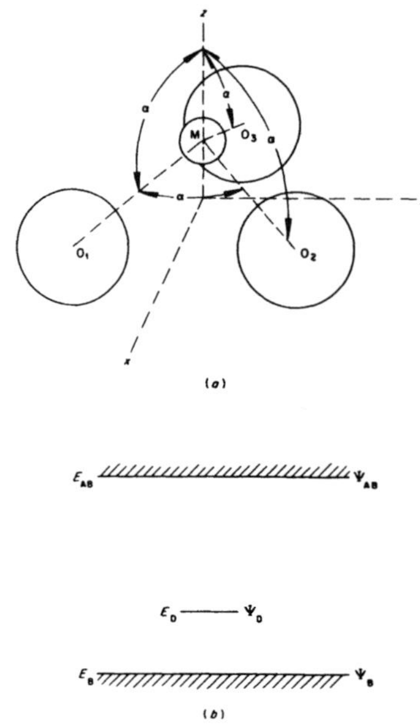

Fig. 2. a) Schematic representation of the non-bridging oxygen vacancy. The metal ion is identified by the letter M.

b) A representation of the energy level of the defect state of the non-bridging oxygen vacancy is shown in relation to the energies of the bonding and anti-bonding states of the metal-oxygen bonds.

that was involved in forming a bonding and anti-bonding state with the missing oxygen. Since the energy levels of defects in many crystals and glasses (21)(22) are observed to fall between the bonding (valence band) and anti-bonding (conduction band) states of the unperturbed crystal, this condition will be assumed here. The atomic configuration (Fig. 2a) is that of an AB_3 type molecule with the angles formed by the atoms B-A-B = 109° for sp^3 hybridization. Walsh (23) has obtained the eigenstates and energy levels for such a molecule by a molecular orbital calculation. The eigenstate in his calculation, corresponding to the defect eigenstate, is labeled "a_1s" and has an energy intermediate between the anti-bonding state and the bonding states of the molecule. It is assumed that the ions in the +z direction (Fig. 2a) are far enough away to have only a small effect on the defect state, $\Psi_D(sp)$. The defect state can be occupied by two electrons with opposite spins, in which case it will be diamagnetic. If it is occupied by one electron, it will have a net spin moment and electron spin resonance (ESR) absorption may be observed.

The second type of oxygen vacancy is shown in Fig. 3a. The complex is now defined as two adjacent MO_4 tetrahedra with the shared oxygen missing and there are eight ions to be considered. The principal axes will be those shown in Fig. 3a. In this case two hybrid orbitals, one from each M ion, will contribute to the formation of the defect states. The separations in energy of the

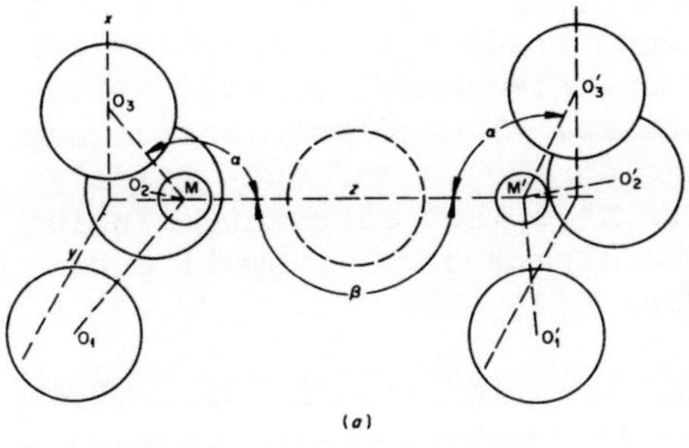

Fig. 3. a) A representation of the bridging oxygen vacancy is shown with the angle β = 180° and thus axial symmetry.

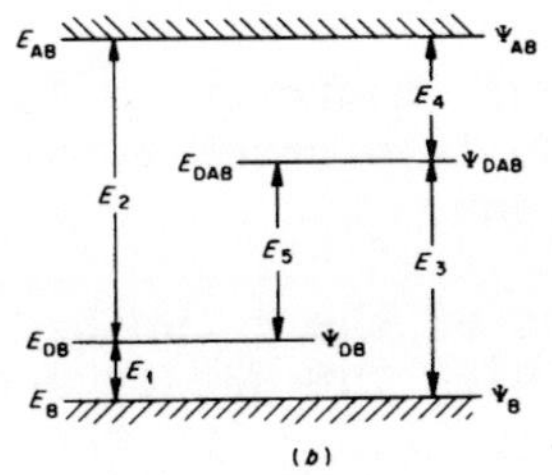

b) A schematic energy level diagram of the defect states is shown in relation to the energy levels of bonding and anti-bonding states of the metal-oxygen bonds.

bonding and anti-bonding states of the elements silicon, germanium, phosphorus, and arsenic are certainly no greater than the separation of bonding and anti-bonding states of the oxides of interest here (<5 eV as compared to >5 eV) (24). The M-M distance is less in the elemental forms than the corresponding distance in their oxides (∿ 2 Å as compared to ∿ 3 Å) (8). The difference in energy of the two defect states resulting from a linear combination of the two hybrid orbitals is expected to be less than that between the bonding and anti-bonding orbitals of elemental M (25) and less than the energy difference of the bonding and anti-bonding states of the oxides. These two defect states will have four electrons when filled. If only three electrons occupy the two states, there will be a net spin moment and ESR may be observed. Similarly, if only one electron occupies the lower energy state, ESR may also be observed.

The M ions which are of interest have either an s^2p^2 or s^2p^3 configuration of their valence electrons. The oxygen ion has a $2s^2p^4$ configuration of its outer electrons. Only sp hybridization is assumed.

B. Spin-Orbit Interactions

Consideration will now be given to the way in which the parameters of the ESR spectra are determined by the defect states. A magnetic field, H, will remove the spin (Kramer's) degeneracy of a single electron in a defect state, and irradiation with polarized electromagnetic radiation of the correct frequency, ν, will stimulate transitions between the spin states. The defect states are orbital singlets to a first approximation (26), and the paramagnetic electron has properties similar to a free electron. Taking into account crystal field effects to higher order, small deviations from free-electron behavior occur (26) (27).

With these conditions a "spin Hamiltonian" can be used which has the form

$$\mathcal{H} = \beta\vec{H} \cdot \overleftrightarrow{g} \cdot \vec{S} \quad , \tag{1}$$

where $\vec{S}$ is the spin operator, β is the Bohr magneton, $\vec{H}$ is the magnetic field and $\overleftrightarrow{g}$ is a tensor which takes into account the admixing of states due to spin-orbit interactions.

Some indication of the principal values of the g-tensor can be gained by a consideration of the way in which the defect states are admixed by the spin-orbit interaction (28), L.S, with the M-O bonds in the complex. The shift in the values of the components, g_{ij}, of the g tensor from the g-value of the free electron, $\Delta g_{ij} = g_{ij} - g_{fe}$, is given to first order in the spin-orbit coupling constant, λ, by

$$\Delta g_{ij} = \sum_n \frac{\langle \Psi_D | H_{SOj} | \Psi_n \rangle \langle \Psi_n | L_i | \Psi_D \rangle}{E_n - E_D} \tag{2}$$

where it is assumed (29) that the spin orbit interaction may be expressed as $H_{SOj} = \lambda \vec{S} \cdot \vec{L}_j$ and the sum is over all the electronic states of the complex. In order to make this calculation a number of additional assumptions are necessary. These have been discussed elsewhere (30) and will not be considered in detail here. The results of the calculation will be given with the comment that by reducing the number of assumptions, e.g. including interactions such as exchange polarization, improvement in the calculated values will probably be obtained.

For the non-bridging oxygen vacancy with one electron and with the three O-M-O angles being equal

$$\Delta g_{zz} = \Delta g_{\parallel} \simeq 0 \quad \Delta g_{xx} = \Delta g_{yy} = \Delta g_{\perp} = \lambda\, \beta^2 \left[\frac{1+\varepsilon}{|E_B - E_D|} \; \frac{1+\varepsilon}{|E_{AB} - E_D|} \right] \tag{3}$$

where β is the coefficient of the p part of the defect sp hybrid wave function, λ is the spin-orbit constant for the neutral M ion and ε is a parameter that is determined by the spin-orbit interaction of the defect electron with the core electrons of the M ion. The energy of the defect orbital relative to the energies of the bonding and anti-bonding orbitals is not known either from theory or experiment with any certainty. In order to obtain an estimate of $\Delta g_\perp$ it is assumed that $|E_B - E_D| = |E_{AB} - E_D|$, the value of $\varepsilon \sim 0.17$ for both Si and P (31) and $\varepsilon \sim 0.4$ for Ge and As, (30) and that $\beta = \sqrt{3/2}$, i.e. the defect state is an sp^3 hybrid. The result is $\Delta g_\perp \simeq 3\lambda\ \varepsilon/2\ |E_B - E_D|$. In Table I the observed band gaps in the oxides of interest here, the spin-orbit coupling constant λ, and the estimate of $\Delta g_\perp$ for this defect are given.

The more complex bridging oxygen vacancy will now be considered. Figure 3a shows the atomic configuration with the z axis parallel to the M-M direction. A linear combination of two hybrid M orbitals, one from each M ion, will form two defect states. A schematic energy level diagram is shown in Fig. 3b where it has

TABLE I

Spin-Orbit Constants, Band Gap, and $\Delta g_\perp$

of the Non-Bridging Oxygen Vacancy in Some Oxides

Ion	λ_{np}[a] (ev)	n	Oxide Band Gap (ev)	ΔE (ev)	$\Delta g_\perp$
Si	0.02	3	8,5[b]	4.3	0.003
Ge	0.07	4	6[c]	3	0.024
P	0.02	3	6.5[d]	3.3	0.003
As	0.04	4	6[d]	3	0.012

[a]These values were taken from *Atomic Energy Levels*,

C. E. Moore, NBS Circular 467 (1958) for the neutral ion.

[b]C. M. Nelson and R. A. Weeks, J. Appl. Phys. 32, 883 (1961).

[c]V. Garina-Corino, Compt. Rend. 247, 593 (1958).

[d]R. A. Weeks, unpublished data.

been assumed that the two defect states have energies intermediate to those of the M-O bonding and anti-bonding states. Ignoring other effects, such as thermal excitation, charge repulsion, and polarization, four electrons could occupy these two states producing a diamagnetic defect. Similarly if only two electrons occupy the lowest defect state the defect will be diamagnetic. Paramagnetic states of the defect occur when there are either one or three electrons in the defect states. Slichter (32) has discussed a method of calculating the spin-orbit interaction for a similar type of defect and using this method (30) the g-shifts are for one electron.

$$\Delta g_{\parallel} \sim \Delta g_{\perp} \sim 0$$

where it has been assumed that the hybrid orbitals from each M ion have the same hybridization and equal admixtures of each form of the defect states. With three electrons the g-shifts are

$$\Delta g_{\parallel} \simeq 0$$

$$\Delta g_{\perp} \simeq 2\lambda\ \beta^2\ N_{D-}^{\ 2} \left[\frac{1+\varepsilon}{|E_3|} - \frac{1-\varepsilon}{|E_4|}\right] \tag{4}$$

where N_{D-} is the normalization constant for the antisymmetric combination of the hybrid orbitals and E_3 and E_4 are defined in Fig. 3b. For a separation of the M ions of $\sim$ 3A, $N_D^{\ 2} \sim 0.5$ (30) with the result that

$$\Delta g_{\perp} \simeq \lambda\ \beta^2 \left[\frac{1+\varepsilon}{|E_3|} - \frac{1-\varepsilon}{|E_4|}\right] \quad , \tag{5}$$

If it is assumed that $|E_3| = |E_4|$, i.e. the antisymmetric state is in the middle of the gap, then this defect with three electrons can not be distinguished from the non-bridging oxygen vacancy with one electron. However, if the energy of the state is close to that of the anti-bonding states a negative value for $\Delta g_{\perp}$ might be expected. Watkins and Corbett (31)(33) have proposed models for two paramagnetic defects in silicon which are similar to the non-bridging and bridging oxygen vacancies. The results of their calculations give $\Delta g_{\parallel} \simeq 0$ and $\Delta g_{\perp} > 0$ for one electron in the defect analogous to the non-bridging vacancy and for three electrons in the defect analogous to the bridging oxygen vacancy. Perturbation calculations to higher order and admixtures of other excited states (34) would result in a further alteration of the g-values. Since it is not expected that the sp hybridization, which was assumed, will account for all of the defect electron density (30), overlap onto other ions will contribute to the g-shifts. For the heavier M ions, such as Ge and As, where the spin orbit interaction is larger (35), a more careful evaluation of ε is necessary.

In Figure 3a the symmetry axes of the two MO_3 groups are assumed to be colinear. In the hexagonal phase of GeO_2 and SiO_2 (36), for example, the M-O-M angle is 135°. An axis of symmetry would not be present for a bridging oxygen vacancy and $\Delta g_{xx} \neq \Delta g_{yy} \neq \Delta g_{zz}$. For small deviations from colinearity (M-O-M = 180°), $|\Delta g_{xx} - \Delta g_{yy}|$ (the z axis is defined as the M-M direction) would be expected to be small compared to $|\Delta g_{xx} - \Delta g_{zz}|$. An ESR powder pattern for these conditions would have a shape similar to that for axial symmetry at a frequency of $\sim$ 10 Gcps.

C. Hyperfine Interactions

The sp hybridization for the M-O bonds is a function of the O-M-O angle which may vary from 120° (sp^2) to 109° 28' (sp^3) to 90° (P_x, P_y, P_z and s). For the non-bridging oxygen vacancy (Fig. 1) the paramagnetic electron would be in a p orbital if the angle M-O-M = 120° and in a s orbital if the angle M-O-M = 90° (23). Some indication of the sp admixture may be obtained from the interaction of the paramagnetic electron with nuclei in the vicinity of the defect which have nuclear magnetic moments. The electronic density at the nuclear positions determines the magnitude and orientation dependence of this interaction. A hyperfine interaction with one nucleus can be described by a Hamiltonian (26)(27),

$$\mathcal{H} = \vec{S} \cdot \overleftrightarrow{A} \cdot \vec{I} \tag{6}$$

where $\vec{S}$ and $\vec{I}$ are the electron and nuclear spins and $\overleftrightarrow{A}$ is a second-order tensor coupling the electron and nuclear spins. $\overleftrightarrow{A}$ is composed of dipole-dipole interaction terms which are angularly dependent and isotropic Fermi contact terms. The spin Hamiltonian is the sum of Eq. (6) and Eq. (1)

$$\mathcal{H} = \beta\vec{H} \cdot \overleftrightarrow{g} \cdot \vec{S} + \vec{S} \cdot \overleftrightarrow{A} \cdot \vec{I} \tag{7}$$

where the nuclear Zeeman term (37) is neglected.

Hyperfine interactions of a paramagnetic defect with the nuclei of surrounding ions can furnish detailed information about the atomic environment of the defect and the nature of the defect wavefunction. For the defect hybrid orbitals the isotropic hyperfine interaction, a, is proportional to the s part of the hybrid and the anisotropic interaction, b, to the p part. Thus for the defect wavefunction of the non-bridging oxygen vacancy, $\Psi = \alpha s + \beta p$, $a \propto \alpha^2$ and $b \propto \beta^2$. The magnitude of the isotropic hyperfine interaction has been calculated for a pure s state and for a pure p state and are given in Table II with the isotropic and anisotropic interactions expected for an sp^3 hybrid orbital. Thus, it is possible to get reasonable values of the hyperfine interaction

TABLE II

Hyperfine Interactions of Electrons in s and p Orbital States

	^{29}Si (I =1/2) (Mcps) (a)	^{31}P (I = 1/2) (Mcps) (b)	^{73}Ge (I = 9/2) (Mcps) (c)	^{75}As (I = 3/2) (Mcps) (d)
s	4000	10190	5850	9582
p	140	288	312	511
sp^3				
a	1250	2550	1460	2395
b	105	216	233	383

(a) G. D. Watkins and J. W. Corbett, Phys. Rev. 134, A 1359 (1964)

(b) G. W. Chantry, et al., Mol. Phys. 5, 233 (1962)

(c) These values were estimates from those given for ^{75}As

(d) W. C. Lin and C. A. McDowell, Mol. Phys. 7, 223 (1964).

constants from theory for various values of α and β. Five per cent of naturally occurring Si has spin I = 1/2 (^{29}Si) and 100% of phosphorous (^{31}P). In GeO_2 the 7% abundant ^{73}Ge has a nuclear spin of 9/2 and ten hyperfine lines would be expected for 7% of the non-bridging oxygen vacancies. Their intensity would be < 1% the intensity of the 93% of such vacancies with even-even isotopes of Ge. Because of the width of each hyperfine line, due to anisotropy of the g-tensor and hyperfine interactions, detection of these hyperfine lines is difficult in a glass. The 100% ^{75}As with spin I = 3/2 should furnish an identifiable quartet for vacancies in arsenic oxide glasses.

Bridging oxygen vacancies with two adjacent M ions would, in the case of SiO_2, have 87.9% of the sites with ^{28}Si in both M sites, 9.6% with ^{29}Si in one of either site and 2.5% with ^{29}Si in both M sites. Approximately the same percentage apply to GeO_2. In the phosphate and arsenate glasses 100% of the sites would have ^{31}P or ^{75}As in both M sites.

IV. DEFECT PARAMAGNETIC STATES OF OXYGEN

In addition to the oxygen vacancy there are many other defects which can have paramagnetic states. The most important of these are those which are due to the paramagnetic states of various configurations of the oxygen ion. These configurations can be divided into three groups: a) An oxygen ion, O^-, in an interstitial site or bonded to only one M ion; b) an oxygen molecule ion, O_2^-, in an interstitial site or bonded to an M ion; and c) complex molecule ions, such as O_3^-, in interstitial sites. The oxygen molecule is also paramagnetic and in glasses with open structures it may be possible for a molecule to exist. However, because of a large zero field splitting (38), electron spin resonance transitions are difficult to observe, and in the usual electron spin resonance (ESR) spectrometer operating at 10 to 35 Gcps, they will not be detected. If the crystal field can be specified and the separation of the energy levels of the ion can be obtained from optical data (or by calculation) then an approximate evaluation of the g-tensor can be made by a second order perturbation calculation of the spin-orbit interaction. Most of the information necessary for this calculation is not available and one has to rely upon informed guesses for the values of the various parameters. To indicate the possible characteristics of the ESR spectra due to these various forms of oxygen in oxide glasses, consideration will be given to calculations and experimental results for the O ions in crystalline solids, in superoxides and in various perborates.

A. O^- Ion

The electronic configuration of the O^- ion is $1s^2\ 2s^2\ 2p^5$ and in the absence of an external crystal field the p_x, p_y and p_z orbital states are degenerate. If a field of axial symmetry is imposed on the ion, this degeneracy is lifted. In Figure 4a a schematic representation of these states is shown. Because of symmetry the matrix elements of L_z in Eq. (2) between the ground state, $\sigma_-(p_z)$, and the other states in Fig. 4a vanish and contributions from higher energy states will be negative. Since these states are several eV above the ground state, their contributions will be small. Thus $\Delta g_{\parallel} \lesssim 0$. L_x and L_y do admix the other states in Fig. 4 with $\sigma_-(p_z)$ giving

$$\Delta g_{\perp} \simeq -\lambda/(E_1 - E_o) = \frac{\lambda}{|E_1 - E_o|} \tag{8}$$

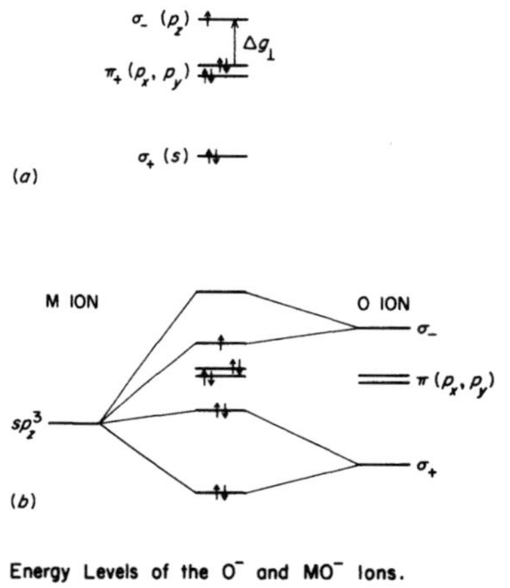

Fig. 4. a) A schematic energy level diagram of the O^- ion in a field of axial symmetry.

b) A schematic energy level diagram of the MO^- radical in an axial field. The field is determined by the metal ion.

where E_1 is the energy of the p_x and p_y states. It is evident that as the strength of the crystal field increases Δg decreases. If the crystal field has rhombic symmetry then the degeneracy of the p_x and p_y states will be removed with the result that $\Delta g_x \neq \Delta g_y$. If the separation of the p_x and p_y states is small then $|\Delta g_x - \Delta g_y| \ll |\Delta g_z|$ and in a powder pattern it might not be possible to resolve the three g-values. As the strength of the rhombic term in the crystal field increases the difference between Δg_x and Δg_y increases and a different ESR powder pattern would be obtained. However, other complications in the ESR spectra could occur. For a small separation of the $\sigma_-(p_z)$ and the p_x state thermal excitation of one of the electrons in the p_x state to the $\sigma_-(p_z)$ state becomes highly probable with a consequent radical alteration of the resonance spectra.

A form of the O^- ion which could be paramagnetic is the non-bridging oxygen which has lost an electron forming a paramagnetic MO^- radical. It will be useful to think of the molecular states of an MO molecule as a linear combination of an sp hybrid orbital of the M ion and the four hybrids of the O ion shown in Fig. 4a. Since the overlap term

$$< \Psi_M(sp) \mid \pi(p_x) > \text{ and } < \Psi_M(sp) \mid \pi(p_y) > \simeq 0 \qquad (9)$$

the largest contributions to the new set of states will be $\Psi_M(sp) \pm \sigma(s)$ and $\Psi_M(sp) \pm \sigma(p_z)$ (normalization has not been included). The state $\Psi_M(sp) + \sigma(p_z)$, will contain the unpaired electron. A schematic diagram of the energy levels (39) resulting is shown in Fig. 4b. The M ion is the source of a strong axial field with the result.

$$\Delta g_{\parallel} \approx 0 \text{ and } \Delta g_{\perp} \approx \lambda/(E_1 - E_o) > 0,\ E_1 < E_o \text{ and } \lambda < 0 \quad (10)$$

Here the field is determined by the M ion and rhombic terms due to other ions should be quite small. The net effect will be to decrease the value of $E_1 - E_o$ compared to that obtained for only the O^- ion in the axial crystal field with the result that $\Delta g_{\perp}(RO^-) > \Delta g_{\perp}(O^-)$. If this decrease is large enough the energy of the state, $\Psi(sp) + \sigma(p_z)$, may be less than that of the π states and the π states will now be occupied by three electrons.

There are two other ionization states of oxygen, O^o and O^+, that may also be present in oxides. In a crystal field of axial symmetry the degeneracy of the P states will be lifted and the energy level scheme shown in Fig. 4a will be assumed to be appropriate. If the splitting of the p_z and $\pi(p_x,p_y)$ states is large compared to thermal energies then no $\Delta M_s = \pm 1$ resonance transitions would be expected for O^o. If the crystal field has rhombic symmetry then the degeneracy of the $\pi(p_x, p_y)$ states will be removed and still no resonance transitions would occur for O^o, but $\Delta M_s = \pm 1$ transitions of the O^+ ion may be observed. The g-values of this transition are a function of the splitting of the p_x and p_y states and the virtual transitions which determine the g values are similar to the O_2^- molecule ion. Since the O_2^- molecule ion will be discussed below, no extensive discussion will be given here.

The g-values will be

$$\Delta g_{\perp} \leq 0 \qquad \Delta g_{\parallel} \approx \frac{2\lambda}{\Delta} > 0 \quad (11)$$

where the spin-orbit coupling constants, $\lambda < o$ and Δ is the energy difference of the p_x and p_y states and is negative. The ordering of the g-values is just reversed to those for the O^- ion. Usually axial terms of the crystal field are larger than rhombic terms, and thus it would be expected that for the O^- ion the energy term in the denominator is larger than for the O^+ ion. Thus g-shifts for the O^- ion will be smaller than for the O^+ ion.

B. O_2^- Ion

A schematic energy level diagram for an O_2^- ion is shown in Fig. 5a with the electrons represented by arrows. In a crystal

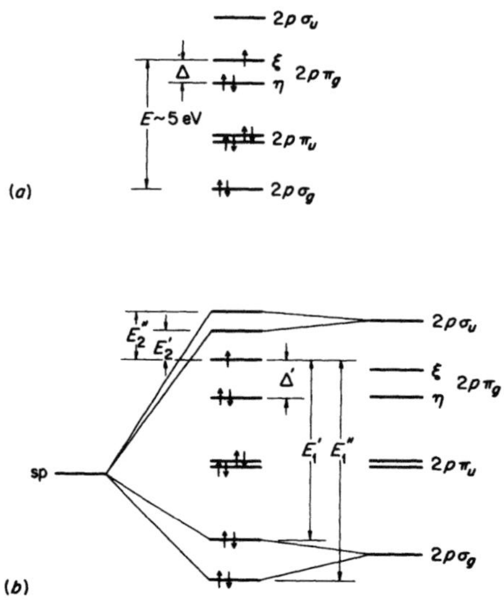

Energy Levels of the O_2^- and MO_2^- Molecule Ion and Radical.

Fig. 5. a) A schematic energy level diagram of the O_2^- molecule ion in a rhombic crystal field. The rhombic field splits the $2p\pi_g$ states.

b) A schematic energy level diagram of the MO_2^- peroxy radical. Admixtures of the sp hybrid of the M ion and the π states of the O_2^- molecule are assumed to be negligible.

field with axial symmetry and with an axis colinear with the axis of the molecule ion the two π_g levels are degenerate. The g values are $g_{\parallel} = 4$ and $g_{\perp} = 0$ and $\Delta M_s = \pm 1$ transitions will not be observed. In a crystal field with rhombic symmetry about the molecular axis this degeneracy will be removed and $\Delta M_s = \pm 1$ transitions are allowed. The g-shifts for this transition have been calculated by Kanzig and Cohen (40) for the ion in an interstitial site in KCl. In the oxides of interest here interstitial sites would be expected with rhombic symmetry and the results of this calculation should be applicable. The molecular axis is chosen as the z axis and the x and y axes will depend upon the particular details of the site. The g-values are

$$g_z = g_{fe} + 2\left(\frac{\lambda^2}{\lambda^2 + \Delta^2}\right)^{\frac{1}{2}} \qquad \ell$$

$$g_y = g_{fe}\left(\frac{\Delta^2}{\lambda^2 + \Delta^2}\right)^{\frac{1}{2}} - \frac{\lambda}{E}\left[\left(\frac{\lambda^2}{\lambda^2 + \Delta^2}\right)^{\frac{1}{2}} - \left(\frac{\Delta^2}{\lambda^2 + \Delta^2}\right)^{\frac{1}{2}} + 1\right] \qquad (12)$$

$$g_y = g_{fe}\left(\frac{\Delta^2}{\lambda^2 + \Delta^2}\right)^{\frac{1}{2}} - \frac{\lambda}{E}\left[\left(\frac{\lambda^2}{\lambda^2 + \Delta^2}\right)^{\frac{1}{2}} - \left(\frac{\Delta^2}{\lambda^2 + \Delta^2}\right)^{\frac{1}{2}} - 1\right]$$

where g_{fe} = free electron g-value, λ is the spin orbit coupling constant (in this case the spin orbit coupling constant of O^-, λ = 0.014 eV (41), is appropriate), E and Δ are the energies shown in

Fig. 5, and ℓ represents a correction to the angular momentum about z due to an interaction with the crystal field ($\ell = 1$ for the free ion and in the crystal field of KCl $\ell \simeq 1.04$).

In this approximation $g_z > g_{fe}$, and $g_x < g_{fe}$, $g_y < g_{fe}$. If $\Delta > 0.1$ eV then the term $[\Delta^2/(\lambda^2+\Delta^2)]^{1/2} \simeq 1$ and with E of the order of several eV (40) the terms multiplied by λ/E will be quite small. It can be seen that $(g_x - g_y)/2 \simeq \lambda/E$ and thus the difference between g_x and g_y will be much smaller than the difference between either one and g_z. Comparing these results with those obtained for an interstitial O^- ion it can be seen that the powder pattern of the interstitial O_2^- would be distinctly different from that of O^-.

As in the case of the O^- ion a form of the O_2^- molecule ion would be expected in which a bond is formed between the molecule ion and a metal ion. Such a molecular configuration is called a peroxy radical. There are two axes which are important in determining the ESR spectrum. These are the M-O and the O-O directions. If these are colinear and the crystal field contributions from neighboring ions are negligible, the degeneracy of the π orbitals will not be lifted and no resonance will be observed. If, on the other hand, the M-O and O-O directions are not colinear the degeneracy will be lifted and a resonance should be observed similar to that observed for interstitial O_2^-. The stable configuration of an MO_2^- molecule ion is one in which the M-O-O angle is between 90 and 180° (21)(22). If it is assumed that the admixture of the orbital states of the M ion with those of the O_2^- molecule ion causes a small perturbation of the σ and π states of the O_2^- ion then the g-values of the resonance should be similar to those of the interstitial ion, i.e. $g_z > g_{fe}$; $g_x < g_{fe}$; $g_y < g_{fe}$. A large shift in the energies of these states coupled with a strong interaction of the π orbitals with the electrons of the M ion (which would be expected as the M-O-O angle approaches 90°) could alter the ordering of the states shown in Fig. 5 with a consequent radical alteration in the resonance spectrum.

C. O_3^- Molecule Ion

The ozonide ion has been observed in a number of compounds (42) and has been found to be stable at room temperature. Since it may be possible to form this molecule ion in the oxide glasses the resonance spectrum expected for this molecule ion will be considered. The molecular orbitals and the most probable energies of this molecule ion and the isoelectronic SO_2^- and ClO_2^- molecules have been determined (41) and are shown in Fig. 6. D. Griscom (42) has utilized the available data on these states in calculating the g-values of the ESR spectrum of the O_3^- ion and in comparing the results with observed spectra. The results of his calculation indicate that $\Delta g_x \simeq 0$, $\Delta g_y \simeq 0.01$ and $\Delta g_z \simeq 0.02$ (see Fig. 6a and b). These results do not take into account crystal field interactions. If this interaction is large enough not only may it alter the ordering of the orbital states of the molecule ion, it may even be sufficient to prevent the formation of the ion. Tentatively, then, such complex molecule ions of oxygen may be found in open network structures such as the glasses in which interactions with surrounding ions are presumably weak rather than in crystals.

More complex molecules of oxygen, such as O_4, could be discussed, but the probability of formation of such complexes by irradiation of a solid with energetic particles would appear to be negligible. There is one report (43) in which an ESR spectrum is interpreted as due to an O_4 molecule ion, but such a molecule ion will not be discussed here.

The only isotope of oxygen with a nuclear magnetic moment is ^{17}O with a natural abundance of 0.03%. Without isotopic enrichment the effects of interactions with ^{17}O on the ESR spectra would be difficult to detect (44). Effects of hyperfine interactions with surrounding metal ions on the ESR spectra may produce small but detectable effects. In single crystals these effects may be resolved, but in SiO_2 and GeO_2 glasses they have not been resolved, while in glasses of P_2O_5 and As_2O_5 some effects may be detected.

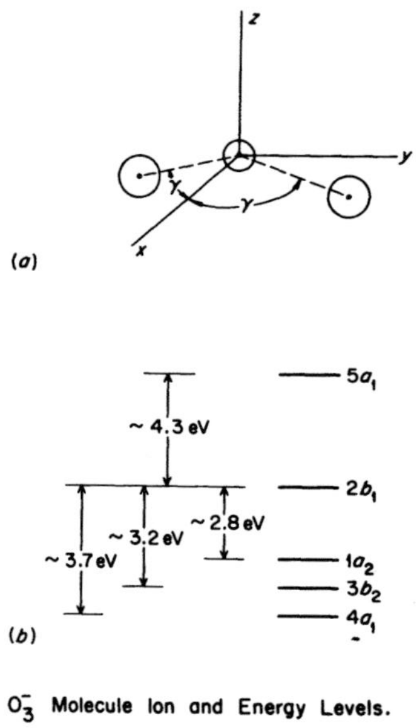

Fig. 6. a) A schematic representation of the O_3^- molecule ion. The axes are those for the principal g-values.

b) Schematic energy level diagram of the O_3^- molecule ion.

V. POWDER PATTERNS

For most of the paramagnetic centers axial or a close approximation to axial symmetry is expected. Ideally, a proper evaluation of the g-tensor of a paramagnetic center should be made in a single crystal. In many cases single crystals are not available and the analysis must be made on the spectra of glasses and powdered material. The analysis of such spectra have been discussed by several authors (45)(46)(47)(48) and show that considerable information can be obtained from powder patterns.

Axial symmetry of the g tensor will result in the powder pattern (here the term "powder pattern" will refer to crystalline powder) shown in Fig. 7a. The theoretical resonance absorption is shown for zero line width. The derivative of the resonance absorption of a paramagnetic center in silica glass is shown in Fig. 7a (50). In Figure 7b a comparison is made between the resonance absorption of this center and the powder pattern calculated from the g-values of a center observed in α-quartz, including the effects of line width, which is thought to be the same center. The agreement between the calculated and observed pattern apparently justifies the identification. Curve 4 in Fig. 7b shows the effect of a large neutron dose. The alteration in the shape is due primarily to other paramagnetic centers with g-values differing slightly from those of the paramagnetic center produced by γ-ray irradiation or a low neutron dose. Heavily neutron irradiated single crystal α-quartz has a spectrum (Fig. 8, Curve I) similar to heavily irradiated glass (51) (Fig. 7, Curve IV). Heat treatments change the shapes similarly in both cases (Curves II and III Fig. 8), and the result is a pattern similar to a low irradiation dose glass.

Rhombic symmetry of a defect site produces the powder pattern shown in Fig. 9. The derivative curve which is shown exhibits three well resolved peaks which are readily associated with the three g-values of the center. For some centers with rhombic symmetry two of the g-values may differ by an amount too small to be resolved at a specific frequency. Such a center is the O_2^- molecule ion when measurements are made at ∿ 10 Gcps.

With hyperfine interactions the powder pattern has added complexity. Contributions from hyperfine interactions are frequency independent in contrast to the frequency dependent Zeeman term. Weil and Hecht (52) and Blinder (53) have discussed the powder patterns expected for some values of $A_{\parallel}$, $A_{\perp}$, $g_{\parallel}$ and $g_{\perp}$. Since a large range of values for these parameters are expected for the centers which have been discussed, powder patterns for only one possible set will be discussed. In Fig. 10 a and b two extreme cases are shown. These extreme cases do not necessarily occur,

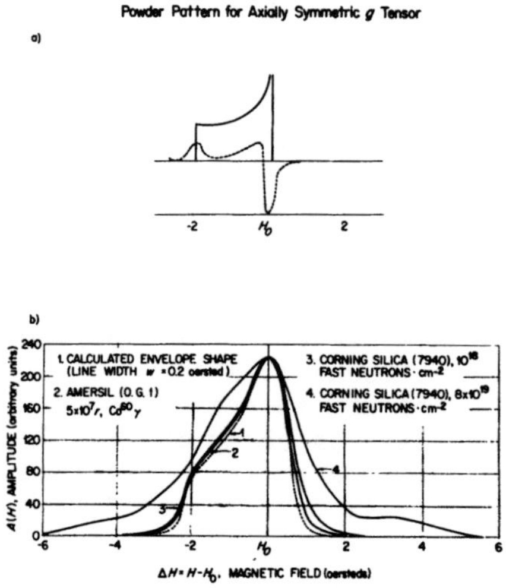

Fig. 7. a) ESR absorption of a S = 1/2 spin state with axial symmetry and zero line width. The first derivative of the resonance absorption is shown with finite "single crystal" line width.

b) Powder pattern of an ESR absorption in irradiated silica glass is compared with a calculated powder pattern using the g-values of a center (E_1') in an irradiated single crystal of α-quartz.

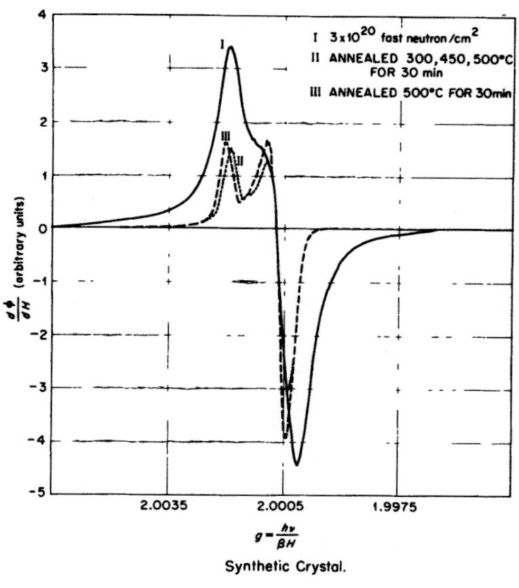

Fig. 8. Effect of heat treatment on the ESR resonance absorption (E_1' center) in a neutron irradiated single crystal of α-quartz. The powder pattern is expected since x-ray diffraction studies show the structure to be similar to silica glass.

but some of the characteristics of actual powder patterns are illustrated by them. The g-values chosen in Fig. 10a are approximately those expected for a non-bridging oxygen vacancy while the isotropic hyperfine interaction is that given in Table II for an sp^3 hybrid orbital. In Fig. 10b the anisotropic hyperfine interaction is approximately that expected for an sp^3 hybrid (Table II) while an isotropic g-value has been assumed. At a frequency of ∿ 10 Gcps an anisotropy in the g-tensor of the magnitude used in Fig. 10a would be only a small perturbation on the powder pattern in Fig. 10b. Even at a frequency of measurement of 35 Gcps the effect of g-anisotropy on the powder line shape will not be large. Fig. 11a and b show the spectra in a phosphate glass measured at two frequencies and show the case for which hyperfine interaction dominates any effect that may be due to a small g-anisotropy. The eigenvalues of the g-tensor and hyperfine tensor for this center are probably of the same magnitude as those used in the theoretical powder patterns in Fig. 10.

In addition to these factors which determine a powder pattern, there are other effects in a glass which are important. There is a distribution of M-O and M-M separations (54), although the mean separation in silica glass is the same as the separation in α-quartz. This variation in separation in a glass will result in a variation of the crystal field from one point to another and hence a variation of the g-tensor of a paramagnetic center, although the atomic environment may be quite similar. For example, for a bridging oxygen vacancy the angle M-vacancy-M, which will change with a change in the M-M separation, may not be the same from one vacancy to the next, and thus small variations in the g-tensor would be present. The effect of these variations will usually be small and not resolvable. However, a frequency dependent width of the "single crystal" line width for a powder pattern, defined by Weil and Hecht (52), will be observed.

VI. SUMMARY OF PREDICTIONS

It is evident from the preceding discussion that a large variety of paramagnetic defect states may be generated by irradiation of oxide glasses. These paramagnetic centers have different symmetries and Δg-values that fall within the range $-0.1 < \Delta g < 0.2$. Contrary to the usual assumptions made about g-shifts of electron and hole traps, electron trapping defects may have g-values $> g_{fe}$ and hole trapping defects may have g-values $< g_{fe}$. In the absence of other interactions, such as hyperfine interactions, the spin-orbit interaction as determined by g-value measurements is not a sufficient basis for constructing a model for a paramagnetic defect.

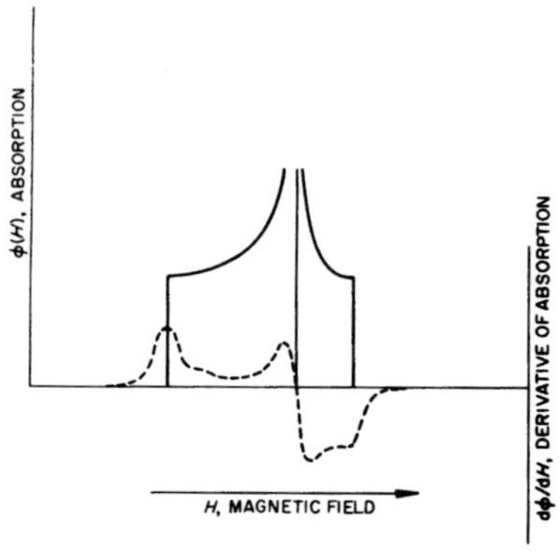

Powder Pattern for Rhombic Symmetry

Fig. 9. ESR powder pattern for a paramagnetic center with S = 1/2 and rhombic symmetry. The absorption curve is shown for zero single crystal line width while the first derivative curve is shown for finite width.

Fig. 11. ESR spectra in X-ray irradiated Lithium phosphate glass (P_2O_5 / Li_2O = 1). The two outer lines have been attributed to the paramagnetic state of a non-bridging oxygen vacancy. This center is similar to the case shown in Fig. 10b.

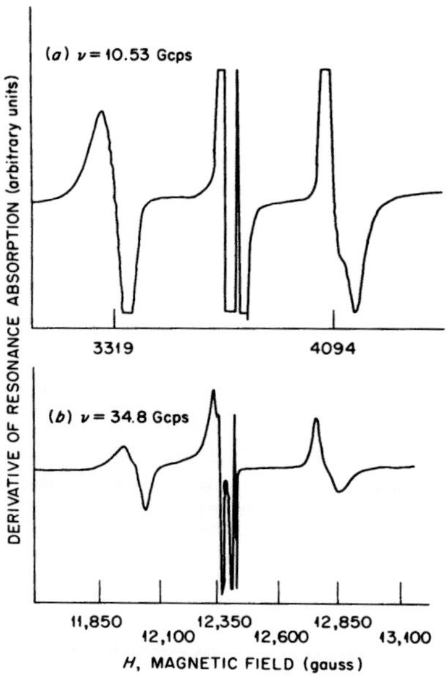

ESR Spectra in X-Ray Irradiated Lithium Phosphate Glass (P_2O_5/Li_2O =1).

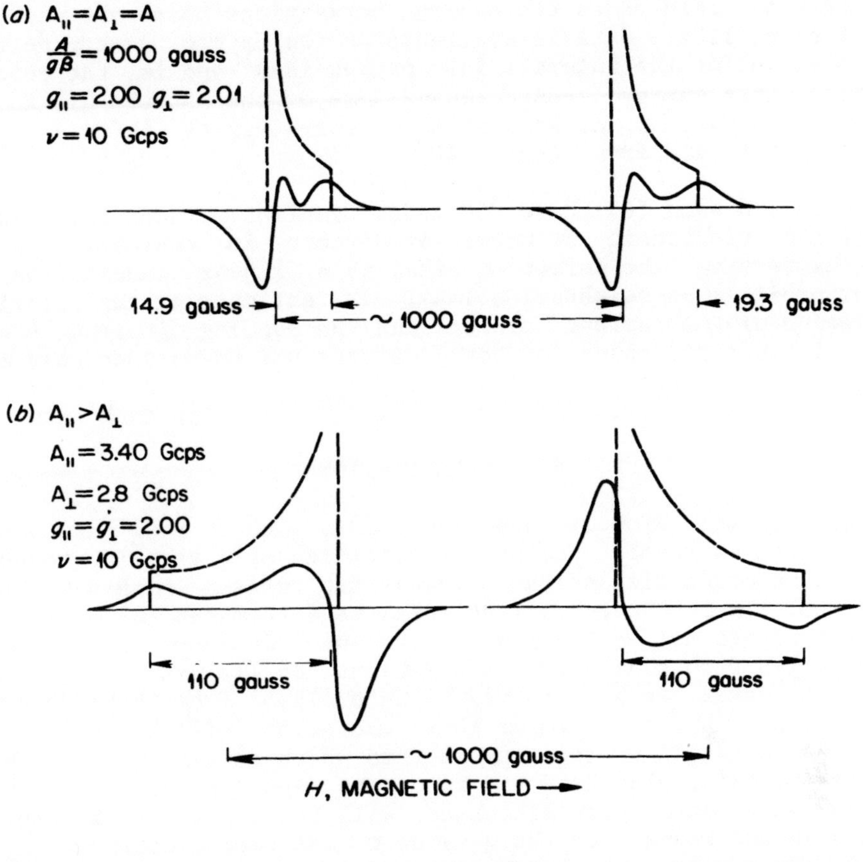

Fig. 10. Powder patterns for ESR absorption with two extreme cases of g-anisotropy and hyperfine anisotropy. a) An isotropic hyperfine interaction and a small axial anisotropy of the g-tensor is shown. b) An isotropic g-tensor and anisotropic hyperfine interaction is shown. In both a) and b) the absorption is shown with zero "single crystal" line width and the first derivative with finite width.

A critical parameter needed for an estimate of g-values is the energy separation of the defect states from those states which are admixed by spin-orbit coupling. These separations are usually not those measured by optical spectroscopy, since the optical transitions between states involved in g-shifts are forbidden or at least partially forbidden (42). Because of the admixing of states by crystal fields in a solid, defect optical absorption bands may be related to the energy separations which are needed. In any case, little data is available on the oxygen vacancy defect states while for the interstitial oxygen ions and for the peroxy radicals, the energy separation of some of the states are known and the separations of some of the others may be inferred from available data and from theory (42).

Öpik and Wood (55) have discussed various methods for calculating the electronic structure of defects in insulators based upon expressing the defect orbital as a linear combination of atomic orbitals on neighboring ions. The estimate or the relative separation of the defect orbital and the bonding (valence bond) and anti-bonding (conduction bond) states was based upon this method. The estimates are indeed very approximate and a detailed calculation is required for more realistic values in each case.

VII. REALIZATION

Irradiation of well annealed SiO_2, GeO_2, P_2O_5 and As_2O_5 glasses with energetic photons or particles at a very low temperature may produce all of the paramagnetic states which have been described. Their concentrations will be determined by many factors which have not been evaluated and which in many cases can not be evaluated with available information. An estimate of the number of vacancies in a glass may be determined from thermodynamic calculations, but the form of these vacancies will not be determined. As a result of the many factors which determine the structure of the oxide glasses, the resonance spectrum of many of the defects which have been discussed will contribute to a complex spectra in the region of the g-value of the free electron, g_{fe} = 2.0023. By varying the temperature of irradiation or by thermal treatments after irradiation, a partial resolution of a complex spectra may be obtained. Such a resolution coupled with a knowledge of the characteristics of these various defects in a single crystal is necessary before the spectra of glasses can be unambiguously interpreted in terms of the models which have been proposed. In a few cases it has been possible to compare the spectra of single crystals with glasses of the same composition.

A. Non-Bridging Oxygen Vacancies

There have been several investigations of the non-bridging oxygen vacancy in phosphate compounds (56) (in these reports it is usually described as a PO_3^{2-} radical) and one report of an AsO_3^{2-} radical (57). There is also one report on the defect in polycrystalline and glassy P_2O_5 and in a variety of polycrystalline and glassy alkali phosphates (58). A portion of this information is summarized in Table I. Because of the large isotropic hyperfine interaction of the defect electron with the ^{31}P nucleus, the interpretation of the ESR spectra in terms of a non-bridging oxygen vacancy model appears to be well founded. On the basis of hyperfine interactions this model is also a reasonable description of a defect in disodium hydrogen arsenate and in α-quartz. The observed isotropic hyperfine interaction given in Table III and the calculated values given in Table II for an sp^3 hybrid orbital are also in approximate agreement with the exception of the ultraphosphate compositions. In these compositions the isotropic hyperfine interaction is indicative of a higher proportion of s character in the defect wave function. Weeks and Bray (10) attribute the variation of isotropic hyperfine interactions in the phosphate glasses to different stable configurations of the PO_3 structure, in which the defect wave functions have different p/s hybridization ratios. This defect state has been identified with the one labeled "a_1s" by Walsh (23) and he notes that the s character of the state increases with a decrease in the oxygen-phosphorous-oxygen angles. An increase in the isotropic hyperfine interaction would be expected.

It is also important to note that the sp part of the defect wave function comprises only a fraction of the defect electron density. It has been assumed that the remainder is distributed over the three oxygens (56). This remainder ranges from ∿ 50% to ∿ 20% and is a significant fraction of the electron density. The anisotropic interaction is smaller than that expected from the model. This effect may be due to exchange polarization (57) which has not been considered in the calculations.

Despite the reasonable agreement found between the hyperfine interactions calculated from the model and the observed values, the agreement between observed and calculated g-values is not so good. In the single crystal phosphate compounds all the g-values are less than g_{fe} while in the glasses $< g > \; > g_{fe}$. Weeks and Bray (10) in the analysis of their data on the glasses were unable to resolve the contributions of g-anisotropy and of hyperfine anisotropy with the result that only an approximate average of the g-values is given. In the single crystal arsenate compounds, the g-values are both slightly less and slightly greater than g_{fe}. No data is available on the glass. In α-quartz they are slightly less

TABLE III

ESR Parameters of Paramagnetic Defects Attributed to Non-Bridging Oxygen Vacancies

COMPOUND OR GLASS (Species)	g_x	g_y	g_z	a (mcps)	b (mcps)	ν (cps)	T (°K)	Ref
$(NH_4)_2H_2P_2O_6$ ($PO_3^{=-}$)	2.0000	2.0012	2.0012	2156	158	∿10	∿300	Morton (59)
$Na_2PHO_3 \cdot 5H_2O$ ($PO_3^{=-}$)	1.9994	2.0004	2.0011	1998	151	8980	∿300	Horsfield et al. (56)
$MgHPO_3 \cdot 6H_2O$ ($PO_3^{=-}$)	1.998	1.999	1.999	1890	160	∿10	∿300	Hanna and Altman (56)
P_2O_5 (PO_3^{--})		$\langle g \rangle \approx 2.03$ $\langle g \rangle \approx 2.01$		∿4000 ∿2800	∿80 ∿80	{10, 35} {10, 35}	∿300 ∿300	Weeks and Bray (10)
$P_2O_5:0.6Na_2O$ (PO_3^{--})		$\langle g \rangle \approx 2.03$		∿4000	∿80	{10, 35}	∿300	Weeks and Bray (10)
$P_2O_5:Na_2O$ ($PO_3^{=-}$)		$\langle g \rangle \approx 2.01$		∿2000	∿80	{10, 35}	{∿300, ∿180}	Weeks and Bray (10)
$P_2O_5:1.6Na_2O$ (PO_3^{--})		$\langle g \rangle \approx 2.01$		∿1875	∿80	10	∿300	Weeks and Bray (10)
$AsO_3^{=-}$ in Calcite	2.00162	2.00195	2.00195	2215	189	10	∿300	Serway and Marshall (68)
$Na_2HAsO_4 \cdot 7H_2O$ ($AsO_3^{=-}$)	2.004	2.004 (±0.002)	2.005	1737	148	∿9	∿300	Lin and McDowell (72)
SiO_2 (α-quartz) $SiO_3^{=}$)	2.0006	2.0006	2.0018	∿1200	∿60	∿10	∿300	Weeks (60)

than g_{fe}, while in silica glass the identification is uncertain. Probably the most significant feature of these values is their small deviation from g_{fe}. The model predicts much larger deviations from g_{fe} for the arsenates than for the phosphates or silicates, but this effect is not observed. Since such a large fraction of the electron density is presumably distributed over the three oxygens and a proper account of contributions from this part of the defect wave function to the g-values has not been made, the disagreement is not surprising. There may also be contributions to the g-values from admixtures of excited d orbitals by spin-orbit interactions. The importance of these orbitals is indicated by the existence of stable six-fold coordinated crystalline phases in the silicates (6) and arsenates (7). The existence of six-fold coordinated Ge in alkali germanate glasses has been deduced from acoustic absorption data (58).

B. Bridging Oxygen Vacancies

There is very little data on this type of vacancy. In single crystals it has only been positively identified in a phosphate compound (59). The data showed that the electron had a high density on one of the two phosphorous ions at the defect site. The g-values are $g_{\parallel} = 2.0034$, $g_{\perp} = 2.0060$ and the hyperfine interaction constants are, for one of the p ions, $a_1 = 1109$ mcps and $b_1 = 161$ mcps, and for other $a_2 = 377$ mcps and $b_2 = 31$ mcps. The principal axes for the g-tensor and the larger hyperfine tensor are the same. The large value of a/b is evidence for a large exchange polarization interaction (57) of the defect electron with the core electrons of not only the near P ion but also the more distant P ion. If this is the case, then the value of the overlap integral of the two hybrid orbitals makes a significant contribution to the shift in g-values.

A spectrum observed in α-quartz and correlated with a powder pattern in silica glass has been interpreted in terms of an oxygen divacancy (60,50). Defect wave functions for an oxygen divacancy would be more complex than for the bridging oxygen vacancy. Silsbee (61) has discussed a number of objections to this model. It is clear from the observed hyperfine interactions that the defect electron is localized on only one Si ion at the defect site, and that the principal axes of the g-tensor and hyperfine tensor (of the ^{29}Si on which the electron is localized) are the same. If a large relaxation of nuclear positions is assumed to take place upon formation of a bridging oxygen vacancy then it is possible to interpret the data in terms of this model. In this case, as for the phosphate compound, the defect electron is associated primarily with only one of the Si ions at the defect site. The g-values are $g_x = 2.0003$, $g_y = 2.0005$, $g_z = 2.0018$ and $a_1 = 1151$ mcps, $b_1 = 60$ mcps.

A number of ESR spectra have been observed in polycrystalline powders of the tetragonal and hexagonal phases of GeO_2 and in glassy GeO_2 (11). One of these in the glass and in the hexagonal phase powder has been tentatively attributed to a bridging oxygen vacancy with either one or three electrons. No parameters on hyperfine interactions with ^{73}Ge are given and the proposed model is based only upon the g-values. These are $g_{\perp} \approx 1.996$, $g_{\parallel} \approx 2.002$. On the basis of the model, the vacancy with one electron would have both g-values $\approx g_{fe}$, while with three electrons $g_{\parallel} \approx g_{fe}$ and $g_{\perp} \neq g_{fe}$. However, as was pointed out for the non-bridging oxygen vacancy, contributions to the g-values from other interactions would be expected and thus no final model can be proposed. No models were proposed for the other spectra which were observed and which were independent of impurities. The g-values of these resonances were all greater than g_{fe} and thus are probably attributable to one or more of the other types of paramagnetic defect states which were discussed above.

Agreement between the observed g-values of centers attributed to bridging oxygen vacancies and those predicted from the model is not as good as for the non-bridging oxygen vacancy. One defect center which can be attributed to a bridging oxygen vacancy because of its hyperfine interactions (observed in ammonium hypophosphite single crystals (50)) has g-values that are clearly not in agreement with the model. It is evident from the hyperfine interaction constants that the one defect electron is localized principally on only one of the two P ions at the defect site. The g-values for the electron are more closely in accord with those expected for a non-bridging vacancy. It is also possible to interpret the resonance data on a center in α-quartz (the E_1' center) (60) as indicating the same kind of localization.

One other interesting possibility due to this localization is apparent. The energy separation between the symmetric and antisymmetric states of the defect center is a function of the separation of the M ions. If this separation is small, thermal excitation of the defect electrons from the lower defect state to upper defect state may occur. The defect may now have S = 1 and another term, $\vec{S} \cdot \overleftrightarrow{D} \cdot \vec{S}$, will be added to the spin Hamiltonian. In silicon (62) and α-quartz (63) S = 1 states have been detected and have been attributed to this effect. Because of the magnitude of the components of the $\overleftrightarrow{D}$ tensor $\Delta M_s = \pm 1$ resonance transition due to S = 1 states would be difficult to detect in glasses. However, it may be possible to detect $\Delta M_s = \pm 2$ transitions since the width of the powder pattern should be small and the peak intensity of the resonance absorption correspondingly larger (64).

C. O^- and MO^-

ESR spectra observed in irradiated non-oxide single crystals have been attributed to O^- ions in interstitial and substititional sites. In calcium fluorophosphate and calcium fluoride the spectra is most consistently interpreted in terms of O^- in a F^- site. The g-values are $g_{\parallel} = 2.0012$, $g_{\perp} = 2.0516$ and $g_{\parallel} = 2.0016$, $g_{\perp} = 2.0458$, respectively (65). The g-values are in reasonable agreement with calculated values based upon the energy level scheme shown in Fig. 4, where the energy difference of the p_x, p_y and p_z orbitals, $E_1 - E_o \sim 0.3$ eV and $-\lambda \sim 100$ cm^{-1}. In two alkali halides a spectrum with $g_{\parallel} \simeq 1.981$, $g_{\perp} \simeq 2.258$ (66) has been attributed to O^- in interstitial sites. The difference in g-values between O^- in the calcium and alkali compounds is probably due to the difference in the cation charges. This difference will be discussed below. Spectra in irradiated oxide glasses have not been reported with g-values similar to those for the O^- ion in the calcium or alkali compounds.

Spectra observed in several glasses have been attributed to the MO^- radical. Karapetyan and Yudin (57) proposed this model to explain a spectrum that is characteristic of most phosphate glasses and several compounds (P. Beekenchamp's doctoral dissertation, technische Hogeschool te Eindhoven, Holland, 1965, provides a summary of the reported data and gives additional results). Lee and Bray (67) have also attributed spectra observed in borate glasses to a defect of this type. In both cases $\Delta g_{\perp} \simeq 0$ and $\Delta g_{\parallel} \neq 0$ in disagreement with the values observed for the O^- ion in calcium fluorophosphate or calcium fluoride. A possible explanation of this discrepancy was suggested in Section IV-A. This explanation requires that the admixing of the sp orbital of the M ion alter the ordering of the energy levels in Fig. 4 with the energy of the p_z orbital less than the p_x and p_y orbitals and that a rhombic term in the crystal field interaction removes the degeneracy of the p_x and p_y orbitals. In the absence of this field resonance would be difficult to observe. One of the reasons for attributing these spectra to the MO^- defect is the observed hyperfine interaction with the M nucleus. In the phosphate glasses a $\approx$ 150 mc and an anisotropic hyperfine interaction is not observed. This isotropic interaction is 3 times larger than that observed for atomic phosphorus where it is $\sim$ 50 mcps (68). It does not appear reasonable to expect the hyperfine interaction of an electron localized on a nearby oxygen ion with ^{31}P to be larger than that for an electron in a p orbital of the phosphorus ion. A mechanism has been proposed (69) by which the isotropic interaction with the ^{31}P of an electron on an oxygen ion would be appreciable but still not comparable with the observed value. A similar argument can be advanced for discounting the model proposed for the resonance in the borate glasses. On the basis of this discussion, it is concluded that the assignment of observed ESR spectra to a MO^- type defect in the oxide glasses is still an open question.

D. O_2^- and MO_2^-

There are a number of reports on the ESR characteristics of the O_2^- molecule ion. These data are summarized in Table IV. In the oxides two of the g-values are almost equal and axial symmetry will be assumed with $g_\perp \gtrsim g_{fe}$ and $g_\parallel > g_{fe}$. The data in this table show that in the oxides $g_\parallel$ varies between $\sim$ 2.04 and 2.18 and $g_\perp$ between 1.98 and 2.007. In two different irradiated zeolites both extremal values are found for the g-values. The variation is particularly interesting since it may be due to an alteration of the charge of the cation. If the cation charge determines the M-O-O angle, it would be expected that in the borate, silicate, phosphate, and arsenate glasses the angle would be a minimum and the MO_2^- radical would have $g_\parallel \sim 2.04$ and $g_\perp \sim 2.002$. In silicate (70,50) and borate (67) glasses an ESR spectrum is observed with approximate axial symmetry and with these g-values. Tentatively, this spectrum may be identified with the MO_2^- defect. Identification of this defect in a single crystal of α-quartz or in any other oxide single crystal in which the cation is multivalent would furnish a suitable test of this hypothesis. For this purpose, and ENDOR measurement could also be used to identify the cation in the glasses.

On the basis of these results it is useful to speculate on the electronic interaction that decreases the g-values of the MO_2^- defect in different host lattices. Kasai has attributed this decrease to an increase in the M ion charge (Table II). One of the spectra observed by Griscom in alkali-perborates (Table II) has g-values comparable to those found by Kasai in Na-zeolite and by others in the alkali superoxides. Thus Kasai's hypothesis seems reasonable on the basis of these data. In the absence of strong crystal field interactions from other ions a decrease in the M-O-O angle will produce the increase in Δ (Fig. 5a) required for the decrease in the g-values. From the data it is apparent that this angle has a value that is almost independent of whether the cation charge is +2, +3, or +4. This minimum value for the angle is probably determined by a coulombic interaction of the π electrons of the oxygen with the electrons of the M ions.

An estimate of the energy difference, Δ, between the ξ and η levels in Fig. 5a can also be made for the extremal values of $g_\parallel$ and $g_\perp$ from the equations given in Section IV-B. Assuming that E is approximately constant for the two conditions and $\sim$ 5 eV, the maximum value of Δ is 0.63 eV for $g_\parallel = 2.04$ and $g_\perp = 2.002$ and the minimum value is 0.14 eV for $g_\parallel = 2.18$ and $g_\perp = 1.997$.

TABLE IV

ESR Parameter of O_2^- Molecule Ion

Glass or Compound	g_x	g_y	g_z	ν (Gcps)	T of obs. (°K)	λ/E[a]	λ/Δ[a]	Ref
O_2^- as impurity in KCl	1.9551	1.9512	2.4359	∿10	<20	0.0025	0.23	b
NaO_2 in Na_2O_2 and KO_2 in K_2O_2	2.002	2.002	2.175	∿91 ∿23 ∿36	90			c
O_2^- in treated alkali perborate monohydrate	2.001 1.981	2.001 1.981	2.116 2.306			0.002		d,e
O_2^- in Ba Zeolite	2.0046	2.0090	2.057	9.10	77	0.0038	0.03	f
O_2^- in Na Zeolite	2.0023	2.0066	2.103	9.10	77	∿0.0038	0.06	f
RO_2^- in Teflon	2.002	2.002	2.04	9.2	77 296			g

[a]These values have been obtained from the g-values of the observed resonance spectra. From these values with $\lambda \simeq 0.014$ eV, E ≈ 4 eV.
[b]W. Kanzig and M.H. Cohen, Phys. Rev. Let. 3, 509 (1959).
[c]J. Bennet, et al., Proc. Phys. Soc. 69A, 556 (1956); J. Bennet, et al., Phil. Mag. 46, 443 (1955).
[d]D. Griscom, Ph.D. Thesis, Brown Univ. (1966); In a careful analysis Griscom has shown that a distribution of g-values between these two limits are found in perborate preparations.
[e]R. Bruce, et al., J. Am. Chem. Soc. 87, 2057 (1965).
[f]P. H. Kasai, J. Chem. Phys. 43, 3322 (1965).
[g]D. W. Ovenall, J. Phys. Chem. of Sol. 26, 81 (1965).

E. O_3^- and Other Complex Molecule Ions

The resonance spectrum of O_3^- has been observed in a number of compounds. Data on this molecule ion are listed in Table V. No spectrum in the glasses or their crystalline forms have been reported for a powder pattern with these g-values. The resonance signal of such a defect would be difficult to resolve from those due to O_2^- and O^-, particularly when a distribution of g-values of these defects is present. Griscom (42) has been able to resolve the contributions from O_2^- and O_3^- in perborate preparation.

VIII. CONCLUSIONS

There is a considerable body of information which shows that the average atomic configuration at any point in a glass is similar to that in one of the crystalline phases of the same composition. X-ray diffraction studies (71) show that the radial coordination in silica glass, for example, is similar to α-quartz. Infrared absorption bands have been found to be similar in silicate, germanate, arsenate and antimonate glasses, to those observed in α-quartz, hexagonal phase (not rutile) GeO_2, As_2O_5, and Sb_2O_5, respectively. Bray and his students have also demonstrated this similarity of structure. The results of these studies have shown that the concept of a structural unit in the glasses has considerable substance and utility. These studies also support the hypothesis that the defect structures of single crystals can be related to defects in a glass. There may not be an exact correspondence of defect structure due to liquid-like fluctuation in structure and to the high concentration of Schottky vacancies. These deviations at paramagnetic defects may be revealed by line widths that are due to small variations in g-values and hyperfine interaction constants from site to site.

Of the various oxides which have been discussed, only α-quartz is available in large single crystals, and α-quartz is the favorite crystal of many students of defects in glass-forming oxides. P_2O_5 single crystals are relatively easy to grow but difficult to use in experimentation, because they dissolve in the moisture of the atmosphere. Apparently no one has yet been successful in growing large single crystals of GeO_2 in either phase. Similarly, large single crystals of As_2O_5 have not yet been successfully grown. It would appear at this moment that only α-quartz single crystals are available for study, and successful growth of crystals of the other oxides would open other fruitful avenues of study.

Despite the difficulties of analyzing the powder patterns of ESR spectra in glasses it is apparent that some progress has been made in constructing models for paramagnetic defects in glasses.

TABLE V

ESR Parameters of the O_3^- Molecule Ion

Host Material	g_x	g_y	g_z	ν (Gcps)	T (°K)	Ref
Treated Alkali						
Perborate	2.0051	2.0105	2.0183	10	300	a
Monohydrate	2.0022	2.0113	2.0184	10	77	a
NaO_3 } NH_3		2.012		10	90	b
$KClO_3$	2.0025	2.0113	2.0174	10	77	c

[a] D. Griscom, Ph.D. Thesis, Brown University (1966).

[b] A. D. McLachlan, et al., J. Chem. Soc. 1959, 952 (1959).

[c] P. W. Atkins, et al., J. Chem. Soc. 1962, 4785 (1962).

A lack of information about the relative energies of the defect states and the bonding and anti-bonding states of the glass prevents a more critical approach to the comparison of models with experiment. Despite this lack, some obvious deficiencies of some of the models are apparent and the necessary steps for improving them have been indicated. Comparison of calculated and experimental g-values of the non-bridging oxygen vacancy showed that the experimental values were usually smaller. In addition to the spin-orbit interaction there are other electronic interactions which affect the g-values. For example inclusion of exchange polarization (57) will probably decrease the calculated values. In spite of the lack of agreement between observed and calculated g-values the validity of the non-bridging vacancy model is attested by the observed hyperfine interactions in silicate and phosphate glasses.

ESR spectra have been observed in the oxide glasses which are attributed to the O_2^- molecule ion. Because of the expected similarity of g-values there is a possibility that the spectra are due to an interstitial O^+ ion. No spectra have been observed which have g-values expected for the O^- ion in an axial or rhombic field.

Since the g-values expected for a hole trapped on a non-bridging oxygen are expected to be similar to those for the O^- ion some questions have been raised about spectra attributed to this type of defect. Spectra which may be due to the other paramagnetic states of oxygen complexes have not been identified.

ACKNOWLEDGMENTS

Discussions with M.M. Abraham, P.J. Bray and D. Griscom have aided in the development of this paper. This research was sponsored by the U. S. Atomic Energy Commission under contract with **Union Carbide Corporation and under AEC Contract AT(30-1)-2024 with Brown University**

REFERENCES

1. *E.R. Lipincott, A.V. Valkenberg, C.E. Weir and E.N. Bunting, J. Res. Nat'l. Bur. Std. 61, 61 (1958); V.P. Cheremisinov, Tr. Fiz. Inst. Akad. Nak (SSSR) 25, 153 (1964).*

2. *The band-to-band transition will be discussed in later sections.*

3. *H.A. Robinson, J. Phys. and Chem. of Sol. 26, 109 (1965) and references contained therein; V.V. Tarasov, New Problems in the Physics of Glass, Chap. I.*

4. *M.C. Wittels and F.A. Sherrill, Phys. Rev. 93, 1117 (1954).*

5. *L. Bragg and C.F. Claringbull, Crystal Structure of Minerals (Cornell Univ. Press, Ithaca, N.Y., 1965).*

6. *E.C.T. Chao, J. Geophys. Res. 67, 419 (1962); Foote Prints 32, 25 (1966); Sci. 133, 882 (1961).*

7. *R.W.C. Wyckoff, Crystal Structures II (Interscience Pub. Inc., N.Y. (1960).*

8. *J.R. Van Wazer, Phosphorus and Its Compounds, Chap. 12 (Interscience Pub. Inc., N.Y. 1958); E. Thilo, Advances in Inorganic Chemistry and Radiochemistry 4, 1 (Editors H.J. Emeleus and A. G. Sharpe, Academic Press, N.Y., 1962).*

9. *R.A. Weeks and E. Lell, J. Appl. Phys. 35, 1932 (1964) and references therein; R.A. Weeks and C.M. Nelson, J. Appl. Phys. 131, 1555 (1960).*

10. *R.A. Weeks and P.J. Bray, to be published.*

11. *R.A. Weeks and T. Purcell, J. Chem. Phys. 43, 483 (1965) and references therein.*

12. J.H. Gibbs, Modern Aspects of the Vitreous State I, p. 152 (J. D. Mackenzie, Ed., Butterworth Inc., 1962).

13. R.W. Gurney, Handbook of Physics, p. 8-43 (E.U. Condon and H. Odishaw, Editors, McGraw-Hill Book Co., N.Y., 1958).

14. R.A. Weeks, Paper 42, Proc. VII Intl. Congress on Glass, Brussels (1965).

15. E.U. Condon and H. Odishaw (Editors) Handbook of Physics, p.8-46, (McGraw-Hill Book Co., N.Y., 1958); J.H. Schulman and W.D. Compton, Color Center in Solids, Chaps. 1 and 10 (Macmillan Co., N.Y., 1962).

16. Richard H. Bube, Solid State Physics II, p. 223 (Edited by F. Seitz and D. Turnbull, Academic Press, N.Y., 1960).

17. O.S. Oen and D.K. Holmes, J. Appl. Phys. 30, 1289 (1959).

18. W.A. Weyl and E.C. Marboe, The Constitution of Glasses II, p. 289 (Interscience Publishers, N.Y., 1962).

19. H.B. Callen, Handbook of Physics, p. 8-24 (E.U. Condon and H. Odishaw, Editors, McGraw-Hill Book Co., N.Y., 1958).

20. S. Karavelas, Thesis, University of Michigan (1964).

21. J.H. Schulman and W.D. Compton, Color Centers in Solids (Macmillan Co., N.Y., 1962).

22. C.A. Coulson, Valence, Chap. IV (Oxford University Press, N.Y. 1957).

23. R.A. Walsh, J. Chem. Soc. 1953, 2301 (1953).

24. R.L. Sproull, "Properties of Semiconductors", p. 5-157, American Institute of Physics Handbook, 1957.

25. For example, the two defect electron states of a Si vacancy in silicon fall between the bonding and anti-bonding orbital states of the normal Si-Si configuration. G.D. Watkins and J. W. Corbett, Report 64-RL (3567-X), G.E. Research Laboratory, Schenectady, N.Y., Feb. 1964.

26. C.P. Slichter, Principles of Magnetic Resonance, p. 66 (Harper N.Y., 1963).

27. H.S. Jarrett, Solid State Physics, Vol. 14, p. 215 (F. Seitz and D. Turnell, Editors, Academic Press, N.Y., 1963).

28. M.H. L. Pryce, Proc. Roy. Soc. (London) A63, 25 (1950); A. Abragam and M.H. L. Pryce, Proc. Roy. Soc. A205, 135 (1951).

29. D.H. Tombulain and W.M. Cadis, Phys. Rev. 59, 422 (1944).

30. R.A. Weeks, Proc. British Cer. Soc. on Point Defects No. 2, 1966.

31. G.D. Watkins and J.W. Corbett, Phys. Rev. 134, A1359 (1965).

32. C.P. Slichter, Principles of Magnetic Resonance, Sec. 7.4 (Harper and Row, N.Y., 1963).

33. G.D. Watkins and J.W. Corbett, Phys. Rev. 138, A543 (1965).

34. G. Feher, Phys. Rev. 114, 1219 (1959); G. Feher, J.C. Hensel and E.A. Gere, Phys. Rev. Let. 5, 309 (1960).

35. C.J. Ballhausen, Introduction to Ligand Field Theory, Chap. 6 (McGraw-Hill Book Co., Inc., N.Y., 1962).

36. G.S. Smith and P.B. Isacs, Acta. Cryst. 17 (7), 842 (1964).

37. G.T. Trammell, H. Zeldes and R. Livingston, Phys. Rev. 110, 630 (1958).

38. H. Meyer, M.C.M. O'Brien and J.H. Van Vleck, Proc. Roy. Soc. A243, 414 (1958).

39. R.S. Mullikin, Rev. Mod. Phys. 4, 1 (1932).

40. W. Kanzig and M.H. Cohen, Phys. Rev. Let. 3, 509 (1959).

41. A.D. Mclachlan, M.C.R. Symons and M.G. Townsend, J. Chem. Soc. 1959, 952 (1959); P.W. Atkins, J.A. Brivotti, N. Keen, M.C.R. Symons and P.A. Tveralion, J. Chem. Soc. 1962, 4785 (1962).

42. D. Griscom, Ph.D. Thesis, Brown University (1966).

43. L.N. Muloy and L.K. Keys, BAPS 10, 75 (1965).

44. J.C. Baird, J. Chem. Phys. 37, 1879 (1962).

45. F.K. Knebuhl, J. Chem. Phys. 33, 1074 (1960).

46. R. Nieman and D. Kivelson, J. Chem. Phys. 35, 149 (1962).

47. J.A. Ibers and J.D. Swalen, Phys. Rev. 127, 1914 (1962).

48. J.W. Searl, R.C. Smith, and S.J. Wyard, Proc. Phys. Soc. 78, 1174, (1961).

49. R.H. Sands, Phys. Rev. 99, 1222 (1955).

50. R.A. Weeks, J. Appl. Phys. 27, 1376 (1956).

51. M.C. Wittels, Phil. Mag. 2, 1445 (1957).

52. J.A. Weil and H.G. Hecht, J. Chem. Phys. 35, 281 (1963).

53. S.M. Blinder, J. Chem. Phys. 33, 748 (1960).

54. B.E. Warren, J. Appl. Phys. 8, 645 (1937).

55. U. Opik and R.F. Wood, Conference on the Calculation of the Properties of Vacancies and Interstitials, May 1, 1966, Washington, D.C., to be published.

56. H. Horsfield, J.R. Morton and D.H. Whiffen, Mol. Phys. 4, 475 (1961); M.W. Hanna and L.J. Altman, J. Chem. Phys. 36, 1788 (1962); W.E. Hughes and W.G. Morton, J. Chem. Phys. 39, 1359 (1963); J.R. Morton, Mol. Phys. 6, 193 (1963).

57. D.A. Goodings, Phys. Rev. 123, 1706 (1961).

58. C.R. Kurkjian and J.E. Krause, J. Am. Cer. Soc. 49, 134 (1966)

59. J.R. Morton, Mol. Phys. 6, 193 (1963).

60. R.A. Weeks, Phys. Rev. 130, 570 (1963).

61. R.H. Silsbee, J. Appl. Phys. 32, 1459 (1961).

62. W. Jung and G.S. Newell, Phys. Rev. 132, 648 (1963).

63. R.A. Weeks and M.M. Abraham, BAPS 10, 374 (1965).

64. E. Wasserman, L.C. Snyder and W.A. Yager, J. Chem. Phys. 41, 1763 (1964).

65. B.S. Segall, G.W. Ludwig, H.H. Woodbury and P.D. Johnson, Phys. Rev. 128, 76 (1962); H. Bill and R. Lacroix, Phys. Let. 21, 257 (1966).

66. W. Sander, Naturwissenschaften 51, 404 (1964).

67. *S. Lee and P.J. Bray, J. Chem. Phys. 39, 2863 (1963); S. Lee and P.J. Bray, Ibid. 40, 2982 (1964); G.M. Muha, J. Phys. Chem 70, 1390 (1966).*

68. *H.G. Dehmelt, Phys. Rev. 99, 527 (1954); R.H. Lampert and F.M. Pipkin, Phys. Rev. 128, 198 (1962); R.A. Serway and S.A. Marshall, BAPS 11, 203 (1966).*

69. *W.H. From, C. Kikuchi and P. Dorain, BAPS 9, 37 (1964).*

70. *S. Lee and P.J. Bray, Phys. and Chem. of Glasses 3, 37 (1962).*

71. *B.E. Warren, J. Appl. Phys. 8, 645 (1937).*

72. *W.C. Lin and C.A. McDowell, Mol. Phys. 7, 223 (1964).*

DISCUSSIONS

A. GUINIER (Faculte des Sciences d'Orsay, France): *What is the number of vacancies in neutron irradiated quartz detected by ESR?*

R.A. WEEKS: *After approximately 10^{20} fast neutrons cm^{-2} and for an ambient temperature of irradiation of approximately 100°C, the concentration of centers which are identified as oxygen vacancies is approximately 10^{19} cm^{-3}. However, γ-ray irradiation will produce a higher concentration in pure silica glass if the glass has a concentration of hydrogen $< 10^{18}$ ions cm^{-3}.*

J.H. CRAWFORD, JR. (Oak Ridge National Laboratory U.S.A.): *If the concentration of unpaired spins can be taken as a measure of the defect content, it might be pointed out that D.K. Stevens found that fast neutron irradiation at about 100°C produced a saturation spin concentration (Measured by paramagnetic susceptibility) of about 2×10^{19} cm^{-3} for both crystalline and glassy silica.*

R.A. WEEKS: *The paramagnetic defects apparently have a distribution of thermal activation energies. Thus an irradiation at a lower temperature will produce a higher concentration of centers.*

J.D. MACKENZIE (Rensselaer Polytechnic Institute U.S.A.): *How do you explain the large differences in the absorption edge shown for silica Quartz and the various SiO_2 glasses and how does this influence your calculations?*

R.A. WEEKS: *Localized regions of strain produce some of the shift; the high concentration of Schottky vacancies probably accounts for some of the shift, and the remainder is due to unknown causes.*

J.D. MACKENZIE: *If strains are used to explain the differences in absorption edge, it is difficult to attribute it to thermal expansion because fused silica has negligible expansion.*

R.A. WEEKS: *The anomalies in the thermal expansion of glasses may be due in part to localized modes of oscillation.*

E. LELL (Bausch and Lomb. Inc., U.S.A.): *Dr. Weeks, would you be willing to give an estimate of the concentration of the number of peroxide groups in glasses?*

R. A. WEEKS: *After neutron irradiation of approximately* 10^{20} cm^{-2} *a resonance is observed which may be due to a peroxy radical; a concentration of approximately* 10^{19} cm^{-3} *is found. For a lower temperature of irradiation there would probably be a higher concentration for the same irradiation dose.*

RADIATION DAMAGE STUDIES OF LIBYAN DESERT SILICA GLASS

A. Bishay and F. Hassan

The American University in Cairo

U.A.R.

The EPR spectrum for a number of Libyan Desert silica glass samples was studied before and after reactor and cobalt 60 irradiation. A resonance line at about g = 4.2660 was observed in the unirradiated samples and increased in intensity with increasing total iron content. This line was attributed to Fe^{3+} ions since Fe^{2+} does not give any EPR spectrum at temperatures above the liquid hydrogen range. Irradiation of different samples resulted in decreasing the intensity of the 4.2660 line and increasing the intensity of the optical absorption band at 1.8 microns which is characteristic for four coordinated ferrous ions. A broad resonance was also observed at about g = 2 in the unirradiated sample. However, this resonance was not observed in the same sample after heating for 30 minutes at 500°C, which indicated that it was mainly due to solarization effects. These observations support the postulate that the g = 4.2660 line corresponds to Fe^{3+} iron in 4-coordinated positions and that practically all the ferric ions in Libyan Desert silica glass enter the silica network substitutionally for silicon. Ferrous ions, on the other hand, take both four and six coordinated positions.

A sharp resonance at about g = 2.0005 was observed in irradiated samples and attributed to the E' center reported by Weeks in pure quartz and pure silica. This resonance was much stronger in reactor irradiated samples as compared to those irradiated with cobalt 60. However, it was noted that splitting becomes more apparent as a result of fading. Furthermore, the intensity of the resonance in this range increased with increasing iron content, which indicated that in addition to the E' center, a ferric defect center contributes to this resonance. A broad resonance was ob-

served in the region of g = 2.009 and was attributed to the well known aluminum center.

INTRODUCTION

A number of studies on the origin and properties of the Libyan Desert silica glass (about 98% SiO_2) have been reported during the last 35 years. Recently, increasing interest in space work attracted our attention to this natural glass found in Egypt.

A party surveying the southern western regions of the desert of Egypt under the leadership of Sir Clayton of the Egyptian Survey Department was the first to report the existence of Libyan Desert silica glass. Later in 1932, Clayton accompanied by L.J. Spencer of the British Museum visited the same area and collected a large number of specimens of this glass (1,2,3). There was no evidence of volcanic action in the area to account for the glass, nor was any volcanic rock or any sign of rupture of the earth (2), which made a wholly terrestial origin of the glass seem improbable. An alternate suggestion was that the glass is due to some meteoric cause, having either fallen from the sky in its present state (i.e. a tektite), or being due to the impact of a gigantic meteorite which generated sufficient heat to fuse the desert sand (i.e. impactite).

Valuable information has been obtained from electron spin resonance studies of irradiated quartz and fused silica (4). The significant iron content of the Libyan Desert silica glass (0.085-0.68% Fe_2O_3 by weight) suggested the use of EPR techniques in determining the structural role of iron in this glass, and studying the color centers induced in different samples by means of reactor and ^{60}Co irradiation.

EXPERIMENTAL

Various specimens of the Libyan Desert silica glass, ranging from clear pale greenish yellow to dark greenish black, were provided by the Geological Museum in Cairo. Samples were ground and polished to a thickness of about one millimeter for spectrophotometric studies. A Beckman DK-2A recording spectrophotometer was used for absorption measurements in the visible, ultraviolet and near infrared regions.

The different samples for EPR measurements were finely powdered and mixed thoroughly. All measurements were conducted at room temperature using a Varian No.V-4502 EPR spectrometer system.

Most of the samples were irradiated in a ^{60}Co Gamma Cell 220 (6300 Curie) at a dose rate of 4.8×10^5 rad/hr. Measurements of gamma irradiated samples were taken immediately after being removed from the radiation field with the exception of EPR measurements which were taken about six hours later. A few specimens were irradiated in one of the horizontal channels of the U.A.R. Atomic Energy Establishment's tank type experimental reactor. The experiments were conducted at the reactor's full power of 2000 Kw, at a distance of 70 cm from the core. The flux at this position was about 10^{10} n/cm^2.sec. of thermal neutrons, 10^8 n/cm^2.sec. of epithermal neutrons, and, 10^8 n/cm^2.sec. of fast neutrons (energy between 3.1-8.2 Mev). The gamma ray intensity was about 10^7 rad/hr. next to the core of the reactor (average energy about 2.0 - 2.15 Mev).

RESULTS AND DISCUSSION

Figure (1) shows the EPR spectrum for sample 1 (0.08% Fe_2O_3) of the Libyan Desert silica glass. The dashed line indicates the spectrum before irradiation, while the continuous line shows the spectrum after irradiation for 169 hours in the reactor. The intensity of the resonance at about g = 4.2660 decreased appreciably after irradiation. In order to demonstrate this observation more clearly the conditions for obtaining the resonance were changed and the signal is shown in Figure (2). A similar observation is shown in Figure (3) for sample 9 (0.68% Fe_2O_3) irradiated by ^{60}Co (dose 2.0 x 10^8 rads/hr.).

It is known that the ferric ion with the spin S=5/2 and orbital angular momentum L = 0 would have a relatively sharp line at an apparent g = 4.28 (5). In addition to this line a broad peak centered at about g = 2.00 may also be observed for ferric ions in a different type of site. On the other hand, no resonance is expected for Fe^{2+} ions at temperatures above the liquid hydrogen range because its integral spin S = 2 and orbital angular momentum L = 2.

The resonance line observed in the unirradiated samples at about g = 4.2660 is expected to be associated with ferric impurities in the Libyan Desert silica glass. This is confirmed in Figure (4) which shows the increase in intensity of this line with increasing iron content in the different samples.

The decrease in the intensity of the 4.2660 resonance line with increasing radiation dose was found to be associated with an increase in the optical absorption due to ferrous ions (Fig. 5). This observation suggests the following reaction during irradiation:

$$Fe^{3+} + e \longrightarrow Fe^{2+}$$

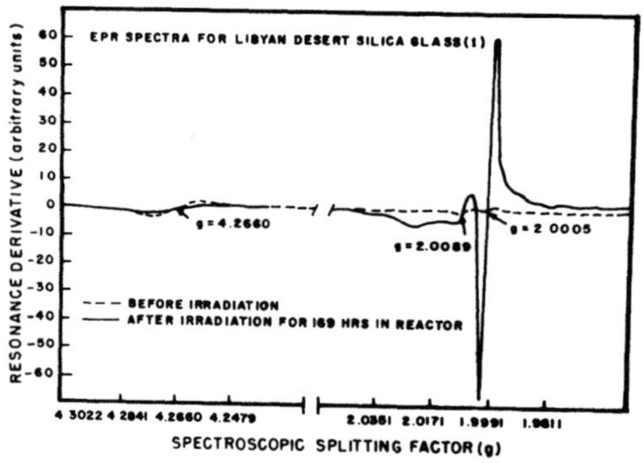

Fig. 1. EPR spectra for Libyan Desert silica glass (1).

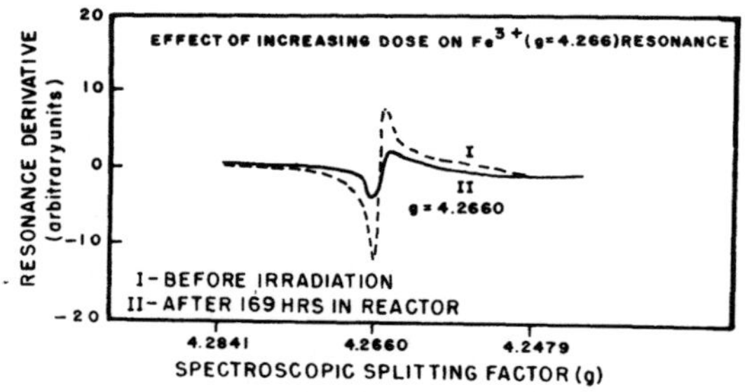

Fig. 2. Effect of increasing dose on Fe^{3+} (g = 4.266) resonance.

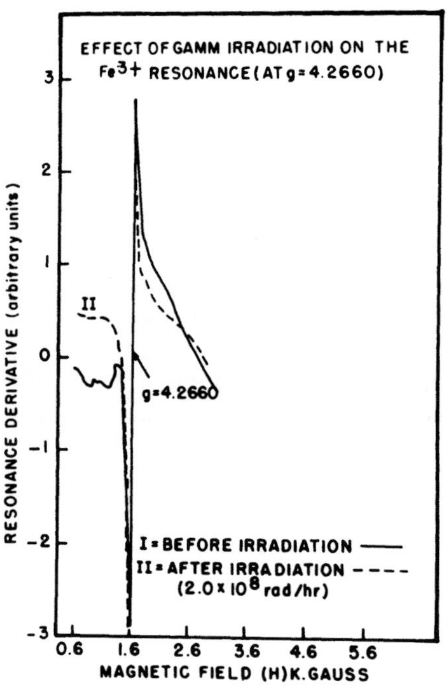

Fig. 3. Effect of gamma irradiation on the Fe^{3+} resonance (AT g = 4.2660).

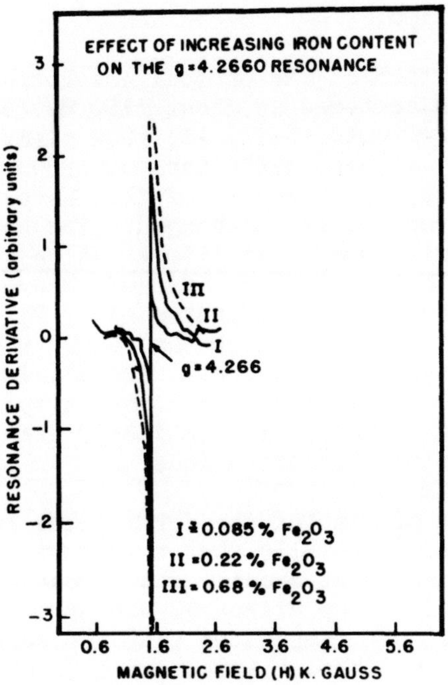

Fig. 4. Effect of increasing iron content on the g = 4.2660 resonance

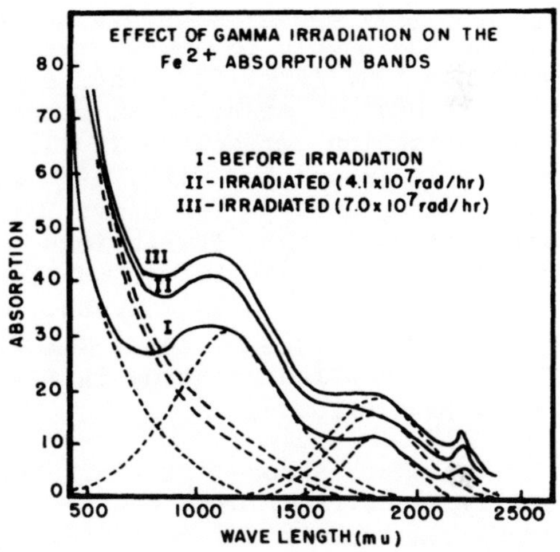

Fig. 5. Effect of Gamma irradiation on the Fe^{2+} absorption bands.

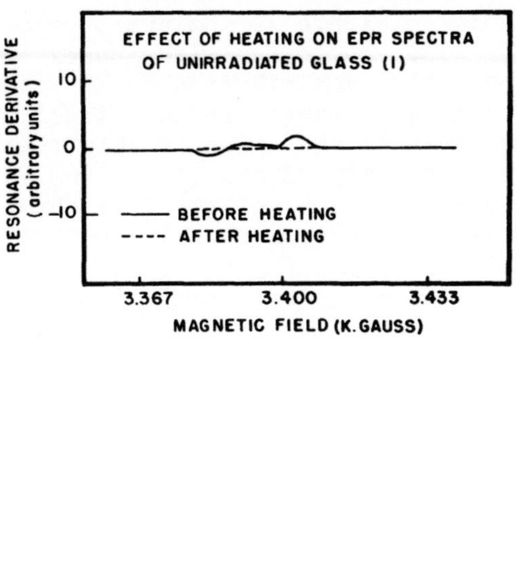

Fig. 6. Effect of heating on EPR spectra of unirradiated glass (1)

which would explain the decrease in the Fe^{3+} resonance and the increase in the Fe^{2+} optical absorption.

An approximate gaussian resolution of the optical absorption spectrum (Fig. 5) for the two bands centered at about 1150 mμ (6-coordinated Fe^{2+}) and 1800 mμ (4-coordinated Fe^{2+}) (6) show clearly that the intensity of the Fe^{2+} 4-coordinated band at 1800 mμ increases with increasing dose while the intensity of the ferrous 6-coordinated band at 1150 mμ is practically unchanged. The observed spectrum, however, shows an increase at 1150 mμ which is mainly attributed to the increase in the visible induced absorption whose tail overlaps with the ferrous 6-coordinated band. This observation strongly suggests that the resonance at about g = 4.2660 is associated with ferric 4-coordinated ions. Irradiation does not change the coordination of the iron atoms but simply changes its valency, resulting in a decrease in the 4.2660 resonance and an increase in the ferrous 4-coordinated ions.

The possibility of the presence of some of the ferric ions in a different site was also considered specially that a weak broad resonance was observed before irradiation at about g = 2. However heating the sample to 500°C resulted in the disappearance of this resonance line (Fig. 6). This experiment suggested that the weak broad resonance observed at about g = 2 before irradiation is due to solarization effects. It also shows clearly that there is no resonance associated with Fe^{3+} ions at about g = 2 in these samples. Accordingly, the above discussion clearly indicates that practically all the ferric ions in the samples of Libyan Desert silica glass are present in substitutional positions for silicon, i.e. 4-coordinated with oxygen. On the other hand, ferrous ions are present in both 6- and 4-coordination sites as proved by the **optical absorption spectrum in Fig. (5) (according to ligand field theory predictions).**

The sharp strong line at about g = 2.0005 (Fig. 1) was originally identified with the E' center reported by Weeks (7,8) for irradiated quartz and fused silica. This sharp line is found to be superimposed on a broad line which may be attributed to the well known Al center ($Al\,O_4$ + hole). Figure 7, however, shows that the intensity of the sharp line increases with increasing iron content in the glass. The possible contribution of an induced iron center to the resonance in this range was confirmed by studying the effect of increasing dose on the shape of this line. Figure 8 shows clearly that the intensity of the sharp line increases with increasing dose. Furthermore, a splitting is observed at the highest dose. It should be noted here that the resonance was obtained under identical conditions for the three doses. The possibility that the sharp resonance in this range is due to two different lines is confirmed in Figure 9 which shows the chan-

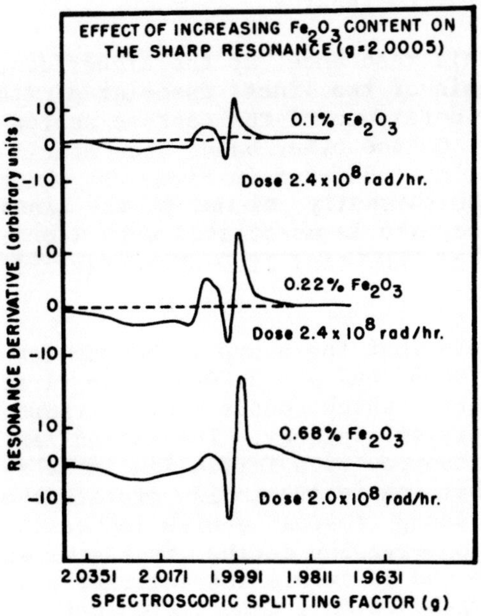

Fig. 7. Effect of increasing Fe_2O_3 content on the sharp resonance (g = 2.0005).

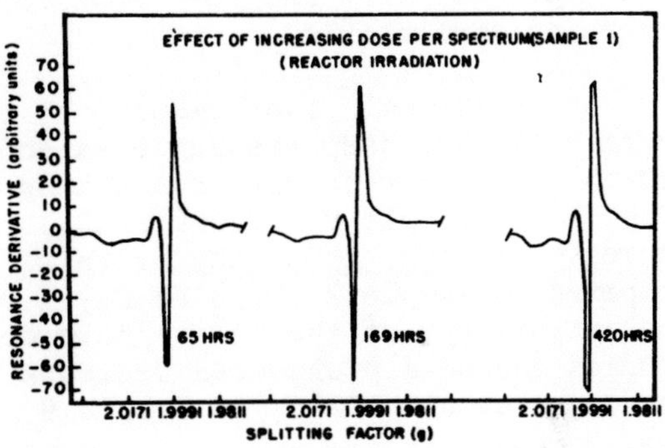

Fig. 8. Effect of increasing dose on EPR spectrum (sample 1) (reactor irradiation).

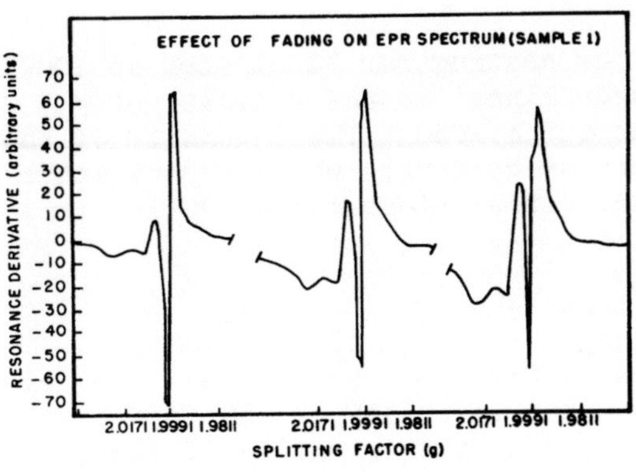

Fig. 9. Effect of fading on EPR spectrum (sample 1).

ge in the shape of this resonance after different periods following the removal of the same sample from reactor.

The splitting observed for this resonance at the higher dose may thus be explained on the basis of two lines associated with two different centers. At higher doses one of the centers approaches saturation before the other. On the other hand, the stability of these lines is different as indicated in Figure 9 which shows a pronounced decrease in the intensity of one of the lines as compared to the other. This decrease is associated with a pronounced increase in the intensity of the broad line characteristic of the Al center.

The above observation suggests that the sharp resonance consists of two lines at about $g = 2.0005$ and $g = 2.0023$. The first line is associated with the E center which consists of a trapped electron at a silicon next to an oxygen vacancy. The second line ($g = 2.0023$) may represent a hole trapped by a ferric ion in substitutional position for Si^{4+}. This postulate would explain the increase in the intensity of the sharp resonance with increasing iron content. Furthermore, the observation during fading of an increase in the intensity of the broad resonance characteristic of the Al^{3+} + hole center, may be attributed to a competition between substitutional Al^{3+} and Fe^{3+} ions for trapping a hole; the Al center being more stable.

An additional confirmation of the effect of iron on the intensity of the sharp resonance is shown from the comparison of the growth curves for samples containing different iron concentrations (Fig. 10). Sample 9 (0.68% Fe_2O_3) shows a much more pronounced increase in the intensity of this resonance at higher doses as compared to sample 2 (0.1% Fe_2O_3). On the other hand, due to the weak stability of the Fe^{3+} + hole center, the thermal bleaching studies showed a pronounced decrease in the intensity of this resonance at about 100°C. in sample 9 as compared to sample 2 (Fig. 11).

Figure 12 shows the effect of heating for 30 minutes at 500°C on the sharp and broad resonance lines in the vicinity of $g = 2$ and on the ferric resonance line $g = 4.2660$. The intensity of the latter line increases as a result of heating; which may be attributed to the unstability of the ferrous ions formed during irradiation:

$$Fe^{2+} \longrightarrow Fe^{3+} + e$$

This postulate is confirmed in Fig. 13 which shows the effect of increasing temperature on the intensity of the ferric resonance line at about $g = 4.2660$ and on the intensity of the opti-

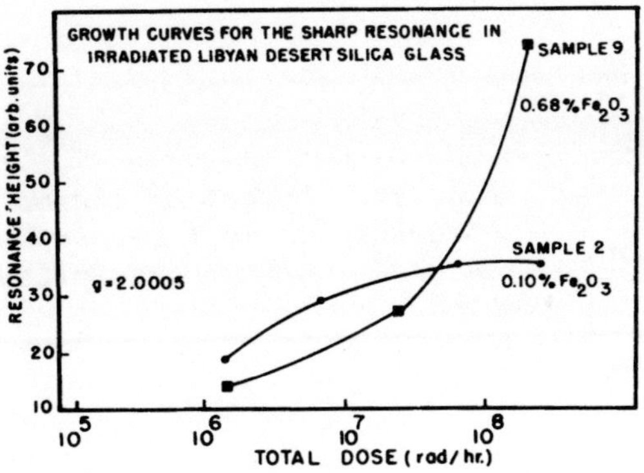

Fig. 10. Growth curves for the sharp resonance in irradiated Libyan Desert silica glass.

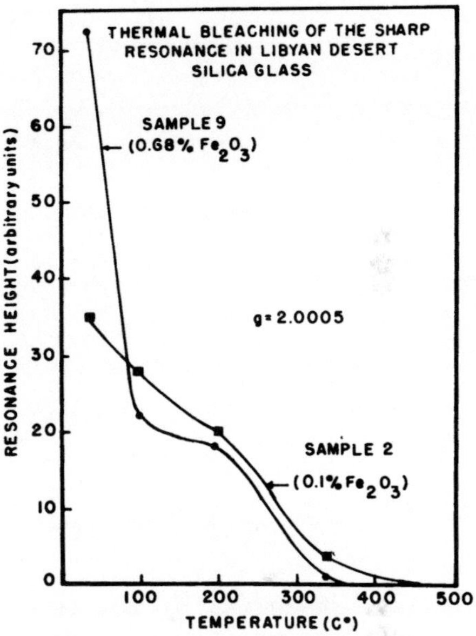

Fig. 11. Thermal bleaching of the sharp resonance in Libyan Desert silica glass.

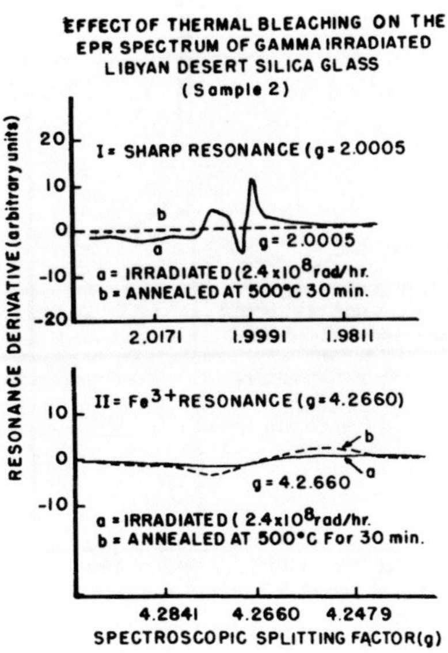

Fig. 12. Effect of Thermal bleaching on the EPR spectrum of gamma irradiated Libyan Desert silica glass (sample 2).

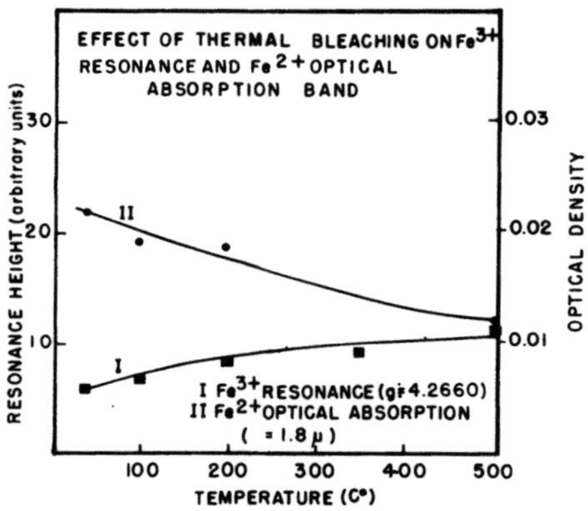

Fig. 13. Effect of thermal bleaching on Fe^{3+} resonance and Fe^{2+} optical absorption band.

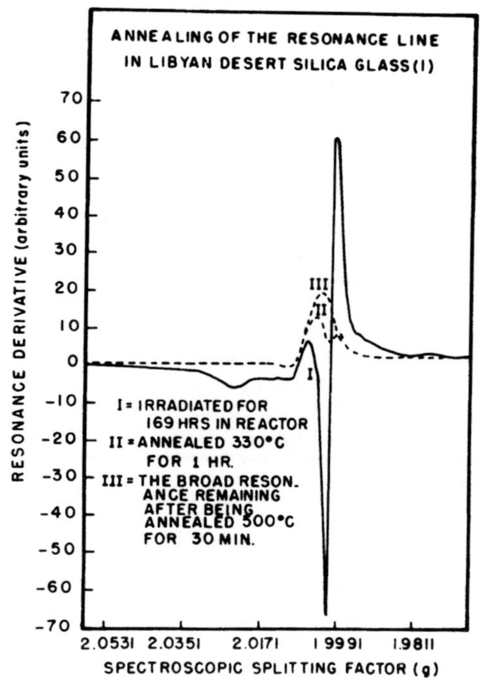

Fig. 14. Annealing of the resonance line in Libyan Desert silica glass (1).

cal absorption band characteristic of the 4-coordinated ferrous ions.

An interesting observation was noted in the effect of thermal bleaching on the resonance in the range of g = 2 for sample I irradiated in the reactor. The continuous line I in Fig. 14 represents the resonance after 169 hours of reactor irradiation. The dashed line II represents the resonance after annealing at 330°C., using the same conditions for EPR measurements. This resonance II shows a drastic decrease in the intensity of the sharp line as well as the broad Al resonance. However, a new broad asymmetric resonance appears at an apparent g value of 2.009. Annealing the sample for another 30 minutes at 500°C resulted in the disappearance of the sharp resonance and the increase in the intensity of the new broad asymmetric resonance (Fig. 14, dotted line III). This latter resonance may be similar to that reported by Bray (9) in alumino silicate glasses after heat treatment. Bray considered the possibility that the broad asymmetric resonance observed in his work arises from a hole associated with a non-bridging oxygen on the basis of VenWriengen and Kats (10) interpretation. Bray suggested that the non-bridging oxygens are associated with aluminium taking network modifying positions. However,. it is more probable that in the samples of the Libyan Desert silica glass the non-bridging oxygens are associated with ferrous ions in 6-coordinated positions. The presence of ferrous ions in these positions is supported by the optical measurements showing the 1150 mμ band characteristic for 6-coordinated ferrous ions.

CONCLUSIONS

A resonance line at about g = 4.2660 was observed and attributed to Fe^{3+} 4-coordinated. The intensity of this line decreased with increasing irradiation, while the intensity of the 4-coordinated Fe^{2+} optical absorption band was increased.

A sharp resonance at about g = 2.0005 was attributed to the existence of two superimposed paramagnetic centers: E' center and Fe^{3+} defect center.

A broad resonance at about g = 2.009 was attributed to the well known Al center. Furthermore, a broad asymmetric resonance remained at about g = 2.009 after thermal bleaching.

ACKNOWLEDGMENTS

Thanks are due to the U.A.R. Atomic Energy Establishment for its cooperation in the irradiation studies and to the Geological Museum of Cairo for providing the glass specimens.

REFERENCES

1. *Nature No. 3405 Vol. 135, Feb. 1935 p. 179.*

2. *Reports Kept in the Geological Museum in CAIRO.*

3. *A.J. Cohen. Space Research III. The Third National Space Science Symposium (1962) Edited by W. Priester.*

4. *E. Lell, N.J. Kreidl, J.R. Hensler, Final Summary Report, U.S. Atomic Energy Commission Contract AT (30-1) 1312, (1965).*

5. *R.F. Tucker, "Advances in Glass Technology" Technical Papers of the VI International Congress on Glass, 103-104, Washington D.C., U.S.A. (1962).*

6. *A.M. Bishay and A. Kinawi. Phys. of Non-Crystalline Solids. (589 - 605). Proceedings of the International Conference Delft, July 1964.*

7. *R.A. Weeks, J. Phys. Rev. 130, 570 (1963).*

8. *R.A. Weeks and C.M. Nelson; J. Am. Ceram. Soc. 43, 401 (1960).*

9. *S. Lee and P.J. Bray. Phys. and Chem. of Glasses Vol. 3 No. 2, April 1962.*

10. *Van Wriengen, J.S. and Kats, A. (1957). Philips Res. Rep. 12, 432.*

DISCUSSION

R.A. WEEKS (Oak Ridge National Laboratory, U.S.A.): *What are the relative concentrations of* Fe^{3+} *and* Fe^{2+} *in the Libyan Desert glass before irradiation and what is the relative decrease in* Fe^{3+} *with irradiation?*

F. HASSAN: *No quantitative measurements were conducted in this part of the work for determining the concentration of* Fe^{3+} *and* Fe^{2+} *or the decrease in* Fe^{3+} *with irradiation. However, this is planned for the near future.*

RADIATION EFFECTS IN SEMICONDUCTORS

James H. Crawford, Jr.

Oak Ridge National Laboratory, Oak Ridge

Tennessee, U.S.A.

Although radiation damage in elemental semiconductors and its consequences for a variety of electronic properties have been intensively investigated for the past 20 years, appreciable progress in understanding the nature of the resultant lattice imperfections has been achieved only in recent years since electron spin resonance experiments have established that stable defects in irradiated silicon for the most part consist of complexes involving point defects and impurity atoms. Positive identification has been made of oxygen-vacancy complexes, donor-vacancy complexes, acceptor-vacancies, divacancies, and interstitial impurity atoms. Except for the divacancies which are a direct product of the irradiation, the remaining complexes are created when point defects (vacancies and interstitials) move through the lattice and interact with impurities. Although equally definitive ESR data are not yet available for germanium, much of the impurity-sensitive response in this material can be understood on the basis of analogous impurity-defect interaction processes demonstrated for silicon. The points of evidence pertaining both to the identification of radiation defects in silicon and germanium and to their electronic behavior will be summarized, and inferences from these studies for other semiconductor systems and the radiation sensitivity of devices will be drawn.

INTRODUCTION

Nearly all of the properties that made semiconductors interesting to study and valuable for applications in solid-state electronics are strongly dependent upon the presence of imperfections in the crystalline structure. Therefore they are excellent materials in which to investigate damage to the crystal lattice produced by high-energy radiation. This modulation of electronic behavior, which we call "Radiation Effects", is a clue to the type of damage introduced and, if we are clever enough to connect specific defect structures with the various qualitative and quantitative changes in properties, we should be able to construct a complete picture of the lattice damage introduced by various types of energetic radiation. Studies of radiation effects in semiconductors have an additional advantage: From them we learn much about radiation sensitivity of properties vital to the operation of semiconductor devices, as well as various factors which affect this sensitivity. Now many potential applications of semiconductor devices may involve radiation environment, space vehicle communication gear for example. Therefore, knowledge pertaining to radiation sensitivity will be invaluable in developing devices with longer service lives in outer space or when used near a nuclear reactor power source.

As a field of study, radiation effects in semiconductors is not new. The first work in this field was done on germanium and silicon nearly 20 years ago at Purdue University and Oak Ridge National Laboratory, and the late Karl Lark-Horovitz (1) was an outstanding leader in these investigations. But although the field is not new, its present complexion is vastly different from what it was six or seven years ago. Before that time the main emphasis was colored by what turned out to be a vain attempt to test quantitatively in terms of defect yield the simple theory of radiation damage, and by a valiant but equally vain attempt to account for the multitude of defect energy levels in terms of simple interstitials and vacancies. Beginning in 1959, several developments occured (2) which revealed without doubt that in the room-temperature range both the yield and nature of radiation defects in silicon and germanium are strongly influenced by the type and amount of impurity present. Since this discovery much effort has been devoted to the determination of the complex structures of defects, involving both impurities and simple vacancies, which are responsible for changes in electronic behavior. These studies have employed properties such as optical absorption of polarized light and electron spin resonance to reveal the symmetry and composition of the photon and microwave absorption centers. In addition, annealing studies have given some insight into the defect-complex structure through the kinetics of the various relaxation processes leading to both their formation and their decomposition.

In this brief summary, I would like to discuss the types of defects that result from irradiation with various types of energetic particles and photons; to indicate briefly the way in which imperfections influence semiconducting behavior; to present evidence for the various defect-impurity interaction products that are produced as a result of the migration of simple defects; and, finally, to mention recent interesting developments in the area of compound semiconductors.

THE NATURE OF RADIATION DEFECTS

In elemental systems such as germanium and silicon, which will be our primary concern, defects are not created in the perfect lattice by ionization through radiochemical processes as they are in the alkali halides. Instead, elastic collisions between the incoming particles and lattice atom are required. Moreover, the recoil energy imparted to the lattice atom must be greater than some minimum value E_d, called the displacement energy, which is required to create a stable interstitial-vacancy pair. In the early, naive damage theory (3), the displacement probability was assumed to be a step function: 0 if $T < E_d$, and unity or greater if $T > E_d$. Using this simple concept E_d is usually measured by exposing a crystal to a prescribed dose of Van de Graaff electrons and increasing the energy until a change in a radiation-sensitive property is first detected. E_d is related to this threshold electron energy by the relativistic expression for the maximum energy imparted in an elastic collision, namely,

$$E_d = 2(E + 2\ mc^2)\ E/Mc^2 \tag{1}$$

where E and m are the energy and rest mass of the electron, and M is the mass of the target atom.

The naive theory does not account for the ordered arrangement of atoms. Obviously, if the lattice structure is considered, the ease of displacement will depend upon the direction of the impulse: in open directions E_d will be much smaller than in the direction of a neighboring ion. Also, in determining E_d care must be taken that the defects created are stable enough for experimental detection. Such a complication is encountered in germanium. Low-temperature electron bombardment experiments (1 to 4.5 MeV electrons at 4^oK) of MacKay and Klontz (4) have shown that conduction electrons are required to stabilize close interstitial-vacancy pairs. These defects, which act as acceptors and hence remove conduction electrons, are therefore produced in higher yield the higher the concentration of conduction electrons. In fact, in p-type specimens where the current is carried by positive holes instead of electrons the yield is almost below detection. From an experimental point of view, then, the electronic state of the semiconductor (whether an electron or hole conductor) would appear to have a strong influence upon the apparent value of E_d for low-temperature ($<10^oK$) irradiations with electrons, since even at liquid helium temperature these close-pair defects are unstable in the absence of conduction electrons.

Another complication which may also be encountered in other solids as well but for which there is direct evidence in silicon is the creation of divacancies by electron bombardment. This occurs when the impulse is along the appropriate crystallographic direction and the recoil energy high enough. Using electron spin resonance techniques, Corbett and Watkins (5) have found that divacancies are formed in silicon during bombardment with 0.7 to 1.5 MeV electrons at 20°K, and that the yield is highest for the bombarding beam parallel to the Si-Si bond direction. These results indicate that the divacancy is formed by a single collision event and that divacancies are a primary radiation defect.

As the energy of the primary recoil increases to many times E_d, the recoil acts as a bombarding particle and secondary, and even tertiary, displacements are created as it collides with other lattice atoms. This results in what is called a displacement cascade in which the defects are closely spaced and have a high concentration. Damage produced by fast neutron bombardment in a nuclear reactor consists predominantly of regions of high defect density (disordered regions) of this type. Typically, these regions are roughly spherical with a diameter of ∿100 Å and these have been resolved in germanium by electron micrography (6). These disordered regions have a rather unusual effect on the electrical behavior of n-type germanium. Since the dominant effect of radiation defects in germanium is acceptor action, the disordered regions are p-type islands in an n-type matrix and as such they must be isolated from the matrix by a potential barrier and an electrical double layer (7), such as shown in Fig. 1. The radial extent of the space-charge zone in the junction surrounding the disordered region is roughly given by the Debye-Hückel length,

$$L_{DH} = \left(\frac{kT\epsilon}{q^2 N_D}\right)^{1/2} \qquad (2)$$

where q is the electronic charge, ϵ is the dielectric constant, and N_D is the donor density. For reasonable purity n-type germanium (donor concentration of 10^{14} to 10^{15}), this distance is between one and two orders of magnitude larger than the radius of the disordered region itself. The potential gradient extending over this region denudes it of mobile charge, so that the entire space-charge zone acts as an insulator, blocking current flow. The situation is shown schematically in Fig.2. Because of this double-layer, disordered regions markedly reduce the conductivity, and for non-overlapping space-charge zones (8) the conductivity is approximately

$$\sigma \simeq \sigma_0 \, (1 - f)/(1 + \tfrac{1}{2}f), \qquad (3)$$

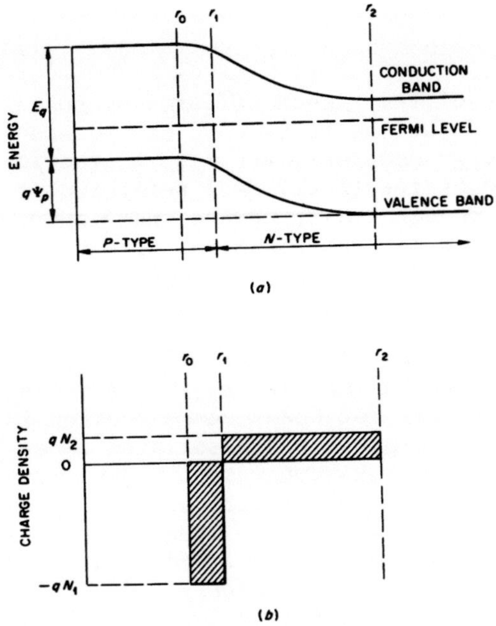

Fig. 1. The structure of the electrical double layers (space-charge zone) associated with a disordered region in germanium.

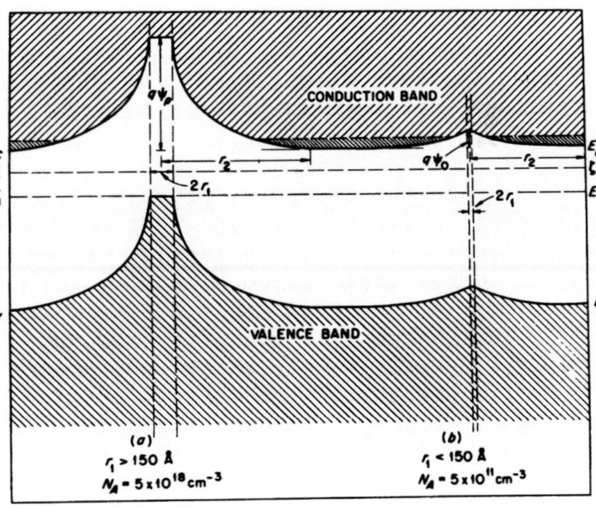

Fig. 2. Schematic drawing of the disordered region and its associated space - charge zone. Part (a) refers to a disordered region large enough to support a double layer of maximum radial extent. Part (b) shows the relatively minor perturbation of the band edges by a small disordered region.

where σ_0 is the true bulk conductivity and f is the fraction of the volume occupied by non-overlapping insulating zones, which is given approximately by

$$f = \overline{V}_{sc}\ N^{o}\ \sigma_s\ \phi_d t, \qquad (4)$$

where $\overline{V}_{sc}$ is the average volume of the region affected by space-charge zones, N^{o} is the density of lattice sites, σ_s is the scattering cross section for high-energy neutrons, and $\phi_d t$ is the integrated flux of neutrons with energies (> 0.7 MeV) sufficient to create a disordered region large enough to support a space-charge zone. Solution of Poisson's equation for the spherical case yields

$$V_{sc} = \psi_p\ \epsilon\ r_1/q\ N_2 \qquad (5)$$

where ψ_p is the difference in electrochemical potential between the disordered region and the unperturbed matrix, r_1 is the radius of the disordered region, and N_2 is the net donor concentration in the matrix. This model of disordered regions and associated space-charge zones has been verified for neutron-bombardment n-type germanium by several investigators in terms of electrical properties alone (7,9,10), and the disordered regions are found to have a radius of 50 to 100 Å. The electrical results have been confirmed by means of thermal conductivity measurements recently made by Van Dong (11). He found the mean disordered region radius to be 70 Å. More recently, Bertolotti and coworkers (12) have identified both the disordered region and the space-charge on electron micrographs of replicas of etched surfaces of germanium after bombardment with 14 MeV neutrons. Figure 3a shows a micrograph on which the space-charge zone (1000 to 1500 Å radius) is clearly visible. Figure 3b shows an enlargement of one of the etch figures, and at the center is detectable a crater of radius 150 to 200 Å which is presumably the disordered region. Interestingly enough, presumably because of its smaller atomic mass and different defect energy level structure, neutron-irradiated silicon shows no visual evidence of disordered regions (13).

Not all the energy of primary recoils is consumed in the displacement cascade. It was mentioned earlier that ionization is not effective in creating lattice defects in elemental solids. In fact, energy lost through "inelastic" ionization and excitation processes is not available for the displacement of lattice atoms (3). The higher the velocity and the larger the charge of the primary recoil, the greater the amount of energy lost in these inelastic processes. Another mechanism of fruitless dissipation of recoil energy is through "focusing" or focused collisions (14) down a close-packed row in a crystal. The succession of energy transfers in a focused chain interaction in effect drains energy away from the region of primary impact or the cascade region. An additional mode of energy loss for energetic recoils or incident

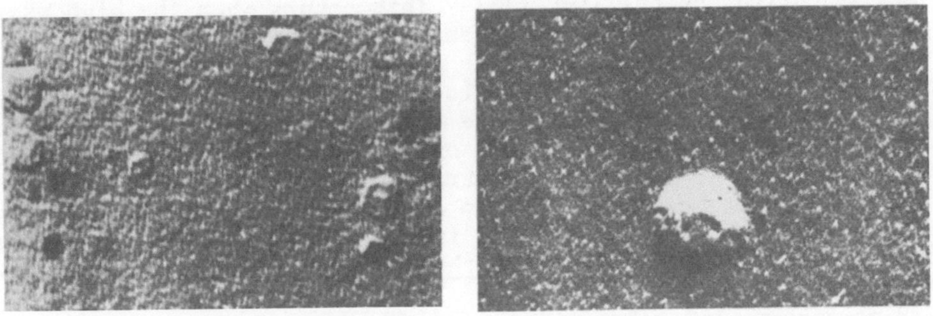

(a) *(b)*

Fig. 3. Replica micrographs of etched structures on fast-neutron bombarded germanium. (a) Low magnification showing the space-charge zone; (b) high magnification showing the disordered region evident as a small pit in the center of the larger etch structure. (After Bertolotti et al., ref. 12).

Fig. 4. A view along a <110> direction of the diamond structure.

Fig. 5. A view along a <100> direction of the diamond structure.

ions is called "channeling" (15). This occurs when the high energy particles find themselves in an open channel or crystallographic direction. Small energy losses occur through ionization or very small angle collisions with the rows of atoms defining the channel. The most open direction and hence the most probable channel in the diamond structure is along any of the <110> directions. This is clearly shown in Fig. 4. In contrast, views along the <100> direction in Fig. 5, and the <111> direction in Fig. 6, show very little open space. That this is indeed the channel direction for the diamond lattice is shown by the dependence of the range of 40 keV xenon ions in silicon in Fig. 7. These measurements by Davies and coworkers (16) show quite dramatically that ions penetrate much more deeply along this direction than along any other low index direction. Also shown for comparison is the range in single-crystal aluminum which has a face-centered cubic structure with a <110> channel direction.

Another very effective demonstration of channeling is afforded by the response of silicon p-n junction fission detectors. One problem that arises in using junction detectors in fission-fragment spectroscopy is the failure of the pulse-height absorbed energy calibration for high energy massive particles, which is designated by the term pulse-height defect. The smaller than expected pulse of current indicates that some other process besides ionization is available for dissipating a portion of the energy. Recently, Moak and coworkers (17) have investigated the influence of crystal orientation upon pulse-height response of a detector to ^{127}I ions with energies in the range of 10 to 100 MeV. When the detector was aligned such that the incident beam was parallel to a <110> direction, the pulse-height defect for the channel ions virtually disappeared. Therefore, one is lead to the conclusion that the pulse-height defect is caused by energy losses in elastic collisions which displace atoms. When the ions are channeled the probability of displacement collisions is much reduced and nearly all of the particle energy appears as ionization current in the p-n junction.

INFLUENCE OF DEFECTS ON SEMICONDUCTING BEHAVIOR

Defects, like impurity atoms, are capable of introducing local energy states into the band structure of semiconductors. Depending upon their nature and the electronic characteristics of the host lattice, defect centers may act as (a) donors, yielding electrons on thermal excitation in n-type material or capturing holes in p-type; (b) acceptors, in which capacity they may thermally ionize to produce holes in p-type conductors and capture electrons in n-type; or (c) as amphoteric centers capable of capturing electrons in n-type and holes in p-type. Moreover, it is common for radiation defects to possess two or more energy levels.

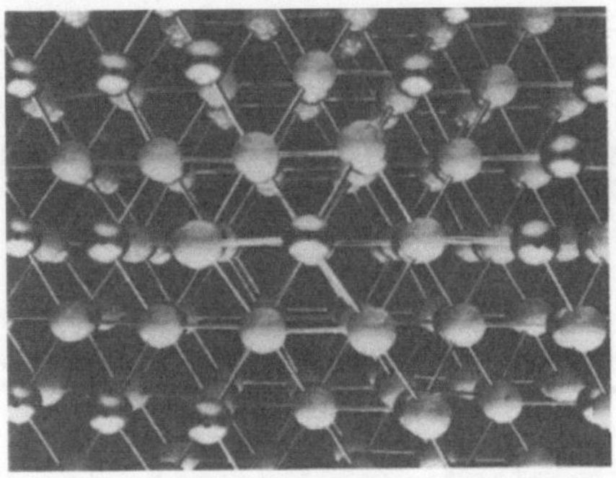

Fig. 6. A view along a <111> direction of the diamond structure.

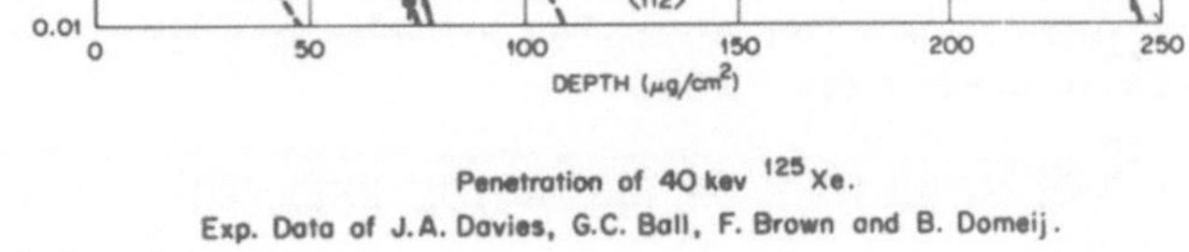

Fig. 7. Penetration of 40 keV ^{125}Xe ions into single crystals of silicon and aluminum. (After Davies et al, ref. 16).

In consequence, introduction of defects by irradiation can produce profound and often complicated changes in the electronic behavior. Not only is the carrier concentration altered even to the extent of inducing conversion from n- to p-type (or conversely), but carrier scattering is affected by the introduction of charged centers and the influence upon non-equilibrium minority-carrier processes such as trapping and recombination by the deep level defects which act both as efficient recombination centers and traps is very large. In short, all semiconducting properties are influenced in some degree by exposure to energetic particles or photons capable of displacing lattice atoms.

In studies of radiation effects (18) use is made of the influence of defects on properties to gain quantitative information about the position and relative numbers of energy levels produced by a given exposure to various types and energies of incident particles. The most common investigation involves the temperature dependence of the carrier concentration and the mobility which are obtained from measurements of Hall coefficient and mobility. This approach yields the position of the thermally accessible energy levels and information on their charge state. Other electronic properties that have been useful in locating energy levels of radiation defects not readily detected by the above approach include excess carrier lifetime, trapping of minority carriers, and the spectral dependence of photoconductivity. Both electron spin resonance and optical absorption have been very valuable tools in gaining detailed information about defect structure.

Additional information on defect stability and the tendency for various defects to interact with other imperfections (point defects, impurities, and dislocations) is obtained from the thermal annealing of radiation defects. By employing a radiation sensitive property as an index of the defect content, the kinetics governing the restoration of initial conditions through some annealing cycle (isochronal or isothermal) may be examined to yield values of defect motion energies or the binding energies for defects in certain complexes (18).

THE STRUCTURE OF RADIATION DEFECTS

As we have already indicated above, defects remaining in germanium or silicon irradiated with electrons in the 0.5 to 5 MeV range, or even more massive particles such as protons or neutrons in the room-temperature range, are not simple interstitials and vacancies but are invariably complexes formed by the interaction of simple point defects with other imperfections, notably impurity atoms and vacancy pairs as well as more complex clusters of defects.

The first clue to this state of affairs came in 1959 when Bemski (19) and Watkins and coworkers (20) observed that irradiation of oxygen containing n-type silicon gave rise to an imperfection with both a spin-resonance spectrum and an ionization energy that are quite different from those observed in specimens containing little oxygen. It was concluded and subsequently confirmed by optical absorption of polarized light (21) that the defect in the oxygen-containing material is a silicon vacancy associated with an interstitial oxygen atom. The model for this defect is shown in Fig. 8, and it is distinguished by the fact that it acts as an acceptor center with an energy level located 0.17 eV below the conduction band. Conclusive identification (21) of this center with an oxygen-vacancy complex was accomplished via an optical absorption band at 12μ, which is produced by electron irradiation under the same conditions as those under which defect resonance is produced. This band is made birefringent by uniaxial pressure in the same way that certain equivalent orientations of the paramagnetic center become more heavily populated by the application of pressure and, significantly, the thermal annealing of both the birefringence and the imbalance in the spin system follows the same path. The final point of evidence that the center responsible for the 12μ band contains oxygen is the observation that specimens doped with oxygen enriched in ^{18}O exhibit a second band shifted to lower frequencies by the amount expected for the increase in isotopic mass to influence an antisymmetric stretching vibration.

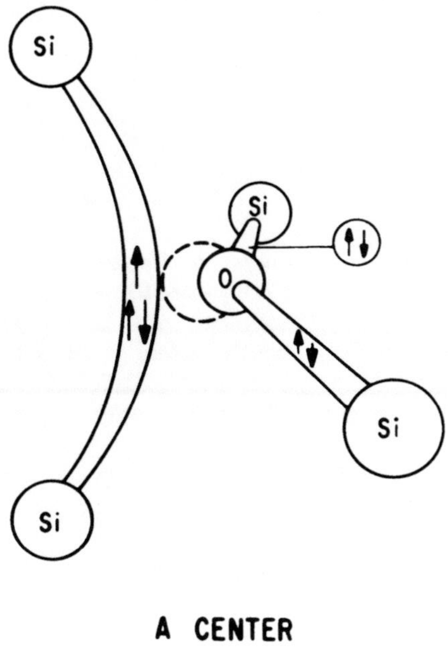

Fig. 8. Model for oxygen-vacancy center in electron-irradiated silicon. (After Watkins and Corbett, ref. 20).

Since this early discovery, a number of defect complexes have been identified in both n- and p-type silicon after irradiation with electrons. Figure 9 shows the model (22) of the vacancy-phosphorus complex which in n-type silicon is responsible for at least one acceptor level located 0.43 eV below the conduction band. This center can be identified by ESR alone, since the hyperfine interaction of the unbalanced spin with the nuclear moment of the phosphorus demonstrates conclusively that phosphorus impurity is a part of the defect structure.

In Fig. 10 we see the model of the divacancy (5), the production of which as a primary defect by electron bombardment along a <111> direction we have already discussed. This rather complex-looking center is responsible for three energy levels: an acceptor state ∿0.4 eV below the conduction band; an acceptor state quite near the center of the forbidden energy gap (∿0.5 eV above the valence band); and a donor level quite near the top of the valence band.

More recent studies (23) on p-type silicon have revealed the existence of a vacancy-acceptor complex (specifically a vacancy attached to an aluminum atom), interstitial acceptor impurity (both gallium and aluminum), and complexes one component of which is an interstitial acceptor atom. An example of the latter is a Ga_2^+ molecular complex, with one of the gallium atoms in a substitutional site and the other occupying an interstitial position. The energy levels that have been attributed to these defects are summarized in Fig. 11.

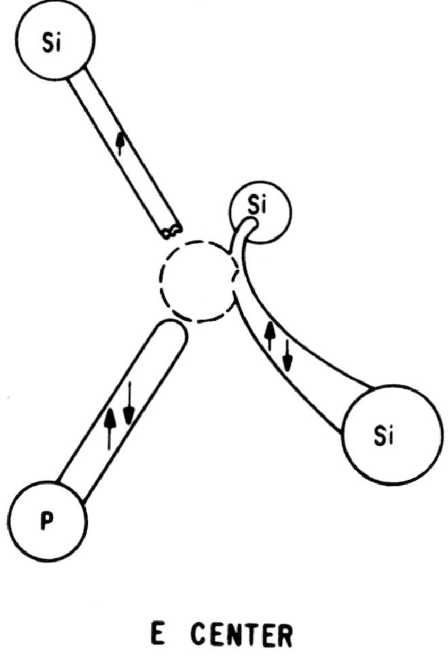

Fig. 9. Model for the phosphorous - vacancy center in electron-irradiated silicon. (After Watkins and Corbett, ref. 22.)

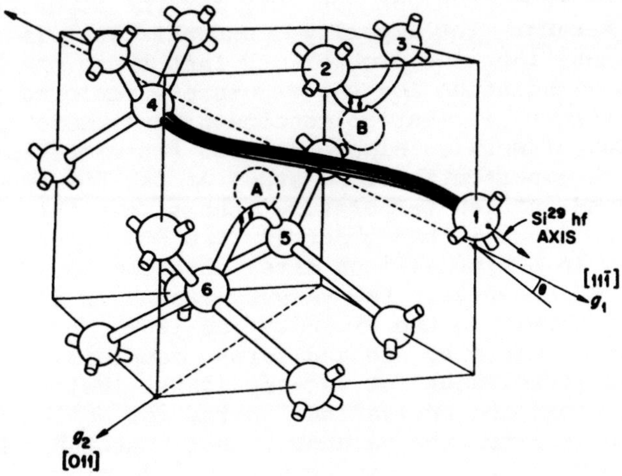

Fig. 10. Model of the divacancy in electron-irradiated silicon. (After Corbett and Watkins, ref. 5).

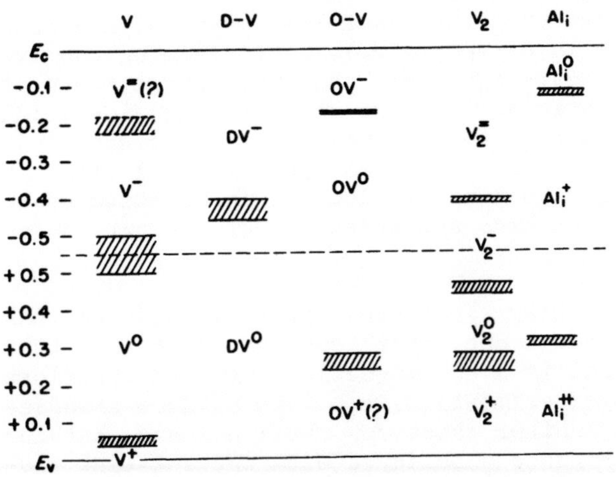

Defect Structures in Irradiated Silicon from ESR.

V - vacancy
D-V - donor-vacancy complex
O-V - oxygen-vacancy complex
V_2 - divacancy
Al_i - interstitial aluminum

Fig. 11. Position in the band gap of energy levels associated with defect complexes as identified by ESR in electron-irradiated silicon.

We must now ask by what mechanism these complexes are formed. The divacancies (5) have already been discussed and it has been shown that they are primary defects formed directly by a single collision. However, the same can not be said for the impurity-containing complexes because the impurity concentration is so small (<10^{-4} mole fraction) that the probability that these can be created directly by the irradiation for the exposures employed is vanishingly small. Therefore, it must be concluded that these are formed by the interaction of mobile simple defects (interstitials and vacancies) with these imperfections. A study of the ESR spectra of silicon irradiated at low temperature shows that this is indeed the case. Watkins (23,24) has observed the ESR spectrum for the isolated vacancy in p-type silicon after electron bombardment at both 4° and 20°K. However, the vacancy resonance disappears as the crystal is warmed to the vicinity of 170°K, presumably because the vacancy migrates to and forms complexes with other imperfections, or annihilates at vacancy sinks. Pulse annealing studies reveal that the activation energy for motion is 0.33 ± 0.03 eV. In n-type silicon the vacancy is not detectable by an ESR signal presumably because it contains only paired electrons when the Fermi level is in the upper half of the forbidden energy gap. However, the motion and trapping of vacancies can be detected through the formation of phosphorus-vacancy pairs which are formed during annealing of a n-type silicon after low-temperature irradiation. It is noteworthy that the vacancies move with a much smaller activation energy in n-type material, confirming the hypothesis put forth many years ago (25) that the charge state of defects should influence their ease of motion. From consideration of the energy level structure of the vacancy, it is concluded that the neutral vacancy is the form that moves in p-type Si with the 0.33 eV motion energy, while both the V^- and V^{--} forms expected to exist in n-type Si (Fig. 11) have appreciably smaller activation energies of migration (23).

All attempts at detecting isolated interstitials by ESR methods have failed. Although one might conclude from this that they may always be present in a nonmagnetic form and therefore escape detection, careful experiments with p-type silicon irradiated at 4°K indicate quite another cause for their absence. Watkins (23) has identified a resonance associated with interstitial aluminum ions (Al^{++}), and measurements of the relative intensities of microwave absorption show that their concentration is virtually the same as the isolated vacancy concentration. The interstitial impurities are tightly lodged in the lattice, and begin to move only upon heating to the vicinity of 200°C, when they also interact with other imperfections to form complexes. In gallium-doped material post-irradiation anneals near 200°C there appear gallium pairs, one member of which is interstitial and adjacent to the other member in a substitutional site (22).

As a result of these observations, it is concluded that the interstitial silicon moves rapidly through the lattice by some type of interstitialcy mechanism (the interstitial propagates by interchange with substitutional atoms), and at temperatures even as low as 4°K, when impurities such as Al, Ga, or perhaps even donor atoms are encountered, these are ejected into interstitial sites, thus breaking the interstitialcy chain process. The interstitial impurities can now migrate only by a normal interstitial diffusion and a much higher activation energy. (Note that the activation energy of interstitial motion for Li^+, the smallest ion studied, is 0.66 eV in silicon (26)).

It should be mentioned here that these defect structures account for only a fraction of the resonances that have been isolated from the complex spectra of electron-irradiated silicon. In addition, other resonances including those for S = 1 states have been obtained in fast neutron-irradiated silicon (27) and an oxygen-containing state in p-type silicon (28) has been reported.

As a result of this outstanding series of ESR investigations, we are now in a position to understand the full complexity of the problem of identifying the imperfections responsible for the various acceptor and donor states not only in irradiated silicon but in germanium and other semiconductors as well. Rather than to attempt to force experimental observations into a conceptual framework in which only the existence of simple defects is admitted, we have learned to accept the principle that the products of irradiation will include nearly every conceivable complex between vacancies and impurities that are present, as well as numerous interstitial impurity complexes. This new point of view has led to quite rapid progress in understanding striking differences in behavior in specimens with slightly different impurity content. For example, we show in Fig. 12 the effect of oxygen impurity on the radiation response of the excess carrier lifetime τ in n-type silicon (29). Although the two specimens have nearly the same initial lifetime and conductivity, the oxygen-free material is much more strongly affected than crystals which contain 10^{17} to 10^{18} oxygen atoms per cm^3. The difference can be readily explained: the deep acceptor state associated with the phosphorus-vacancy complex, which is the dominant radiation defect in the oxygen-free crystal, is a much more efficient recombination center than the shallow state associated with the oxygen-vacancy complex. Interestingly, the higher purity specimen is more sensitive to radiation with respect to this particular property than the less pure one, and since excess carrier lifetime is so important to the operation of most devices, it would appear that, other things being equal, a diode or transistor fabricated from quartz-crucible-grown silicon would have a better radiation resistance than the less highly oxygen-contaminated floating-zone-grown crystals.

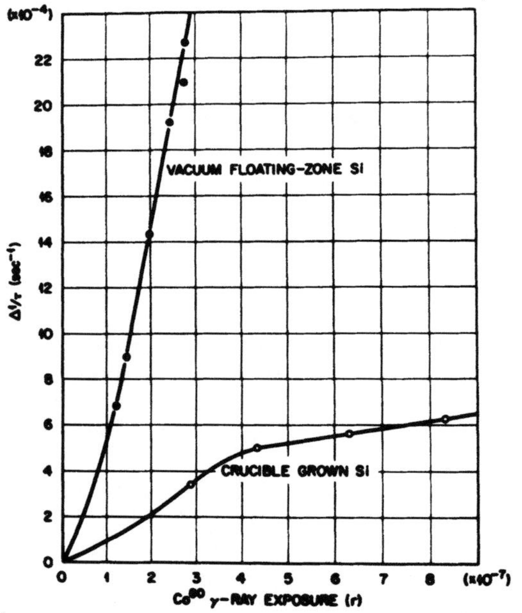

Fig. 12. Change in reciprocal of lifetime of excess carriers in n-type silicon exposed to γ rays showing the difference in response of oxygen - free (vacuum floating zone) and oxygen containing (crucible grown) material. (After Matsuura and Inuishi, ref. 29).

Let us now see what the results obtained on silicon can tell us about germanium. Unfortunately, defects in germanium are not so amenable to study by ESR techniques and one is therefore forced for the most part to infer the nature of the processes associated with electrical property changes resulting from irradiation and subsequent annealing, in terms of defect structures analogous to those found in silicon. The possibility of defect-impurity complexes in irradiated germanium was first pointed out by Brown and coworkers (30), who found that radiation defects were thermally more stable in n-type specimens doped with small donor atoms (As and P) than those doped with antimony. The recovery is much more rapid in the antimony-doped specimens than in the others. A naive interpretation of this observation would be that the smaller donor atoms, i.e. arsenic or phosphorus, trap and stabilize a simple defect. Subsequent work (31,32) has shown this view to be untenable since the rate of recovery in this temperature range increases as the antimony concentration increases, indicating that the impurity must either be directly involved in the unstable defects or participate in the annealing stages. The influence of antimony concentration on the recovery of excess carrier lifetime is shown in Fig. 13. Arsenic-doped material shows only a very slight influence of concentration.

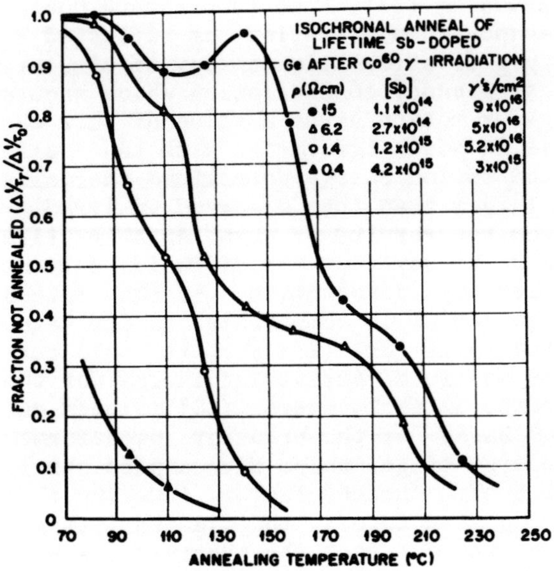

Fig. 13. The influence of antimony concentration on the recovery of excess carrier lifetime in γ-irradiated n-type germanium. (After Curtis and Crawford, ref. 31).

Until recently there was no direct evidence that point defects might exhibit a similar behavior and form similar complexes with impurities in germanium as they do in silicon. However, Whan (33) has presented conclusive evidence that vacancies begin to move in germanium near 65°K and form a series of oxygen complexes which can be detected through their vibronic infrared absorption spectra. Armed with this vital piece of evidence, analysis of annealing data in terms of vacancy-impurity complexes and perhaps even interstitial-impurity complexes can be carried out with reasonable confidence. Indeed such defects have been invoked in interpreting the isochronal recovery of electron concentration and reciprocal mobility in n-type germanium which had been exposed to ^{60}Co γ rays at 78°K. Evidently, even at this temperature, vacancies are able to migrate and form donor-vacancy complexes, with the mutual annihilation of interstitial donors and vacancies.

It should be pointed out that Whan's results as well as those of Watkins on the ease of vacancy motion confront us with a serious paradox concerning the mechanism underlying self-diffusion in these materials. The self-diffusion energy for germanium as measured by Letaw and coworkers (34) is 3 eV, while recent studies on silicon (35,36) yield 5.1 eV. If the diffusion is assumed to proceed through the motion of vacancies, this energy is the sum of the energy of formation and the energy of motion of the vacancies.

Now these component energy values can be separately determined by quenching and subsequent annealing experiments which have been carried out in the case of germanium (37). The temperature dependence of the quenched-in vacancy concentration as reflected by changes in Hall coefficient yields a formation energy of $\sim$1.9 eV, and the annealing of the excess vacancy concentration which should reflect the activation energy of motion gives a value of $\sim$1.2 eV. The sum of these values is in good agreement with the self-diffusion energy. However, the motion energy determined thermally is at least a factor of five larger than that observed in irradiated specimens, as inferred from the results of Whan. An attractive solution to the paradox would be to say that the diffusion-current at high temperature is carried by divacancies. If the motion energy for the divacancy in germanium is comparable to its value in silicon determined experimentally (22) (1.3 eV), it would certainly be consistent with the observed activation energy for the annealing of quenched-in defects. In fact, Seeger (38) pointed out a number of years ago, on the basis of the behavior of deformed germanium, that the motion energy of germanium divacancies should be in the neighborhood of 1.5 eV. Another factor favoring this viewpoint is the dislocation jog structure in the diamond lattice. Haasen and Seeger (39) have pointed out that the shortest jog produced by the displacement of an edge dislocation from one glide plane to the next adjacent one is two atoms high. Now, in many solids vacancies are known to be created in the bulk thermally by evaporation from dislocation jogs (evaporation of a vacancy causes the jog to advance a unit spacing along the dislocation) and these jogs also act as sinks for the deposition of defects as the crystal is cooled from high temperature. Consequently, if in the diamond lattice the energy to move a jog by removing a pair of vacancies one at a time is much larger than that for removing the pair as a unit, one would expect them to be produced bound together in pairs. If now the binding energy holding this pair together is large enough to prevent thermal breakup, the predominant thermal defects will be divacancies rather than single vacancies. This is just another way of stating that the formation energy of divacancies is smaller than that of single vacancies. In order for divacancy motion to dominate self-diffusion, the following relation must hold:

$$E_F'' + E_M'' < E_F' + E_M' , \qquad (5)$$

where the double-primes refer to the divacancies and the single primes to single vacancies. In view of the small value of E_M' ($\sim$0.1 eV) indicated by the experiments of Watkins (23) and Whan (33), the self-diffusion energies would then require that $E_F' > 3$ eV for germanium and 5 eV for silicon. It should be pointed out, however that theoretical calculations of E_F (40-42) all yield values for germanium that are consistent with the accepted single vacancy model for self-diffusion, i.e., ∿2 eV. Consequently, unless it can be shown that the excellent agreement between these calculations and experiment is fortuitous, it appears doubtful that the paradox can be resolved on the basis of divacancies being responsible for self-diffusion.

Another possible way of accounting for the disparity between the motion energies of radiation vacancies and thermal vacancies is the influence of charge state. However, inspection of the energy level structure of vacancies as shown in Fig. 11 indicates that this suggestion would hardly be valid, since at high temperature the vacancy would have an appreciable probability of being negative (the Fermi level lies in the center of the gap near the position of the acceptor level), and this is the charge state of highest mobility (23). Evidently, a careful re-examination of our ideas concerning the mechanism of self-diffusion in the diamond structure is required.

COMPOUND SEMICONDUCTORS

Because of the binary nature of III-V or II-IV semiconductors, the complexity of defect structures that might be expected to result from irradiation is considerably greater than that encountered in elemental semiconductors. Because of mutual interaction between radiation defects, interaction of two types of interstitials and vacancies with chemical impurity and a new type of defect—the misplaced atom (a Group III atom on a Group V lattice site and conversely)—the number of defect structures in these materials is perhaps nearer N^2 than 2N, where N represents the number of prominent defect structures in the elemental case. In view of this seemingly overwhelming complexity, one might question whether the experimental results could be interpreted in terms of any realistic defect models or yield any physical insights into the nature of radiation effects in these materials. For a review of the variety of results on the III-V compounds, the reader is directed to Aukerman's comprehensive review (43). Although in many instances complexity is indeed an obstacle, in some cases it works to our advantage and I would like to present briefly two such cases.

The first of these concerns a peculiarity of the zinc-blende structure in a material such as indium antimonide. Each In atom is covalently bound to four Sb atoms, each being situated at the corners of a regular tetrahedron, and the inverse, of course, holds for each Sb atom. If these atoms could be transformed into identical atoms without any change in position or binding, the lattice would assume the diamond structure, which indeed is a highly symmetrical structure. But let us now examine the environment of each In atom or each Sb atom situated on a given (111) plane. It is immediately evident that the lattice is unsymmetrical with respect to this plane. For example, one covalent bond extends from each In atom to the left and three to the right, whereas this configuration is inverted from the Sb atoms. This suggests that there are two possible (111) surfaces, each containing a single type of atom, and a polarization direction, namely, the <111>. The existence of this unsymmetrical situation has important consequences for many properties of III-V compounds: cleavage, piezoelectric effect, chemical behavior both with respect to oxidation and etching, a variety of electronic transport effects, and even crystal growth (44).

As might be expected this polarized structure is also reflected in the ease of defect creation by fast neutrons. Eisen (45) has determined the threshold energy and the defect yields in indium with the electron beam in the two senses of <111> direction. In the "minus" direction, examining those atom pairs parallel to the beam, the electron sees first the In atom immediately behind which is the Sb atom, and immediately behind that there is an open interstitial site. The "plus" direction is the inverse, i.e., the Sb atom, followed by the In atom, followed by the interstitial site. The results indicate that not only is the displacement energy E_d different for the two directions, but that the thermal stability of the defects is quite different. For the beam direction into the (111) face (the plus direction), it is seen from Fig. 14 that there is an open interstitial site immediately behind the In atom; hence this should be the direction for easy In-atom displacement. On the other hand, an Sb atom receiving an impulse in this direction would have to collide with an indium atom before reaching the interstitial site. The situation is, of course, inverted for the beam into the $(\bar{1}\bar{1}\bar{1})$ face, which corresponds to the direction for easy Sb-atom displacement. An analysis of the effect of irradiation on the Hall coefficient reveals that the displacement energy $E_d(In)$ for In in the easy direction is 6.4 eV and the resulting defect is thermally more stable than the structure created by displacement of Sb atoms for which $8.5 < E_d(Sb) < 9.9$ eV. Although the precise structure of these defects which decrease the electron concentration by capturing electrons can not be described, this investigation indicates that a displaced In atom plays a dominant role in the one which is thermally stable up to 128°K, and Sb-atom displacement is involved in the one that is stable only up to 83°K.

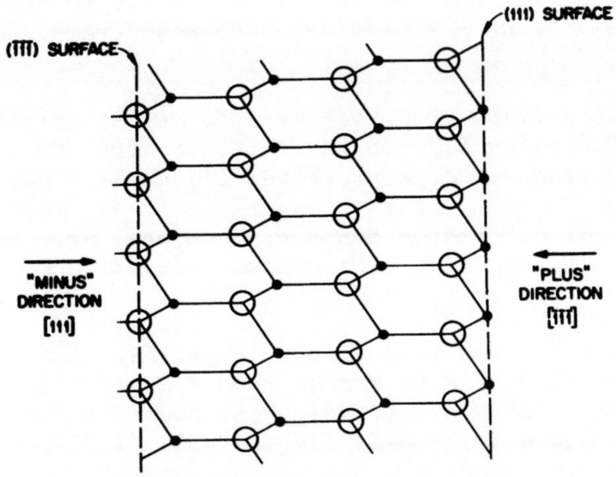

Projection of the Zinc-Blende Lattice on a (110) Plane. (●-In; ○-Sb).

Fig. 14. A projection of the zinc-blende lattice onto a (110) plane.

A second series of experiments by R. O. Chester (46) has employed nuclear characteristics of the atoms comprising the compound semiconductor to throw light upon the defects responsible for acceptor and donor action. She studied the response of cadmium sulfide and cadmium telluride to gamma irradiation employing different γ-ray sources, namely, ^{60}Co with E_γ = 1.17 and 1.33 MeV, and ^{137}Cs with E_γ = 0.66 MeV. Because of the markedly different mass and atomic numbers of the Cd and S atoms, it can be demonstrated from simple collision theory that the 0.66 MeV photon from ^{137}Cs decay displaces S atoms and the ^{60}Co photons predominantly displace Cd atoms. The situation is reversed in cadmium telluride for which the smaller energy photon preferentially displaces Cd atoms, but because of the greater similarity of atom mass the distinction is not as great as in cadmium sulfide. Investigation of the behavior of Hall coefficient of gamma-irradiated n-type cadmium sulfide and both n- and p-type cadmium telluride shows that preferential Cd-displacement leads to the introduction of acceptor states while displacement of the electro-negative constituent leads to donor action. Because of the propensity of simple defects for forming complexes with other imperfections, one can not infer from these results the precise structure of the defects responsible for the dominant electronic states. Nevertheless, this study indicates what are the dominant ingredients of the radiation acceptors and donors in the II-VI compounds.

REFERENCES

1. *For an early review, see K. Lark-Horovitz, pp. 47-69 in Semiconducting Materials, Ed. by H.K. Henisch (Academic Press, New York) 1951.*

2. *These developments were first reported in 1959 at the Gatlinburg Conference on Radiation Effects in Semiconductors and the Proceedings were published in J. Appl. Phys. 30, No. 8 (1959).*

3. *For a review, see D.S. Billington and J.H. Crawford, Jr., Radiation Damage in Solids, pp 11-54 (Princeton University Press New Jersey) 1961.*

4. *J.W. MacKay, E.E. Klontz, and G.W. Gobell, Phys. Rev. Letters 2, No. 8, 146 - 147 (1959); J.W. MacKay and E.E. Klontz, J. Appl. Phys. 30, No. 8, 1269-1274 (1959); E.E. Klontz and J.W. MacKay,J. Phys. Soc. Japan 18, Suppl. III, 216-221 (1963).*

5. *J.W. Corbett and G.D. Watkins, Phys. Rev. Letters 7, No. 8, 314-315 (1961).*

6. *J.R. Parsons, R.W. Balluffi, and J. S. Koehler, Appl. Phys. Letters 1, No. 3, 57-58 (1962).*

7. *B.R. Gossick, J. Appl. Phys. 30, No. 8, 1214-1218 (1959); J.H. Crawford, Jr., and J. W. Cleland, J. Appl. Phys. 30, No. 8, 1204-1213 (1959); J.W. Cleland and J.H. Crawford, Jr., pp 299-302 in Proc.International Conference on Semiconductors Physics (Czechslovakia Academy of Sciences, Prague) 1961.*

8. *H.J. Juretschke, R. Landauer, and J.A. Swanson, J. Appl. Phys. 27, No. 7, 838-839 (1956).*

9. *H.J. Stein, J. Appl. Phys. 31, No. 8, 1309-1313 (1960).*

10. *W.H. Closser, J. Appl. Phys. 31, No. 9, 1693 (1960).*

11. *N. Van Dong, J. Appl. Phys. 36, No. 11, 3450-3452 (1965). See also, N. Van Dong, P. Ngu Tung, and M. Vanderyner, Compt. Rend. 236, 1722-1723 (1963).*

12. *M. Bertolotti, T. Papa, D. Sette, V. Grosso, and G. Vitali, Nuovo Cimento 29, 1200-1204 (1963); M. Bertolotti, T. Papa, D. Sette, and G. Vitali, J. Appl. Phys. 36, No. 11, 3506-3512 (1965).*

13. *For a summary of recent development, see F.L. Vook, pp 51-63 in Radiation Damage in Semiconductors, Ed. by P. Baruch (Academic Press, New York) 1965.*

14. *R.H. Silsbee, J. Appl. Phys. 28, No. 11, 1246-1250 (1957).*

15. *M.T. Robinson and O.S. Owen, Appl. Phys. Letters 2, No. 2, 30-32 (1963); Phys. Rev. 132, No. 6, 2385-2398 (1963)*

16. *J.A. Davies, G.D. Ball, F. Brown, and B. Domeij, Can. J. Phys. 42, No. 6, 1070-1080 (1964).*

17. *C.D. Moak, J.W.T. Dabbs, and W.W. Walker, Bull. Am. Phys. Soc. 11, Series 11, No. 1, 101 (1966); to be published in Rev. Sci. Inst.*

18. *For a coverage of Radiation-sensitive properties and thermal annealing of radiation defects in semiconductors, the reader is directed to recent review papers: J.H. Crawford, Jr., pp. 421-470 in the Interaction of Radiation with Solids, Ed. by R. Strumane, J.Nihoul, R. Gevers, and S. Amelinckx (North Holland Publishing Co., Amsterdam) 1964; V.S. Vavilov, Phys. Stat. Sol. 11, No. 2, 447-497 (1965). An excellent coverage of recent work is also to be found in Radiation Damage in Solids, Ed. by P. Baruch (Academic Press, New York) 1965.*

19. *G. Bemski, J. Appl. Phys. 30,No. 8, 1195-1198 (1959).*

20. *G.D. Watkins, J.W. Corbett, and R.M. Walker, J. Appl. Phys. 30 No. 8, 1198-1203 (1959).*

21. *J.W. Corbett, G.D. Watkins, R.M. Chrenko, and R.S. McDonald, Phys. Rev. 121, No. 4, 1015-1022 (1961).*

22. *G.D. Watkins and J.W. Corbett, General Disc. Faraday Soc. 31, 86 (1961).*

23. *G.D. Watkins, pp 97-113 in Radiation Damage in Semiconductors, Ed. by P. Baruch (Academic Press, New York) 1965.*

24. *G.D. Watkins, J. Phys. Soc. Japan 18, Suppl. II, 22-27 (1963).*

25. *H.Y. Fan and K. Lark-Horovitz, pp 113-138 in Semiconductors and Phosphors, Ed. by M. Schon and H. Welker (Interscience Publishers, New York) 1958.*

26. *J. Maita, J. Phys. Chem. Solids 4, 68-70 (1958).*

27. *Wun Jung and G.S.Newell, Phys. Rev. 132, No. 2, 648-662 (1963).*

28. N. Almeleh and B. Goldstein, Bull. Am. Phys. Soc. 11, Series 11, No. 1, 17 (1966).

29. K. Matsuura and Y. Inuishi, J. Phys. Soc. Japan 16, No. 2, 339 (1961). Y. Inuishi and K. Matsuura, J. Phys. Soc. Japan 18, Suppl. III, 240-245 (1963).

30. W.L. Brown, W.M. Augustyniak, and T.R. Waite, J. Appl. Phys. 30, No. 8, 1258-1268 (1959).

31. O.L. Curtis, Jr., and J.H. Crawford, Jr., Phys. Rev. 124, No.6 1731-1740 (1961); Phys. Rev. 126, No. 4, 1342-1349 (1962).

32. J.C. Pigg and J.H. Crawford, Jr., Phys. Rev. 135, A1141-A1150 (1964).

33. R.E. Whan, Appl. Phys. Letters 6, No.11, 221-223 (1965), Phys. Rev. 140, No. 2A, A690-A698 (1965).

34. H. Letaw, W.M. Portnoy, and L. Slifkin, Phys. Rev. 102, No. 3, 636-639 (1956).

35. W. R. Wilcox and T.J. LaChapelle, J. Appl. Phys. 35, No. 1, 240-246 (1964).

36. B.J. Masters and J.M. Fairfield, Appl. Phys. Letters 8, No.11, 280 (1966).

37. For a recent assessment of this matter, see A.Hiraki, J. Phys. Soc. Japan 21, No. 1, 34-41 (1966).

38. A. Seeger, pp 61-71 in Vol. 1, Solid State Physics in Electronics and Telecommunications, Ed. by M. Desirant and J.L. Michiels (Academic Press, New York) 1960.

39. P. Haasen and A. Seeger, pp 68-114 in Halbleiterprobleme, Vol. 1, Ed. by W. Schottky (Vieweg und Sohn, Braunschweig) 1958.

40. R.A. Swalin, J. Phys. Chem. Solids 18, No. 4, 190-196 (1961).

41. A. Scholz and A. Seeger, Phys. Stat. Sol. 3, No. 8, 1480-1490 (1963); pp 315-322 in Radiation Damage in Solids, Ed. by P. Baruch (Academic Press, New York) 1965.

42. K. Bennemann, Phys. Rev. 137, No. 5A, A1497-A1514 (1965).

43. L.Aukerman, in Semiconductors and Semimetals, Ed. by R.K. Willardson and A.C. Beer (Academic Press, New York) Vol. 4 (to be published).

44. *J.W. Faust, pp 445-468 in Compound Semiconductors, Ed. by R.K. Willardson and H. L. Coering (Reinhold Publishing Corp., New York) 1962.*

45. *F.H. Eisen, Phys. Rev. 135, No. 5A, A1394-A1399 (1964).*

46. *R.O. Chester, to be published. See also, PH.D. Thesis from the University of Tennessee, "Radiation Damage in Cadmium Sulfide and Cadmium Telluride", ORNL-3767 (unpublished.*

DISCUSSIONS

R.A. WEEKS (Oak Ridge National Laboratory, U.S.A.): *Would you comment on the effect of the directional properties of displacement events on the number of displaced atoms per incident particle in an amorphous solid?*

J.H. CRAWFORD: *The random arrangement of structural units in a glass will present all local orientations of such units to the electron beam. Therefore, a good average of the displacement energy would be obtained. A similar effect occurs in electron irradiation of thick crystalline specimens since successive scattering events cause the incident electron beam direction to be altered through a wide range of angles.*

D. DAUTREPPE (Centre d'Etudes Nucleaires, Grenobles, France): *Is there any argument why there are less successful ESR experiments in Ge than in Si?*

J.H. CRAWFORD: *Although ESR signals from radiation defects in Germanium have been observed, the structures are not sharp and are difficult to analyse because spin lattice interactions tend to smear out the resonance lines.*

A. CHARLESBY (Royal Military College of Science, England): *From a purely practical view, could you give some estimates of the minimum radiation levels likely to affect the performance of semiconductors? How are these levels affected by the type of radiation?*

J.H. CRAWFORD: *A dose of 10^6r ^{60}Co gamma rays will decrease the amplification factor of conventional germanium or silicon P-N-P transistors by a large factor. The rate of carrier removal at room temperature in n-type germanium by 1.5 MeV electrons is $\sim$ 0.2 per incident electron.*

O. HAUSER (Academy of Science, NRC, Dresden, Germany): *Can you explain the maximum in the annealing of irradiated Ge?*

J.H. CRAWFORD: *Unfortunately, we do not understand the details of the early annealing stages (Fig. 13) which produce the maximum is the curve for high purity Ge. Evidently, they involve subtle defect impurity reactions.*

O. HAUSER: *I understand from your paper that the radiation effects were created mainly by the interaction of the point defects with impurity atoms. Were any experiments conducted with Si samples containing a low concentration of different impurities but irradiated in the same manner. I would expect that you will observe some differences.*

J.H. CRAWFORD: *Work using photoconductivity and lifetime experiments has shown that the yield of electrically active deep level defects is indeed smaller in high purity silicon after the impurity atoms have been saturated in the defect-impurity interaction process.*

LOTFIA EL NADI (U.A.R. Atomic Energy Establishment): *Are the changes in the resolution of Semiconductor detectors, e.g. Li drifted Si detectors, due to the diffusion of Li through Si or due to the changes in the Si junction itself?*

J.H.CRAWFORD: *In view of the high rate of diffusion of Lithium ions in germanium and silicon, I would judge that the change in Lithium distribution, which can readily occur if the crystals are warmed above liquid nitrogen temperature, is responsible for changes in junction configuration and hence the calibration of these detectors.*

PHOTOCONDUCTIVITY AND PHOTOTROPY IN NON-CRYSTALLINE SOLIDS

J. D. Mackenzie

Rensselaer Polytechnic Institute

New York, U.S.A.

This survey paper is concerned with two reversible phenomena arising from the interaction of light with non-crystalline solids. The photoelectric properties of solids in the chalcogenide family are reviewed and a comparison is made between the photoconductivity of crystalline and non-crystalline phases of materials such as selenium and arsenic selenide. Phototropy refers to the darkening of materials under irradiation and their subsequent fading when the radiation is removed. Results for inorganic and organic non-crystalline systems are reviewed and the effects of crystallinity discussed.

INTRODUCTION

The reversible phenomena of photoconductivity and phototropy (or photochromism) have been known for a long time. Most of the studies reported have been on crystalline solids and until recently relatively little is known concerning these two interesting electronic properties for non-crystalline solids. Although both organic and inorganic materials can exhibit photoconductivity and phototropy, this survey paper will be restricted to inorganic systems. Most published work on non-crystalline solids, in particular those on photoconductivity, have been concentrated on the understanding of the mechanisms only rather than a correlation of the phenomena with the structures of the materials involved. In this review, attempts will be made to indicate the equal importance of the understanding of material structures. Before we proceed to review the observed properties, therefore, it is pertinent to clarify the meaning of "non-crystalline" solids.

NON-CRYSTALLINE SOLIDS

Non-crystalline solids can be prepared by a large number of grossly different methods (1). The two most important methods, particular for the present review, are (a) by cooling the melt until it is rigid in the absence of crystallization, and (b) by condensation of the vapor on a solid substrate at some temperature at which the condensate is rigid. In either case, absence of crystallinity is usually assumed if diffuse x-ray or electron diffraction patterns similar to that given by a liquid are obtained. It is generally accepted by workers in the field of glass science and technology that the non-crystalline solid obtained by method (a) is termed a *glass* or a *vitreous solid* (2). This glassy solid is a metastable phase whose frozen-in structure corresponds to the structure of a liquid or a supercooled liquid at some higher temperatures. A non-crystalline solid obtained by vapor deposition is generally referred to as an *amorphous solid*. The structure of an amorphous solid is dependent on many factors some of which are interrelated. For example, the structures of the vapor species, the substrate temperature, the rate of deposition and the duration of deposition. If a substance can be rendered into a non-crystalline state by both methods (a) and (b), for instance selenium, it is unlikely that the structures of the two non-crystalline solids will be the same. This will be further clarified when selenium is discussed below. Although there appears to be some evidence that non-crystalline oxides obtained by different methods are structurally similar, this is to be regarded as anomalous at present (3). In any study of the properties of non-crystalline solids, therefore, the preparation history must be considered as an important factor.

PHOTOCONDUCTIVITY

Semiconducting Non-Crystalline Solids

Electronic conduction in semiconductors is not confined to crystalline solids. Ioffe and Regel (4) have presented extensive data which has shown conclusively that semiconduction is primarily governed by short-range order and not long-range order. Thus the mechanism of semiconduction in a molten semiconductor may be similar to that in the corresponding crystal providing the short-range order in the two phases is similar. It is of course frequently difficult to define exactly the meaning of short-range order. Semiconducting behavior was reported long ago for amorphous films of elementary solids such as antimony, germanium, silicon and tellurium. More recently, semiconducting behavior has been found in many oxide glasses (5) as well as non-oxide glasses (6) based on mixtures of elements such as S, Se, Te, P, As, Tl, Si and Ge. Photoconductivity has only been reported for the latter materials.

In general, these semiconducting glasses have relatively low electrical conductivity (10^{-2} ohm^{-1} cm^{-1} or less) and the mobility of the carriers is small (1 cm^2 V^{-1} sec^{-1} or less). There is at present little quantitative understanding concerning such low-mobility solids. Both n-type and p-type conduction have been reported. It is interesting to note that for some of these glasses, whereas thermoelectric power measurements indicate that the major carriers are positive holes, Hall measurements suggest that electrons are the primary carriers. It was suggested that this anomaly could be the result of inhomogeneities in the glass (7). Recently, however, it was reported that this anomaly persisted when glasses of the systems As_2Se_3 and As_2Te_3 were heated to temperatures corresponding to the liquid state (8). It is unlikely, therefore, that inhomogeneities are the dominant factor.

From a practical point of view, glasses are more flexible as an electronic material, since their chemical compositions are not limited by stoichiometry and also there are no grain boundaries as for polycrystalline solids. Further, unlike semiconducting crystals, the electrical conductivity of the glass is apparently insensitive to the presence of impurities. In Table I, the effect of zone-refining on the variation of electrical properties for the crystalline and glassy phases of the same material is shown.

TABLE I

Effect of Impurities on Glass and Crystal Conductivity (9)

Material	Crystal				Glass	
	σ (Ω^{-1}cm^{-1})	n (cm^{-3})	μ (cm^2/Vs)	Type of conduction	σ (Ω^{-1}cm^{-1})	Type of conduction
As_2SeTe_2 before zone refining	60	-	-	P	1×10^{-6}	P
As_2SeTe_2 after zone refining	4×10^{-2}	10^{16}	15	n	0.9×10^{-6}	P

Photoconductivity of Non-Crystalline Selenium

Photoconductivity was first observed for non-crystalline selenium by Weimer in 1950 (10). The sample was prepared by condensing the vapor on a substrate. The spectral response of the non-crystalline sample as compared to that of the hexagonal crystal is shown in Fig. 1. Although the photocurrents are normally very small, because of the high dark resistivity (greater than 10^{14} ohm cm at 20°C), they are many times the dark current. From a scientific point of view, after Weimer's discovery, non-crystalline selenium immediately became a challenge with many unexplained properties; among them were:

1) Difference in photon energy for absorption and photoconductivity.

2) Origin of the carriers which are predominantly positive holes.

3) Mobility of the carriers.

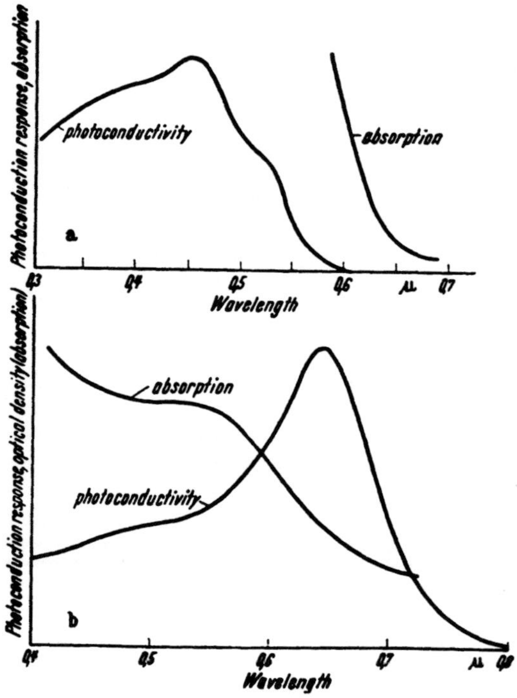

Fig. 1. Spectral response of photoconduction and optical absorption in selenium: (a) amorphous, and (b) hexagonal crystal. (After Garlick (11)).

From an industrial point of view, the high resistivity of non-crystalline selenium was desirable for the Xerographic printing process (12). In the past 15 years it has, therefore, been widely studied. An excellent summary of published work has been made by Pell (13). Almost without exception, research has been concentrated on the mechanism of photoconduction and/or attempts to explain electrical and optical properties by band theory. Important experimental data have been obtained from drift mobility measurements, particularly by Spear (14), and from studies of space-charge limited currents (15). The electronic states in non-crystalline selenium at room temperature as indicated by optical, photoconductivity, and charge-transport measurements have been summarized recently by Hartke and Regensburger (16), and are reproduced in Figure 2.

Structure of Non-Crystalline Selenium

Selenium exists in three known crystalline modifications (17, 18):

1) A common hexagonal form which is known as grey or metallic selenium. The structure of this is made up of parallel infinite spiral chains.

2) α -monoclinic; and

3) β -monoclinic.

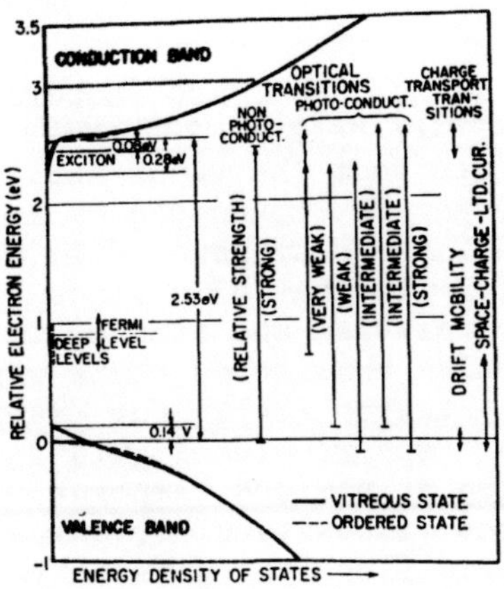

Fig. 2. Electronic states in vitreous selenium at room temperature as indicated by optical, photoconductivity, and charge - transport measurements. (After Hartke and Regensburger (16)).

These latter two modifications are made up of puckered Se_8 rings. The hexagonal form is the high temperature stable form and is insoluble in carbon disulphide whereas the monoclinic forms will dissolve in CS_2. Molten selenium probably consists of a mixture of Se_8 rings, Se_6 rings, larger rings and chains of varying lengths (19,20,21). The distribution of these various units is dependent on temperature and controversy exists between different workers regarding the polymeric units as well as their distribution. The melting temperature of selenium is 217°C. A glass obtained by quenching the melt from different temperatures will therefore have different structures. It is also to be noted that if a Se_8 ring were to open there will be two unpaired electrons at the two ends of the resultant chain. Thus the number of unpaired electrons in a glass, say, will depend on the equilibrium distribution of the chain lengths for that particular sample. Unambiguous information on the detailed structures of glasses is not available from X-ray diffraction experiments. This is exemplified by the controversy which exists, for example, between Krebs (19) who suggested Se_8 rings and larger rings of about 1000 Se atoms to be the basic units of glassy selenium and Richter (21) who postulated Se_6 rings. The glass is metastable with respect to the crystalline phases and will devitrify through nucleation and subsequent crystal growth. The temperature at which devitrification will occur and the particular crystalline phase which forms are both sensitive to impurities, atmosphere and mechanical stresses. The glass transition temperature for selenium is approximately 30°C; which means that above this temperature viscous flow will be perceptible.

The vapor species of selenium are temperature dependent. At low temperatures, molecular weight measurements (22) and electron diffraction (23) indicate that the vapor species are likely to be Se_6 rings. When the vapor is condensed on a substrate at 20°C, say, it is unlikely that the amorphous film which is formed will have a structure similar to that of a glass formed by quenching the melt to 20°C. Many workers who studied the photoconductivity of non-crystalline selenium considered their vapor-formed films to be glassy or vitreous. Lanyon (24), for example, considered his vapor-formed films as a "supercooled liquid". Amorphous selenium films have been deposited at various temperatures as high as 90°C. At this temperature, the viscosity of selenium is 1×10^6 poise. It is a supercooled liquid since the glass transition temperature is only 30°C. On the other hand, many films were deposited on substrates at 20°C. This will of course make the meaningful correlation of film properties even more difficult. As for the glassy selenium, amorphous films will undergo transformation. Weimer (10), for instance, mentioned that his vapor-formed amorphous films contained *some* monoclinic crystals. Keck (25,26) deposited amorphous films at temperatures between 30° and 75°C and reported that at lower temperatures, monoclinic crystals were present but at above 65°C, some hexagonal crystals were formed. Further, crystal growth could occur even at 20°C, the rate being 2700A per day. It is known that the electrical properties of the film are dependent on temperature duration and rate of deposition (12). Figure 3 is a reproduction of the recent work of Griffiths. The bottom figure shows that the electron diffraction pattern of selenium film deposited on the (100) cleavage face of MgO at 90°C is diffuse. However, the transmission electron-micrograph shows widely separated spherical particles of about 2000A in diameter. Presumably these are non-crystalline droplets. Their separation certainly does not resemble the transmission picture of an ordinary glassy sample.

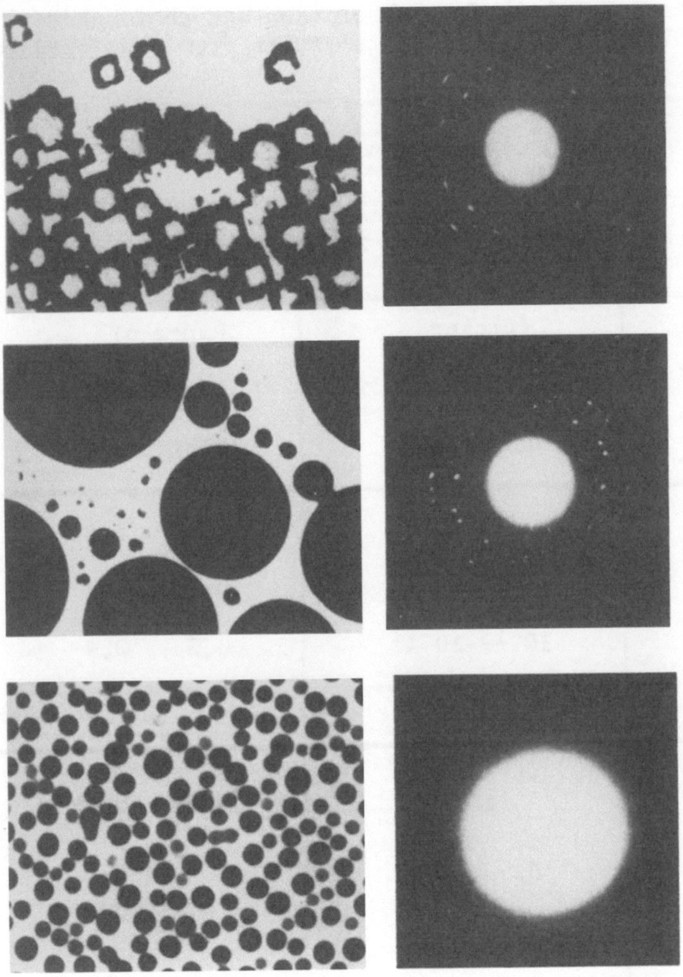

Fig. 3. Transmission electron micrograph and electron diffraction pattern of selenium films deposited on various substrates. (After Griffiths (27)).

Further progress in understanding the electrical properties of non-crystalline semiconductors will be difficult if the structures of the materials themselves are neglected.

Photoconductivity of Chalcogenide Glasses

Besides selenium, many chalcogenide glasses have been demonstrated to be photoconductors. Most of the pioneer work in this field has been carried out by B. T. Kolomiets of Russia (28). Examples of some such glasses are shown in Table II. The results showed that the conductivity of a glass increases when the content of the heavier element is increased. In the glass-forming system of As, Se and Te, for example, the activation energy also decreases with increasing As_2Te_3. In Table III, it is seen that the activation energies determined from absorption and photoconductivity are similar although they differ somewhat from that obtained from conductivity.

TABLE II

Conductivity ranges and Photosensitivity limits of some Chalcogenide glasses. (After Kolomiets et al. (28)).

Systems	Limiting values of conductivity, $ohm^{-1}cm^{-1}$	Range of observed spectral peak positions, μ
As_2Se_3 - As_2Te_3	10^{-13}-10^{-6}	0.6 - 1.25
As_2Se_3 - Tl_2Se	10^{-13}-10^{-7}	0.6 - 1.1
As_2Se_3 - Sb_2Se_3	10^{-13}-10^{-12}	0.6 - 0.9
$Tl_2Se \cdot As_2Se_3$ - $Tl_2Se \cdot As_2Te_3$	10^{-7}-10^{-3}	1.0 - 1.2
$Tl_2Se \cdot As_2Se_3$ - $Tl_2Se \cdot Sb_2Se_3$	10-8	0.95 - 1.15

TABLE III

Carrier Activation Energies of Glassy Semiconductors (28).

Composition	Activation energy (in eV), determined from		
	Absorption band edge	Photocon-ductivity	Temperature dependence of conduc-tivity
As_2Se_3	1.57	1.64	1.96
$4As_2Se_3.As_2Te_3$	1.1	1.3	1.46
$3As_2Se_3 \cdot As_2Te_3$	1.05	1.20	1.37
$2As_2Se_3 \cdot As_2Te_3$	1.01	1.04	1.30
$As_2Se_3 \cdot As_2Te_3$	0.94	0.9	1.17
$9As_2Se_3 \cdot 11As_2Te_3$	0.92	0.93	1.17
$2As_2Se_3 \cdot 3As_2Te_3$	0.9	0.83	1.04
$As_2Se_3 \cdot 2As_2Te_3$	0.89	0.83	0.95
$As_2Se_3 \cdot 3As_2Te_3$	0.86	0.74	0.94
$As_2Se_3 \cdot 4As_2Te_3$	0.85	-	0.91

In Figure 4, it is seen that the conductivity of glasses of the system As, Se and Te varies smoothly from the glassy state to the supercooled liquid state. In Figure 5, the photoconductivity of glassy and liquid $4As_2Se_3 \cdot As_2Te_3$ is seen to be quite similar. Photoconducting amorphous As_2Se_3 films have been made by vapor deposition (29). The vapor deposition of a polycomponent system can generate an extra degree of complexity since the composition of the condensate may deviate from that of the original material. In Table IV it is seen that the composition of condensed amorphous films is influenced by evaporation temperature as well as substrate temperature (30).

TABLE IV
Composition of Arsenic Sulfide Non-Crystalline Films as a Function of Deposition History. (After Tanaka et Al. (30)).

Composition (mole ratio)	Temperature Control of distillation: Temperature of batch container (°C)	Temperature of condensate depositer (°C)	Composition of Glass Sheet: Weight % As	Formula
As_2S_3	530 ± 10	410 ± 5	61.0 - 65.0	$As_2S_{3.0}-As_2S_{2.5}$
As_2S_3+S	530 ± 10	410 ± 5	61.0 - 65.0	$As_2S_{3.0}-As_2S_{2.5}$
$As_2S_3+1.5S$	530 ± 10	380 ± 5	57.0 - 61.0	$As_2S_{3.5}-As_2S_{3.0}$
As_2S_3+2S	530 ± 10	360 ± 5	54.0 - 61.0	$As_2S_{4.0}-As_2S_{3.0}$
As_2S_3+4S	530 ± 10	340 ± 5	48.5 - 54.0	$As_2S_{5.0}-As_2S_{4.0}$
As_2S_3+8S	530 ± 10	360 ± 5	48.5 - 54.0	$As_2S_{5.0}-As_2S_{4.0}$
As_2S_3+8S	490 ± 10	360 ± 5	43.5 - 48.5	$As_2S_{6.0}-As_2S_{5.0}$
As_2S_3+8S	490 ± 10	350 ± 5	41.5 - 46.0	$As_2S_{6.5}-As_2S_{5.5}$
As_2S_3+8S	490 ± 10	330 ± 5	34.0 - 40.0	$As_2S_{9.0}-As_2S_{7.0}$
As+3S	520 ± 10	360 ± 5	57.0 - 65.0	$As_2S_{3.5}-As_2S_{2.5}$
As(anal) + 3.55(anal)	520 ± 10	360 ± 5	51.0 - 57.0	$As_2S_{4.5}-As_2S_{3.5}$

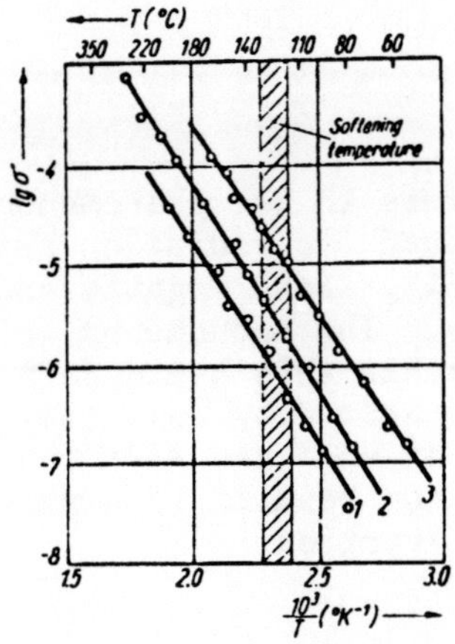

Fig. 4. Temperature dependence of the conductivity of glasses and supercooled liquids:

1) $4As_2Se_3 \cdot As_2Te_3$; (2) $3As_2Se_3 \cdot As_2Te_3$;

3) $2As_2Se_3 \cdot As_2Te_3$. (After Kolomiets [9]).

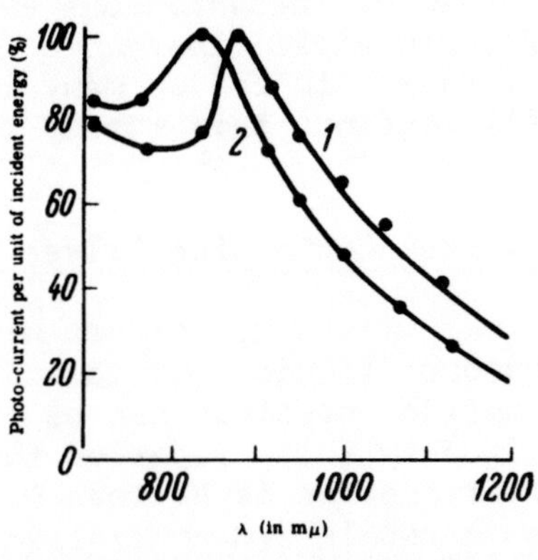

Fig. 5. Spectral distribution of the internal photoelectric effect in a glass of the composition $4As_2Se_3 \cdot As_2Te_3$ in the solid (1) and molten (2) states.

PHOTOTROPY

Arbitrary Definition

Phototropy or photochromism generally refers to the phenomenon whereby a material changes color when exposed to radiation but reverts to its original color in the dark (31). Both organic and inorganic materials can exhibit phototropy. The phenomenon of solarization, for example, can be included within this broad definition. Results on solarization have been summarized by Weyl (32). In a silicate glass containing small amounts of Mn^{2+} and Fe^{3+} ions, for instance, the absorption of a light quantum results in a purple coloration which can be explained by the reactions:

$$Mn^{2+} + h\nu = Mn^{3+} + e$$

and

$$Fe^{3+} + e = Fe^{2+}$$

Such processes are extremely slow. For practical applications, both the coloration and fading should be controllable. The coloration, in particular, should be instantaneous and be sensitive to sunlight. Two types of phototropic inorganic glasses have been developed recently. They have both raised many interesting scientific questions. We will confine this survey to these two types of glasses.

Phototropic Glasses Containing Silver Halides

Glasses can be melted with some silver nitrate and sodium halides to give a homogeneous liquid. On controlled cooling and heat treatment, silver halide crystals can be precipitated. Recently, Armistead and Stookey (33) reported that for some glass compositions, if the silver content is between 0.2 and 0.7 percent, and if the silver halide crystals are controlled to 50 to 100Å in diameter, a transparent glass is obtained. On exposure to sunlight or radiations in the UV, darkening will occur rapidly. This so-called activation and subsequent fading of the glass is exemplified by the results shown in Fig. 6 and Fig. 7, according to Smith (34). The activity of the glasses, typical compositions of which are shown in Table V, is enhanced by small amounts of copper ions.

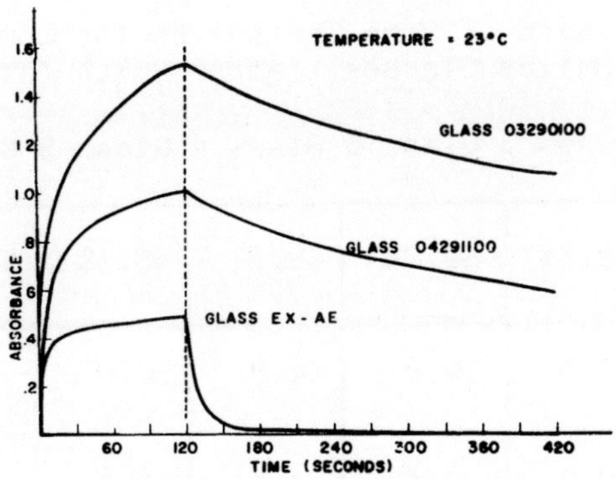

Fig. 6. Darkening and fading of typical phototropic glasses containing silver chloride at 23°C. (After Smith (34)).

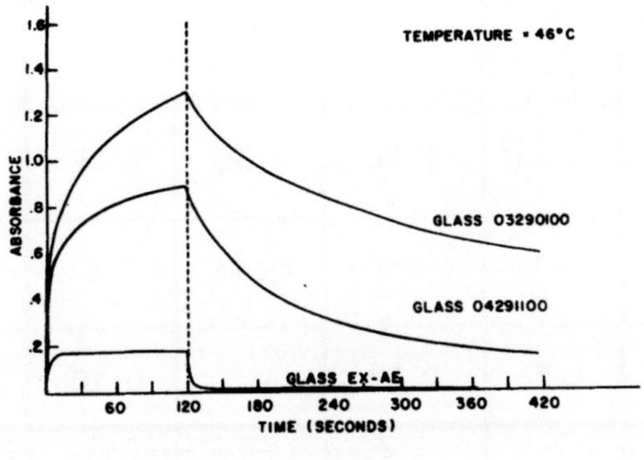

Fig. 7. Darkening and fading of typical phototropic glasses containing silver chloride at 46°C. (After Smith (34)).

TABLE V

Composition of Some Typical Photochromic Silver Halide Glasses (after Smith (34)).

Constituent	Glass 1	Glass 2	Glass 3	Glass 4	Glass 5	Glass 6	Glass 7
SiO_2	60.1%	62.8%	59.2%	59.2%	60.1%	52.4%	51.0%
Na_2O	10.0	10.0	10.9	14.9	10.0	1.8	1.7
Al_2O_3	9.5	10.0	9.4	9.4	9.5	6.9	6.8
B_2O_3	20.0	15.9	20.0	16.0	20.0	20.0	19.5
Li_2O	-	-	-	-	-	2.6	2.5
PbO	-	-	-	-	-	4.8	4.7
BaO	-	-	-	-	-	8.2	8.0
ZrO_2	-	-	-	-	-	2.1	4.6
Ag	0.40	0.38	0.50	0.50	0.40	0.31	0.30
Br	0.17	-	-	0.60	0.17	0.23	0.11
Cl	0.10	1.7	0.39	-	0.10	0.66	0.69
F	0.84	2.5	1.45	1.45	0.84	-	-
CuO	-	0.016	0.016	0.015	0.016	0.016	0.016

Note: Compositions in weight percent; halogens are given as weight percent additions to that of the base glass.

Apparently, under irradiation, three processes occur simultaneously: darkening or creation of color centers; optical bleaching; and thermal bleaching. The darkening rates, and the optical components of fading rates, are considered to be described by first-order kinetics (33). The spectral sensitivity is dependent on the particular halide. Thus a glass with AgCl is sensitive in the 3000-4000Å region whereas one with a mixture of AgCl and AgBr will be sensitive from 3000 to 5500Å and one with a mixture of AgCl and AgI will be sensitive to about 6500Å. The darkening rate is apparently not very sensitive to temperature, but the fading rate increases rapidly with increasing temperature. The advantage of this inorganic phototropic system is that even after 300,000 cycles, no fatigue is noticed (35). This is far superior to organic systems.

A tentative model to explain this interesting reversible phenomenon has been given by Armistead and Stookey (33). The primary photolytic reaction can be represented by:

$$Ag^{+} + Cl^{-} + h\nu = Ag^{o} + Cl^{o}.$$

Thus an electron is freed from the Cl^{-} and is trapped by the Ag^{+}. Such silver atoms, or very small aggregates of them, are the color centers. Sensitization by copper ions is attributed to:

$$Ag^{+} + Cu^{+} + h\nu = Ag^{o} + Cu^{++}$$

The reaction products are metastable and will return to the original state when the incident radiation is terminated. This suggests that the Cl atoms cannot diffuse away nor will extensive aggregation of the silver atoms occur. The phenomenon is thus reversible in contrast to the normal photographic process which is irreversible. Aggregation of silver and the loss of chloride do occur in the latter process. Further support of this model is available from the similarity of the optical absorption curves of such glasses to those for crystalline silver halides (34). The model is admittedly over-simplified, and many observations are not yet explainable. For example, the composition of the base glass and the thermal history are critical factors. Their roles are not yet understood. In addition to silver halides, it has been reported that silver molybdate and silver tungstate crystals can also yield phototropic glasses similar to those described above (36).

Phototropic Glasses Prepared Under Highly Reducing Conditions.

Soda-silica glasses of high purity and melted under highly reducing conditions also show phototropic behavior when exposed to UV radiations (37). It appears that the reducing conditions are necessary and that traces of impurities can have a great effect on the optical properties. Very careful experiments have been reported recently by Swarts and Pressau (38). For soda-silicate glass, the color center generated is located at 5700Å and no deliberately added impurities are necessary although cerium and europium will enhance phototropy. Inhibition of phototropy can also be caused by about 200 ppm of titanium, vanadium or iron. The effects of cerium and europium are shown in Table VI. The parameter K^1 is defined by:

$$K^1 = 1/d \log (T_o/T) \quad ,$$

where d is the thickness of the sample, T is the percent transmission of colored glass at the peak of the absorption band, and T_o is percent transmission of uncolored glass. Under sustained irradiation, the saturation value of K^1 is designated $K_o{}^1$. Although the rare-earth ions can supply photoelectrons, their exact role is unclear since phototropic coloring is not accompanied by a reversible change in absorption of the reduced rare earth ions.

The change in transmittance of a typical soda-silica glass on UV irradiation is shown in Fig. 8. In Fig. 9, the effects of changing the alkali ions are shown. The position of the peak is apparently unaffected by the concentration of the alkali ions. For Na_2O-SiO_2 glasses, for instance, it is constant even when the Na_2O content is varied by as much as 22%. This observation suggests that the color centers in question are dependent on the structure of the glass. It is possible that the reduction reaction during melting generates oxygen-deficient sites which subsequently trap electrons and that the nature of this site is dependent on the nature of the alkali ion.

TABLE VI

Equilibrium Phototropic Response of Soda-Silica Glass to 3660Å Radiation. (After Swarts and Pressau (38)).

Additions	K_o^1 (cm^{-1})
None	0.025
0.1% Ce	0.28
0.005% Eu	0.70

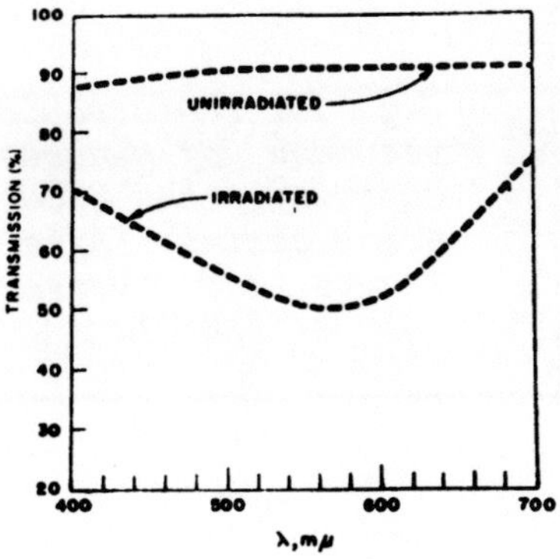

Fig. 8. Spectral change in soda-silica glass irradiated by 2537A UV. (After Swarts and Pressau (38)).

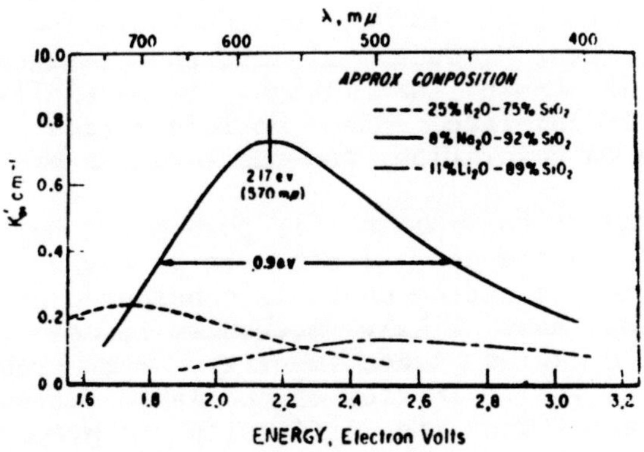

Fig. 9. The 5700A Color center in alkali-silicate glasses. (After Swarts and Pressau (38)).

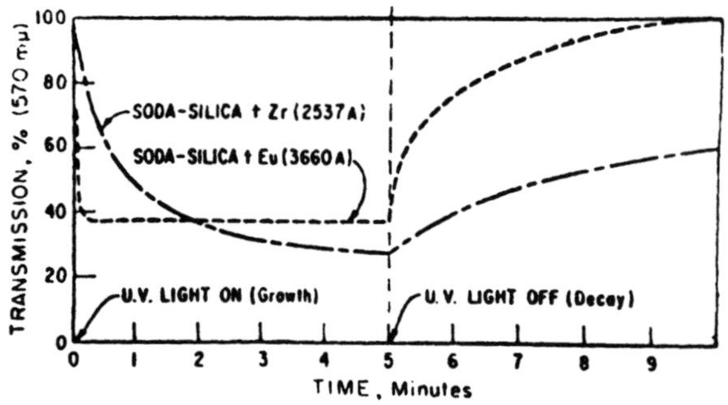

Fig. 10. Phototropic response of glasses containing the 5700A color center. (After Swarts and Pressau (38)).

Typical photoresponse of this type of glass is shown in Figure 10. The formation of the color center is described by first-order kinetics whereas the fading mechanism follows a second-order relation (38). The coloration was not studied as a function of temperature although the fading rate is very temperature-sensitive. For instance, the fading rate at room temperature is about thirty times greater than that at -150°C. Unlike the silver halide phototropic glasses, these "reduced" glasses show fatigue on repeated irradiation. It is known that fatigue is partially due to an acceleration of fading rate. Since no information exists on the effect of fatigue on coloration, the exact causes are unknown.

A number of models to explain this phenomenon has been discussed by Cohen (39). A trapped electron in an oxygen vacancy near an alkali ion or near a cation donor is considered unlikely. Cohen considers that the most likely mechanism is one involving the photoreduction of a cation trace impurity, most probably the reduction of Ti^{4+} to Ti^{3+}. In view of the large effects of minute traces of impurities and the difficulty in preparing glasses totally void of impurities, the validity of any model must await further research.

ACKNOWLEDGMENTS

I am grateful to Dr. S. Sakka and Mr. Roger Haskell for their assistance in the preparation of this review and to the National Aeronautics and Space Administration for support under Contract No. NSG-100-60.

REFERENCES

1. D.R. Secrist and J.D. Mackenzie, Chap. 6 in Modern Aspects of the Vitreous State, Vol. 3, Ed. by J.D.Mackenzie, Butterworths (Washington) 1964.

2. J.D. Mackenzie, The Encyclopedia of Physics, p. 769, Ed. by R. M. Besancon, Reinhold (New York) 1966.

3. D.R. Secrist and J.D. Mackenzie, J. Am. Ceram. Soc., 48, 487-91 (1965).

4. A.F. Ioffe and A.R.Regel, p. 238 in Progress in Semiconductors Vol. 4, Ed. by A.F. Gibson, Wiley (New York) 1960.

5. J.D.Mackenzie, Chap. 5 in Modern Aspects of the Vitreous State Vol. 3, Ed. by J.D. Mackenzie, Butterworths (Washington) 1964.

6. B.T. Kolomiets, Phys. Stat. Sol. 7, 359-72 (1964).

7. A.D. Pearson, Chap. 2 in Modern Aspects of the Vitreous State, Vol. 3, Ed. by J.D. Mackenzie, Butterworths (Washington) 1964.

8. J. Male, "Hall Mobility in Chalcogenide Glasses", presented at Conference on Electronic Processes in Low Mobility Solids, April 18-20, 1966, Sheffield, England.

9. B.T. Kolomiets, Phys. Stat. Sol. 7, 713-31 (1964).

10. P.K. Weimer, Phys. Rev. 79, 171 (1950).

11. G.F.J. Garlick, p. 350, Handbuch der Physik, Vol. 19.

12. Xerography and Related Processes, Ed. by J.A. Dessauer and H.E. Clark, Focal Press (London and New York) 1965.

13. E.M. Pell, Chap. 3, in ref. 12.

14. W.E. Spear, Proc. Phys. Soc. B 76, 826-32 (1960).

15. J.L. Hartke, Phys. Rev. 125, 1177-92 (1962).

16. J.L. Hartke and P.J. Regensburger, Phys. Rev., 139, A970-A980 (1965).

17. A. Jenkins, Chap. 4 in Materials Used in Semiconductor Devices, Ed. by C.A. Hogarth, Interscience (New York) 1965.

18. *W. E. Addison, The allotropy of the Elements, Chap. 4, Oldbourne Press (London) 1964.*

19. *H. Krebs, p. 246 in Semiconducting Materials, Ed. by H.K. Henisch, Academic Press (New York) 1951.*

20. *A. Eisenberg and A.V. Tobolsky, J. Polymer Sci., 46, 19 (1960)*

21. *H. Richter and F. Herre, Z. Naturforsch. 13a, 874-85 (1958).*

22. *F. Metzger, Helv. Phys. Acta 16, 323-42 (1943).*

23. *J.D. Howe and K. Lark-Horovitz, Phys. Rev. 51, 380 (1937).*

24. *H.P.D. Lanyon, Phys. Rev., 130, 134-43 (1963).*

25. *P.H. Keck, J. Op. Soc. Am., 41, 53-5 (1951).*

26. *P.H. Keck, J. Op. Soc. Am., 42, 221-5 (1952).*

27. *C.H. Griffiths, Chap. 8 in Recent Advances in Selenium Physics Ed. by ESTC, Pergamon Press (Oxford) 1965.*

28. *B.T. Kolomiets, T.N. Mamontova and T.F. Nazarova, p. 418-22 in The Structure of Glass, Consultants Bureau (New York) 1960.*

29. *E.J. Felty and M.B. Myers, "Preparation and Properties of Vitreous As_2Se_3", to be published.*

30. *M. Tanaka, O. Mukai, and O. Kamiike, J. Ceram. Assoc. Japan, 73, 51-5 (1965).*

31. *G.H. Brown and W.G. Shaw, Rev. Pure and Appl. Chem. 11, 1-32, (1961).*

32. *W.A. Weyl, Colored Glasses, Dawson's (London) 1959.*

33. *W.H. Armistead and S.D. Stookey, Science, 144, 150-4 (1964).*

34. *G.P. Smith, Proc. Int. Cong. on Glass, p. 108.1, Brussels, (1965).*

35. *G.K. Megla, Appl. Optics 5, 945-60 (1966).*

36. *British Patent No. 1,005,335, Sept. 22, 1965, issued to Corning Glass Works entitled "Phototropic Glass Article and Method of Making it".*

37. *A.J. Cohen and H.L. Smith, Science, 137, 981 (1962).*

38. E.L. Swarts and J.P. Pressau, *J. Am. Ceram. Soc.* 48, 333-8 (1965).

39. A.J. Cohen, *Proc. Int. Cong. on Glass*, p. 111.1, Brussels (1965).

DISCUSSIONS

CHARLES GREENE (Alfred University, U.S.A.): *Is phototropy of the Eu sensitized variety confined to sodium silicate glasses or may soda lime glasses and borosilicates be used as bases for this type of phototropy?*

J.D. MACKENZIE: *I believe Swarts and Pressau (J. Am. Ceram. Soc. 48, 333 (1965)) have mentioned that Plate Glass compositions have been studied, e.g.* Na_2O *- 14 w/o, CaO - 12 w/o, MgO - 2 w/o,* SiO_2 *- 72 w/o.*

CHARLES GREENE: *Are there no references to work on photoconductive selenium prior to 1950? Reference may be made to the Electrical Experiments which were published since 1920; Hugo Cerusback editor.*

J.D. MACKENZIE: *Thank you for your comment.*

V.D. FRECHETTE (Alfred University, U.S.A.): *In the case of AgCl phototropic glasses, are there elements which quench the effect or "poison" the phototropy?*

J.D. MACKENZIE: *From published work of Corning Glass Works, I do not think the effects of "poison" are discussed. I can only "guess" that impurities, at least in concentrations of ppm., are unimportant.*

CHARLES DELBECQ (Argonne National Laboratory, U.S.A.): *I remember a lecture by Dr. Stookey from Corning Glass Works about a year ago, in which he mentioned that they were not able to detect the* Ag^+ *- ion with an electron trapped to it. Has anyone made ESR or photoconductivity measurements on these phototropic glasses?*

J.D. MACKENZIE: *Not that I am aware of.*

E. LELL (Bausch and Lomb., U.S.A.): *Is there proof that the observed crystals in Ag - phototropic glasses are actual Ag-halide crystals and not a - for instance - boro-silicate phase that has crystallized?*

J.D. MACKENZIE: *I think there is evidence from X-ray diffraction that AgCl crystals are found in the glass.*

H. PEISL (T.H. Darmstadt, Germany): *Phototropy in AgCl doped glasses could be understood from annealing behavior of AgCl single crystals after UV irradiation, which is essentially well known and shows very similar properties.*

J.D. MACKENZIE: *Yes, this has been emphasized by Corning workers. On the other hand, the phototropic properties are sensitive to thermal history such as heat treatment which says that the glassy matrix must affect the silver chloride present.*

OSCILLATORY PHOTOCONDUCTIVITY AND LUMINESCENCE IN $SrTiO_3$

L. H. Grabner

National Bureau of Standards

Washington, D.C., U.S.A.

Photocurrents and luminescence in $SrTiO_3$ oscillate at low temperatures ($\sim 4^oK$) at a threshold of about 100 volts/cm. Voltage-current curves show negative differential resistance in the region of oscillation. It is suggested that these effects are due to the field ionization, into the valence band, of a hole trapped 1 eV from the valence band. The hole is ionized by the local field ($\sim 10^6$ V/cm) which differs from the externally applied field of 10^2 V/cm because of the high dielectric constant of $SrTiO_3$ ($\sim 10^4$) at low temperatures.

INTRODUCTION

In semiconductors and insulators an increase in the resistivity with increasing electric field -- negative differential resistivity (NDR) -- and current oscillations has been observed in a variety of materials. The following mechanisms have been suggested for NDR:

1. redistribution of charge carriers over recombination centers resulting in a decrease of the number of free charge carriers; "field quenching (1),"

2. redistribution of free charge carriers over bands with different effective masses (2,3),

3. field-dependent capture cross-sections of recombination centers for free charge carriers or field-dependent scattering mechanisms (4,5).

In this preliminary report we present an example of NDR in which the mechanism is attributable to "field quenching". This example, however, differs from previous reports in that the electric field responsible for the NDR and the oscillation of photocurrent and luminescence is the "local" electric field induced by the applied field rather than the applied field itself. The two differ by a factor of $\sim 10^4$ in $SrTiO_3$ because of the large static dielectric constant of this material at low temperatures ($\sim 10^4$ at 4^oK).

EXPERIMENTAL

The samples are single crystals of undoped $SrTiO_3$. They have dark resistivities larger than 10^{10} Ω-cm and are cut into platelets of dimensions 10 x 5 x 1 mm. Electrodes of molten indium are applied to a single large face of the sample leaving a slit of 2 x 10 mm. Exciting radiation is supplied with a monochromator; luminescence is measured with a second monochromator and an appropriate photomultiplier; photocurrents, with a micro-microammeter; voltages are supplied with a battery. Measurements at 4^oK and 77^oK are made by immersing the sample in liquid helium or nitrogen, respectively.

RESULTS AND DISCUSSION

Figure 1 shows the luminescence and the excitation of luminescence vs. photon energy at 4^oK and 77^oK. The luminescence band at 4^oK is centered at 2.25 eV and has a full width at half maximum of 0.20 eV. At 77^oK the corresponding values are 2.10 eV and 0.20 eV. The excitation spectrum is identical for both temperatures. Figure 2 shows the spectral response of photoconductivity and optical quenching of photoconductivity at 4^oK and 77^oK; neither is normalized. As shown by a comparison of the spectral response of photoconductivity and the excitation spectrum of the luminescence, the onset of luminescence coincides with production of free charge carriers. Thus we conclude that the luminescence is due to the recombination radiation of a free carrier with a trapped carrier. From our data we deduce a band-gap of about 3.35 eV. P-type conductivity has not been observed in $SrTiO_3$ and we shall assume that under excitation the conductivity is predominantly n-type and that the luminescence is due to the recombination of a free electron with a trapped hole about 1.1 eV from the valence band. The hole trap is due to a defect unknown to us; it could be intrinsic or extrinsic. The voltage-current curve, when the sample is simultaneously irradiated with band-gap light, is shown in Fig. 3. After an initially ohmic portion it enters a region of NDR. Oscillations of the photocurrent and luminescence begin where the curve ends. The oscillations have the following characteristics:

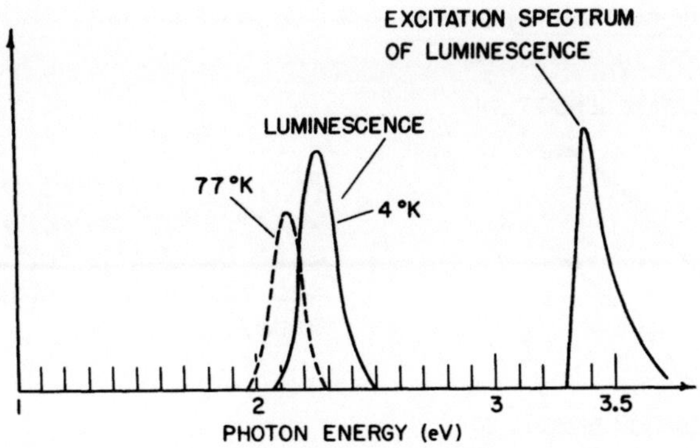

Fig. 1. Spectrum of luminescence and the excitation spectrum of luminescence at 4°K and 77°K.

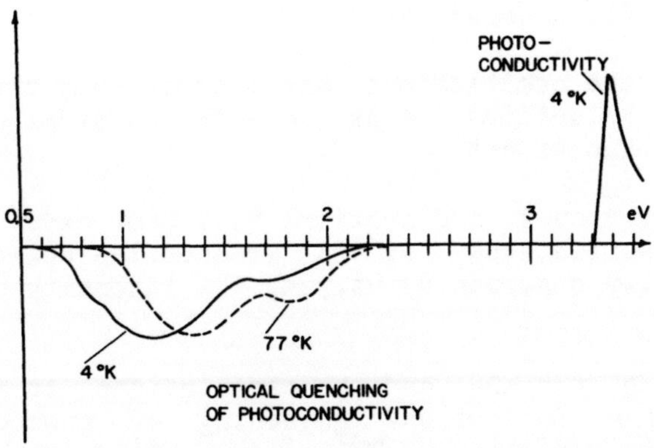

Fig. 2. Spectral response of photoconductivity and the spectrum of the optical quenching of photoconductivity at 4°K and 77°K. Neither spectrum is normalized.

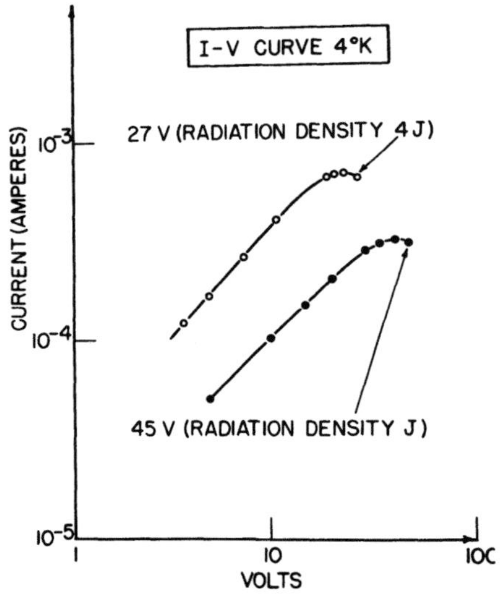

Fig. 3. Current voltage curve under simultaneous irradiation with band-gap radiation. J is the radiation density (relative).

1. The threshold for oscillations decreases as the intensity of the band-gap radiation increases as shown in Fig. 3.

2. The frequency of the oscillations increases with the intensity of the band-gap radiation.

3. The frequency of the oscillations is independent of the applied field above threshold.

4. At 77°K the oscillations are less marked; the threshold occurs at applied fields roughly an order of magnitude larger than at 4°K.

So far only visual observations have been made on the frequency of the oscillations: at 4°K frequencies between 0.1 and 10 hertz are easily observed by varying the intensity of the band-gap radiation.

In explaining the NDR and the oscillations it is important that the mechanism for optical quenching of photoconductivity (luminescence) be understood. This occurs when a trapped hole is optically ionized from its ground state (see Fig. 4) into the valence band. The free hole may then recombine with a trapped electron, emptying an electron trap. The empty electron trap is then able to capture a free electron leading to a decrease of photocurrent. It is clear that this mechanism will also lead to a decrease of luminescence if the radiation transition is due to the recombination of a free electron with a trapped hole.

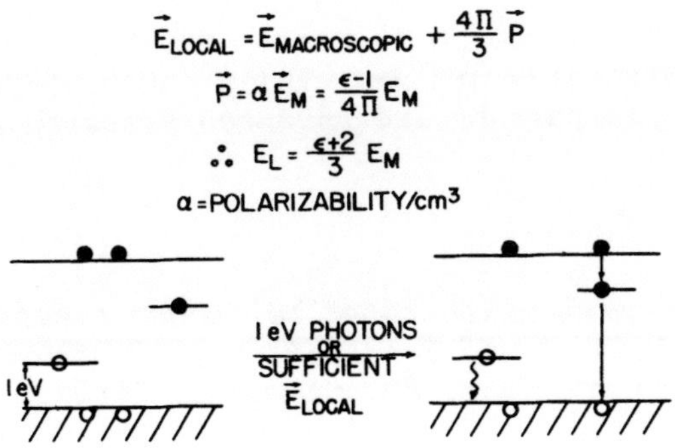

Fig. 4. Model illustrating the quenching of photocurrent or luminescence. Trapped hole at 1 eV can be ionized either optically or by the local field E_L leading to quenching as shown. E_M is the applied field, P the polarization.

In order to explain the NDR and oscillations we propose that the trapped hole is ionized by the electric field with consequences that duplicate the optical ionization of a trapped hole described above. Since the trapped hole is localized (presumably at a lattice point) in a polarizable medium, the electric field effective in the ionization is the local electric field rather than the applied electric field. The local electric field E_L is related to the applied electric field E_M by

$$E_L = \frac{\epsilon + 2}{3} E_M$$

where ϵ is the static dielectric constant. Thus for applied fields of only 10^2 V/cm the local field is $\sim 10^6$ V/cm ($\epsilon = 2 \times 10^4$ at 4^oK). In fact, in order to ionize a hole trap 1 eV from the valence band will require fields of the order of 10^6 V/cm (5).

We visualize the oscillations to occur by the following mechanism: Because any real crystal is physically inhomogeneous the distribution of applied field and hence the local field will be non-uniform. Field ionization of a trapped hole will first occur in a region of maximum field. Because of the decrease in the number of free electrons (quenching) following the field ionization of the trapped hole, the resistivity of this region increases, thus further increasing the applied field (local field) of this region. The electric field gradients are such that free charge carriers from contiguous regions move into the region of high resistivity which thus becomes mobile. Thermodynamically the most stable situation is one in which the region of high resistivity is expelled from the crystal (via one of the electrodes). Once expelled from the crystal, the cycle starts again and hence the oscillations.

ACKNOWLEDGMENTS

Discussions with members of the Solid State Physics Section of the National Bureau of Standards are gratefully acknowledged. This research was supported in part by the Advanced Research projects Agency.

REFERENCES

1. *K.W. Böer, Feld- und Strominhomogenitäten bei hohen electrischen Belastungen in Insulatoren und Photoleitern. Festkorperprobleme (Friedrich Vieweg und Sohn, Braunschweig, 1962), Vol. 1, pp. 38-71.*

2. *J.B. Gunn, International Conference on Semiconductors, Plasma Effects in Solids, Paris 1964 (Academic Press Inc., New York, 1965), p. 199.*

3. *H. Krömer, Phys. Rev. 109, 1856 (1958).*

4. *B.K. Ridley and R.G. Pratt, J. Phys. Chem. Solids 26, 21 (1965)*

5. *Walter Franz, Ann. d. Phys. (6) 11, 17 (1953).*

DISCUSSIONS

A.A. ARAFA (Assiut University, U.A.R.): *Oscillations in electrical conductivity were observed in thin films of SiO_2 evaporated in vacuum. Repeated conductivity measurements in both this material and Sb_2S_3 as the field applied across the field was varried, gave low values of σ when E decreased and high values when E increased.*

L.H. GRABNER: *Oscillations of current are observed in a number of substances; the phenomenon is not uncommon.*

PHOTOMECHANICAL EFFECT IN SEMICONDUCTORS

G.C. Kuczynski

Department of Metallurgical Engineering and Materials Science and the Radiation Laboratory*, University of Notre Dame, Notre Dame

Indiana, U.S.A.

It has been found that light of wavelength between 2.0μ and 4.0μ or shorter than 0.4μ decreases the hardness of the surface layer of n and p type Ge or Si crystals by 10 to 60%. Oxidation of the surface reduces this effect appreciably.

Experiments with Ge plates bent in darkness and under illumination revealed photomultiplication of dislocations. This effect seems to be affected by the chemistry of the surface.

* The Radiation Laboratory of the University of Notre Dame is operated under contract with the U.S. Atomic Energy Commission. This is AEC Document No. COO - 38 - 506.

INTRODUCTION

The science of solid materials recognizes a general correlation between mechanical properties of solids and their electrical conductivities. Mechanical hardness of the material is usually associated with relatively high electrical resistivity and vice versa. Thus, good insulators like aluminum oxide are very hard and brittle while the best metallic conductors belong to the class of ductile materials. Furthermore, any process which hardens the material usually, at the same time, increases electrical resistivity. This is especially true for metals, where the electrical conductivity decreases upon cold work or alloying which increases their hardness. This speculation suggests that the photoconductors may become mechanically softer upon illumination.

EXPERIMENTAL

A simple experiment (1) to test this hypothesis was not difficult to perform. Into a polished surface of a germanium crystal, a wedge-shaped diamond indenter was pressed by a force of a few grams, in darkness and under illumination with a tungsten-filament or mercury lamp. The effect was so large that it was spotted at once. A photograph of two rows of such indentations; one obtained in the dark, an other under illumination, is reproduced in Fig. 1. The length of the indentations produced under illumination are almost twice as large as those taken in darkness. As the hardness is inversely proportional to the square of the indentation length, the decrease of hardness, due to illumination in this case, is about 65%. The effect is reversible, that is, disappears immediately after the source of light is removed. The effect was found to occur in both n and p type germanium and silicon crystals. Application of various loads to the indenter revealed that only a surface layer about 1 - 3μ deep was affected. This is illustrated graphically in Figure 2. The same effect was observed in other semiconductors such as indium antimonide and indium arsenide. It is interesting to note that, in NaCl crystals, the decrease of hardness upon illumination which was about 20% was found not to be confined to the surface layer but was observable at large distances from the surface (2). In LiF, a slight increase of hardness was observed upon illumination.

The photomechanical effect in germanium was found to be sensitive only to certain bands in the optical spectrum (1). Only infrared radiation with wavelength between 2.0μ and ultraviolet shorter than 0.4μ seem to produce this effect. The hardness decreased proportionally to light intensity (3).

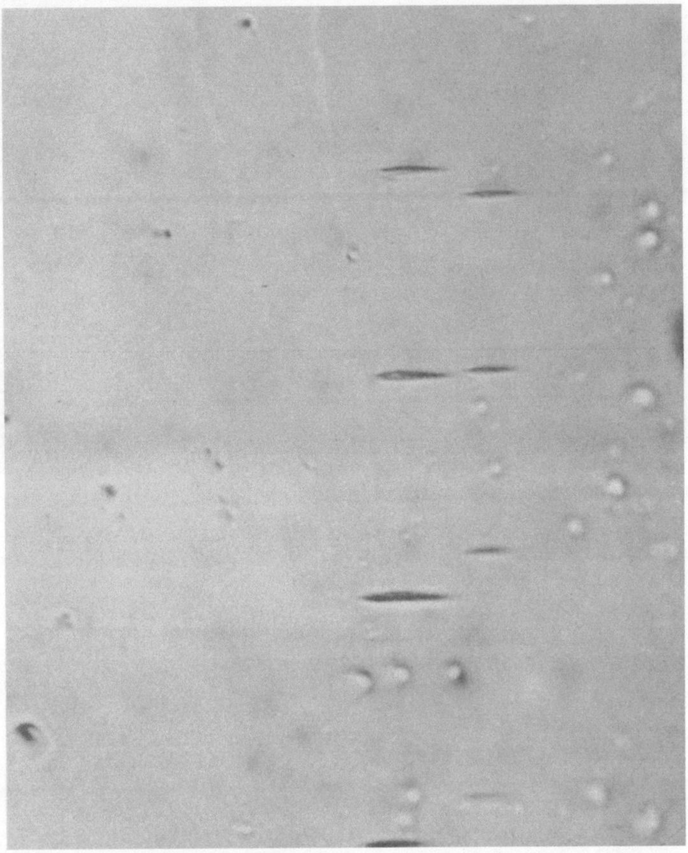

Fig. 1. The Knoop hardness indentation on Ge crystal in dark (left row) and under illumination (right row) 1000X.

(a)

Fig. 2. Hardness of Ge in dark as a function of depth.

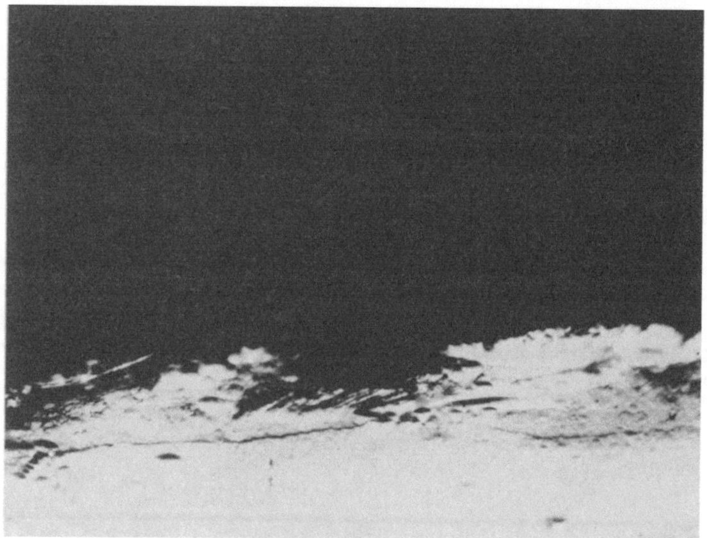

(b)

Fig. 2. Hardness of Ge under illumination as a function of depth.

The photomechanical effect in germanium was confirmed by Westbrook and Gilman (4) and Goridko et al (5). The latter have found that the hardness decrease is, for all practical purposes, proportional to the increase of the electrical conductivity during illumination which, as was mentioned in the introductory remarks, was to be expected. The effect seems to be quite common in semiconductors and semimetals. Apart of n and p germanium, it has been found in silicon (1,2,3) antimony (6,7), indium antimonide (1,7), indium arsenide (1) cadmium sulfide (8), gallium arsenide (9), gallium phosphide (9), and lead sulfide (9). The last three compounds exhibited hardening upon illumination.

Nadeau (10) reported appreciable changes in flow stress in alkali halide crystals upon illumination. This effect as we have remarked above is not limited to the surface. There are in reality two effects. One at low temperatures (in KCl below $100^{o}K$) and the other close to room temperature. No effect was observed at intermediate temperatures. Almost all these changes in flow stress at low temperatures, which he observed to be sometimes of the order of 30%, were positive, that is, the flow stress increased upon illumination. In some crystals, (NaCl, NaF) at room temperature, a decrease of the flow stress on illumination was observed which corroborates the earlier observations of Hochman (2).

It was logical to investigate the effect of light upon the elastic constants of germanium. None, however, was found. This and the reversibility of the effect seem to indicate that the mechanism of generation and/or movement of dislocations is affected by some photon electron reaction. This was borne out by experiment (3,11). Dislocations in germanium are easy to detect by a suitable etchant. Those which end at the surface form pits after etching. Also, it is possible to obtain germanium crystals with very low dislocation density. When the crystals containing about 10^{2} dislocations per square cm. were bent to fracture under illumination, the number of dislocations increased, often to as much as 10^{6} per square cm. Crystals bent in the dark, even to breaking have not exhibited any increase in dislocation concentration. In Figure 3a, a photomicrograph of a section through a crystal bent under light is represented. The etch pits formed at the dislocations ending at the surface are present only in a layer a few microns thick adjacent to the surface, which during the test was under tension. In Figure 3b, a similar section of a crystal bent in darkness is shown with no visible dislocation pits. This photomultiplication of dislocations was confirmed by Kikuchi and Saito (12) by a very ingenious experiment. They measured the photo response due to the change of minority carrier lifetime after a germanium crystal was bent in the dark and under illumination with infrared light. They observed a large decrease in the lifetime of these carriers and an increase in the rate of surface recombina-

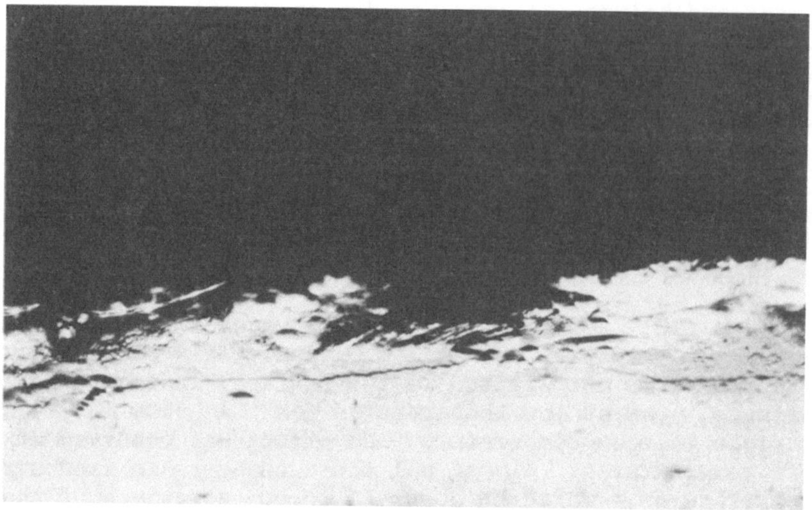

(a)

Fig. 3. Section through Ge crystal fractured under illumination. Etched to reveal dislocation pits. Original dislocation density $10^2/cm^2$. 250X.

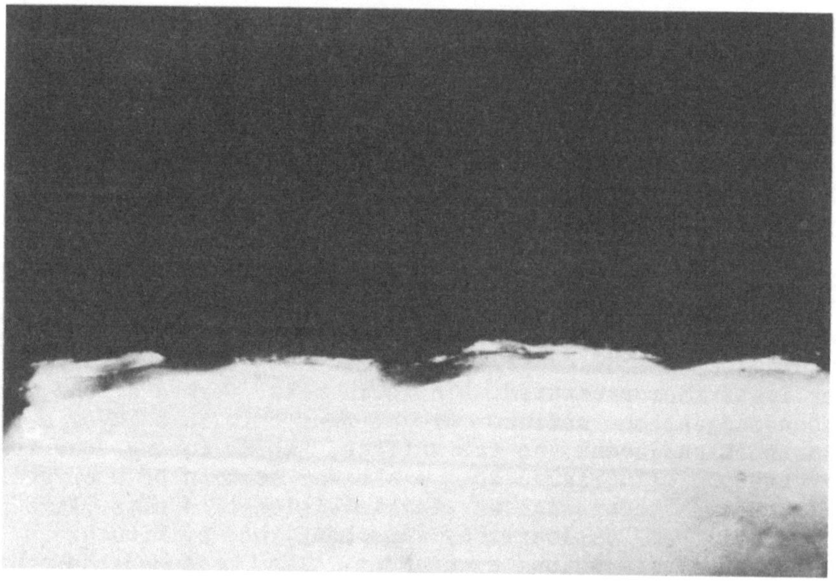

(b)

Fig. 3. Section through Ge crystal fractured in darkness. Dislocation density $10^2/cm^2$. 250X.

tion, both effects caused by dislocations produced during bending under illumination. No effect was observed in specimens deformed in darkness.

Recent investigations of the author of this report indicate that the photomechanical effect is insensitive to the dislocation density. Large indentations in germanium produced by a load in excess of 100 grams, after etching revealed a ring of dislocations as shown in Figure 4. The decrease of hardness upon illumination was found to be the same in the vicinity of these dislocations as in the undeformed regions of the crystal surface.

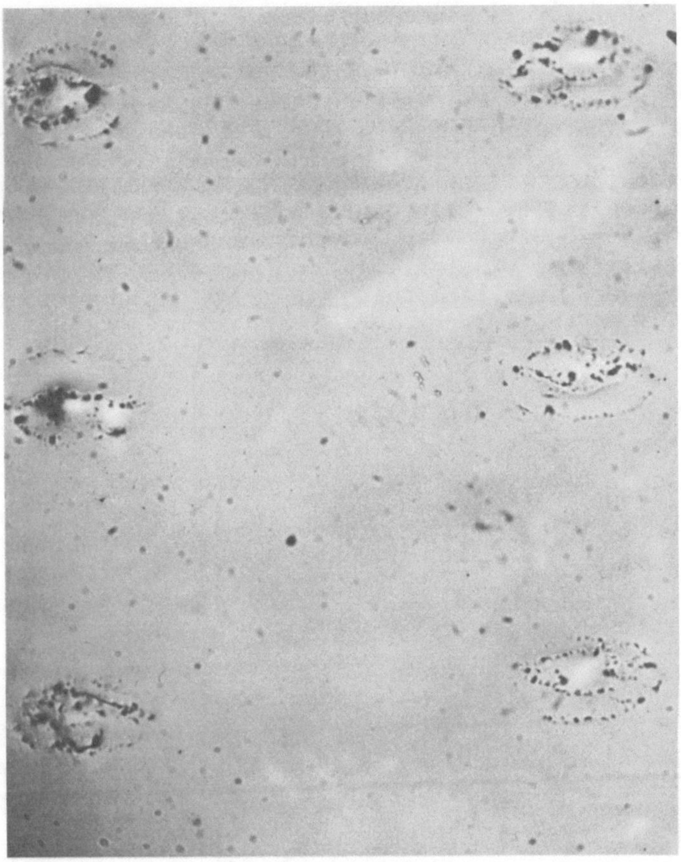

Fig. 4. The impressions produced by 100 grams load on the surface of germanium after etching with CP-4 solution. 100X.

Experiments of Westbrook (13) seem to indicate that the softening of germanium surface occurs only in the presence of absorbed water on the surface of the crystal. He found no effect in germanium crystals dried and tested in toluene. The effects reappeared as soon as the surface of the semiconductor was exposed to air. He also found that in the presence of moisture, GaAs surface softens under illuminations in contrast with observations of Holt (9). The latter experiments were performed on the crystals etched with solution containing iodine ions on the semiconductor surface. Kuczynski et al (3) investigated the effect of oxidation on the photomechanical effect. Germanium crystals were oxidized in air at 400°C for time intervals ranging from 0 to 180 minutes. Figure 5 shows the results of one of these experiments. Surface hardness measured in the dark and the magnitude of the photomechanical effect both decreased with increasing exposure time. After 1.5 to 2 hours of oxidation the effect disappeared completely. In general, the difference between the hardness measured in the dark and under illumination decreased linearly with the square root of time as shown in Figure 5. There are some indications (14) that, at temperatures below 500°C, the growth of the oxide layer on germanium follows a parabolic law. This would mean that the hardness difference decreases proportionally to the increase in the oxide layer thickness. All these observations indicate that the photomechanical effect is very sensitive to the presence of absorbed molecules and oxide film on the surface.

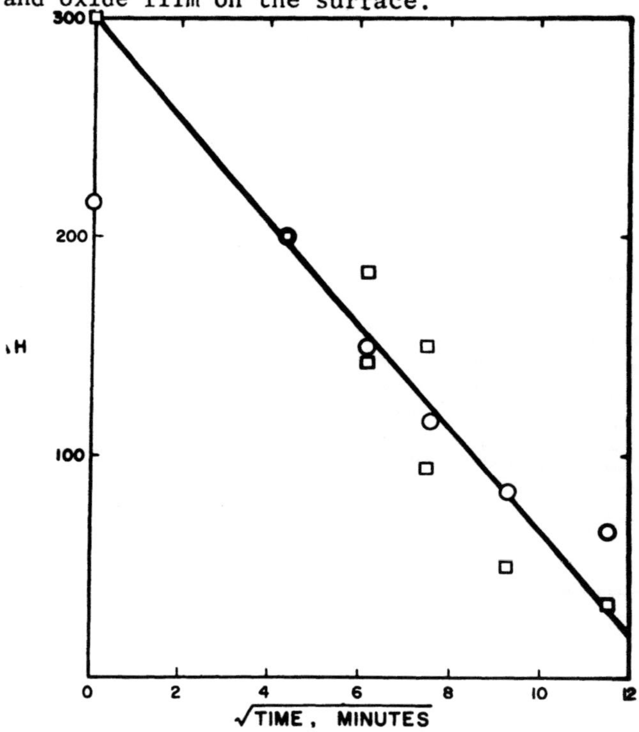

Fig. 5. Effect of oxidation time upon the difference of dark and illuminated hardness.

Hardness increase ΔH in germanium with donor concentration

○ $2 \times 10^{13}/cm^3$

□ $10^{15}/cm^3$.

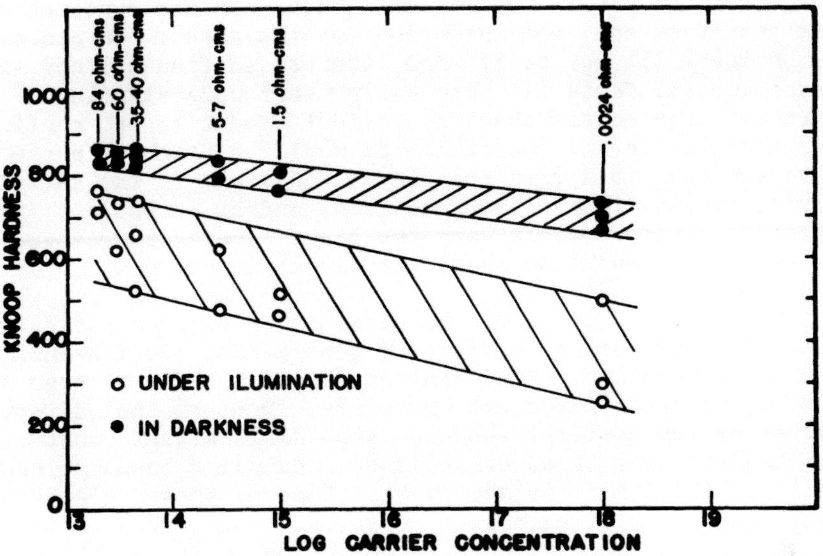

Fig. 6. The dark and illuminated hardness at constant depth of 0.5μ for donor type germanium crystals, as a function of donor concentration.

Finally, the influence of impurity concentration upon hardness of germanium crystals was investigated. Figure 6 summarizes the results obtained on various crystals of germanium with donor concentrations varying from $2x10^{13}$ (84 ohm-cm) to $7x10^{17}$ ($2.4x10^{-3}$ ohm-cm). The hardness measured in the dark decreases only slightly with increasing donor concentration while the magnitude of photomechanical effect increases with donor concentration.

Westbrook and Gilman (4) found that the surface hardness of semiconductors is lowered significally (60%) by the simultaneous presence of a small potential between the indenter and the crystal surface. This effect is also confined to a layer 2 - 3μ deep. They also investigated the possibility of a common origin of photo and electro-mechanical effects. They found that the photomechanical effect is apparently not connected with a potential difference since they could not affect it by either the presence of a polonium discharger or shorting the indenter to the crystal.

DISCUSSION OF RESULTS

There is no single consistent theory of the photomechanical effect and it is not the intention of the author to propose one here. However, it may be of some interest to find out how some of the experimental facts fit into our present understanding of semiconductors. The photomechanical effect seems to be confined to the crystals which are partially or wholly covalently bonded. It is also limited to a very thin layer adjacent to the surface and highly sensitive to the absorbed ions and/or molecules. Its instantaneous disappearance when the light source is removed indicates that the phenomenon is electronic in nature.

In principle, it should be possible to rationalize the plastic behavior of solids in terms of generation, motion and interaction of dislocation. Read (15) developed a model of edge dislocation in diamond structure crystals. Due to the directional character of the covalent bonds, such dislocations must contain broken bonds, or as Read called them, dangling bonds or unpaired electrons, which act as acceptors. Some of these electrons are paired, thus simulating normal covalent bonds. Due to coulombic repulsion the distance between two electron pairs on the dislocation line in germanium is on the average 100 atomic distances. The experiments of Pearson, Read and Morin (16) seem to substantiate this model. These paired electrons act as dragging points, obstructing severely the mobility of dislocations. Johnson (17) found no detectable motion of dislocation in germanium below 200°. He attributed this to very high Peierls force. However, in view of the observed photomultiplication of dislocation, it seems that at least part of this lack of mobility is due to the dragging points which we identified with paired electrons. In this model, it is obvious that any decrease of the density of paired electrons of the dislocation line should produce increase in mobility. This can be accomplished by excitation of a paired electron with a photon of appropriate energy. According to Read, the energy levels E_d of the extra electrons in a dislocation line in germanium lie about 0.4 ev below the bottom of the conduction band. This energy corresponds to the wavelength of about 3μ, which lies within the 2 - 4μ band for which the photomechanical effect in germanium was observed (1). Ultraviolet radiation does not penetrate the crystal, but is capable of exciting surface electrons or those from the impurities absorbed on the surface, to higher energy levels. At certain photon energy the surface electrons may be emitted back into the crystal. When this happens, assuming steady state conditions, the Fermi level E_F would split into two levels, the Fermi level for electrons E_{Fn} and that for holes E_{Fp}. The latter will decrease with the electron population, causing the number of holes to increase, which is equivalent to the decrease of the concentration of paired electrons in the dislocation lines. This, according to our hypothesis, would decrease the number of dragging points and therefore increase the dislocation mobility. The work function of germanium is 4.8 eV and that of silicon 3.6 eV (18), corresponding to about 0.26μ and 0.34μ wavelengths respectively. These wavelengths are shorter than the observed wavelengths associated with the appearance of the photomechanical effect in these semiconductors on the ultraviolet side of the spectrum. However, due to the positive space charge in the crystal layer adjacent to the surface, the energies required to emit back electrons from the surface to the interior of the crystal are expected to be lower than the work functions quoted above.

Far more data is required before a detailed theory of the photomechanical effect can be developed. It would be particularly valuable to be able to detect this effect by means of tests which are easier to interpret than the indentation microhardness. The efforts to bend the plates of germanium about 100μ thick in the dark and under illumination yielded inconclusive results, chiefly because the specimens were too thick. So far, all the efforts to produce thinner plates or wiskers of uniform cross section ended in failure. Also the whole field of the influence of the chemical composition of the surface upon the photomechanical effect awaits thorough experimental investigation.

REFERENCES

1. *G.C. Kuczynski and R.F.Hochman, Phys. Rev. 108, No. 4 pp. 946-948 (1957).*

2. *R.F. Hochman, "light Induced Plasticity in Semiconductors" Ph. D. Thesis Department of Metallurgy, University of Notre Dame, Notre Dame, Indiana (1954).*

3. *G.C. Kuczynski, R.F. Hochman and C.W. Allen, Mechanical Properties of Engineering Ceramics. Proc. Conf. North Carolina State College, pp. 495-505. Interscience Publishers (1961).*

4. *J.H. Westbrook and J.J. Gilman, J. Appl. Phys. 33, No. 7 pp. 2360-2369 (1962).*

5. *N. Goridko, P.P. Kuzmenko and N.N. Novikov, Fiz. Tverd. Tela, (Solid State Phys.) 3, No. 12 pp. 3650-3656 (1961).*

6. *P.P. Kuzmenko, N.N. Novikov and N. Goridko, Fiz. Tverd. Tela. (Solid State Phys.) 4, No. 10, pp. 2656-1659 (1962).*

7. *V.N.Lange and T.I.Lange, Fiz. Tverd. Tela. (Solid State Phys.) 6, No. 4, pp. 1220-1223 (1964).*

8. *N.N. Novikov, N. Ya. Goridko and A.G. Rudenko, Izv. Vysshikh. Uchebn. Zavod. Fiz. 4, pp. 22-24 (1964).*

9. *D.B. Holt to be published in Proc. Symp. on Environmental Effects on Mechanical Properties. Baltimore (1965).*

10. *J.S. Nadeau, J. Appl. Phys. 38 No. 3 pp. 669-677 (1964).*

11. *G.C. Kuczynski and R.F. Hochman, J. Appl. Phys. 30 No. 2, p. 267 (1959).*

12. *M. Kikuchi and M. Saito, J. Phys. Soc. Japan 14 No. 8 p. 1642 (1959).*

13. *J.H. Westbrook to be published in Proc. Symp. on Environmental Effects on Mechanical Properties. Baltimore (1965).*

14. *J.T. Low and P.S.Meigs,J. Electrochem. Soc. 104 No. 3 pp. 154-159 (1957).*

15. *W.T. Read Jr., Phil. Mag. 45 pp. 775-796 (1954).*

16. *G.L. Pearson, W.T. Read and F.J. Morin, Phys. Rev. 93 pp. 666-667 (1954).*

17. *O.W. Johnson, J. Appl. Phys. 36, No. 10 pp. 3247-3250 (1965).*

18. *L. Apker, E.Taft and J. Dickey, Phys. Rev. 74 No. 10, pp.1462-1474 (1948).*

DISCUSSIONS

A. GUINIER (University of Paris): *What is the thickness of the layer affected by ultraviolet light in Ge?*

G.C. KUCZYNSKI: *Only about half wavelength, hence the effect cannot be due to the radiation which penetrated, but rather by the interaction of ultraviolet light with the surface electrons.*

F.H. HAMMAD (U.A.R. Atomic Energy Establishment): *How does the model presented by you explain the increase in dislocation density in bent Ge crystals under illumination?*

G.C. KUCZYNSKI: *This question is difficult to answer. Probably the new dislocations in Ge are produced by collision of moving dislocations. If this is assumed, then the increase of dislocation mobility should also increase the dislocation concentration. However, we cannot exclude the possibility of activation of small dislocation loops existing in the crystals.*

CHARLES GREENE (Alfred University, New York): *How did you measure knoop hardness as a function of depth?*

G.C. KUCZYNSKI: *By changing the load on the indentor, the indentations of various depth can be produced. These depths can be estimated from the measurement of linear dimensions of the indentations assuming that hardness is constant, or by polishing off the surface electrolytically and measuring the amount of material taken off before the indentation disappears.*

J.H. CRAWFORD (Oak Ridge National Laboratory, U.S.A): *In what direction is the effect of light on crack propagation?*

G.C. KUCZYNSKI: *Increase of the stress is necessary to propagate the crack.*

THE EFFECT OF ILLUMINATION ON THE SURFACE POTENTIAL OF CADMIUM SULPHIDE

Y.L. Youssef, A. Mishriki, and H. Mikhail

Faculty of Science, University of Cairo

Cairo, U.A.R.

The surface potential of dark-insulating photoconducting CdS single crystals was measured at various illumination intensities, using white light and also monochromatic light in the wavelength range 0.36 to 1.0 μ. The impurity contents of the specimens were a few parts per million of Ag, Cu, Fe and Mg. The technique employed was based on the electrostatic attraction due to surface potential, appropriately modulated by a low frequency sinusoidal voltage. This technique imposes on insulating specimens an alternating electric field of some hundred volts/cm, which manifests novel aspects of the interaction of light with semiconductors. Optical quenching of the surface photopotential by a bias light as well as its growth and decay characteristics were also investigated. The results suggest that even pure CdS may contain negatively charged surface states which, though necessarily of small density, can be readily detected in a small AC field as a voltage rectification which may be destroyed by photodesorption.

INTRODUCTION

Over the last two decades, several papers (1-9) were published on surface photopotential (SPP) of semiconductors, i.e., the change ΔV_S in contact potential upon illumination. Some of these papers (2,3,4,6,7,9) deal specifically with SPP of CdS. Recently, the present authors (9) reported that the spectral response of SPP of CdS crystals exhibits in the visible region two abnormally large peaks; a broad long-wave peak, and a sharp short-wave peak close to the fundamental absorption edge. Somewhat similar, though much smaller, peaks have often been observed by Strakhov (4) 1963, but their exact origin was not identified.

The method used in this work for measuring the surface potential is based on the electrostatic force experienced by an attractable reference electrode as a result of the surface potential of the specimen, appropriately modulated by a low frequency sinusoidal voltage. This force induces in the sensitive reference electrode noticeable resonant vibrations which can be balanced by a compensating DC voltage equal in magnitude and similar in sign to the surface potential. Further details of this method have already been given by the authors elsewhere (11). This technique may allow an alternating voltage of several volts to be dropped on the specimen, and this has been found to reveal interesting magnified features of some photo-induced processes.

It was noticed that the peak heights depend markedly on the value of the modulating voltage imposed on the crystal, suggesting that they must be associated with voltage rectification. This in turn implies the existence of a Mott-Schottky-Bardeen (10) barrier at the crystal surface due to chemisorbed atoms. This view point is supported by the observation that SPP can be quenched by the action of light or heat, both of which, according to Williams (6), cause desorption.

The appearance of peaks in SPP and of humps in its time dependence plainly point out to the presence of two competing processes occuring at the opposite surfaces of the crystal. Accordingly, a rectifying barrier must reside at each surface, and the observed surface potential is essentially due to the net rectified component of the modulating voltage. The crystal, if n-type, may be regarded as a p-n-p junction traversed by a small constant AC current, and thus acts as a differential voltage rectifier, the direction of voltage rectification being the high-resistance direction. Hence, the resultant polarity of the front surface would be positive if light can destroy the potential barrier at the front more completely than at the rear. This is not only due to promotion of electrons in negative surface states to the conduction bands, but mainly due to dominance of voltage rectification at the back. This situation would be normally expected for frontal illumination.

Sign anomalies have often been observed (9) and will be discussed in the present paper, which describes SPP experiments conducted over a wide intensity range, using either monochromatic light from a monochromator, or a continuous spectrum of white light from a tungsten lamp operating between zero and normal rating.

EXPERIMENTAL

The experimental arrangement is shown in Fig. 1, in which S is the crystal under test (dark insulating photoconducting CdS 0.3 x 0.2 x 0.03 cm^3) stuck by Aquadag to a fixed electrode Q connected to a modulating LF oscillator O. The oscillator can be tuned to the natural transverse frequency f (30 cps) of a thin elastic reference electrode B which is thereby electrostatically excited to vibrate laterally. The component force with the fundamental frequency f is proportional to the surface potential. The plate D is a capacity electrode followed by a vibration detector VD, a low pass filter LPF and a cathode - ray oscilloscope CRO to display the null point which is obtained by varying the opposing voltage V_b supplied by the potential divider P. The sensitivity is 0.01 V.

In some experiments, unfocussed white light (full spectrum) was obtained from a small tungsten lamp 10 cm from the front surface of S. In other experiments the surface potential was determined at different wavelengths and at different monochromatic intensities.

Photoconductivity was measured on one sample separately in a subsidiary experiment by attaching with Aquadag a thin Pt wire to the upper end of S, the second lead being the electrode Q, now disconnected from O.

The crystals used in this work were a gift from Dr. I. Broser of the Fritz - Haber Institute, Berlin. The dark resistivity was often as high as 10^{11} ohm-cm. The impurity contents in a typical crystal were a few parts per million of Ag, Cu, Fe and Mg. The surface exposed to radiation was a plane clean face parallel to the c-axis. Special care was taken to guard against stray light and reflections to the rear. Possible interactions at the contact between the crystal and the supporting electrode Q were avoided by protecting the contact from the action of the exciting light by means of a black paper C covering the lower end of the front surface of S.

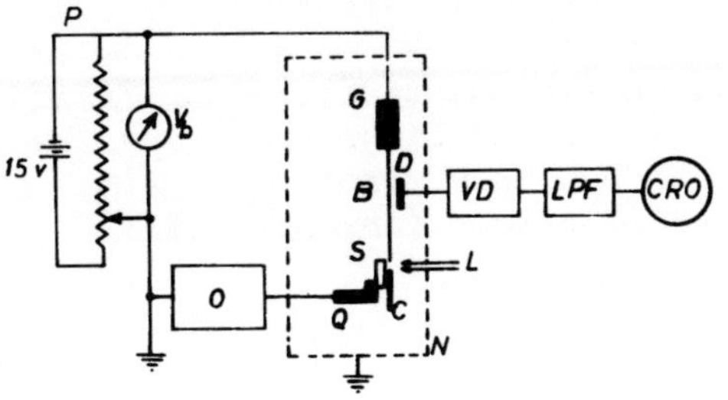

Fig. 1. Diagramatic representation of the experimental set-up.

RESULTS

Dependence of SPP on Intensity

Fig. 2 curve (a) represents for crystal IBX3 (the light screen C being removed) the variation of surface potential in an alternating electric field of several hundred volts/cm with the power consumed in the filament of the tungsten lamp (continuous spectrum) used to irradiate the surface. Three stages ending with saturation (which may be negative or positive) were distinguished; namely the stages OB, BD and DE. These stages appear to be largely an intensity induced effect, although the change of the spectral characteristics of the lamp with the rating may have a secondary role. This view point is supported by Fig. 3-a, which gives the actual intensity dependence for crystal IBL using a fixed electric power in the lamp and reducing the intensity by a neutral wedge. Fig. 3-b shows the monochromatic intensity dependence of SPP for crystal IBX3. Yellow light appears to behave in much the same way as the continuum of a tungsten lamp. Differential interaction at the two surfaces of the crystal may account for the general shape of the curves.

The SPP peak in Fig. 2 is not attended with a peak in photoconductivity, as we may see from curve (b). This should not be surprising if the SPP peak is the resultant of two opposing effects, i.e., the outcome of differential desorption, whereas the photocurrent depends on the sum of the photo carriers released from both surfaces.

Fig. 4-a gives the wavelength dependence of SPP, no account being taken of the change of the incident energy at various wavelengths. Fig. 4-b represents the spectral response at constant incident energy of $15\mu W/cm^2$.

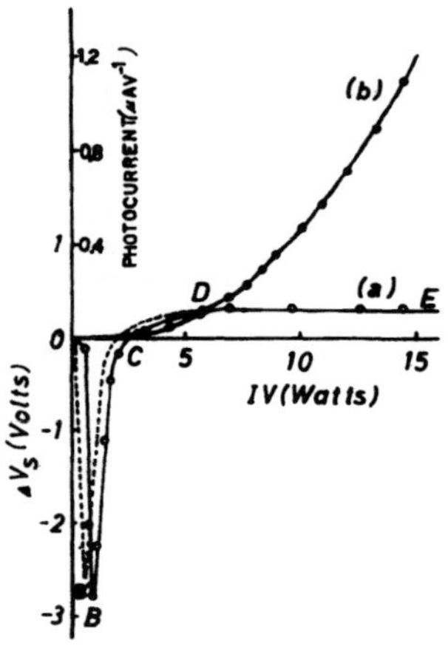

Fig. 2. (a) Surface photopotential in an alternating electric field using white light (full spectrum) from a tungsten lamp operating at different powers. (Crystal IBX3).

(b) DC photoconductivity under irradiation with white light (full spectrum) from a tungsten lamp operating at various ratings. (Crystal IBX3).

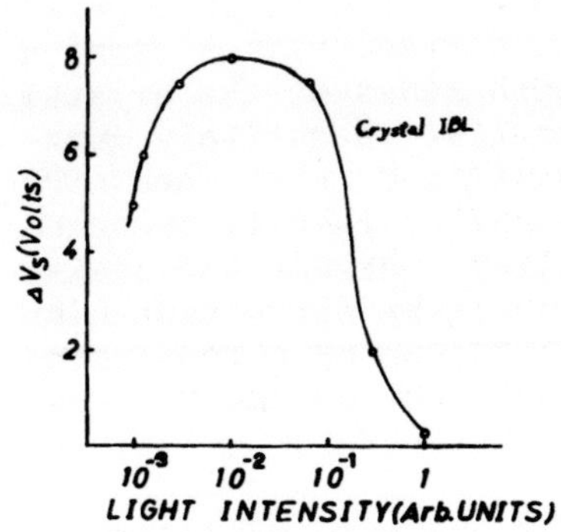

Fig. 3-a Actual intensity dependence using a fixed electric power in the tungsten lamp. The intensity is changed by a neutral wedge. (Crystal IBL).

Fig. 3-b Intensity dependence of SPP using monochromatic light. The abcissa represents the power in the monochromator tungsten lamp. For λ = 6000 Å, the curve between IV = 5 W to IV = 500 W covers a photometric intensity range of 4 decades. The incident energy range is ∿1 to 10^4 μ W/cm^2. (Crystal IBX3).

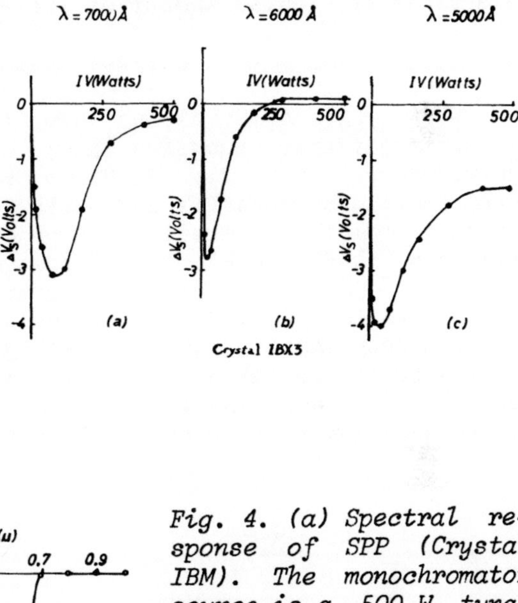

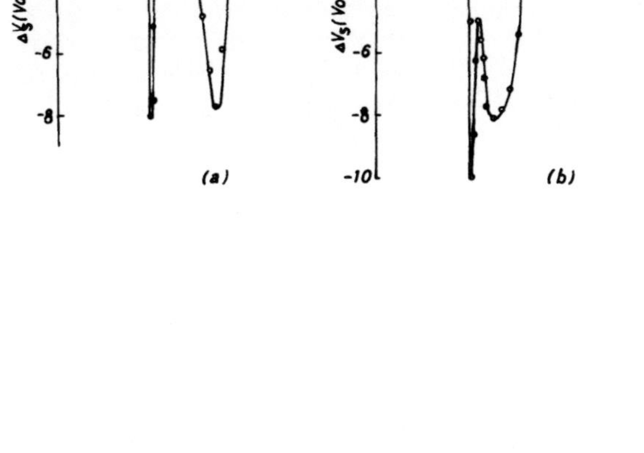

Fig. 4. (a) Spectral response of SPP (Crystal IBM). The monochromator source is a 500 W tungsten lamp. Curve is not corrected for energy distribution.

(b) Spectral response at constant incident energy of 15 μ W/cm^2. (Crystal IBM).

Quenching

The SPP induced by white light in an AC field is in general partially quenched at high intensities when surface states forming the rectifying barriers are removed from both sides of the crystal. At the point C, Fig. 2-a, the surface potential is entirely quenched. The authors (9) have earlier demonstrated that when SPP displays two peaks of the same sign, both may be readily quenched by a yellow range of wavelengths in a valley between the peaks. Fig. 5 gives two further examples of quenching by a constant bias light. The process is regarded as a manifestation of photodesorption with a consequent reduction of barrier height on the two surfaces.

Optical quenching is also possible by the mere variation of intensity keeping the wavelength constant. Evidence for this statement is found in Fig. 3-b at λ = 6000 Å, at the point where ΔV_S changes sign. In this case, it is postulated that light is able to destroy equal number of states on both surfaces.

Optical Hysteresis

When cyclic observations are made of SPP against the rating of the irradiating tungsten lamp, optical hysteresis was noticed (dotted curve, Fig. 2). Findings of a similar nature have also been reported by Schaaffs and Woelk (3) but were not accounted for. Starting from low intensities, then in the opposite direction, the peak at B shifts to B', i.e., towards the direction of lower intensities. This behavior is apparently related to the hysteresis in ΔV_S manifested by CdS under cyclic monochromatic illumination from a monochromator with a wavelength sweep (9), a red shift in the spectrum then taking place as the wavelength traverse is made from short to long waves instead of *vice versa*. This hysteresis is probably connected with the long time constant of the process.

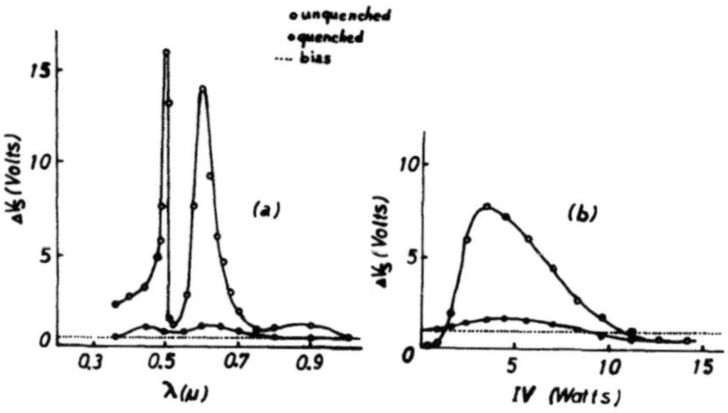

Fig. 5. (a) Quenching of SPP spectrum by strong white light.

(b) Quenching of surface potential due to white light by monochromatic illumination (λ = 5200 Å). (Crystal IBL).

Growth and Decay Characteristics

The growth and decay curves of SPP (Fig. 6) exhibit interesting features, for white light illumination with a continuous spectrum. At small ratings of the tungsten lamp, more specifically, for illuminations by intensities below that corresponding to the peak (Fig. 2, curve a), the time dependence of SPP is almost exponential, with a time constant for growth of the order of a fraction of one minute. This constant decreases markedly with increase of rating IV. The time constant for decay is usually several times longer. At higher ratings, the curves assume a complex character; a hump suggests that they arise from two competing tendencies, one of which predominates at low intensities. With monochromatic illumination, especially by weakly absorbed light, the behavior is essentially unaltered.

DISCUSSION OF RESULTS

The intensity dependence curve of SPP (see, for example, Fig. 3-b curve b) can be analysed into two components acting in opposite directions. It may thus be assumed that at equilibrium

$$\Delta V_s = V_1(1 - e^{-I/I_1}) - V_2(1 - e^{-I/I_2}) \tag{1}$$

where I is the intensity of the incident light and V_1, V_2, I_1, I_2 are constants depending on the wavelength for a given specimen.

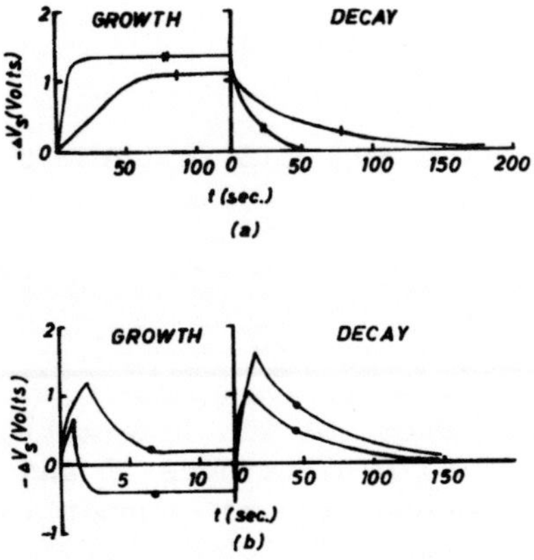

Fig. 6. Growth and decay characteristics using white light:

a- weak excitation corresponding to the stage OB of Fig. 2 (a). The sharper curves (||), correspond to higher intensity.

b- strong excitation:
- *just below the quenching intensity at C.*
- *at saturation (Crystal IBX3).*

A negative peak at low intensity and saturation with a positive value at high intensity may be explained if

$$V_1 > V_2 \quad \text{and} \quad I_1 > I_2$$

The inequality signs have to be reversed if the polarity of the surface potential is reversed.

Further, at low intensities where the effect of the first term in (1) may be neglected, the growth and decay curves are almost exponential (Fig. 6,a). At much higher intensities, however, both the growth and decay curves (Fig. 6,b) show a maximum indicating again the superposition of opposing tendencies. The build-up of SPP at constant intensity can be regarded as arising from two competing processes associated with the two surfaces and acting simultaneously. For inscance, one process may lead to a positive photopotential of the type

$$+ V_1' (1 - e^{-t/\tau_1})$$

and the other to a negative photopotential of the type

$$- V_2' (1 - e^{-t/\tau_2})$$

The photopotential at any time during growth is thus

$$\Delta V_s(t) = V_1' (1 - e^{-t/\tau_1}) - V_2' (1 - e^{-t/\tau_2}) \qquad (2)$$

where τ_1 and τ_2 are the time constants of the two component processes during growth for a specified radiation. Consequently, the resultant growth shows a maximum if

$$V_1' > V_2' \quad \text{and} \quad \tau_1 > \tau_2.$$

The initial rise to a maximum during decay suggests that the observed decay may be represented by

$$\Delta V_s(t) = V_1' e^{-t/\tau_1'} - V_2' e^{-t/\tau_2'} \qquad (3)$$

where $\tau_1' > \tau_2'$

It is easy to demonstrate the importance of the interaction at the rear using a crystal S (Fig. 1) with electrodes Q and B arranged as in the figure, S being attached to the supporting electrode Q by Aquadag (carbon dag). It has been found that full illumination (shield C removed) may give an SPP spectrum with two negative peaks (cf. Fig. 4). The negative sign of the long wave peak may be explained if we assume that at the crystal - metal contact, the density of surface states is larger than that at the free front surface. Thus frontal illumination (even allowing for reduced intensity at the rear by bulk absorption) may cause more desorption at the rear than at the front, and the net voltage rectification imparts a negative polarity to the front where photo-interactions are less favourable.

The negative sign of the short wave peak may be interpreted in a different way, since now lowering of the barrier height at the rear is not due to the direct influence of light which is strongly absorbed at the front surface and so cannot be transmitted through the bulk. Short wave radiation produces electron-hole pairs from the main lattice, and partially eliminates the few states at the front by the union of holes with the negative surface charge. It is likely that the photo electrons have in pure CdS crystals a range of the order of a fraction of a mm, which is comparable with the crystal thickness. These electrons can, therefore, profoundly affect the positive space charge at the rear, to the extent that will make the front surface voltage negative, i.e., there is still predominance of interaction at the rear due to diffusion and drift of photoelectrons. The drift is largely due to the field set up by the rectified voltage at the back barrier. According to Rittner (12), drift is unimportant in an intrinsic semiconductor.

Effective screening of the lower end of S from the action of light would make the distance too long for photoelectrons to reach the contact, and so SPP would be positive and arising entirely from the destruction of states at the front, and rectification at the rear.

By properly supporting the sample and illuminating it, the study can be confined practically to the front surface. It is interesting to note that on raising the light shield C gradually upwards, the spectrum changes sign from negative to positive (cf. Fig. 7). This is expected to happen when the distance L in the crystal between the top of C and the top of Q (Fig. 1) is comparable with the range of electrons in the specimen. This appears to suggest the principle of a direct method of measuring the carrier range. It follows that a convenient study of the interaction of light with the front surface of a crystal may be realized by irradiating the surface of a thick crystal with monochromatic light whose wavelength lies in the absorption region. This radiation is highly absorbed at the front and so cannot directly affect the rear. Fig. 7 gives $\Delta V_S(I)$, the intensity dependence of SPP (a) without using a light screen C, (b) with ample screening. Sign reversal is a beautiful demonstration of the indirect action of light on the contact. Transition from curve (a) to curve (b) is attended by a change of the growth-decay characteristics from the humped form to the simple exponential form.

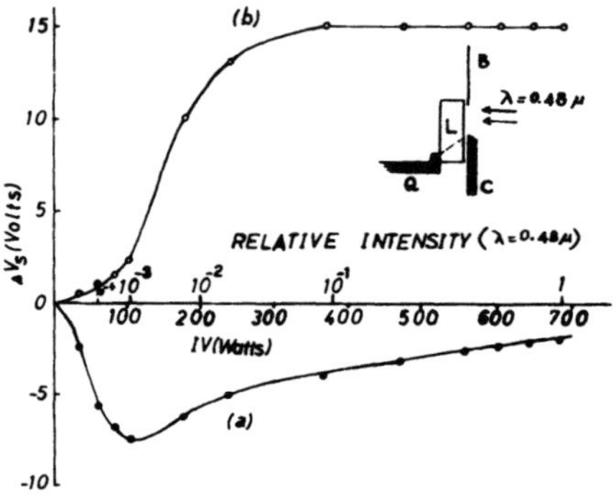

Fig. 7. Intensity dependence of SPP: (Crystal IBL).

a- When the two surfaces interact with the incident radiation.

b- When interaction is restricted to the front surface.

It is probable that for adequate screening, the shape of the curve would be approximated by a relation of the form,

$$\Delta V_s = V\ (1 - e^{-aI}) \tag{4}$$

where V depends on the AC voltage redistribution following the change of illumination.

Equation (4) may be regarded as representing the decrease by light of the rectification at the front from that at the rear. Saturation suggests complete desorption, but it is another problem to correlate (4) with the number of desorbed atoms.

REFERENCES

1. *W.H. Brattain, "Semiconducting Materials", p. 37, Ed. H.K. Henisch (London: Butterworths 1951).*

2. *G. Wlerick, Physica 20, 1099 - 102 (1954).*

3. *W. Schaaffs, and H. Woelk, Z. angew. Phys. 10, 456 - 8 (1958).*

4. *L.P. Strakhov, Fiz. Tver. Tela 1, 583 - 8 (1959); Soviet Phys. Solid State 3, 1169 - 70 (1961); 4, 2505 - 8 (1963).*

5. *I.A. Akimov, Sov. Phys.-Doklady 4, 1046 (1959 - 60).*

6. *R. Williams, J. Phys. Chem. Solids 23, 1057 (1962).*

7. *G. Nazhdakov, and S. Balbanov, Fiz. Tver. Tela 7, 1193 (1965).*

8. *I.A. Akimov, A.M. Meshkov, and A.N. Ternin, Phys. Stat. Sol. 14, 135 (1966).*

9. *Y.L. Yousef, et al. To be published, Brit. J. Appl. Phys. (1966)*

10. *N.F. Mott, Proc. Roy. Soc., Lond. 171, 27 (1939).*
W. Schottky, Z. Phys. 118, 539 (1942).
J. Bardeen, Phys. Rev. 71, 717 (1947).

11. *Y.L. Yousef, et al. J. Sci. Instrum. 42, 873 - 5 (1965).*

12. *E.S. Rittner, Photoconductivity Conference, Atlantic City 1954 p. 215, Ed. R.G. Breckenridge (New York: Wiley, 1956).*

DISCUSSIONS

J.H. CRAWFORD (Oak Ridge National Laboratory, U.S.A.): *Was an affect due to various surface treatments on the spectral sensitivity of the photopotential observed?*

Y.L. YOUSEF: *The spectral response of surface photopotential is characteristic of the band structure, and its general shape is little affected by surface treatment, although fine details may be.*

THE INFLUENCE OF NEUTRON IRRADIATION ON THE THERMOELECTRIC POWER AND THE ELECTRICAL CONDUCTIVITY OF NEARLY STOICHIOMETRIC SINGLE CRYSTALS OF URANIUM DIOXIDE.

R. De Coninck, J. Devreese and P. Nagels

Solid State Physics Department

SCK-CEN, MOL, BELGIUM

Nearly stoichiometric uranium dioxide single crystals have been irradiated by different doses of thermal neutrons.

Before and after irradiation, the electrical conductivity (σ) and the thermoelectric power (S) were measured as a function of temperature. These two properties provide complementary information about neutron irradiation damage of uranium dioxide.

The effect of a small dose on S is most pronounced at low temperature, while a larger dose also causes a decrease of S at higher temperatures. The electrical conductivity also decreases upon neutron irradiation and its activation energy increases. These effects on the electrical conductivity were observed earlier and attributed to a decrease in the number of holes (p) in UO_{2+x}. *However, a decrease in p only would imply an increase of S, whereas a decrease was observed. Therefore, it was necessary to introduce a further assumption.*

The bending of the $\ln \sigma T^{3/2}$ *versus* $1/T$ *curves at low temperature suggests the possibility of n-type impurity conduction. This hypothesis could not be confirmed from thermoelectric power measurements on unirradiated crystals because of experimental difficulties. However, it was supported by an analysis of* S_t *and σ as functions of temperature after neutron irradiation. A maximum in* S_t *appears after irradiation. This maximum shifts to higher temperatures with increasing dose because the irradiation causes an increase in the number of compensating donors (hole-traps) and so enhances the effect of the "impurity conduction".*

INTRODUCTION

In the studies published before 1961, on the conduction mechanism of uranium dioxide, this material is treated as a normal band-type semiconductor. In 1961 Aronson, Rulli and Schaner (1) proposed another mechanism, based on the ideas of Heikes and Johnston (2). They suggest that the electronic conduction in UO_{2+x} is due to a thermally activated jumping of holes from U^{5+} to U^{4+} ions, the U^{5+} ions initially arising at O^{--} interstitials. This conclusion was mainly deduced from measurements of the electrical conductivity.

All investigations before 1962 were performed on sintered or pressed polycrystalline material.

In 1962 single crystals of UO_2 were prepared by Van Lierde et al. (3). Nagels, Devreese and Denayer (4) studied the electronic conduction of these single crystals. Measurements of both the electrical conductivity and the Hall mobility for which an upper limit was given, provided some confirmation of Aronson's conclusions.

A better understanding of the conduction mechanism could be obtained after measurements of the thermoelectric power of UO_2 by R. De Coninck (5), as these measurements provide an upper limit for the activation energy of the number of carriers.

A recent paper by Devreese, De Coninck and Pollak (6), analyzing results on both thermoelectric power and electrical conductivity of UO_2, supports the idea of the small polaron character of the carriers in a quantitative way. Therefore the hopping type behavior (typical for small polarons, at relatively high temperatures) of the charge carriers in UO_{2+x} can now be regarded as a valid assumption.

The study of the influence of neutron irradiation on the thermoelectric power and the electrical conductivity might provide further insight into the conduction mechanism for UO_2.

Nagels et al. (7) studied the behavior of the electrical conductivity after neutron irradiation. The creation of hole traps was suggested as a consequence of irradiation. Annealing experiments led to the supposition that the defect centers which act as hole traps are uranium vacancies.

Again a combined investigation of the variation of the thermoelectric power as well as the electrical conductivity could give more information. The present paper deals with such an investigation; measurements of the thermoelectric power and the electrical conductivity have been performed on the same crystals, before and after irradiation.

Rather surprising results were found after irradiation. It was possible to give an explanation by considering the role played by impurity conduction. On the bases of electrical conductivity measurements alone, "impurity conduction" in unirradiated UO_2 was suggested by Acket (8). (The term "impurity" will also be used in this paper when referring to uranium vacancies and other defects. One could perhaps speak here about "structural defect conduction"). However, "impurity conduction" could not be confirmed by our thermoelectric power measurements due to the high resistance of the samples at low temperatures. After irradiation, the effect of impurity conduction is shifted to higher temperatures and becomes apparent both in the electrical conductivity and in the thermoelectric power measurements.

EXPERIMENTAL

UO_2 is a material which oxidizes easily to UO_{2+x} or other higher oxides. The O/U ratio has an important influence on most properties. Therefore measurements here were limited to nearly stoichiometric crystals. These crystals offer the advantage that easily reproducible measurements can be made on them. Oxidation was avoided by taking all measurements under vacuum (10^{-5} to 10^{-6} mm Hg) and using a temperature not higher than 50°C. The low temperatures used were governed by the resistance of the sample (5×10^9 ohm).

All measurements were carried out in an all metal cryostat cooled by liquid nitrogen. The sample holder (Fig. 1) was placed in the central tube of the cryostat. It consists of two cylinders which can be heated or cooled differentially. The sample was pressed by the spring loaded contacts of two chromel P-alumel thermocouples against two copper plates insulated from each other. These are direct pressure contacts which were found to be ohmic, while the copper plates were in good thermal contact with the two cylinders.

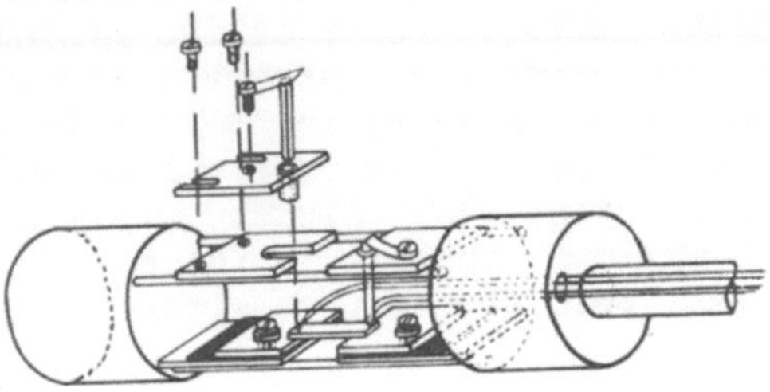

Fig. 1. Sample holder for measurement of thermoelectric power and electrical conductivity.

The technique adopted for measuring the thermoelectric power consists of slowly varying the average absolute temperature of the sample and maintaining the temperature difference as constant as possible.

A switch system permits measurement of the two thermocouples separately and of their difference by a Tinsley type potentiometer; of the thermoelectric voltage by a Wayne Kerr electrometer type M141; and of the resistance by a Lemouzy electrometer-ohmeter.

The reproducibility and precision of these measurements reaches 2 to 3%. The thermocouples were calibrated against tables of standard chromelalumel thermocouples (9). Corrections for the measuring wires were taken into consideration when calculating the absolute thermoelectric power.

To obtain nearly stoichiometric crystals, reduction with pure hydrogen under a pressure of 10 to 20 cm Hg was carried out in an UO_2 crucible placed in an HF furnace. The temperature used for the experiment was 1400°C for periods varying between 10 and 20 minutes.

The samples were irradiated in evacuated quartz capsules in the BR1 reactor (10). Integrated thermal neutron dose measurements were made by Co detectors adjacent to the crystal.

RESULTS AND DISCUSSION

Figures 2 and 3 show the thermoelectric power and electrical conductivity as a function of reciprocal temperature for a nearly stoichiometric UO_2 crystal before irradiation and after successive thermal neutron exposures. The measurements of the electrical conductivity could be taken at lower temperatures than the thermoelectric power measurements.

Figures 4 and 5 represent the same type of measurements for another UO_2 reduced crystal.

It can be noticed that before irradiation, a linear relationship was obtained when the thermoelectric power of the specimens was plotted versus 1/T. The corresponding activation energy deduced from the slope of these lines is equal to 0.057 and 0.072 eV. After short irradiations the thermoelectric power is decreased, especially at the lower temperatures. For higher exposures a larger decrease is found and a maximum appears in the curve; the low temperature part showing less, zero or negative slope.

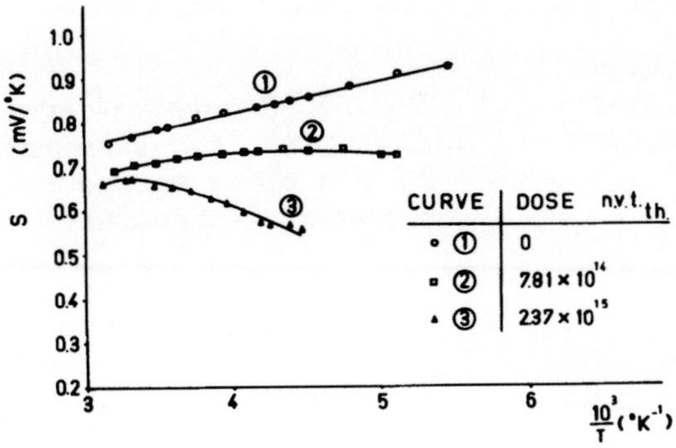

Fig. 2. Thermoelectric power as a function of reciprocal temperature for a nearly stoichiometric UO_2 crystal (specimen I) before irradiation and after successive thermal neutron exposures.

Fig. 3. Electrical conductivity of specimen I (ln $\sigma T^{3/2}$ versus reciprocal temperature).

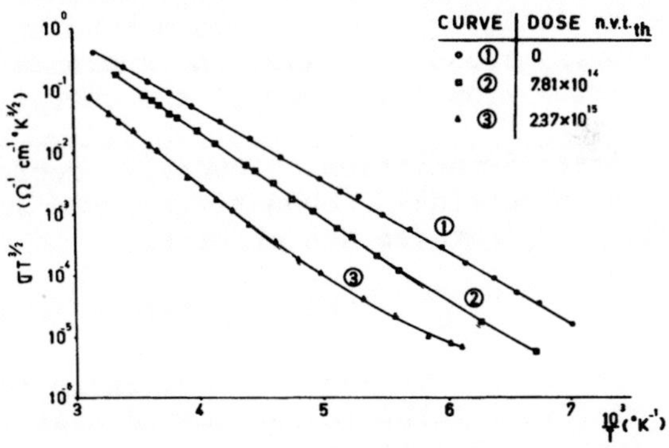

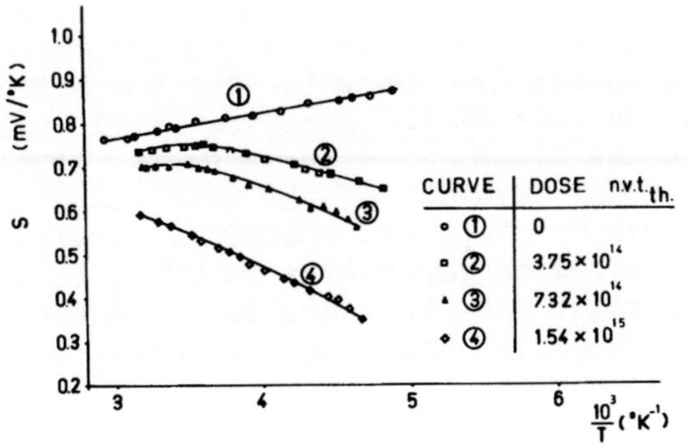

Fig. 4. Thermoelectric power as a function of reciprocal temperature for a nearly stoichiometric UO_2 crystal (specimen II) before irradiation and after successive thermal neutron exposures.

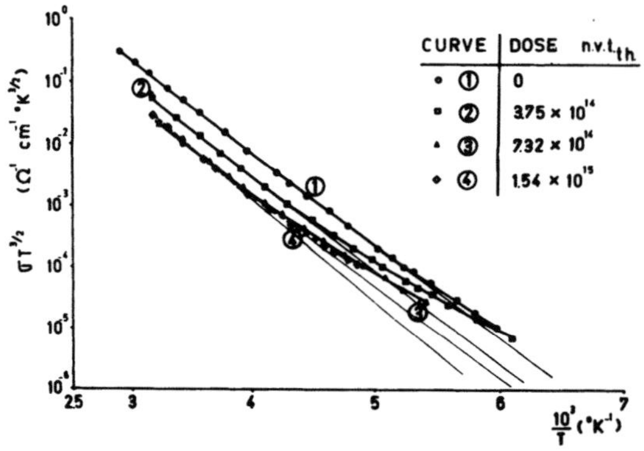

Fig. 5. Electrical conductivity of specimen II (ln $\sigma T^{3/2}$ versus reciprocal temperature).

Before irradiation, the electrical conductivity versus 1/T plot is a straight line over 5 decades in the higher temperature region. At lower temperatures a bending to higher values is (nearly always) found. Activation energies of 0.296 and 0.229 eV were calculated from the slope of the curves.

After irradiation an overall decrease of the absolute values of the electrical conductivity and an increase in the slope at higher temperatures was observed.

Behavior Before Irradiation

The ln σT versus 1/T curves in (4) showed a bending at low temperatures. Acket (8) suggested that this bending is due to conduction via the acceptor centers, a mechanism described by Mott and Twose (11). Let us consider the energy level scheme in Fig. 6: In nearly stoichiometric UO_{2+x} one observes, at relatively high temperatures, p-type conduction dominated by the holes originating from the neighbourhood of O^{--} interstitials (acceptors). At low temperature, n-type conduction in the acceptor level can become important if some compensating donors are present. The holes will fill up all the donors at T=o and some of the acceptor levels will be empty. Electrons will be transferred from those empty levels to the occupied levels if a small overlap of the wave function exists. This conduction is also by hopping. Chemical analysis shows that the presence of compensating impurities in UO_{2+x} is not probable. However, hole-traps might play the same role. We therefore suggest the presence of a number of hole traps before irradiation.

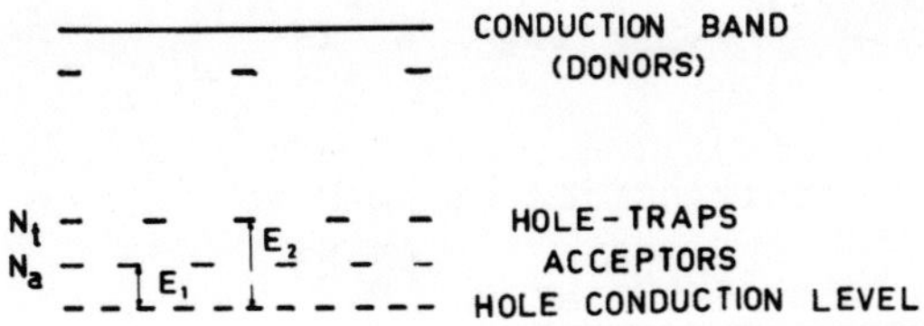

Fig. 6. Energy level scheme for nearly stoichiometric UO_2.

The following conduction mechanism before irradiation is postulated: at high temperatures, conduction occurs by the hopping of polarons. The activation energy for conduction is $(E_1 + E_p^{\mu})$ where E_1 is the separation between the acceptor level and the hole conduction level (~ 0.07 eV (6)), and E_p^{μ} is the activation energy for the drift mobility (~ 0.2 eV (6)).

At low temperatures, the holes "fall" back into the acceptor level except for about N_t which fall into the hole traps. In the acceptor level n-type conduction is then observed:

$$\sigma_n = e\, N_t \left[1 - e^{-(E_2 - E_1)/kT}\right] \mu_o^n\, e^{-E_n^{\mu}/kT} \tag{1}$$

E_n^{μ} is the activation energy for mobility in the acceptor level (~ 0.06 eV). This explains the observed bending at low temperatures, starting approximately at $10^3/T = 5.5$, before irradiation. E_2, the separation between the hole-trap level and the hole conduction level, is assumed to be larger than E_1. This will be confirmed experimentally.

The thermoelectric power will increase with decreasing temperature because of the decrease in the number of p-type carriers. However, a maximum is to be expected as the temperature decreases further because of the increasing importance of the n-type impurity conduction. This maximum was not reached experimentally because of the difficulty of measuring the thermoelectric power at sufficiently low temperatures.

Effect of Irradiation

The main result of irradiation is to increase the number of hole traps. It will be assumed that the mobilities are not noticeably affected.

The effect of the electrical conductivity is easy to understand: in the ln $\sigma T^{3/2}$ versus 1/T diagram (12), the curve can be decomposed into two straight lines obeying the relation

$$\sigma T^{3/2} = N_a\, e\, e^{-E_1/kT}\, \mu_o^p\, e^{-E_p^\mu/kT} + N_t e\left[1-e^{-(E_2-E_1)/kT}\right]\mu_o^n e^{-E_n^\mu/kT} \quad (2)$$

After irradiation the p-type contribution changes to

$$N_\alpha e\, e^{-E_1/kT}\, \mu_o^p e^{-E_p^\mu/kT} + N_\beta\, e\, e^{-E_2/kT}\, \mu_o^p\, e^{-E_p^\mu/kT} \quad (3)$$

If the creation of hole traps does not affect the number of acceptors we expect $N_\alpha + N_\beta = N_a$. From this picture it can be concluded that the effect of irradiation is an increase of the "activation energy" in the high temperature range, an effect which has been observed. The low temperature part of the electrical conductivity curve (the second term in (2)) has a slope E_n^μ and tends to zero for high temperatures. The fact that the second term of (2) results in a straight line at low T provides evidence for the thermal activation of the mobility in the "impurity" region. (See Figs. 7 and 8).

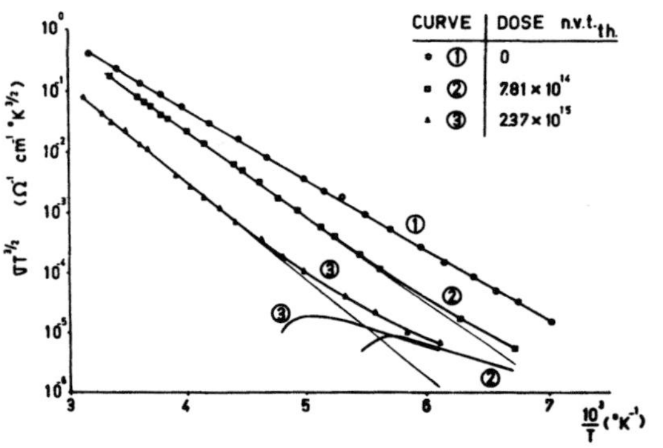

Fig. 7. Electrical conductivity curve of specimen I divided into a p-type and an n-type part.

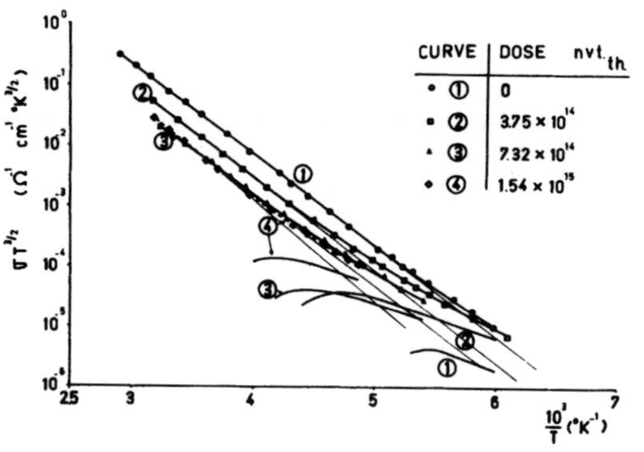

Fig. 8. Electrical conductivity curve of specimen II divided into a p-type and an n-type part.

After irradiation, as N_t increases, the low temperature part of (2) increases. This effect is partly confirmed by the present measurements which also reveal that E_n^{μ} is not very sensitive to irradiation.

From the foregoing discussion it follows that the intersection point of the p-type and n-type lines shifts to larger values of both temperature and electrical conductivity after irradiation. This is also experimentally observed.

In the preceding sections, it was assumed that $N_t < N_a$. This is perhaps not true for higher doses. When $N_t > N_a$, n-type "impurity conduction" will occur in the "hole trap band" and not in the acceptor band. This offers an interesting possibility, as it corresponds to a drastic change in the conduction mechanism. However, this effect will not be observed for the electrical conductivity if the activation energy for the mobility is about the same in the "hole-trap band" as in the acceptor band.

The thermoelectric power of a single conductor which has two independent groups of charge carriers, e.g. p-type and n-type, with conductivities σ_p and σ_n respectively, and such that each group, if acting alone, would have a thermoelectric power S_p and S_n respectively is (see e.g. Putley (13)):

$$S_t = \frac{S_p\sigma_p + S_n\sigma_n}{\sigma_p + \sigma_n} \quad (4)$$

with

$$S_p = \frac{k}{e} \left(A - \ln \frac{p}{P}\right) \quad (5)$$

$$S_n = -\frac{k}{e} \left(B - \ln \frac{n}{N}\right) \quad (6)$$

$$\sigma_p = pe\,\mu_p \quad (7)$$

$$\sigma_n = ne\,\mu_n \quad (8)$$

p and n are the numbers of holes and electrons, P and N the corresponding density of states, and A and B are constants connected with the kinetic energy of the charge carriers.

At low temperatures some of the holes are trapped after neutron irradiation and the "impurity" conduction becomes relatively more important. As the impurity conduction is n-type, the thermoelectric power tends to lower values. The maximum in S_t will shift to higher temperatures: indeed the temperature at which it occurs is related to the temperature at which $\sigma_n = \sigma_p$. As was shown above, the effect of irradiation realizes this equality at higher temperatures. This explains why a maximum is observed after irradiation.

In the high temperature region one would expect an increase of S_t because of the decrease in the number of p-type charge carriers. However, it turns out that the influence of the n-type conduction of S_t is not negligible even at relatively high temperatures.

Consider the change of S_t at a fixed temperature if a hole is trapped

$$\frac{\partial S_t}{\partial(-p)} = \frac{e\mu_p}{\sigma_t} \left[(S^* - A) - (E^p_{irr}/kT - 1) \right] \tag{9}$$

with $S^* = \frac{S_t}{k/e}$ and E^p_{irr} is the activation energy for the number of p-type carriers after irradiation.

If the data from Fig. 3 and Fig. 5 are introduced in equation (9) a negative value of $\frac{\partial S_t}{\partial(-p)}$ is found. This shows that the n-type carriers still have an influence on S_t after irradiation.

CONCLUSION

The bending of the ln $\sigma T^{3/2}$ versus 1/T curves at low temperature suggests the possibility of n-type "impurity" conduction (8). This hypothesis could not be confirmed from thermoelectric power measurements on unirradiated crystals, because the apparatus used does not permit this measurement for nearly stoichiometric UO_2 at sufficiently low temperatures.

In the present paper this hypothesis is confirmed by an analysis of S_t and σ as a function of temperature after neutron irradiation. After irradiation a maximum in S_t appears. This is characteristic for impurity conduction. This maximum shifts to higher temperatures because the irradiation causes an increase of the number of compensating donors (hole-traps). This results in a decrease in the number of holes and enhances the effect of the "impurity conduction". By irradiation the maximum was shifted from an inaccessible to an accessible temperature region. An interesting fact is the sensitivity of the thermoelectric power to the n-type carriers even at higher temperatures. Finally it is not certain that the compensating donors are the same before and after irradiation.

ACKNOWLEDGMENTS

The authors are grateful to the European Atomic Energy Community and to the SCK-CEN for permission to publish this paper. The authors wish to thank Mr. A. Gijs for his help in the construction of the apparatus and in performing the measurements. Thanks are also due to Dr. W. Van Lierde for supplying the single crystals.

REFERENCES

1. *S. Aronson, J.E. Rulli and B.E. Schaner, J. Chem. Phys. 35, 4, 1382 - 1388 (1961).*

2. *R.R. Heikes and W.D. Johnston, J. Chem. Phys. 26, 3, 582 - 587 (1957).*

3. *W. Van Lierde, R.Strumane, E. Smets and S. Amelincks, J. Nucl. Mat. 5, 2, 250 - 253 (1962).*

4. *P. Nagels, J. Devreese and M. Denayer, J. Appl. Phys. 35, 4, 1175 - 1180 (1964).*

5. *R. De Coninck in EURAEC 1476, p. 13 (1965).*

6. *J. Devreese, R. De Coninck and H. Pollak, "On the Conduction Mechanism in Uranium Dioxide", to be published, Phys. Stat. Sol. 2, 17 (1966).*

7. *P. Nagels, Proceedings of the Third International Conference on the Peaceful Uses of Atomic Energy, 11, p. 353 - 361 Geneva 1964.*

8. *G.A. Acket, "On the Conduction Mechanism of Some Transition Metal Oxides". Thesis, University of Utrecht, 1965. (and priv. Comm.).*

9. *R.K. Adams and E.G. Davisson, ORNL - 3649, 2, section 2,3, Chromelalumel Thermocouples (1965).*

10. *R. De Coninck and M. Heerschap in "Experience with Irradiation Facilities for Solid State Experiments in BR1. Section 2. Furnace". Communication presented at Colloque de la Societe Europeene d'Energie Atomique-Grenoble (France) 25-27 Juin, 1962.*

11. *N.F. Mott and W.D. Twose, Advances in Physics, 10, 107-163 (1961).*

12. *I.G. Lang and Yu.A. Firsov, Sov. Phys. Jetp., 16, 5, 1301-1312 (1963).*

13. *E.H. Putley, "The Hall Effect and Related Phenomena", p. 88, Butterworths, London (1960).*

DISCUSSIONS

CHARLES GREENE (Alfred University, U.S.A.): *Why does the effect of neutron irradiation on the conductivity of one specimen appears to saturate at about* $8x10^{14}$ *neutrons/cm*^{2}*.sec while in the other specimen the effect increases up to* $25x10^{14}$ *nvt?*

R. DE CONINCK: *We assumed that* $N_e < N_a$. *For higher doses, n-type impurity conduction can occur in the hole-trap band instead of in the acceptor band; in that case* $N_t > N_a$. *For still higher doses it is possible that a conversion to p-type conduction will be found in the hole-trap level.*

J.H. CRAWFORD (Oak Ridge National Laboratory, U.S.A.): *Is excess oxygen incorporated into* UO_2 *interstitially or accompanied by uranium vacancies?*

R. DE CONINCK: *Only interstitially as far as we know.*

I. EL SHANSHOURY (U.A.R. Atomic Energy Establishment): *Do you expect fission fragments to play a role in the electrical conductivity and thermoelectric power?*

R. DE CONINCK: *Fission fragments are formed in* UO_2 *by thermal neutron irradiation. Along the fission tracks the damage is of course very high.*

C. ERGINSOY (Brookhaven National Laboratory, New York): *The low temperature range you investigated appears to be about 150°K. Presumably other effects could exist at lower temperatures.*

R. DE CONINCK: *The described effects will probably extend to temperatures which are e.g. 50°K lower. At still lower temperatures we do not know what occurs.*

C. ERGINSOY: *Did you look at the Hall effect?*

R. DE CONINCK: *Measurements of the Hall effect are difficult on account of the high resistivity of the samples and the low Hall coefficient. An upper limit of about 0.015* $cm^2\ V^{-1}\ sec^{-1}$ *was deduced.*

RADIATION EFFECTS ON COMPLEX GLASSES

E. Lell and N.J. Kreidl

University of Missouri at Rolla

Rolla, Missouri, U.S.A.

While the complexity and specificity of radiation induced centers in complex glasses -- more than 50 have been described in detail in the literature -- is not surprising because of the variability in the environment in vitreous substances, it has become increasingly possible to characterize several centers in a manner to elucidate principles of structures and radiation effects. The role of non-bridging oxygen and of the systematic variation of alkali near these oxygens in electron trap centers as well as a series of specific centers around technologically important additives is now capable of reasonable interpretation. Special reference is given to the different centers induced in phosphate glasses containing silver.

INTRODUCTION

This broad review refers to that published in "Progress in Ceramic Science" (1) which is in turn a revision of an earlier review by the present authors (2). At the present pace of progress, this material in turn will soon require substantial revision. The following text merely records remarks made at the Cairo Solid State Conference regarding revision of concepts, new and future experiments in the field covered by this review.

Work on the effects of high energy radiation on complex glasses so far has yielded considerable information about percursor centers as well as radiation-induced centers.

At the present time; it appears that this increase in information has, with some elegant generalizations, raised many new problems, the evidence of definite complexity or specificity - an experience not at all an unusual occurrence in many fields of science.

In the past, we have tried to study, in detail, pure and doped silica in the hope of obtaining clues on silicate, borate and phosphate glasses of complex but industrially applicable composition. It turned out that silica behaves quite differently; a characteristic consequence, as R. Weeks (3) has shown conclusively, of the precise nature and concentration of certain defects in silica. Conversely, as was suggested by Stevels (4) and Bishay (5), as well as a few others, the "non-bridging oxygen" is the defect which dominates the nature of radiation-induced centers in complex glasses. Tucker (6) and others have quantitatively elaborated the hole character of the most characteristic defect induced in complex oxide glasses. The general similarity between these centers in many alkali containing oxide glasses as well as the sensitivity or insensitivity of the centers to the amount and kind of alkali have suggested their generally accepted assignment to non-bridging oxygen and neighboring alkali regions.

SILVER IN OXIDE GLASSES

The unique electronic behavior of silver, which is mobile like alkali, but acts as an electron or hole trap like transition elements, has further elucidated the complexity of these centers, and again introduced considerable difficulties in interpretation (7).

Transparent solid materials which contain silver ions in their structure normally show absorption bands which are not present in the pure material. Also, an ultraviolet excited fluorescence is frequently observed, which must be associated with the silver ions since it is absent from the silver-free material. When the silver-containing solid is irradiated with X- or gamma rays, the resulting radiation-induced absorption is quite different from that induced in the material containing no silver. It is, therefore, apparent that silver ions strongly influence the electronic structure of the irradiated material and affect the distribution of electrons and holes at existing defect-trapping centers. This was shown to be true for silver containing alkali halides and glasses.

Experiments with electron spin resonance have shown that silver is able to trap both types of charge carrier, electrons as well as holes (7,8,9). In this respect, silver does not seem to be different from many other transition metal ions as, for instance, iron or cerium. However, while Ce^{3+} captures holes and Ce^{4+} captures electrons (10,11), and Fe^{2+} captures holes and Fe^{3+} captures electrons (13), silver (Ag^{+}) seems to be one of the few ions that is a capture center for electrons - becoming a neutral species, $Ag^{+,-}$, as well as for holes - becoming $Ag^{+,+}$.

EXPERIMENTAL

Relatively large amounts of silver can be introduced during the melting of phosphate glasses, up to an Ag/P ratio of 1, but it is difficult to melt silicate glasses containing appreciable amounts of silver. On the other hand, substantial amounts of silver may be diffused into silicate glasses. Glasses studied were of a sodium-aluminum-silicate type. Silver was diffused by treating the glass with molten AgCl at about 1000°F. In this case, silver presumably replaces sodium.

RESULTS

Luminescence

Phosphate Glasses. A number of luminescence emission bands are associated with silver and can be observed when excited with ultraviolet light. Some of these bands are present in the unexposed glass, others are induced by gamma-irradiation. Studying the growth and bleaching of the emission bands, their correlation to absorption bands, to the composition of the host glass and the silver concentration has brought some clarity into a confusing and complex picture of luminescence emission.

Some of the emission bands associated with silver in phosphate glass are shown in Fig. 1.

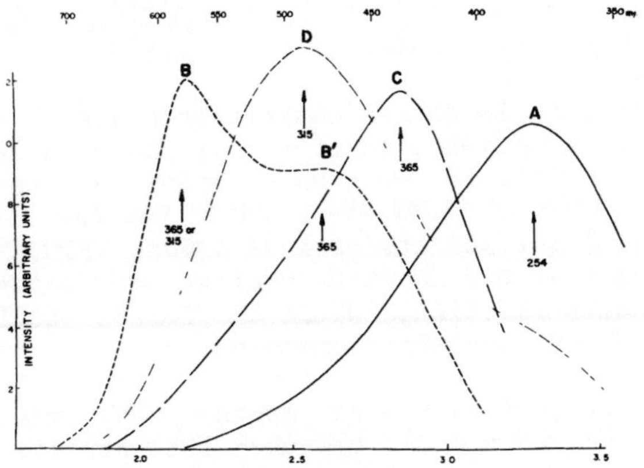

Fig. 1. Luminescence bands in silver activated phosphate glass. The number under the arrows indicates the excitation wavelength in mμ for each band.

The individual bands will be discussed subsequently:

1. AgA Centers.

The high energy side emission band (Fig. 1), denoted AgA, is not radiation induced. It is excited with 254 mμ Hg-light and its peak is found at 2900 A° (3.25 eV). This band does not seem to be affected by gamma-radiation or heat treatment. Since it is present in virtually every silver-containing glass and in some crystals, it is believed to be associated with the Ag^+ ion.

2. AgB Centers.

These centers are excited with 365 mμ and 315 mμ Hg-light, giving an emission peak around 5900 A° (2.1 eV). This band is induced readily by small doses of irradiation, and grows linearly with dose over a wide dose range. It is currently being used for dosimetry. Warming the glass to about 100°C - 200°C for 30 minutes increases the intensity sometimes up to 50%. The center begins to bleach notably on heating to 200°C. Also, exposing to high doses (above 10^6r) bleaches the center notably.

3. AgB^1 Center.

The AgB emission when excited by the 365 mμ Hg-light has a long tail on its short wavelength side which becomes a shoulder at higher doses. This emission band, denoted AgB^1, becomes a prominent band at doses above 2 x 10^6 rads when the AgB centers are radiation-bleached. Heating the sample to 150°C bleaches AgB^1 centers completely.

4. and 5. AgC and AgD Centers.

When a sample containing AgB centers is heated, the AgB centers bleach and two additional centers, not observed before, begin to grow. The AgC center is excited with 365 mμ light and has a peak emission at 4350 A° (2.85 eV), while the AgD center is excited with 315 mμ light and its peak is around 4850 A° (2.55 eV). It should be noted that the shorter excitation wavelength excites the longer wavelength emission, while the longer excitation wavelength excites shorter wavelength emission.

The AgC and AgD centers always appear to be present simultaneously. Heating to above 400°C bleaches both centers.

In addition to the four radiation-induced luminescence bands and the one band present before irradiation, there are two bands found in unexposed samples with emission peaks around 4400 A° and 4950 A°. The long wavelength tail of these bands interferes with the measurement of the orange AgB luminescence; these bands therefore are known as "predose" in dosimetry. Compared to the radiation-induced bands, the predose emission is weak but the wavelength is identical to the AgC and AgD emission.

Other properties such as, for instance, the bleaching of these bands under γ - irradiation at room temperature indicate that the predose luminescence and the radiation-induced AgC and AgD luminescence have their origin in identical centers.

It should be mentioned that the wavelength of the AgA and AgB emissions is essentially independent of the glass composition and melting schedule, while the emission wavelength and intensity ratio of the two predose peaks as well as the AgC and AgD emissions vary with composition and melting temperature. The figures given in this paper are for a glass of the composition 50 $Al(PO_3)_3$ 50 $LiPO_3$ 4 $AgPO_3$ (by weight).

The series in Fig. 2 shows the relationship between AgB and AgC centers. Fig. 2a shows the emission of a glass exposed to three different dose levels. The figure demonstrates the radiation bleaching of the AgB center at higher doses and the growth of the AgB^1 centers. Fig. 2b shows these glasses after heating to 120°C for 10 minutes. The AgB center concentration has increased while the AgB' centers are bleached. Fig. 2c shows the bands after heating to 250°C for 10 minutes. As the AgB centers bleach, the AgC centers grow. The ratio of AgB/AgC center concentration is the same for all three glasses and independent of dose. After heating to 300°C, the AgB centers have almost completely disappeared while the AgC centers have reached their maximum concentration (Fig. 2d). Heating to still higher temperatures will bleach AgC centers. However, if the glasses whose emission is shown in Fig. 2d are re-exposed, AgC centers are bleached completely while the AgB centers reappear as shown in Fig. 2e.

It is apparent from these figures that the intensity of the AgC band depends on the intensity of the AgB band rather than on the received dose. One may conclude, therefore, that AgC centers grow at the expense of the AgB centers and not from other centers which are not observed by luminescence.

A further observation, made also by other authors (7,13), is the dependence between silver concentration and the increase in the AgB band upon heating - that is, the difference between Figs. 2a and b. The lower the silver concentration, the larger is the increase of the AgB after heating to 150°C. A glass with high amounts of silver (30% $AgPO_3$) does not develop the AgB emission upon exposure but gives strong AgC and D luminescence.

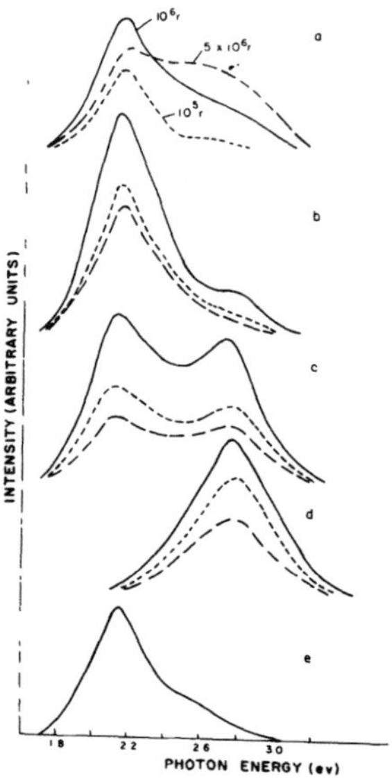

Fig. 2. (a) Luminescence of the B-band after three different doses,

(b) after heating to 100°C

(c) after heating to 250°C

(d) after heating to 300°C

(e) after reexposure to 10^6r.

Irradiation of glasses at elevated temperature yields AgC and AgD centers without any indication of AgB centers.

Silicate Glasses. The luminescence of silver diffused into silicate glasses differs from silver molten into phosphate glasses. The emission appears to have a continuum between 4400 - 6000 A° and the wavelength of the emission depends on the wavelength of the excitation. Four emission bands of this type of glass are shown in Fig. 3. Irradiation with X-rays affects these bands very little, but subsequent heating increases the luminescence intensity of each band without changing its wavelength. New bands have not been observed after heating.

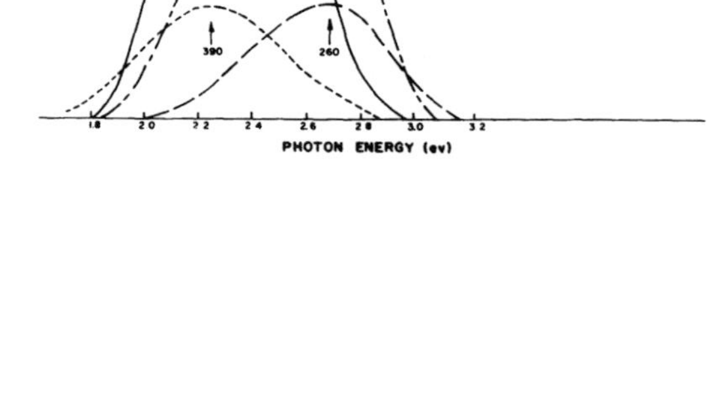

Fig. 3. Luminescence of silver diffused into a silicate glass. The number under the arrows indicates the excitation wavelength for each band.

ABSORPTION

Fig. 4 compares the radiation-induced absorption of the phosphate base glass with the silver-containing glass. This figure demonstrates most clearly that the silver ion captures charges before they can be trapped at network sites normally present in the glass. The band at 2.3 eV is now believed to be associated with a trapped hole (5,12). The absence of this band in the silver containing glass suggests that Ag^+ is an efficient hole trap.

Fig. 5 shows the radiation-induced absorption in a silver-containing phosphate glass exposed to different dose levels. The absorption band associated with the orange AgB emission is located at 3900 A° (3.2 eV).

Fig. 6 gives the radiation-induced absorption after several bleaching temperatures. The absorption curve obtained after heating to 300°C belongs to a glass that shows the strong blue AgC and AgD emission. It appears from this figure that the AgC luminescence is not associated with a new absorption band - a band that develops during bleaching just as the AgC and D emission develops during bleaching. The absorption band associated with AgC and AgD emission is not identified conclusively since there is no apparent correlation between emission and absorption intensities. Excitation into the centers with absorption peaks around 3.9eV (3180 A°) and 4.35 eV (2850 A°) seems to be primarily responsible for the AgD and AgC emission

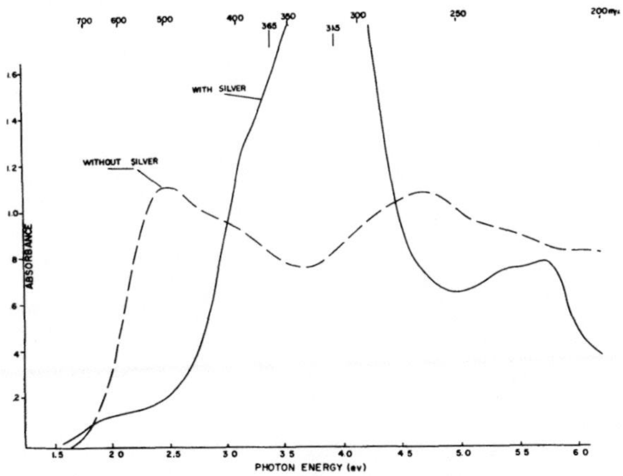

Fig. 4. Radiation-induced absorption of a phosphate base glass and a phosphate glass containing silver.

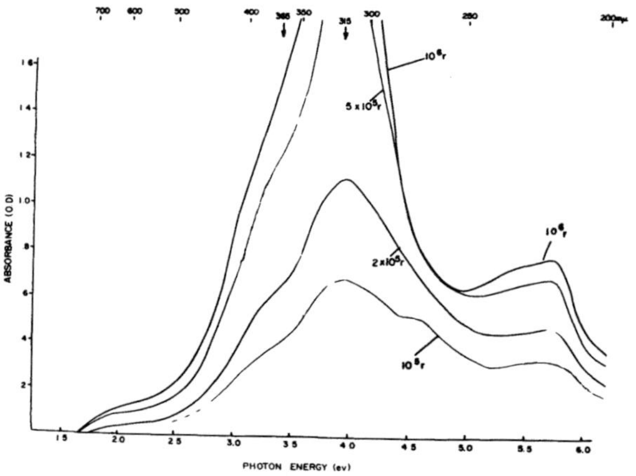

Fig. 5. Radiation-induced absorption of a silver containing phosphate glass exposed to different doses.

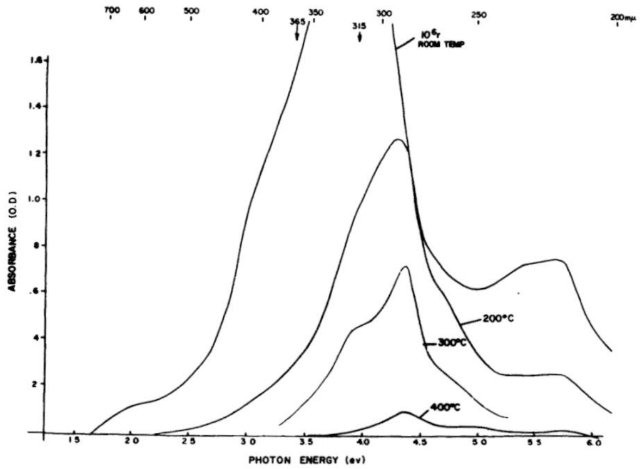

Fig. 6. Radiation-induced absorption of a silver containing phosphate glass exposed to 10^6r and bleached at various temperatures.

DISCUSSION OF RESULTS

It is obvious that it is not possible to propose defect models for the observed luminescence centers on the basis of the present experiments. However, the experiments allow a re-examination of existing and previously proposed models.

It has been mentioned earlier that silver ions may act as traps for electrons as well as holes. This has been shown by spin-resonance experiments in phosphate glasses (6,7,9), and in silver-doped alkali halides (8). Unfortunately, the studies in glass make no attempt, or present sufficient evidence to correlate spin-resonance signals to absorption or luminescence bands.

Models suggested for the orange AgB luminescence include a silver ion Ag^+ with a trapped electron (6,14), and, a silver ion with a trapped hole (7,9). For the orange radiation-induced luminescence in silver-doped alkali halides, Etzel and Schulman (15) proposed a more complex center comprising a silver ion, a hole and a positive ion vacancy. Other absorption and luminescence centers in silver-doped alkali halides and phosphate glasses have been assigned to pairs of silver ions by these authors.

The evidence is strong that the two long wavelength absorption bands in the undoped phosphate glass are associated with trapped holes (12). The absence of these bands in the silver-containing glass is strong indication that the Ag^+ ions indeed do capture holes and, therefore, prevent the formation of these bands. Consequently, the orange AgB luminescence is for the moment attributed to a trapped hole in the vicinity of an Ag^+ ion $(Ag^+,^+)$. The fact that the band decreases if the silver concentration exceeds a certain amount would exclude silver pairs as responsible for this center. If silver pairs play a role in the luminescence of phosphate glasses, they would be most likely to be detected in glasses with a high concentration of silver. The more prominent band in these glasses is the AgB^1 band which is strong in a glass containing 30% $AgPO_3$ and is the only band in a $AgPO_3$ glass. This center is not very stable, that is, it bleaches fast at 100°C, indicating that the energy for destroying it is low. This, however, is in contradiction to Yokota and Imagowa (7,9) who attribute an electron spin-resonance signal which had formed after the exposed glass was allowed to stay at room temperature for several days to silver pair formation.

Silver pairs can also be excluded from being the cause of the AgC and AgD luminescence centers. Although these pairs form at temperatures between 200 and 300°C, at which silver ions are probably quite mobile, heating to higher temperatures brings the glass back to its pre-irradiation state and re-exposure at room temperature forms orange AgB centers but not blue AgC or AgD centers. This suggests that the precursors of the AgC and D centers are present in the glass from the beginning and are not formed during the heating of the exposed glass.

If silver pairs had formed during heating, they would not be expected to dissociate at still higher temperatures. Moreover, it was shown in Fig. 2e that if a glass which contains a high concentration of AgC centers is re-exposed to irradiation, AgB centers are formed while the AgC centers are bleached simultaneously. It is evident from these experiments and from exposures at elevated temperatures that there exists a highly temperature-dependent equilibrium between the concentration of AgB and AgC centers. In other words, AgC and AgB compete for the trapping of holes. However, at lower temperatures, the production of AgB centers is favored, while at higher temperatures, AgC centers are favored. The same equilibrium is established if the glass is heated subsequent to irradiation. The competition between AgB and AgC centers implies that both centers trap the same type of charge carrier. Indeed, electrolysis experiments indicate that both the orange and blue luminescence begin to form on the anode before they move to the cathode, suggesting that both centers are associated with a trapped hole.

An alternative explanation would assume a reorientation and, therefore, a slightly different position of the $Ag^{+},^{+}$ ions at higher temperature. This new position then gives rise to the AgC and AgD emissions. This mechanism would not require the assumption of a release and subsequent retrapping of charge.

It may be significant to mention here that Etzel and Schulman (15) find a band in bleached silver-doped alkali halides that was not observed in unbleached crystals.

There can be little question that the luminescence centers discussed here and shown in Fig. 1, are associated with silver. The large differences in the AgB center concentration in various glasses probably have to be explained on the basis of the structure of phosphate glasses. Unfortunately, our knowledge of the structure of phosphate glasses is rather limited, although it is known that melts and glasses contain a distribution of metaphosphates (ring structures), polyphosphates (chain structures) of various chain lengths and also some ultraphosphates (cross-linked chains) (16). The distribution and chain length is controlled by the metal-to-phosphorus ratio and by the melting and annealing temperatures. The structures and chains of various lengths could be isolated by paper chromatography (17). In addition, six different types of chains have been identified, depending on the orientation and the way in which the PO_4 tetrahedra are linked to produce straight chains or helices (18). It is probably this difference in structure and the predominance of one structure which accommodates silver in special sites that may explain the difference in the concentration of the AgB and AgC centers.

REVISION OF CONCEPTS ON DEFECTS IN COMPLEX GLASSES AND PROJECTION OF FUTURE APPROACHES

Most recent evidence, illustrated for instance by some of the results on silver centers (7) discussed here, suggests some revisions in the conclusions reached in any earlier review (1), as well as further attacks on the understanding of defects in complex glasses of technological importance.

It has been found experimentally that the 5.7 eV absorption ascribed by a recently proposed (12) *band model* to a fundamental transition from an electronic level of the non-bridging oxygen stops, in actuality, at some shorter wavelength. This seems to indicate a different defect mechanism, much more in line with R. Week's (3) description of certain defects in pure silica, but present here in a much higher concentration. This description involves orbit interactions at defective locations substantiated by ESR signals.

In the discussion of centers, as exemplified in the case of silver, to date too little attention has been paid to the fine structure of the complex technologically synthesized glasses. The phosphate glasses, for instance, in which silver has been studied because of its solubility in them, have very complex structures. Westman (17) and Thilo (18), who have made a good start, describe at least six quite different types of chains that may exist in their structures. Progress in the analysis of defect centers, not limited at all to silver, will depend on detailed progress in the description of the structure of the phosphate base glasses at least as detailed as in the work of Bray (19) and co-workers on borate glasses. In both borate and phosphate glasses, different probes should be used to study the fine structure such as chain shape and length, nascent immiscibility or nucleation, alkali order - disorder, etc.

In other sets of experiments, glasses should be prepared under *extreme reducing conditions* and compared with those containing normal or *excess quantities of oxygen*. Centers depending on oxygen vacancies, oxygen ions or peroxide (3,20) (superoxide) groups might be discovered and described. Glasses heavily reduced have great theoretical and industrial significance; the absorption centers induced in these glasses differ greatly from those obtained in conventional syntheses.

The intensity of bands ascribed to non-bridging oxygens in complex glasses is often found quite independent of calculated concentration and varies with alkali type. In all such cases, it will be necessary, as in Bray's interpretation of the simple borate glasses, to ascertain *definite groups* in the structure which will, of course, show significantly different fields of stability as one varies R^+ from Li^+ to Cs^+.

REFERENCES

1. *E. Lell, N.J. Kreidl, J.R. Hensler, "Radiation Effects in Quartz, Silica and Glasses", Progress in Ceramic Science Vol.4 Edited by J. Burke, Pergamon Press, Oxford, New York 1966.*

2. *E. Lell, N.J. Kreidl, J.R. Hensler, Final Summary Report, Atomic Energy Commission Contract AT (30-1) 1312, (1965).*

3. *R.A. Weeks Phys. Rev. 130,570 - 576 (1963), R.A. Weeks Proceedings Cairo Solid State Conference (1966).*

4. *A.Kats and J.M. Stevels, Philips Res. Rep. 11,115 - 156 (1956).*

5. *A. Bishay and K.R. Ferguson, "Advances in Glass Technology" Technical Papers of the VI International Congress on Glass, Washington 1962, 133 - 148 published: American Ceramic Society.*

6. *R.F. Tucker, "Advances in Glass Technology" Technical Papers of the VI International Congress on Glass 103 - 114 Washington 1962.*

7. *R. Yokota and H. Imagawa, Proceedings of the Symposium on Defects in Glass, Tokyo - Kyoto, September 1966.*

8. *C.J. Delbecq, W.Hayes, M.C.M. O'Brien, P.H. Yuster, Proc. Roy. Soc. A 271,243 (1963).*

9. *R. Yokota and H. Imagawa, J.Phys. Soc. Japan 20,1537-38 (1965).*

10. *J.S. Stroud, J. Chem. Phys. 37,836 - 41 (1962).*

11. *A.M. Bishay, J. Am. Ceram. Soc. 45,389 - 393 (1962).*

12. *J.S. Stroud, R.F. Tucker, W.H. Schreurs, Proceedings of the VII International Congress on Glass, Brussels 1965.*

13. *K. Becker, IAEA - Symposium on Solid State and Chemical Radiation Dosimetry, Vienna 1966.*

14. *W.A.Weyl, J.H. Schulman, R.J. Ginther, L.W. Evans, J. Electrochem. 95,70 - 79 (1950).*

15. *H.W. Etzel, J.H. Schulman, J. Chem. Phys. 22,1549 (1954).*

16. *J.R. VanWazer, "Phosphorus and its Compounds" Interscience,New York 1958,Vol. 1.*

17. *A.E.R. Westman, "Modern Aspects of the Vitreous State" Vol. 1 pp. 63 - 91 Edited by J.D.Mackenzie, Butterworths,London 1960.*

18. *E. Thilo, Angew. Chemie, Int. Edition 4,1061 - 1071 (1965).*

19. *P.J. Bray in "Modern Aspects of the Vitreous State" Vol. 1 pp. 92-119 Edited by J.D. Mackenzie, Butterworths,London 1960.*

20. *J.R. Hensler "Advances in Glass Technology" p. 25, Technical Papers of the VI International Congress on Glass, Washington 1962.*

DISCUSSIONS

CHARLES GREENE (Alfred University, U.S.A.): *Might not the rather precise information which Westmans work on phosphate glasses has given us on these structures aid in formulating "band theory" for such glasses?*

N.J. KREIDL: *Yes, with the limitation that lower alkali systems are of interest here, and these lack the solubility needed to exploit chromatographic techniques.*

G.C. KUCZYNSKI (University of Notre Dame, Indiana, U.S.A.): *P. Flinn treated a problem similar to that discussed here. He found that in metallic alloys with strong short range order the band model can still be usefully applied. If we identify superlattice with crystalline structures and disordered alloy with glass, Flinn's model could be perhaps applied to the latter structure at least as the first approximation.*

N.J. KREIDL: *Prof. DEBYE also proposed a "Wiggled" band model at the Alfred Conference on Non-Crystalline Solids.*

R.A. WEEKS (Oak Ridge National Laboratory, U.S.A.): *From a theoretical view point the electronic structure of a glass cannot be considered as a band model since the structure of the glass is not periodic in the sense of a single crystal. However, by taking a region of the glass and assuming it to be ordered, on approximation to a band model might be made.*

N.J. KREIDL: *This question has also been examined by KEYSER. (VII Glass Congress 1965).*

E. DEEG (American University in Cairo, U.A.R.): *In connection with the application of the band model to glass, I would like to add that the existence of forbidden energy zones in non-crystalline materials has been shown by different authors. This means that we are allowed to use the band model at least in principle. However, details have to be worked out for each special glass separately. A survey of theoretical work has been given by R. KRONIG in the book "Physikartag Stuttgart, 1962".*

GAMMA INDUCED COLORATION IN RELATION TO STRUCTURAL CHANGES IN GLASS

Adli Bishay, American University in Cairo, U.A.R.

M. Maklad, I. Gomaa and S. Arafa

U.A.R. Atomic Energy Establishment

This paper presents a number of examples in which induced coloration resulting from the interaction of high energy radiation with different glasses can be used to detect structural changes. Infrared absorption and X-ray diffraction studies of the same glasses are used to confirm these structural changes.

The changes in the intensity of the 1.6 and 2.4 eV bands induced in lead borate glasses suggested structural changes at compositions containing about 25, 33 and 45 mole% PbO. These changes were confirmed by X-ray diffraction studies of the heat treated devitrified samples which showed the appearance or disappearance of certain lead borate compounds at these compositions. Infrared absorption studies of the glass samples of this series showed maxima in the intensity of the BO_4/PbO_x bands at the above mentioned compositions.

Similar studies were made on arsenic borate and barium silicate glasses. The changes in As^{5+}/As^{3+} as well as the changes in the role of arsenic and boron in structure were indicated in the induced absorption and confirmed by other physical properties. Barium silicate glasses showed changes at 22.5 and 27.5 mole% BaO in the induced optical absorption and infrared absorption spectra. This was confirmed by changes in the X-ray diffraction patterns of the heat treated devitrified samples of the same series.

INTRODUCTION

A general review of the use of radiation induced optical absorption as a tool for studying different glass structures was reported earlier by Bishay and Ferguson (1). This paper will deal with three systems; namely: lead borates, alkali borates containing arsenic, and, alkali-barium silicate glasses. The effect of increasing PbO, As_2O_3, or BaO was studied, respectively. Abrupt changes in the intensity of some of the gamma induced bands were in many cases related to structural changes resulting from changes in composition. This was confirmed by using some of the classic tools for studying glass structure (infrared, X-ray, molar volume etc.).

EXPERIMENTAL

Specimen Preparation

Reagent grade chemicals were used whenever possible for preparing the different glasses. Boric oxide was added as boric acid, and, SiO_2 as chemically washed sand from Sinai. Alkali oxides were introduced as their respective carbonates, while As_2O_3, As_2O_5 and PbO were introduced as oxides. All samples were melted in platinum crucibles with the exception of those containing arsenic. The latter were melted in porcelain crucibles due to the corrosive effect of arsenic on platinum. The melting and annealing temperatures were varied according to composition. Most of the samples were ground and polished to a thickness of about one millimeter.

Spectrophotometric and Irradiation Studies

A 700 SP Unicam and a DK-2A Beckman recording spectrophotometers were used for abosrption measurements in the visible, U.V. and near infrared regions of the spectra before and after irradiation. Most of the specimens were irradiated at a rate of 4×10^5r/hr using a 6000 Curie ^{60}Co "Gamma Cell 220". Some arsenic borate samples were irradiated at a rate of about 10^5r/hr using a 500 Curie ^{60}Co "Agromat" source.

Samples needed for infrared studies were prepared using the KBr pellet method. All infrared studies (2.5-25μ) were performed using a 337 Perkin-Elmer infrared grating spectrophotometer.

X-Ray Diffraction Studies

The specimens were heat treated at the temperature and for the period needed for devitrification. The devitrified samples were then finely powdered and X-ray diffraction (XRD-6 General Electric Apparatus) was used to identify the different products of devitrification. These studies were based on the postulation that the structure of the identified compounds in the devitrified samples will be similar to that of the corresponding structural groups in the glass.

LEAD BORATE GLASSES

It was reported earlier (2), that a gamma induced absorption band was observed at about 1.5 eV (825 mμ) in lead aluminoborate and lead borate glasses. This band was also observed when a potassium aluminoborate glass containing Tl^{+} was gamma irradiated.

A similar band was induced in an alkali alumino lead phosphate glass (3) ($Al_2O_3 \cdot 2PbO \cdot 5P_2O_5$) and in an alkali lead silicate glass (1) ($SiO_2 \cdot 0.22Na_2O \cdot 0.15PbO$).

These results indicate that the 1.5 eV band is associated with the configuration of Pb^{2+} ions since it was only observed in glasses containing Pb^{2+} or Tl^{+} which have a similar electronic configuration. (This band will be later called the 1.6 eV band).

The effect of addition of cerium on the intensity of this band in glasses melted under different conditions suggested strongly that it is associated with an electron trap in the vicinity of Pb^{2+} ions (4).

Induced Absorption

Figure 1 shows a gaussian resolution of the induced absorption of the lead borate glass containing 27.4 mole% PbO after ^{60}Co irradiation (3×10^{6} rad). The bands in this glass have maxima at E_o = 1.59, 2.4, and 3.46 eV, respectively. A similar resolution was carried out for the rest of the glasses in this series.

The number of color centers (N_i) giving rise to an induced absorption band can be considered proportional to the product of $\alpha_m U$ where α_m is the intensity at the peak of each band, and, U is the width at half maximum. Figure 2 shows the effect of increasing PbO concentration on $\alpha_m U$ for the 1.6 and 2.4 eV induced bands. The figure shows an increase in the intensity of the 1.6 eV center till about 25 mole% PbO. Further increase in PbO concentration is associated with a decrease in the intensity of this center. However, a change in the rate of decrease is observed at about 45 mole% PbO. The intensity of the 2.4 eV center, on the other hand, shows two maxima at about 33 and 45 mole% PbO respectively.

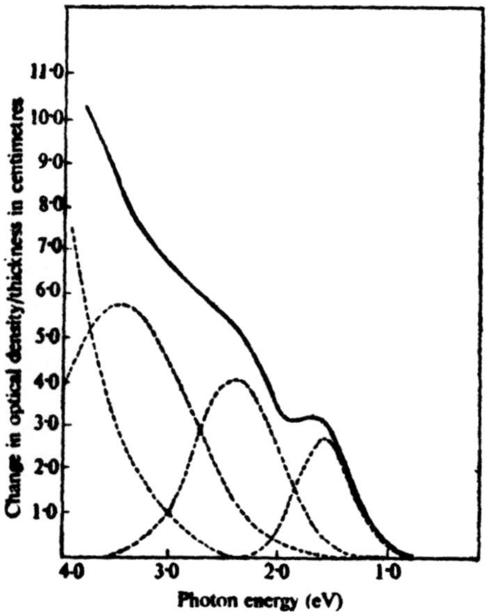

Fig. 1. The resolution of induced absorption spectrum of lead borate glasses containing 27.4 mole% PbO (^{60}Co irradiation 3×10^6 rad).

E_o	U	α_m
1.59	0.57	2.7
2.4	0.89	4.05
3.46	1.42	5.75

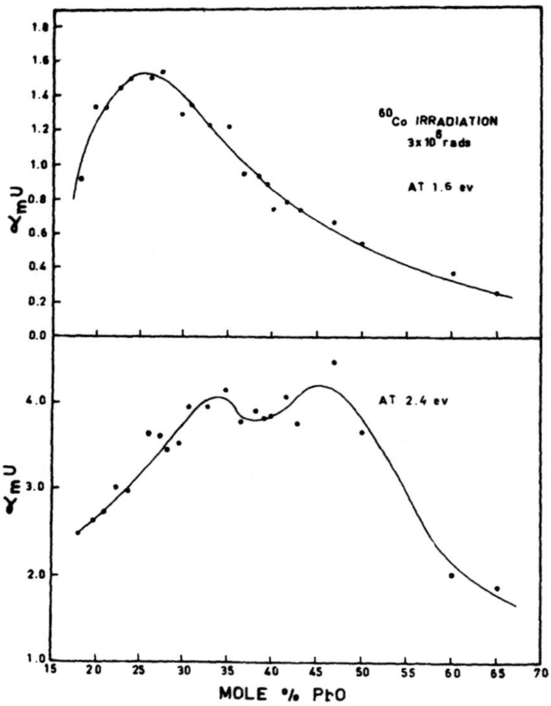

Fig. 2. The effect of increasing PbO content on $\alpha_m U$ for the 1.6 and 2.4 eV induced bands (^{60}Co irradiation: 3×10^6 rad).

The increase in the intensity of the 1.6 eV band in the range from 18 to about 25 mole% PbO supports the suggestion that lead is present in the ionic form in this range of compositions (5) since this band was attributed to electron traps associated with Pb^{2+} ions (4). On the other hand, the decrease in the intensity of this band in glasses containing more than 25 mole% PbO suggests a decrease in the ionic character of lead in these glasses. This is supported by the chemical shift data for Pb^{207} reported by Bray (5).

Infrared Absorption

Figure 3 shows the infrared spectra for lead borate glasses of increasing mole% PbO. Since many of the bands characteristic for PbO(450, 525, 700, 850 and 1450 cm^{-1}) (4) overlap with most of those of boron (BO_4 and BO_3), it was found necessary to restrict our semi-quantitative studies to one of the triply degenerate frequencies of BO_4(ν_3 at 1100 cm^{-1}) and to the absorption at 450 cm^{-1} characteristic of PbO. This ratio between the optical density at 1100 cm^{-1} and 450 cm^{-1} is shown in Figure 4 for the series of lead borate glasses. Three maxima are observed in the curve at about 25,33 and 45 mole% PbO. Recent NMR results by Bray (6) showed a maximum N_4 at about 50 mole% PbO and a change in the rate of increase of N_4 at about 33 mole% PbO where N_4 was plotted versus x/1-x for glasses of molar composition x PbO (1-x)B_2O_3.

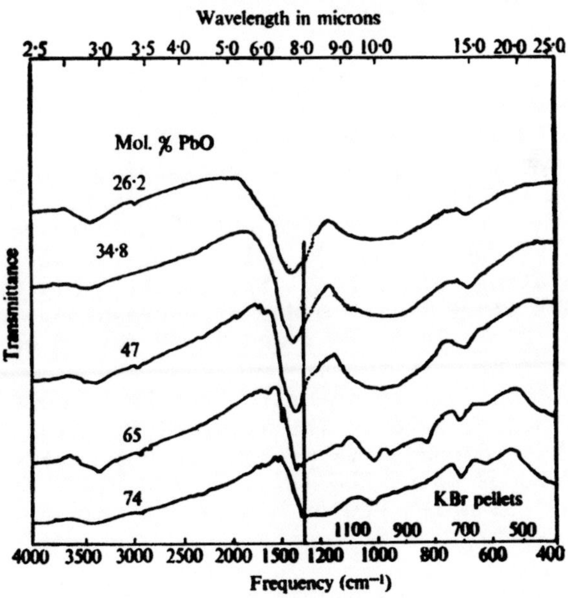

Fig. 3. The effect of increasing mole% PbO on the infrared transmittance of lead borate glasses (note displacement of curves).

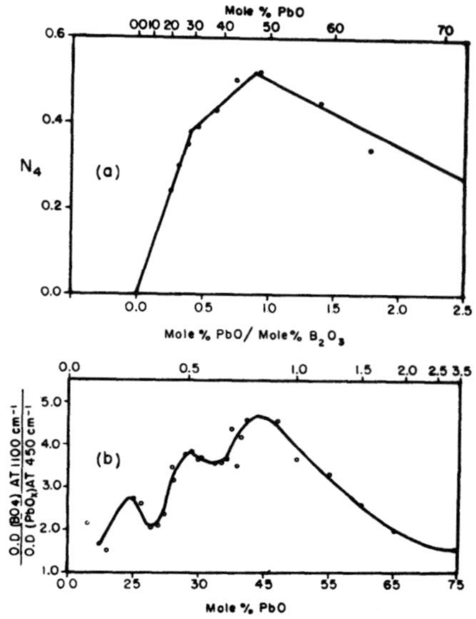

Fig. 4. The effect of increasing mole% PbO on:

(a) $\dfrac{\textit{optical density at 1100 cm}^{-1}\ (BO_4\ \textit{band})}{\textit{optical density at 450 cm}^{-1}\ (PbO_x\ \textit{band})}$

(b) Fraction of four-coordinated boron, N_4 *(after Bray).*

Heat Treatment Studies

The x-ray diffraction patterns of the heat treated devitrified samples of the compositions containing 18 to 24 mole% PbO showed crystals of $PbO{\cdot}2B_2O_3$ as the predominant phase. The compound $5PbO{\cdot}4B_2O_3$ is slightly indicated in these samples (Fig. 5a and b). Glasses containing 25 mole% PbO, showed a new type of crystals separating after heat treatment. This type is characterized by peaks at 2Θ in the order of decreasing intensity: 5.5, 22.8, 28.6, 17, 29.7, 34.5, 24.1 and 18.6° (using copper target).

It was not possible to identify the above mentioned compound by using the A.S.T.M. cards. The x-ray diffraction patterns show clearly that this compound is present in the devitrified samples containing 25 to 32 mole% PbO (Fig. 5c and Figs. 6a and b). Samples containing 33 mole% PbO or higher do not show this compound (Fig. 6c and Fig. 7).

On the other hand, the $PbO{\cdot}2B_2O_3$ and $5PbO{\cdot}4B_2O_3$ compounds are again clearly indicated in the samples containing 33.8 and 40

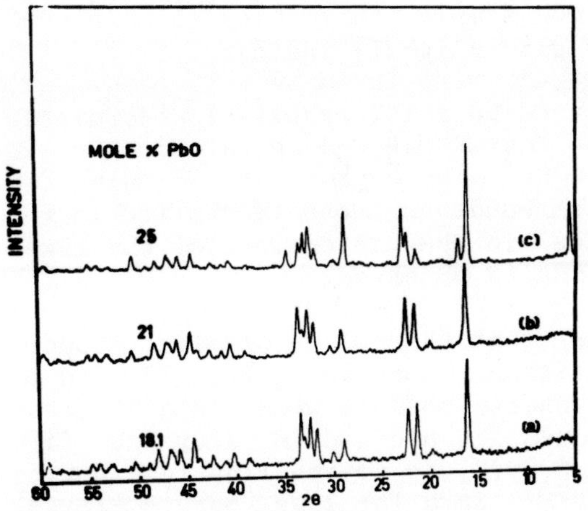

Fig. 5. X-ray diffraction patterns for heat-treated lead borate glasses (devitrified) (18.1, 21 and 25 mole% PbO).

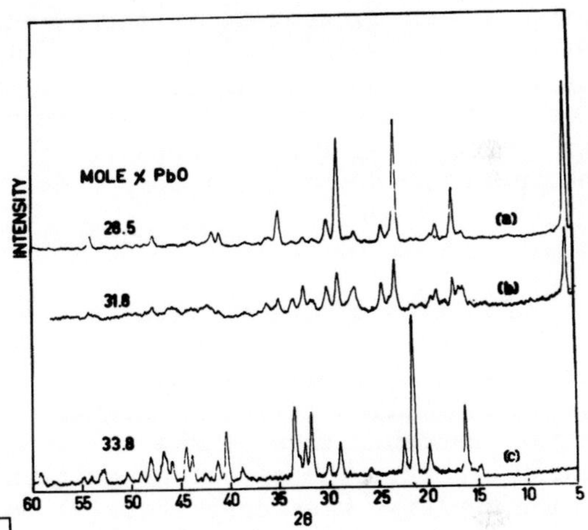

Fig. 6. X-ray diffraction patterns for heat-treated lead borate glasses (devitrified) (28.5, 31.8 and 33.8 mole% PbO).

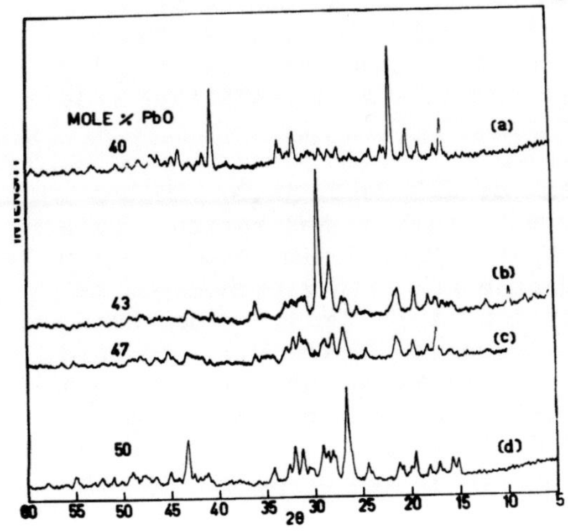

Fig. 7. X-ray diffraction patterns for heat-treated lead borate glasses (devitrified) (40, 43, 47 and 50 mole% PbO).

mole% PbO (Fig. 6c and Fig. 7a). The compound $\beta 2PbO \cdot B_2O_3$ starts to appear clearly in the devitrified sample containing 43 mole% PbO (Fig. 7b). $\alpha PbO \cdot B_2O_3$ compound is also slightly indicated.

Increasing the PbO content to 50 mole% results in increasing the intensity of the patterns for $\alpha PbO \cdot B_2O_3$ and a decrease in $\beta 2PbO \cdot B_2O_3$ patterns. Samples containing higher than 50 mole% PbO show the presence of the same compounds as those identified in the latter sample, though a change in the intensity of the lines characteristic for these compounds is observed.

The above results indicate that changes in physical properties may be expected at compositions containing 25, 33 and 43 mole% PbO, respectively. Corresponding changes were actually found in the physical properties discussed above and in reference (4). Furthermore, recent NMR work by Bray(7) on crystalline lead borate compounds indicated that the Pb^{207} data for lead borate glasses are, without exception, in accord with the presence in the glasses of the lead-oxygen configurations which occur in lead-borate crystalline compounds. These results are also in line with Krogh-Moe's structural model (8) for borate glasses, in which the glass is viewed as a random network of the structural groupings that occur in the crystalline compounds.

ALKALI ARSENIC BORATE GLASSES

Addition of one multivalent element to a glass may result in an increase or decrease of induced absorption or produce new color centers. These effects have been found to depend on the state of oxidation (i.e. ratio of higher to lower valence states), the total concentration of the multivalent element, and the type of color center. The effect of multivalent ions on the intensity of color centers was considered (9) to be due to: a) providing electrons which may annihilate the positive hole centers responsible for the visible and near U.V. induced absorption, or, b) trapping electrons which would have otherwise annhilated the positive hole centers.

Bishay reported that addition of 2.0 mole% As_2O_3 causes a pronounced decrease in the induced visible absorption for a $CaO \cdot P_2O_5$ glass (3), while the addition of 2.0 mole% As_2O_3 to 1.0 $Al_2O_3 \cdot 4.5\ B_2O_3 \cdot 1.0\ K_2O$ glass was associated with a pronounced increase in the induced visible absorption (10). On the other hand, a decrease in the visible and an increase in the U.V. induced absorption was observed when 0.5 mole% As_2O_3 was added to a $CaO \cdot P_2O_5 \cdot GeO_2$ glass (3). Furthermore, the addition of 2.0 mole% As_2O_3 to a 1.0 $Al_2O_3 \cdot 4.5\ B_2O_3 \cdot 1.0$ PbO glass was associated with an increase in the absorption band induced at 1.6 eV as well as the U.V. induced absorption (10).

This section will deal mainly with the effect of addition of increasing amounts of As_2O_3 on the induced absorption of alkali borate glasses, with special reference to its relation to the As^{5+}/As^{3+} ratio and to the structural role of arsenic and boron in the glass.

Induced Optical Absorption

Effect of arsenic on induced absorption: Figure 8 shows the effect of 1.8 mole% As_2O_3 on the induced absorption of a $4.5B_2O_3 \cdot 1.0Na_2O$ glass. Gaussian resolution showed that the induced absorption spectrum of the base glass consisted mainly of absorption bands at about 2.5 eV (500 mμ) 3.5 eV (345 mμ) and 4.9 eV (260 mμ). A fourth band can also be predicted at about 6.4 eV (195 mμ). These bands are the characteristic induced absorption bands for borate glasses (11,12). The position and intensity of some of these bands varied with the borate glass composition (1,12). It is clear (Fig. 8) that 1.8 mole% As_2O_3 markedly increases the induced absorption; beyond 3.6 eV for a 2.0-mm sample of the glass containing arsenic it was too high to be measured by the spectrophotometer.

Figure 9 shows that increasing As_2O_3 up to 0.5 mole% increases induced absorption in the visible and ultraviolet regions of the spectrum. On the other hand, in glasses containing high concentrations of As_2O_3, increasing As_2O_3 from 21 to 45 mole% decreased the general induced absorption.

The observed induced absorption at 2.5 and 3.5 eV is plotted versus the concentration of As_2O_3 in Figure 10. There is a slight increase in the 2.5 eV band up to about 5.0 mole% As_2O_3 followed by a slight decrease in induced absorption. On the other hand, the 3.5 eV band shows a pronounced increase up to about 5.0 mole% As_2O_3 followed by a general decrease in intensity. The rate of decrease of the intensity at 3.5 eV changed at about 10 and 25 mole% As_2O_3. From Figures 9 and 10 it may be concluded that increasing As_2O_3 in the system $4.5B_2O_3 \cdot 1.0Na_2O \cdot xAs_2O_3$ increases the induced absorption up to about 5.0 mole%. Further increase in As_2O_3 is accompanied by a general decrease in the induced absorption.

Effect of oxidation state of arsenic on induced absorption: Two glasses of the molar composition $4.5B_2O_3 \cdot 1.0Na_2O \cdot 0.1As_2O_3$ were prepared; arsenic was added as As_2O_3 in one and as As_2O_5 in the other. It was found that the glass with As_2O_5 has only a slightly higher induced absorption than the corresponding As_2O_3 glass, indicating that the ratio of As^{5+}/As^{3+} in these glasses is not strongly dependent on the form in which arsenic is introduced. Increasing As_2O_5 from 1.79 mole% to 8.33 mole%, however, decreased

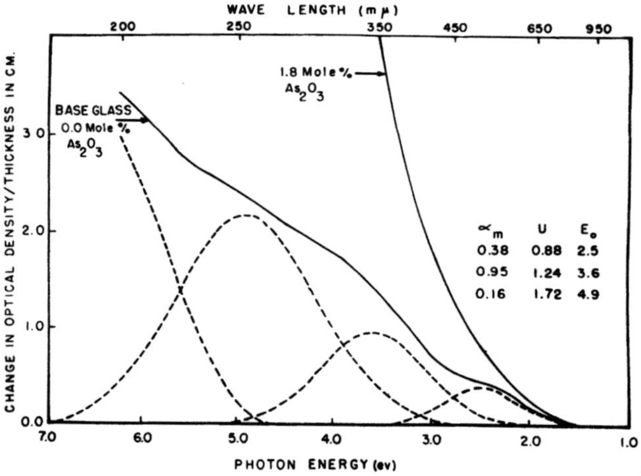

Fig. 8. Effect of addition of 1.8 mole% As_2O_3 on the induced absorption in $4.5B_2O_3 \cdot 1.0Na_2O$ glass (Dose=$1.2x10^6$ rad).

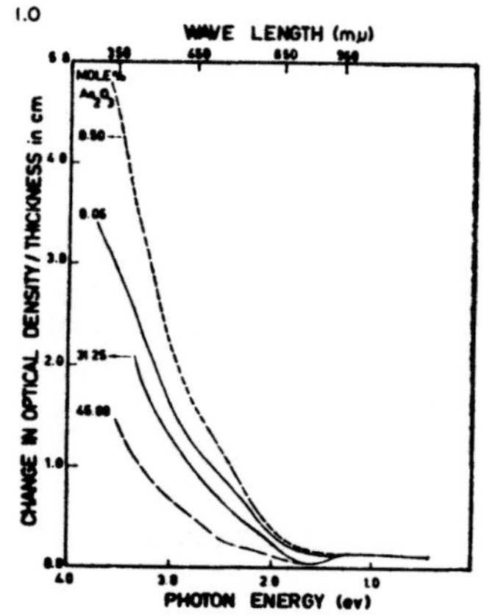

Fig. 9. Effect of increasing As_2O_3 on the induced absorption in $4.5B_2O_3 \cdot 1.0Na_2O$ glass (Dose = $1.2x10^6$ rad).

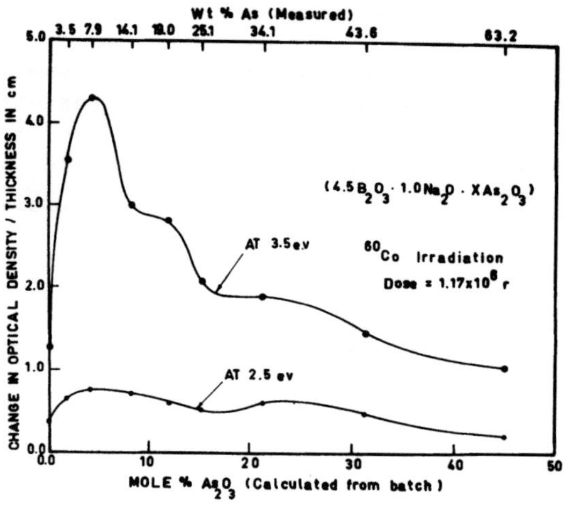

Fig. 10. Effect of increasing arsenic content on induced absorption at 2.5 and 3.5 eV (Dose = $1.17x10^6$ rad).

the visible and increased the ultraviolet induced absorption.

Effect of replacing Na_2O by Li_2O or K_2O on induced absorption. The induced absorption spectrum in the visible and near ultraviolet for three glasses of molar composition $4.5B_2O_3 \cdot 1.0R_2O \cdot 0.1As_2O_3$ where R_2O represents Li_2O, Na_2O, or K_2O was also studied. The potash glass showed the highest induced absorption and the soda glass showed the lowest induced absorption.

In glasses containing arsenic the induced absorption is strongly influenced by the following reactions occurring during the irradiation process:

$$As^{3+} + h\nu = As^{5+} + 2e \quad \text{(mechanism a)}$$

$$As^{5+} + 2e = As^{3+} \quad \text{(mechanism b)}$$

The kinetics of these reactions will be controlled by: (a) the ratio of As^{5+}/As^{3+} in the glass and (b) total arsenic content (which in turn affects the ratio of As in network-modifying to As in network-forming positions). Values for As^{5+}/As^{3+} and for total arsenic for glasses containing 1.79 mole% As_2O_3 (as calculated from batch) but containing different alkalies are given in Table I. The high ratio of As^{5+}/As^{3+} in the K^+ glass favors mechanism b which is accompanied by an increase in the 2.5 and 3.5 eV bands. The Li^+ glass, containing the lowest As^{5+}/As^{3+} ratio, would be expected to have less induced absorption in the visible and near ultraviolet (mechanism a). The X-ray fluorescence analysis, however, gave a much higher value for total arsenic in the Li^+ glass than for the Na^+ glass.

Figure 10 shows that in low As_2O_3 concentration, increasing arsenic is associated with an increase in the induced absorption up to a certain arsenic content. This may explain the higher induced absorption for the Li^+ glass containing 6.4 wt.% total As compared to the Na^+ glass containing only 3.5 wt.% total As. Addition of 1.79 mole% As_2O_3 to a lithium borate glass of the molar composition $4.5B_2O_3 \cdot 1.0Li_2O$ decreased the induced absorption (9).

The difference in the effect of arsenic on the induced coloration of different glass compositions (3, 9, 10) may also be explained on the basis of the As^{5+}/As^{3+} ratio and total arsenic.

The results discussed in the previous sections show that there is a pronounced increase in the induced absorption on increasing the arsenic content, reaching a maximum at about 5.0 mole % As_2O_3. Other weak maxima can be detected at about 10 and 25 mole% As_2O_3 (Fig. 10).

TABLE I

Weight Percent Loss of As, As^{5+}/As^{3+} Ratio and Total Arsenic in Different Glasses (by Analysis)

As_2O_3* (mole%)		Total As* (wt%)	Total As# (wt%)	Loss of As by melting (wt%)	As^{3+} ≠ (wt%)	As^{5+}/As^{3+} ratio
Na	0.00	0.000	0.00	0.00	0.00	0.00
Na	1.79	3.790	3.50	7.65	1.30	1.70
Na	4.34	8.813	7.90	10.36	1.80	3.30
Na	8.33	15.787	14.10	10.69	5.70	1.48
Na	12.00	21.440	19.00	11.40	7.00	1.70
Na	15.33	26.130	25.10	3.93	10.00	1.50
Na	21.43	33.420	34.10	-2.03	12.00	1.61
Na	31.25	43.040	43.60	-1.31	15.28	1.85
Na	45.00	53.250	63.20	-18.69	23.93	1.64
Li	1.79	4.120	6.40	-35.54	3.38	0.88
K	1.79	3.500	4.20	-16.57	1.18	2.56

* Calculated from batch.

\# Determined by X-ray fluorescence.

≠ Determined chemically.

The induced absorption of these glasses is expected to be very sensitive to the As^{5+}/As^{3+} ratio. Since the 2.5 and 3.5 eV bands are due to positive hole centers (13), it is expected that these bands would decrease in intensity with increasing As^{3+} ions according to mechanism (a), which provides electrons for the annihilation of these positive hole centers. On the other hand, increasing As^{5+} ions would be expected to be associated with an increase in the intensity of these bands because of trapping electrons which would otherwise have annihilated the positive hole centers according to mechanism (b). An increase in the As^{5+}/As^{3+}

ratio in the glass should, therefore, be accompanied with an increase in the intensity of these positive hole bands. Conversely, an increase in the intensity of these bands may be taken as an indication of an increase in the As^{5+}/As^{3+} ratio.

The induced absorption in the visible may also be affected by a change in the structural role of arsenic in the glass. Furthermore, if arsenic takes network-modifying positions in the structure, the ratio of BO_4/BO_3 is increased in a way similar to that occurring when an alkali oxide is added to a borate glass. It was reported earlier (11) that the ratio of BO_4/BO_3 strongly affected the intensity of the 3.5 eV band. A change in the structural position taken by arsenic was considered to affect the ratio of BO_4/BO_3 which would, therefore, be expected to affect the intensity of the induced visible absorption. The change at about 10.0 mole% As_2O_3 in the induced absorption curve (Fig. 10) may be attributed to a change in the BO_4/BO_3 ratio as well as to a change in the structural position taken by arsenic.

The formation of certain groups has been postulated (8) in borate glasses. The possibility of similar groups occurring in glasses containing different amounts of As_2O_3, B_2O_3, and Na_2O is quite reasonable. A change in the type or concentration of these groups may also be accompanied by a change in the intensity of induced absorption which may be due to a change in the kinetics of formation or annihilation of the induced centers. The change detected at about 25.0 mole% As_2O_3 (Fig. 10) may be associated with the formation of certain structural groups at this composition.

To confirm the foregoing hypothesis suggested from the work on induced optical absorption, additional studies were conducted (14).

Total Arsenic and As^{5+}/As^{3+} Ratio

The total arsenic content was determined by the X-ray fluorescence technique. The concentration of As^{3+} ions was determined by chemical analysis using the potassium permanganate method. The As^{5+}/As^{3+} ratio was calculated from these two values. The loss in total arsenic due to melting was determined by subtracting the calculated values (from batch composition) from the values obtained by analysis. Table I shows the wt. % loss of total As as a result of melting and the As^{5+}/As^{3+} ratio in the glasses studied.

The percentage loss of total arsenic is plotted versus mole% As_2O_3 (calculated from batch composition in Figure 11a). There was a pronounced increase in percentage loss of total arsenic as the arsenic content was increased from 0.0 to 5.0 mole%, followed by

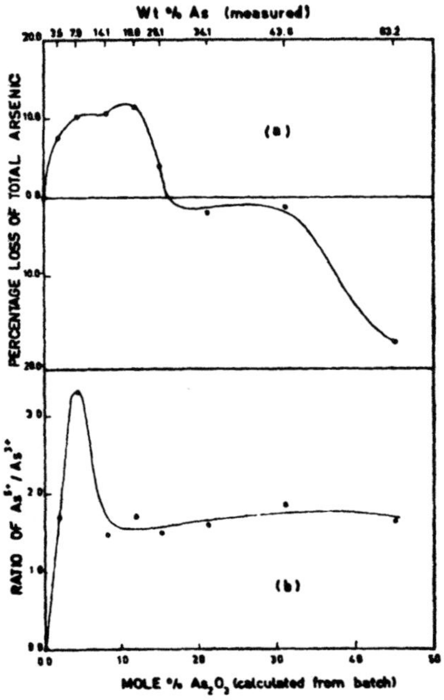

Fig. 11. Effect of increasing arsenic content in 4.5 B_2O_3· 1.0 Na_2O glass on (a) percentage loss of total arsenic and (b) As^{5+}/As^{3+} ratio.

a slight increase in percentage loss reaching a maximum at about 10.0 mole% As_2O_3. A general decrease was then observed as the arsenic content was increased beyond 10.0 mole% As_2O_3 reaching nearly zero % loss at about 15.0 mole% As_2O_3. Glasses containing greater than 25.0 mole% As_2O_3 showed negative values.

The As-O bond in the three-coordinated configuration is a much stronger bond than that in the six-coordinated configuration. The decrease in the percentage loss of total arsenic with increasing As_2O_3 content above 10.0 mole% may be explained by assuming that at this concentration arsenic changes its coordination number from six to three (i.e. from network-modifying to network-forming positions).

The negative percentage loss of total arsenic shown for compositions between 25.0 to 45.0 mole% As_2O_3 may be explained on the basis of an arsenic-oxygen-boron configuration in which the boron is held less strongly than the arsenic in the structure,i.e. boron evaporates at a higher rate than arsenic. This property may be characteristic of certain arsenic borate compounds. On the other hand, in other compounds in which the arsenic content is less than 15 mole%, the boron is held more strongly to the structure than the arsenic, resulting in a positive percentage loss of total arsenic. Compositions with As_2O_3 between 15 and 25 mole% may correspond to a mixture of the two types of compounds discussed in the foregoing, which may explain the negligible percentage loss of total arsenic shown for this composition range (Fig. 11a).

Figure 11 (b) shows the change in the ratio of As^{5+}/As^{3+} (as determined from analysis) with increasing mole% of total As_2O_3. The ratio As^{5+}/As^{3+} increased with increasing As_2O_3 reaching a maximum at about 5.0 mole% As_2O_3. Further increase in As_2O_3 may decrease the rate of oxidation of As^{3+} ions during the melting process. The reducing atmosphere formed by the arsenic vapors above the melt in glasses containing more than 5.0 mole% As_2O_3 (Fig. 11b) may partly explain the decrease in the rate of oxidation and accordingly the decrease in the As^{5+}/As^{3+} ratio. The low As^{5+}/As^{3+} ratio for glasses containing more than 10 mole% As_2O_3 may be attributed to the possibility that a major part of the arsenic in these compositions takes network-forming positions.

Infrared Studies

The results obtained from infrared absorption studies of the system $4.5B_2O_3 \cdot 1.0Na_2O \cdot xAs_2O_3$ are as follows:

Composition between 0.0 and 10.0 mole% As_2O_3. The increase in the BO_4/BO_3 ratio (Fig. 12a), as well as the increase of the BO_4/AsO_3 ratio (Fig. 12b), indicates an increase in the BO_4 concentration. On the other hand, the decrease in the $(BO_4)_I/(BO_4)_{II}$ ratio (Fig. 12d) may indicate the introduction of non-bridging oxygens (14). These observations suggest that in this composition range, the As_2O_3 enters chiefly as a network modifier giving part of its oxygen to four-coordinate the boron whereas the other part introduces non-bridging oxygens in the structure.

Composition between 10.0 and 15.0 mole% As_2O_3. In this range, the ratio of BO_4/BO_3 decreases appreciably (Fig. 12a). A similar decrease was obtained in the BO_4/AsO_3 ratios (Fig. 12b and e). This definitely proves that the total number of BO_4 groups decreases. In other words, the maximum four-coordination of the boron in this ternary system was reached at the composition containing about 10.0 mole% As_2O_3. The fundamental bands due to the AsO_3 groups increase in intensity, as shown from the decrease in both BO_4/AsO_3 and BO_3/AsO_3 ratios (Figs. 12 b and c). In other words, the arsenic becomes strongly bonded to three oxygens (14). There is also an indication of a decrease in the number of non-bridging oxygens, as shown from the increase in the $(BO_4)_I/(BO_4)_{II}$ ratio (Fig. 12d).

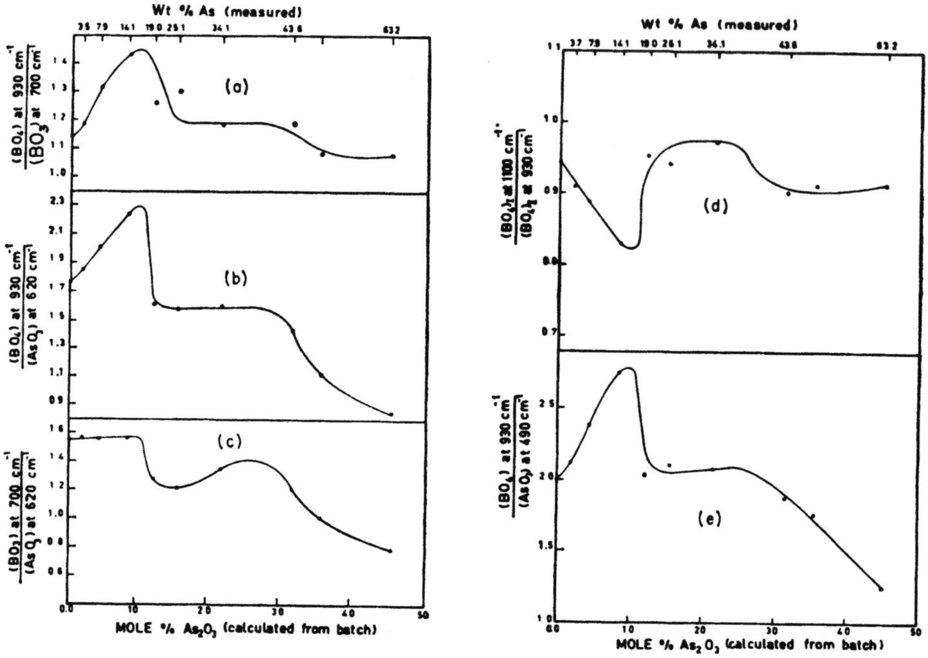

Fig. 12. Effect of increasing arsenic content in $4.5B_2O_3 \cdot 1.0Na_2O$ glasses on the infrared band intensities (KBr pellets).

Compositions between 15.0 and 25.0 mole% As_2O_3. This range seems to be characterized by little change since the ratios of BO_4/AsO_3, BO_4/BO_3, and $(BO_4)_I/(BO_4)_{II}$ show nearly no change in this composition range. On the other hand, an increase in the BO_3/AsO_3 (Fig. 12c) ratio was observed. This observation may indicate the formation of compounds of different configuration from those between 10.0 and 15.0 mole% As_2O_3.

Composition between 25.0 and 45.0 mole% As_2O_3. This range is associated with a decrease in the ratios BO_4/BO_3, BO_4/AsO_3, BO_3/AsO_3, and also a decrease in the $(BO_4)_I/(BO_4)_{II}$ ratio. This indicates a general decrease in the number of BO_4 and BO_3 groups which was seen clearly from their low band intensities. In this composition range arsenic continues to take part in the network. In these high-arsenic glasses the network may be based mainly on AsO_3 units, whereas from 0.0 to 10.0 mole% As_2O_3, the network is based mainly on BO_3 and BO_4 units. Compositions between 15 and 25 mole% As_2O_3 may contain a mixture of the two structures.

The structural changes which were suggested from gamma-irradiation studies are, therefore, in line with the results obtained from chemical analysis (total As and As^{5+}/As^{3+} ratio) and infrared studies.

BARIUM SILICATE GLASSES

The coloration induced by ionizing radiation in silicate glasses was studied by a large number of authors (1,15-18). Gaussian resolution of the observed induced absorption in silicate glasses showed four absorption bands at 2.0, 2.7, 4.0 and 5.5 eV, respectively (1). Recent electron spin resonance and optical studies of cerium-doped glasses by Stroud (17) suggested that the 2.0 and 2.7 eV bands are associated with positive holes and the ultra-violet band with electron traps. This is in line with Bishay's earlier postulation for the origin of these bands in borate (13) and silicate (18) glasses.

This section deals mainly with the effect of gamma irradiation on silicate glasses containing high concentrations of BaO ($1.0SiO_2 \cdot 0.08K_2O \cdot xBaO$) where x ranges from 0.23 to 0.5 mole (17.5 to 31.5 mole% BaO).

Induced Optical Absorption

Figure 13 shows the induced optical absorption for two glasses containing 17.5 and 31.5 mole% BaO respectively. Each glass was irradiated for two doses: 9×10^6 rads, and 8×10^7 rads. Three peaks were observed at about 2.1, 2.75 and 4.0 eV in the sample containing 17.5 mole% BaO. However, the 2.1 and 2.75 eV bands overlap completely giving a maximum at about 2.4 eV in the sample containing 31.5 mole% BaO. Increasing the radiation dose from 9×10^6 to 8×10^7 rads was not accompanied by any observable shift in the position of the different induced bands. On the other hand, increasing BaO from 17.5 to 31.5 mole% was accompanied by an observed shift in the position of the induced bands to lower energies as well as a general decrease in induced absorption.

Figure 14 shows the effect of increasing BaO on the intensity of induced absorption at 2.1, 2.75 and 4.0 eV. A general decrease in the intensity of these bands was observed with increasing BaO content. However, the rate of decrease of intensity with increasing of BaO showed two main changes at about 22.5 and 27.5 mole% BaO.

Corresponding changes were observed at these compositions in other physical properties for the same series of glasses (19).

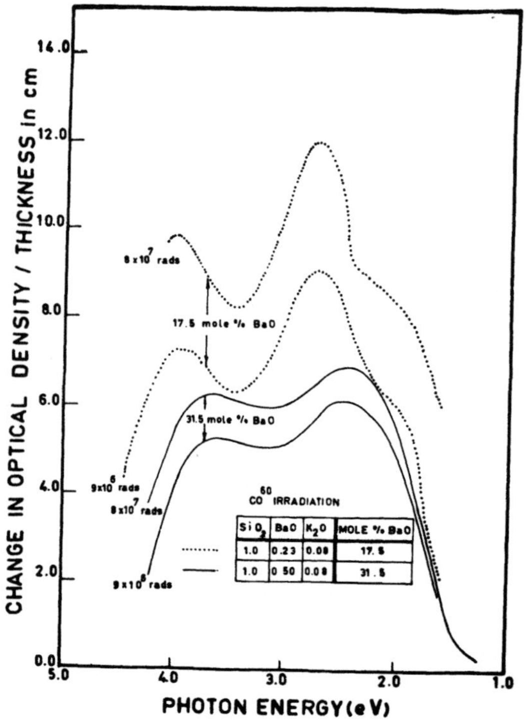

Fig. 13. Gamma induced optical absorption for two barium silicate glasses containing 17.5 and 31.5 mole% BaO irradiated for 9x10^6 and 8x10^7 rads.

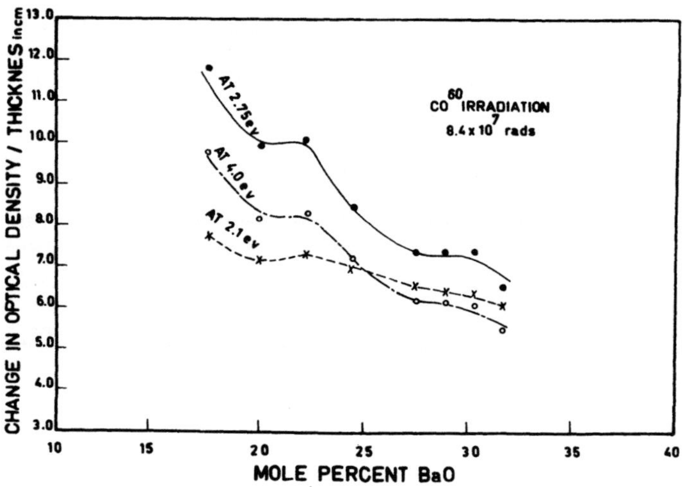

Fig. 14. Effect of increasing BaO on the intensity of three absorption bands induced at 2.1, 2.75 and 4.0 eV.

Infrared Studies

Figure 15 shows the infrared transmittance curves for three barium silicate glasses containing 17.5, 27.5 and 31.5 mole% BaO, respectively. The transmittance curve for fused silica is included for comparison. The fundamental vibrations for SiO_4 in fused silica are clearly shown at about 1100 cm^{-1} (9.1 μ) for the triply degenerate ν_3, 800 cm^{-1} (12.5 μ) for ν_1 and 470 cm^{-1}(21.3μ) for the triply degenerate ν_4. The doubly degenerate ν_2 frequency is beyond the range of the spectrophotometer. The infrared transmittance curves of the barium silicate glasses are characterized by a shift to lower frequencies when compared to fused silica.

The relative intensity of the ν_1 band was obtained by comparing the optical density at the center of the band for each glass to that observed for a glass containing the lowest BaO content (17.5 mole% BaO). The changes in this relative intensity with changing BaO are shown in Figure 16. This figure also shows the changes in molar volume for the same series of glasses, as well as changes in induced absorption at 4.0 eV.

It may be postulated that the changes in physical properties observed at about 22.5 and 27.5 mole% BaO for this series of glasses are associated with the formation or disappearance of certain structural groups in the glass at these compositions. In order to confirm this postulated structural change, the glasses were heat treated followed by x-ray diffraction studies of the devitrified samples.

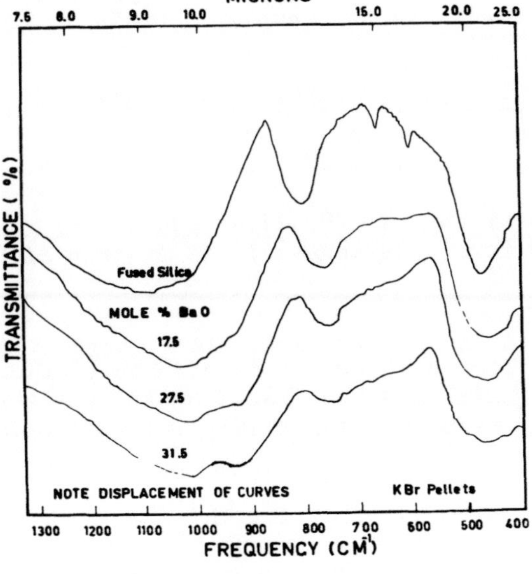

Fig. 15. Infrared transmittance for three barium silicate glasses (17.5, 27.5 and 31.5 mole% BaO) and fused silica.

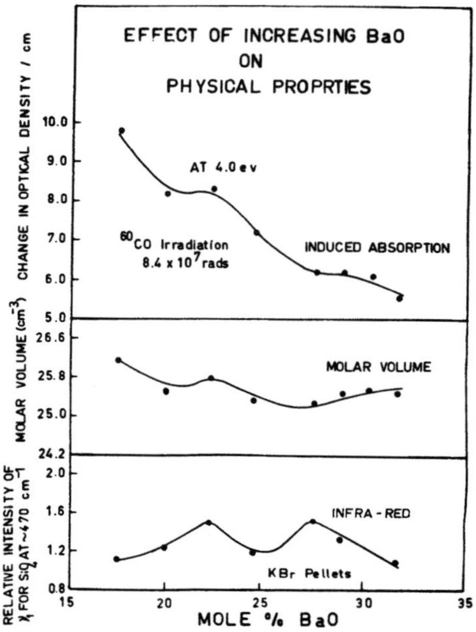

Fig. 16. Effect of increasing BaO on induced optical absorption at 4.0 eV, molar volume, and the relative intensity of the ν_1 band for SiO_4($\sim$470cm-1).

X-Ray Diffraction Studies

Figure 17 shows the x-ray diffraction patterns for finely powdered devitrified samples containing 17.5, 22.5, 24.5, 27.5, and 31.5 mole% BaO. Using the A.S.T.M. Cards, it was possible to identify the following compounds in the sample containing 17.5 mole% BaO (Fig. 17a).

1. alpha-quartz (strongest lines at 2θ = 26.6° and 40.3°)
2. $Ba_5Si_8O_{21}$ (strongest lines at 2θ = 23.7°, 27.4° and 32.3°)
3. $Ba_3Si_5O_{13}$ (strongest lines at 2θ = 23.2°, 27.4°, 32.1° and 40.6°)
4. $Ba_2Si_3O_8$ (strongest lines at 2θ = 23.7°, 27.4° and 32.4°)
5. Traces of tridymite (low form, strongest lines at 2θ = 21.6, 35.6 and 47.0).

This identification is in lines with Mackenzie's suggestion (20) for the presence at this composition of islets of vitreous silica (represented by the α-Quartz and the traces of Tridyimite), discrete ions (represented by $Ba_5Si_8O_{21}$ and $Ba_3Si_5O_{13}$) and rings as ($Ba_2Si_3O_8$).

When comparing the diffraction patterns for the five devitrified samples, it was observed that the above compounds are present in all the samples. However, some changes occur at the composi-

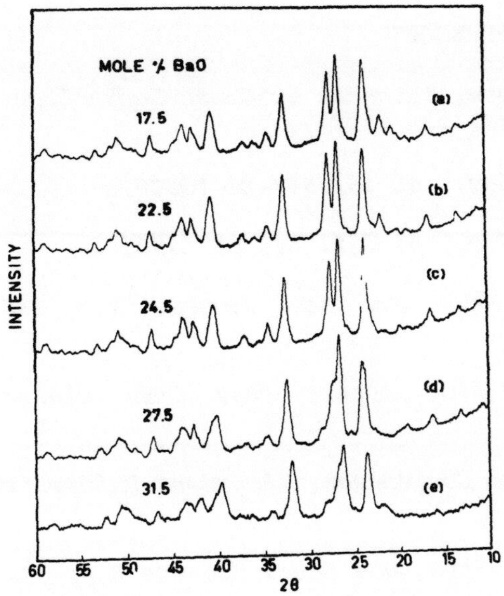

Fig. 17. X-ray diffraction patterns for five heat treated devitrified samples containing 17.5, 22.5, 24.5, 27.5 and 31.5 mole% BaO.

tions containing about 22.5 and 27.5 mole% BaO (Fig. 17). The first major change is the disappearance of the strongest line of tridymite at $2\theta = 21.6^{o}$ in those devitrified samples containing more than 22.5 mole% BaO. The second major change which is observed at compositions containing 27.5 mole% BaO or higher, is the pronounced decrease in the intensity of the ($2\theta = 27.4^{o}$) line characteristic for $Ba_5Si_8O_{21}$, $Ba_2Si_3O_8$ and $Ba_3Si_5O_{13}$.

The changes observed in the x-ray diffraction curves of the devitrified samples confirm the suggestion that changes in the structural groupings occur at these compositions (22.5 and 27.5 mole% BaO).

CONCLUSIONS

Changes in optical absorption induced by gamma irradiation of a number of lead borate, alkali-arsenic-borate, and alkali-barium-silicate glasses may be used as a valuable tool for detecting some of the structural changes in these glasses.

It is postulated that the structural groupings in glass are similar to those in corresponding crystals. Accordingly, crystals which have these groupings are separated on devitrification of the glasses by heat treatment. The appearance or disappearance of these crystals at certain compositions was found to coincide with changes in physical properties like molar volume, infrared and induced absorption.

REFERENCES

1. *A.M. Bishay, and K.R. Ferguson, Advances in Glass Technology, Plenum Press, New York pp. 133-138 (1962).*

2. *A.M. Bishay, J. Am. Ceram. Soc., 43 (8) 417-21 (1960).*

3. *A.M. Bishay, J. Am. Ceram. Soc., 44 (11) 545-52 (1961).*

4. *A.M. Bishay and M. Maklad, Phys. and Chem. Glasses, 7,5, 149-156 (1966).*

5. *P.J. Bray, M. Leventhal and H.O. Hooper, Phys. Chem. Glasses 4,2, 47-66 (1963).*

6. *P.J. Bray, Cairo Solid State Conference, American University in Cairo, September 3, 1966.*

7. *M. Leventhal and P.J. Bray, Phys. and Chem. Glasses 6,4, 113-125 (1965).*

8. *J. Krogh-Moe, Phys. and Chem. Glasses, 6,2, 46-54 (1965).*

9. *A.M. Bishay and S. Arafa, Phys. and Chem. Glasses, 6,4, 134-42 (1965).*

10. *A.M. Bishay, Phys. and Chem. Glasses, 2, 2, 33-38 (1961).*

11. *A.M. Bishay, J. Am. Ceram. Soc., 44, 6, 289-96 (1961).*

12. *K.R.Ferguson, pp. 360-66 in Proceeding of the IV International Glass Congress, Paris, 1956. Imprimerie Chaix, Paris, France, 1957.*

13. *A.M. Bishay, J. Am. Ceram. Soc., 45,8,389-93 (1962).*

14. *A.M. Bishay and S. Arafa, J. Am. Ceram. Soc., 49,8, 423 - 30 (1966).*

15. *R. Yokota, Phys. Rev., 93,4, 896-897 (1954).*

16. *A. Kats and J.M. Stevels, Philips Research Reports, Vol. 11, April 1956, pp. 115-156.*

17. *J.S. Stroud, The Journal of Chemical Physics, 37,4, 836-841 (1962).*

18. *A.M. Bishay, proceedings of the VII Conference of the silicate Industry (Budapest, Hungary) June 1963.*

19. *A.M. Bishay and I. Gomaa, "Radiation Induced Defects and Structure of Barium Silicate Glasses", accepted for publication, J. Amer. Ceram. Soc.*

20. *J.D. Mackenzie, Modern Aspects of the Vitreous State, Vol. 1, P. 6, Butterworths, London, 1960.*

DISCUSSIONS

W. HALLER (National Bureau of Standard, U.S.A.): *Are you aware that your lead borate glasses may be heterogeneous? You are in the middle of a subliquidus immiscibility gap. This may not only give you the changes in your curves but the devitrification may start in one of the two glassy phases.*

A.M. BISHAY: *This is quite possible. Thank you for your comment.*

CHARLES GREENE (Alfred University, New York, U.S.A.): *Were the melts in air? Was* As_2O_3 *the batch raw material? Was the melting time sufficient to reach equilibrium between* As_2O_3 *and* As_2O_5 *in the glass?*

A.M. BISHAY: *The answer is "Yes" to all three questions. In one case* As_2O_5 *was introduced as the batch raw material. However, the induced optical absorption was practically the same as when* As_2O_3 *was used thus indicating that the time of melting was enough to reach equilibrium between* As_2O_3 *and* As_2O_5 *in the glass.*

J.D. MACKENZIE (Rensselaer Polytechnic Institute, New York, U.S.A.): *I see that there are sharp breaks at certain compositions. This reminds me of the old crystallite theory of glass. If one were to put error limits on the values shown then we can be really sure if there are sharp breaks.*

A.M. BISHAY: *The sharp breaks in the* N_4 *figure of Bray may be somewhat exagerated. However, the fact that many physical properties show pronounced changes at certain compositions cannot be doubted. Duplicate samples confirmed these changes.*

GAMMA RADIATION INDUCED OPTICAL ABSORPTION

IN LEAD BORATE GLASSES

R.S. Barker

Research and Development Laboratories

Pilkington Brothers Ltd., Lathom, Ormskirk, England

The gamma radiation induced absorption spectra between 3.45 eV and 0.81 eV in lead borate glasses containing between 36% molar and 64% molar lead oxide have been measured after radiation doses up to 10^8 rads.

The spectra have been resolved into three Gaussian-shaped absorption bands centered at 1.61 eV, 2.40 eV and 3.28 eV.

For each glass the variation in intensity of each band with radiation dose consists of a saturating component and a linear component.

After any given dose, the intensity of the 2.40 eV band decreases continuously with increasing lead content, but the intensity of the 3.28 eV band passes through a minimum corresponding to about 55% molar PbO.

INTRODUCTION

Absorption bands induced in lead borate glasses by gamma radiation have been the subject of studies by Bishay who extensively investigated radiation induced absorption in borate glass systems. In the calcium-alumino-borate system, Bishay reported bands centered at 2.3 eV, 3.5 eV and 5.0 eV (1); he also reported an additional band at 1.5 eV in lead borate glasses (2).

Previous work by the author and others has shown that the growth, with increase in radiation dose, of the induced bands at 3.31 eV and 2.36 eV in lead silicate glass can be related to two components (3). One component, the saturating part, was attributed to trapping of charge at defect sites present in the glass; the other, the linear part, was attributed to the formation by the radiation of similar defect sites and the subsequent trapping of charge at these sites.

The present work was undertaken to examine the growth of the radiation induced absorption bands in lead borate glasses of different molar compositions.

EXPERIMENTAL

Batches prepared from A.R. materials were melted under normal atmospheric conditions in an electric furnace to produce glasses of the molar compositions given in Table I.

TABLE I

MOLAR COMPOSITON AND REFRACTIVE INDICES

OF THE GLASSES

Glass	Mol. % B_2O_3	Mol. % PbO	Refractive Index
A	63.2	36.8	1.765
B	57.9	42.1	1.810
C	51.6	48.4	1.863
D	44.4	55.6	1.930
E	34.1	63.9	2.000

After casting into plates, the glasses were carefully annealed so that they were strain free.

From each glass plate specimens, having a cross-section of 3 cm by 1 cm and thickness of either 1 cm or 0.33 cm, were prepared with opposite faces flat and parallel.

The samples were irradiated in the dark at room temperature in the gamma radiation field from a 4,000 curie cobalt 60 source, which gave a dose rate in air of 10^5 rads per hour. The samples received doses varying between 10^4 rads and 10^8 rads.

As soon as possible after each irradiation, the radiation-induced optical absorption spectrum for each sample was measured, using a Uvispek spectrophotometer, to compare the absorption of the irradiated specimen with that of a previously matched sample.

The absorption measurements were repeated after given intervals of time during which the samples were kept in the dark at room temperature.

RESULTS

Following high radiation doses, intense darkening of the samples was produced and it was not possible to measure complete absorption spectra of these samples immediately after removal from the radiation field. In certain cases reliable measurements could not be obtained until after the samples had undergone five hours thermal bleaching in the dark at room temperature.

To compare the absorption induced by different radiation doses, it was necessary to compare measurements obtained after more than five hours of thermal bleaching.

Glasses were exposed to radiation doses between 10^4 rads and 10^8 rads. The induced absorption spectra for glasses A, B and C showed evidence for three bands in the range 0.35 to 1 μ. The spectra of glasses D and E, however, showed evidence for only two absorption bands in this range.

Assuming that the absorption can be resolved into a number of Gaussian shaped absorption bands, each complete spectrum was resolved into either two or three such bands by means of an iterative process based on the equation,

$$\alpha = \sum \alpha_{i_{max}} \exp B_i \left({}^1/\lambda - {}^1/\lambda_{i_{max}} \right)^2$$

where α = absorption at wavelength λ

B_i = a function of the width of the i'th band at ½ max absorption

$\alpha_{i\,max}$ = the peak absorption of the i'th band

and $\lambda_{i\,max}$ = the wavelength corresponding to the peak of the i'th band.

The computed bands for each glass after radiation doses of 10^4 and 10^8 rads and after 400 minutes thermal bleaching are shown in Figures 1 to 5. Between 0.4 and 1 μ, glasses A, B and C show

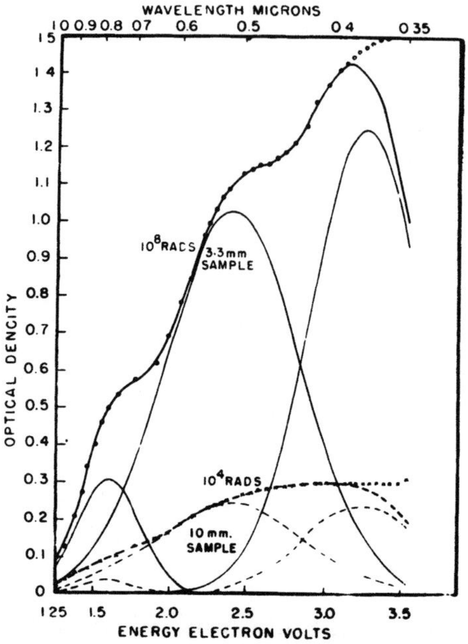

Fig. 1. Absorption Spectra Induced in Glass 'A' by Gamma Radiation.

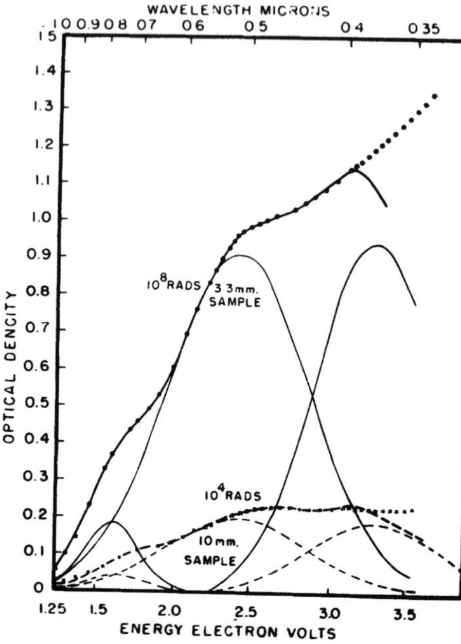

Fig. 2. Absorption Spectra Induced in Glass 'B' by Gamma Radiation.

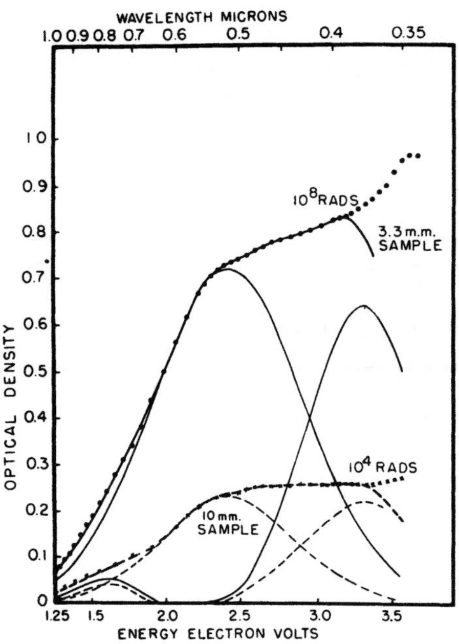

Fig. 3. Absorption Spectra Induced in Glass 'C' by Gamma Radiation.

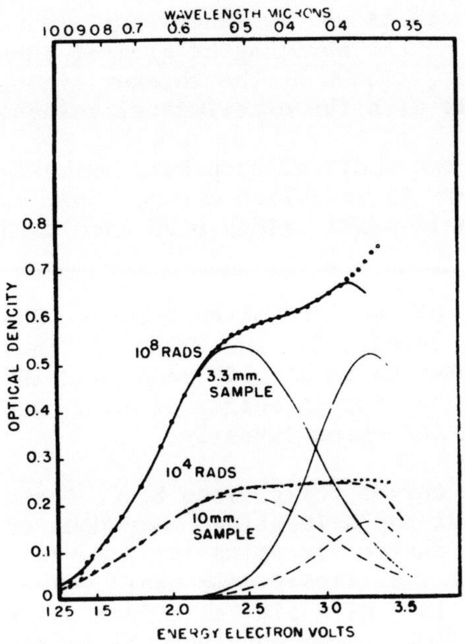

Fig. 4. Absorption Spectra Induced in Glass 'D' by Gamma Radiation.

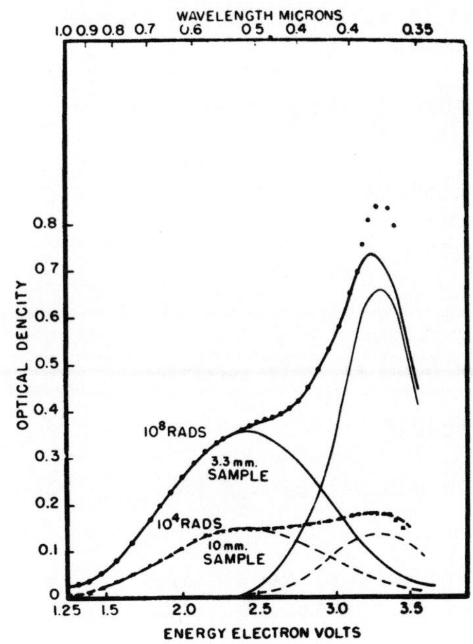

Fig. 5. Absorption Spectra Induced in Glass 'E' by Gamma Radiation.

three absorption bands at 1.61 eV, 2.4 eV and 3.28 eV for all radiation doses up to 10^8 rads. There is no evidence for a band induced at 1.6 eV in glasses D and E. At wavelengths greater than 0.4 μ, the sum of the computed bands, shown by the thicker lines, can be seen to be in close agreement with the experimental values.

For a given glass composition the width of each band at half-maximum absorption was constant for all radiation doses, but, as can be seen from Table II, the 'half-peak' width U of each band changed with changing composition.

In all glasses the intensity of each radiation-induced absorption band increases with dose. Figure 6a and Figure 6b show the increases for glass A after doses up to 2×10^6 rads and 10^8 rads respectively. It can be seen that the intensity of each band increases rapidly at first and then increases linearly.

Figures 7a to 10b show similar curves for glasses B, C, D and E. It has been postulated (3) that the saturating component of each growth curve corresponds to charge trapping at a certain defect site in the glass, and that the linear component corresponds to the creation by the radiation of a similar defect which rapidly traps charge. (The absorption value obtained by extrapolation of the linear part of the curve to zero dose should thus give some indication of the concentration of that defect in the glass which gives rise to the absorption).

The color center concentration is related to the absorption maximum of a band by the Smakula formula;

$$N_e f = 0.87 \times 10^{17} \frac{n}{(n^2 + 2)^2} U \alpha_m$$

where, N_e = number of color centers

f = oscillator strength

n = refractive index

U = full width of the band at half maximum

α_m = absorption coefficient

Table II gives the values of the product of $N_e f$ for each band calculated from the above equation.

The values of α_m are those obtained by extrapolation to zero dose of the linear part of the curves in Figures 6b - 10b.

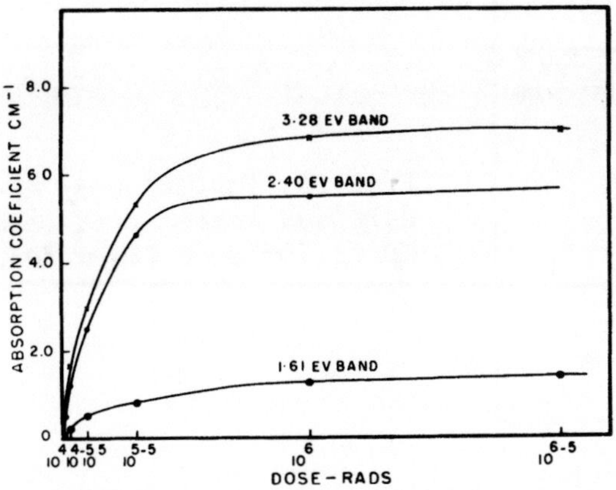

Fig. 6a. Variation of Absorption Maxima with Dose up to 2x10^6 rads Glass 'A'

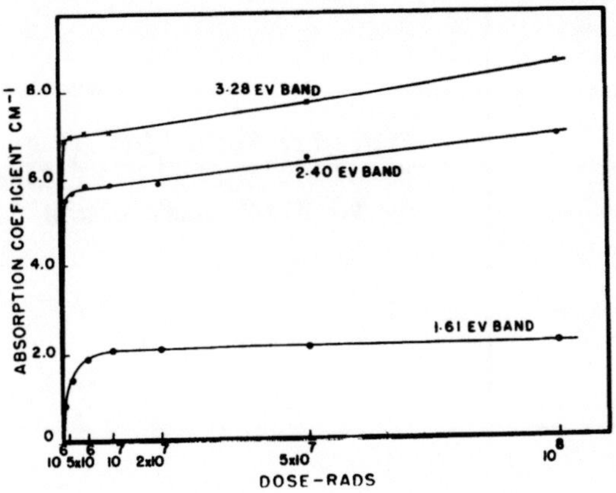

Fig. 6b. Variation of Absorption Maxima with dose up to 10^8 rads glass 'A'.

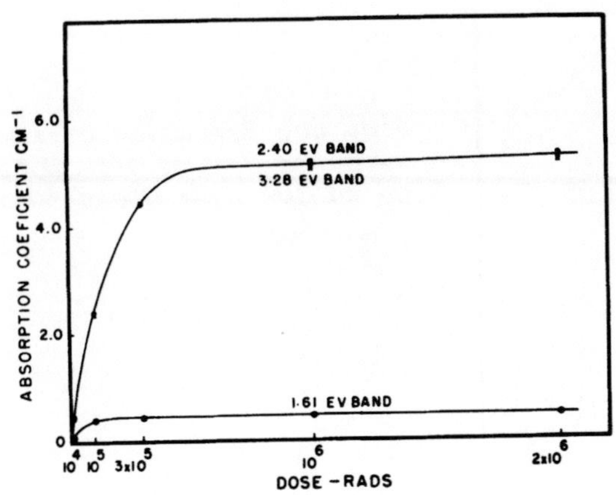

Fig. 7a. Variation of Absorption Maxima with dose up to 2x10^6 rads Glass 'B'

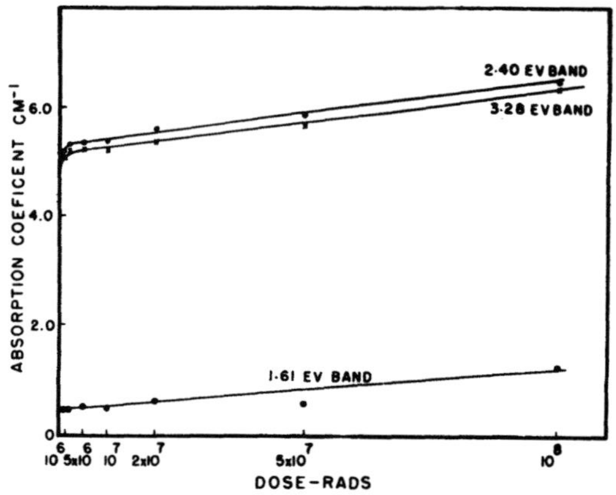

Fig. 7b. Variation of Absorption Maxima with Dose up to 10^8 Rads Glass 'B'.

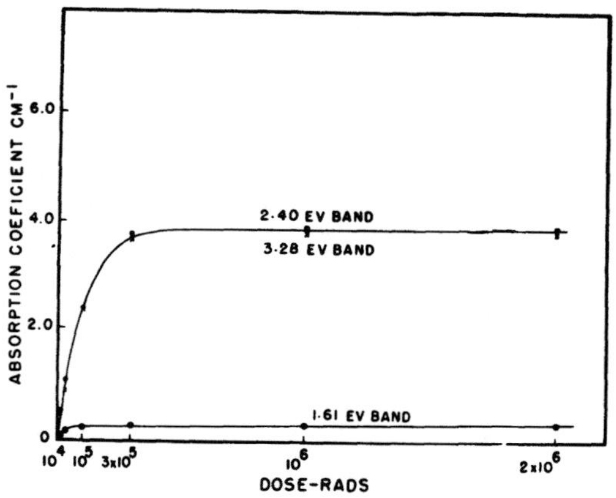

Fig. 8a. Variation of Absorption Maxima with Dose up to $2x10^6$ Rads Glass 'C'

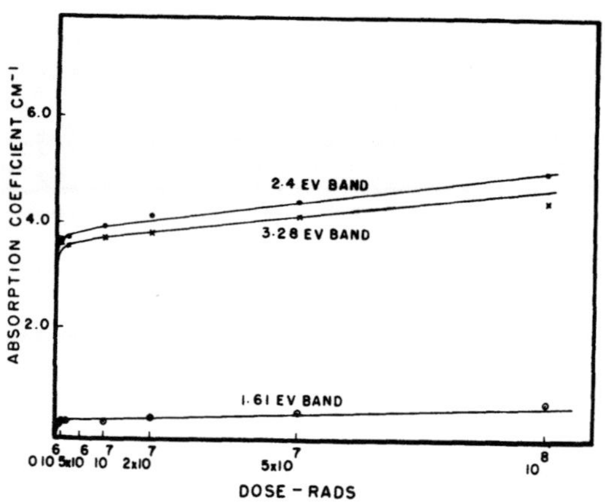

Fig. 8b. Variation of Absorption Maxima with Dose up to 10^8 Rads Glass 'C'.

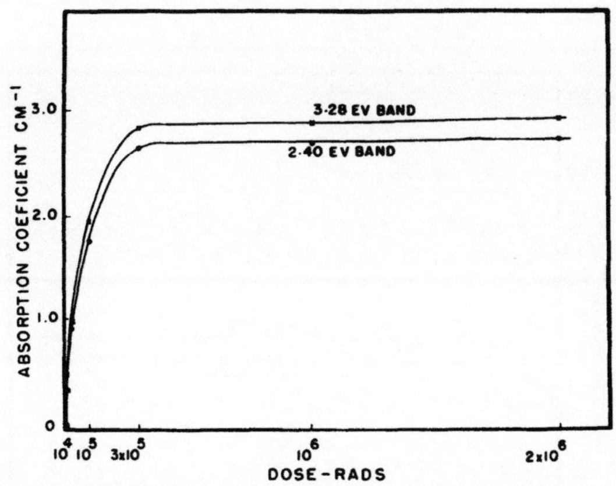

Fig 9a. Variation of Absorption Maxima with Dose up to 2x10⁶ Rads Glass 'D'

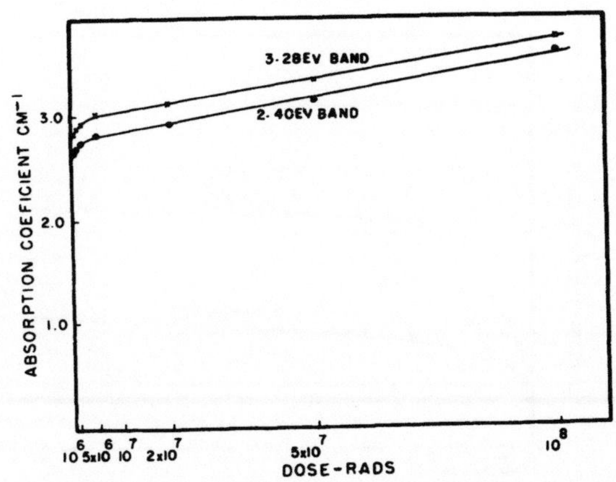

Fig. 9b. Variation of Absorption Maxima with Dose up to 10^8 Rads Glass 'D'.

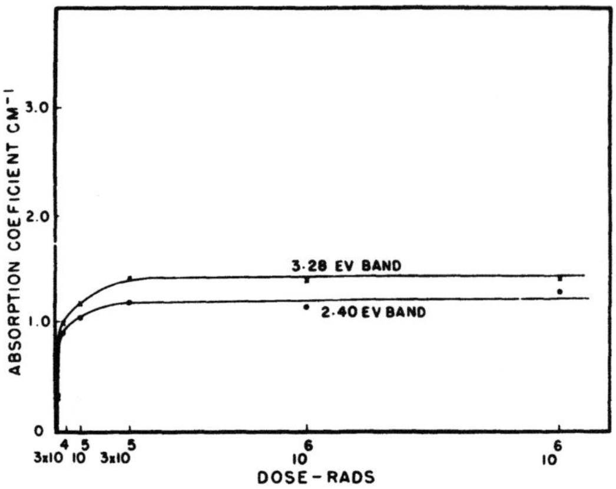

Fig. 10a. Variation of Absorption Maxima with Dose up to 2×10^6 Rads Glass 'E'.

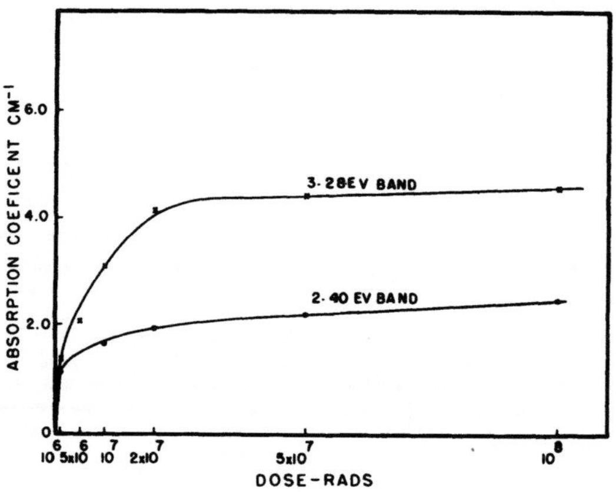

Fig. 10b. Variation of Absorption Maxima with Dose up to 10^8 Rads Glass 'E'.

TABLE II

PARAMETERS OF THE INDUCED ABSORPTION BANDS

Glass	Ref. Index n	1.61 eV band			2.40 eV band			3.28 eV band		
		α_m	U(eV)	Nf x 10^{16}	α_m	U(eV)	Nf x10^{16}	α_m	U(eV)	Nf x 10^{16}
A	1.765	2.05	0.5	.6	5.82	1.05	3.6	6.90	0.83	3.37
B	1.810	0.48	0.43	.112	5.25	1.06	3.15	5.16	0.84	2.47
C	1.863	0.33	0.45	.081	3.75	1.10	2.25	3.55	0.83	1.60
D	1.930				2.80	1.10	1.59	2.95	0.79	1.2
E	2.000				1.80	1.20	1.04	4.3	0.63	1.31

The values of Nf should give an indication of the concentration of those defects in the glass which give rise to the absorption.

The variation with lead oxide content of the glass of the values of Nf for each of the three centers is shown in Figure 11.

It is clear that insufficient data have been obtained to describe the change in concentration of the 1.61 eV centers. The concentration of the 2.4 eV centers is seen to decrease steadily with increasing lead content. The concentration of the 3.28 eV centers passes through a minimum corresponding to about 56 molar % lead oxide.

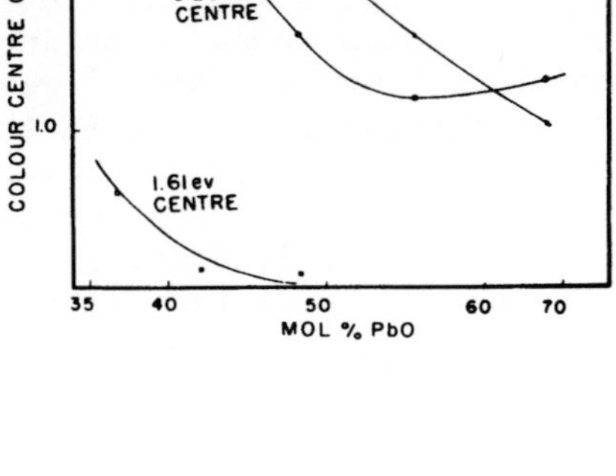

Fig. 11. Variation of Color Center Concentration with Composition.

DISCUSSION OF RESULTS

1.6 eV Band

Bishay (2) reported a gamma-radiation induced absorption band centered at about 1.5 eV in lead borate glasses. It is probable that this band is the same as the one we locate at 1.61 eV. From Figures 1 to 5, for radiation doses of 10^4 and 10^8 rads, and also from Figure 11, it can be seen that the intensity of this band decreases with increasing lead content. This band is not detected in the case of the two glasses of highest lead content.

Since it would not be expected that the Pb^{++} concentration would decrease as the lead content of the glass increased it seems unlikely that the 1.5 eV band is associated with the Pb^{++} ion.

2.48 eV Band

It seems probable that this band is the same as the radiation induced band at 2.36 eV found in calcium-alumino-borate glasses and attributed by Bishay (1) to holes trapped at sites of non-bridging oxygen ions. The concentration of the centers at 2.4 eV can be seen from Figures 5 to 10 and from Figure 11 to decrease with increasing lead content throughout the range of compositions studied, i.e. 36.8 molar % to 63.9 molar %. Bray (4) has shown that, in lead borate glasses containing more than 15 to 20 molar % lead oxide, some of the lead enters the network to form PbO_4 units and it would be expected, therefore, that increasing the lead content beyond 20 molar % would decrease the number of non-bridging oxygen atoms in the glass. Our observations are thus in agreement with the suggestion made by Bishay that the center responsible for this band is associated with non-bridging oxygen ions.

3.28 eV Band

It can be seen from Figures 6 to 10 and also from Figure 11 that the intensity of the 3.28 eV band is a minimum for a lead borate glass containing 55 molar % lead oxide. From N.M.R. studies of lead borate glasses, Bray (4) has shown that the ratio of the concentration of four co-ordinated boron to that of three co-ordinated boron reached a maximum at about 50 molar % lead oxide. The fact that the concentration of the centers at 3.28 eV is a minimum at about the same composition suggests that this band is connected with three co-ordinated boron. Further support for this postulate can be found from Bishay's work on calcium-alumino-borate glasses which showed that the intensity of the induced band at 3.5 eV decreased with increasing lime content but increased with increasing alumina content. Addition of Al_2O_3, which can form AlO_4 units, would increase the ratio B_3/B_4, whereas addition of CaO would decrease the ratio.

REFERENCES

1. *A.M. Bishay, J. Am. Ceram. Soc., 44, No. 6, (1961), 289 - 296.*

2. *A.M. Bishay, J. Am. Ceram. Soc., 43, No. 8, (1960), 417 - 421.*

3. *R.S. Barker, D.A. Richardson and E.A.G. McConkey, Phys. and Chem. of Glasses, 6, No. 1, (1965), 24 - 29.*

4. *P. Bray, M. Leventhall and H.O.Hooper, Phys. and Chem. of Glasses, 4, No. 2, (1963), 47 - 66.*

DISCUSSIONS

ADLI BISHAY (American University in Cairo, UAR): *I am afraid that your conclusion regarding the origin of the 1.6 eV band is not convincing. You must remember that your lowest PbO content was 36.8 mole %. In our paper, presented in this conference, we have clearly shown that the intensity of this band increased with increasing lead content, reaching a maximum at about 25 mole % PbO. Furthermore, since this band is only observed in glasses containing Pb^{2+}, or ions of similar electronic configuration (Tl^{+}, Bi^{3+}), and from the observation that addition of cerium in the ceric state strongly suppresses the absorption in this vicinity, we have suggested elsewhere that this band can be attributed to an electron trap center associated with Pb^{2+} ions in the glass.*

R. S. BARKER: *We did not study glasses of low lead content since they are not of good quality. Furthermore, the decrease in the intensity of this band at high lead concentrations does not favor your postulate that this band is associated with Pb^{2+} ions.*

ADLI BISHAY: *I am surprised to hear that you could not obtain good quality samples of low lead content. We have made very good specimens of glasses containing as low as 18.1 mole % PbO. Furthermore, I would like to stress that we are associating the 1.6 eV band with Pb^{2+} in the ionic state and not with total lead content. The NMR work of Bray (determination of Pb^{207} chemical shift) for the same system has shown that all or most of the lead is present as Pb^{2+} ions in lead borate glasses containing less than about 25 mole % PbO. .On the other hand, lead forms structural groupingsin glasses containing higher PbO contents. This change in the structural role of Pb would explain the decrease in the intensity of this band in glasses containing high concentrations of PbO (> 25 mole %).*

LASER DAMAGE IN TRANSPARENT GLASSES AND CRYSTALS

V.D. Frechette and Carl Cline

College of Ceramics

Alfred University, New York, U.S.A.

The focused beam from a giant-pulse ruby laser effects internal damage in optically transparent crystals and glasses. Bubble formation and incandescence indicate that high temperatures were involved in the process, while open cracks and cavities which were unconnected with free surfaces represent yielding to the action of high mechanical stresses. The fact that E and O rays in calcite were equally damaging although their vibrations correspond to directions of very different atomic packing density and bonding strength is taken to mean that no direct coupling occurs between atoms or ions and the electric field of the beam. Of other mechanisms suggested to account for the conversion of laser light energy to thermal and mechanical energy, only that involving electron avalanche seems to be without objection. Weight is lent to this explanation by the observation that LiF damaged by a laser pulse at 77°K showed thermoluminescence peaks at 100° and 140°K. Such peaks have been related to the annealing of self-trapped electrons and holes and their presence suggests that electrons had been transported by action of the laser beam.

Damage in materials by laser light of wavelengths to which they are normally transparent presents an intriguing aspect of interaction between radiation and solids. While the laser beams in question are of high intensity and by virtue of their sharply monochromatic character, their coherence and their small divergence are capable of focusing to yield an extremely high power density, it is not at all clear what mechanism of energy conversion leads to the high temperatures inferred and to the mechanical displacements observed. A number of suggestions have been offered (1), including (a) normal optical absorption with attendant heating, (b) absorption by included foreign particles, (c) phonon generation by stimulated Brillouin or Raman scattering, (d) acoustic phonons driven at the beat frequency between component frequencies of the beam, (e) direct displacement of electrons in response to the electric field of the beam, possibly precipitating an avalanche in which accelerated freed electrons ionize other sites and are joined by additional electrons thus freed, and (f) direct action of the electrical field to move atoms or ions of the solid from their equilibrium sites.

The present work began with a study of a number of optically clear materials damaged internally by a focused giant pulse ruby laser beam operated slightly above the damage threshold as this was determined (but not quantitatively evaluated) by preliminary trials. The experiments were conducted with a ruby laser 15 cm. long and 1.2 cm. in diameter with Brewster cut, "pumped" by discharge of a condenser bank through Xenon lamps and controlled by a Kerr cell. A 25 mm lens focused the beam within the specimen. The focal spot area was approximately 50 microns in diameter and the pulse duration 30 nsec.

Damage to polished surfaces of inorganic glasses and crystals and to surfaces exposed by cleavage occurs at comparatively low power densities (2). Near threshold it takes the form of a network of shallow open cracks whose courses are dominated by pertinent cleavage directions but are not limited to these. Coloration of the area is sometimes observed. Figure 1 is typical. At higher pulse intensities, cracks are wider and deeper and there is material loss by spalling and volatilization. While such surface damage is enormously important, representing as it does the power-limiting factor in laser operation and in the intensity which can be applied through optical systems without harm to the components, its understanding is complicated by the reduced symmetry of the damage arena, by singularities in chemical bonding and by the likelihood of preexisting flaws and adsorbed species. Attention to internal damage centers is therefore a logical prerequisite to understanding the surface case.

Brought to focus within a glass or crystal a laser beam just intense enough generates a single damage center or a series of unconnected centers along the beam axis. Figure 2 illustrates this in quartz. The most obvious feature of these is a family of interpenetrating crack surfaces. Newton rings observed in transmission show these cracks to be open, their walls separated by as much as several microns. As in the case of surface damage these prefer the cleavage directions but even in crystals with excellent cubic and poor dodecahedral cleavage random directions are also represented in the minor cracks. Prominent hackle further suggests the violence of the event (Fig. 3).

At the cores of these fracture systems, bubbles have sometimes been formed, indicating development of rather high temperatures. Bubbles at damage centers in halite are shown in Figure 3. While these are characteristically elongated and are often larger in diameter at one end, such geometry is consistent with the symmetry of the beam and with stress and thermal interference between neighboring centers. Bubbling has been noted previously in glasses, and Martinelli (2) has seen glass fibers drawn during fragmentation by heavy pulses. Olness (3) has reported that light having the line spectrum of sodium was emitted from damage centers in NaCl crystals.

Irregular patches of reddish brown stained the damaged regions of optical quality LiF crystals as shown in Figure 4. The coloration markedly resembled that caused by gamma irradiation in the same material and required a temperature of 550°C for complete annealing.

A single crystal of LiF immersed in liquid nitrogen was subjected to a focused laser pulse sufficient to cause internal damage and was transferred immediately to a thermoluminescence reader unit where it was allowed to warm, reaching 200°K in ten minutes. The glow curves showed two distinct peaks at 100° and 140°K with half-widths of 15°K. In three runs, brightness values between 2×10^{-6} and 4×10^{-6} lamberts were recorded for the more intense 140°K peak, the lower value being observed when a longer time intervened before the glow curve was run. LiF subjected to irradiation by gamma rays or X-rays at liquid nitrogen temperature and allowed to warm have also been shown to exhibit thermoluminescence with peaks at 100 and 135°K (4,5). These peaks, corresponding to discontinuities in optical absorption, electrical conductivity and paramagnetic resonance (6) appear to be uniquely attributable to the release of self-trapped electrons and to subsequent electron transitions and recombination processes.

In all of the optically isotropic materials studied, relative radial compression existed in the region bordering the damage center. In the case of NaCl crystals this compression exhibited sharp maxima in the 110 directions along which no slip could occur to facilitate relaxation (Fig. 5).

In no case was there evidence of any directional effect related to the polarization direction of the laser beam. It is possible that such an effect might be confined to too small a region to be easily seen, being obscured by the gross fracture effects following the primary event. In order to explore the possibility that the vibration direction of the beam might be critical in generating damage within strongly anisotropic materials, an unpolarized beam was focused in a calcite crystal. Traveling

Fig. 1. Polished 110 prism surface of a synthetic rutile crystal damaged by impingement of the ruby laser beam at normal incidence. Development of the good 110 cleavage (vertical) is evident together with cross fracture and dark coloration (blue-green) in contrast to the faintly yellow region at the top left where the damaged layer has scaled off. Transmitted light 600X.

Fig. 4. Reddish-brown coloration (central dark patches) at internal damage site in a LiF crystal. Laser beam was normally incident on the 100 cleavage. Small bubble-like details are dispersed about the center. Transmitted light, 450X.

FIG. 1

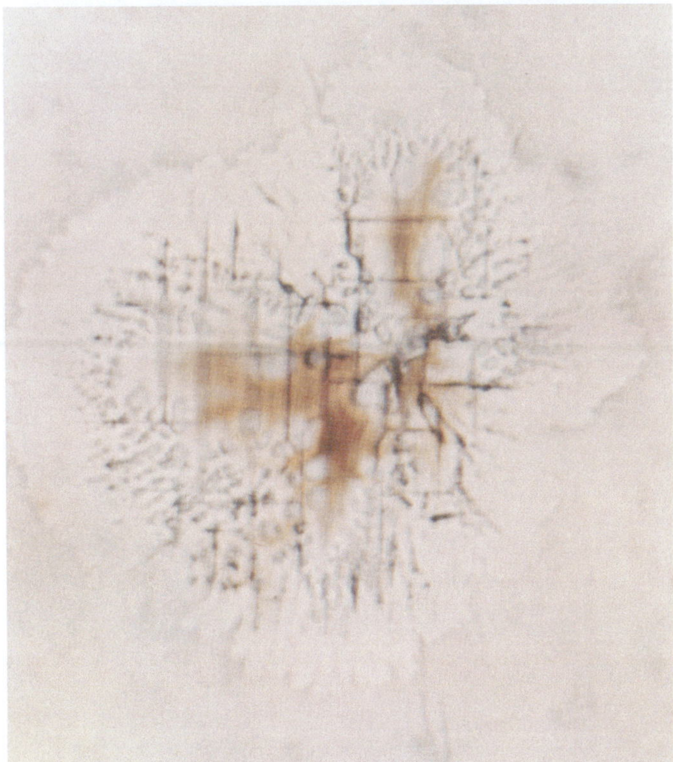

FIG. 4

Fig. 2. Series of internal damage centers along the axis of a focused laser beam travelling normal to the c-axis in a synthetic quartz crystal. Development of the usually unobserved rhombohedral cleavage is evident. 40X.

Fig. 3. Bubble developed in fused zone at an internal damage center in a melt-grown NaCl crystal. Major crack surfaces follow the cubic cleavage but are marked by extreme hackle beam direction right to left. Decentered dark field incident illumination, 400X.

Fig. 5. NaCl crystal internal damage center observed between crossed nicols and crossed quarter-wave plates to show the high concentration of radial compression along the 110 directions (bright cross). The stresses are proportioned to the brightness. 100X.

perpendicular to the rhombohedral cleavage in calcite, an unpolarized light ray is split into two of equal intensity which travel along diverging paths at different velocities and vibrate in directions at right angles to one another. The faster E-ray vibrates in the plane of high thermal expansion and the O-ray in the plane of low expansion and correspondingly tighter bonding. Figure 6 shows the E- and O-ray damage tracks which in a number of trials show essentially identical severity of damage.

Perhaps it is still too early to be able to decide the question of the fundamental mechanism of energy conversion responsible for laser-induced damage to transparent materials near the damage threshold. The following points appear pertinent, however:

(a) The very short length of the individual primary damage centers (ca 10^{-3}cm) makes normal optical absorption highly unlikely as a primary mechanism because of the correspondingly short optical path and the high degree of transparency involved.

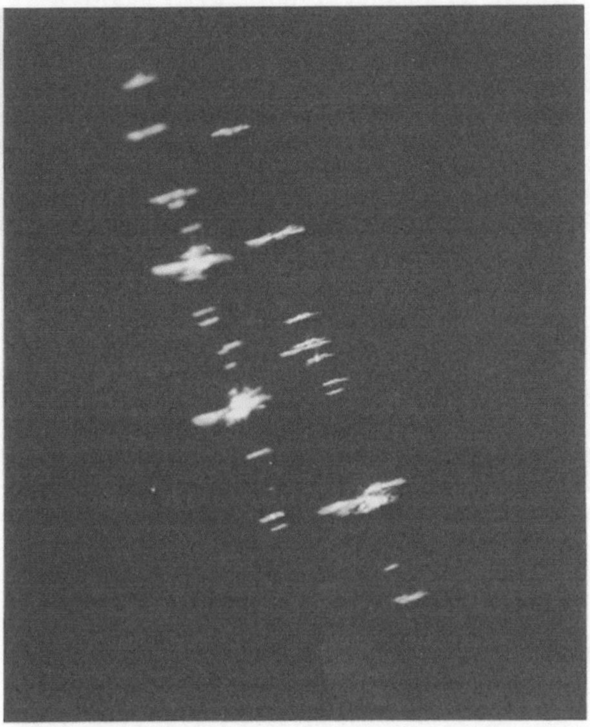

Fig. 6. E-ray (left) and O-ray damage tracks in calcite. The O-ray traveled perpendicular to the 1011 cleavage. Damage is essentially the same in both although vibration directions are perpendicular to one another and represent directions of very different bonding. Incident dark field illumination with polarizer, 120X.

(b) Absorption by included particles may undoubtedly be decisive when these are encountered. However, in careful studies involving high-magnification photography, Olness (3) has shown that damage regularly occured at sites where no inclusions sufficient to scatter a sub-threshold laser beam were present. The necessity for inclusions must therefore apparently be rejected.

(c) Stimulated Brillouin scattering (7) is invariably associated with heavy damage, but Chiao reports (8) that the threshold for such scattering is far above that for damage in many materials. Indeed he was unable to produce stimulated Brillouin scattering in calcite.

(d) Giuliano (1) has calculated that acoustic phonons which can be driven at the beat frequency between the nearly colinear light waves of the laser beam can have a frequency of no more than 260 Mc/sec. He noted that they would have a mean free path of some centimeters, which is far larger than the damaged areas observed.

(e) Direct action of the electrical field of the laser beam on atoms or ions of the solid appears to be untenable as a fundamental damage mechanism because of observations with calcite in which the E- and O-rays were equally damaging although the electric vectors in them are in directions of very different atom packing densities and bonding strengths.

(f) No observations thus far appear to be in conflict with the suggestion that electron avalanche may occur and may initiate damage. Indeed the observation that LiF damaged at liquid nitrogen temperature by a ruby laser beam exhibited the same thermoluminescence peaks as reported for LiF exposed to gamma- or X-rays and attributed to electron transport appears to support the suggestion.

ACKNOWLEDGMENT

The kindnesses of Dolores Olness and V. Garner in assisting with the laser experiments, of D.E. Jones and K.F. Petrock who ran the thermoluminescence curves and of J. Martinelli of the Eastman Kodak Company who irradiated the calcite, are gratefully acknowledged. This work was supported by the U.S. Atomic Energy Commission.

REFERENCES

1. *Concetto R. Giuliano, Applied Phys. Letters 5, 137-139 (1964).*

2. *John Martinelli, J. Appl. Phys. Communications, 37, 1939-1940 (1966).*

3. *D. Olness, Univ. Calif. Lawrence Radiation Lab. Report UCRL-14699, Feb. 16, 1966.*

4. *J.A. Ghormley and H.A. Levy, J. Phys. Chem. 56, 548-554 (1952).*

5. *C.J. Belbecq, P. Pringsheim and P.H. Yuster, Z. Physik. 138, 266 - 275 (1964).*

6. *Werner Kanzig, J. Phys. Chem. Solids 17, 88-92 (1960).*

7. *R.Y. Chiao, C.H. Townes and B.P. Stoicheff, Phys. Rev. Letters 12, 592 - 595 (1964).*

8. *R.Y. Chiao, Lecture at Livermore, Calif. July 1966.*

DISCUSSION

MOSTAFA IBRAHIM EL-AZAB (The American University in Cairo, U.A.R.): *What was the peak power and the integrated power of the laser pulse which produce such kinds of damage; did you measure or estimate the temperature at the damage center during the laser pulse?*

V.D. FRECHETTE: *In the case of NaCl, our colleague Mrs. Olness (3) has measured a damage threshold of 2 GW/cm^2 with a range between 0.6 and 2.6 GW/cm^2 and 50% of all values lying between 1.8 and 2.2 GW/cm^2. The pulse was Gaussian with a half-width of 7 nsec. No measurement of the temperature at the damage center has been attempted but the formation of bubbles in glass and NaCl together with the observation of incandescence suggest that temperatures of probably not less than 1000°C were involved.*

C. ERGINSOY (Brookhaven National Laboratory, Upton, New York, U.S.A.): *Do repeated pulses below threshold affect the value of the threshold?*

V.D. FRECHETTE: *The experiments of Mrs. Olness suggest that they do not.*

H.PEISL (I.Physikalisches Institut der Technischen Hochschule, Darmstadt, Germany): *Have you an idea by what mechanism color centers can be formed in LiF by this extremely low energetic radiation?*

V.D. FRECHETTE: *No definite mechanism is proposed at present. As noted we feel that several of the mechanisms suggested in the literature must be rejected as unnecessary although in special cases they may be sufficient. The displacement of electrons in LiF which seems to be implied by the thermoluminescence peaks observed after laser irradiation above damage threshold in LiF at 77°K does appear suggestive since this could be expected to lead to large stresses and to temperature rise upon relaxation. Such relaxation should be largely prompt, only a fraction of the electron defects remaining to show in the thermoluminescence measurement. The mechanism by which electrons are displaced may involve foreign ions or may possibly involve a multiphoton process. The probability of multiphoton action is of course much enhanced over usual irradiation by the tremendous intensity of the laser pulse.*

F.H. HAMMAD (Atomic Energy Establishment, U.A.R.): *Most of the materials mentioned are suitable for etch pit techniques. Are you aware of any studies on dislocation configurations around the damaged area? Do you think that the understanding of the mechanism of laser damage would give us more understanding of the spike mechanism in radiation damage?*

V. D. FRECHETTE: *We ourselves have not made such studies. Professor Gilman has studied etch pit configurations in laser damaged LiF. We do not have any ideas at the moment of what bearing the laser damage mechanism may have on spikes in radiation damage.*

R.F.WOODCOCK (American Optical Co., Research Division, Southbridge, Mass., U.S.A.): *Was fracture in glass as shown in the slide caused at an unusually low energy - it looks similar to damage observed when a known metallic inclusion is present?*

V.D. FRECHETTE: *The fracture was not caused at abnormally low pulse intensity, nor was any inclusion observed by inspection in a glass laser beam along the same path prior to pulsing.*

RADIATION EFFECTS IN GLASS LASERS

F. Tittel and N. Kamel

The American University in Cairo

Cairo, Egypt, U.A.R.

Experiments are in progress to study the effects of gamma ray irradiation on the stimulated emission properties of laser glasses.

The effects of high energy radiation upon glass are complex. The two principal processes are ionization (leading to color or defect center formation) and displacement of atoms from their position in the lattice. Of particular interest are induced optical absorption effects in commercially available laser glasses of different composition and dopant ions, starting with neodymium doped silicate glass (Eastman Kodak ND 11). Such changes as light absorption in glass produced by ionizing radiation may improve the optical efficiency for a laser by increased effective coupling between an optical pump source and the active medium. Improved laser efficiency was reported for irradiated crystal systems. However, a degradation of laser performance may also occur depending on the glass network and modifier composition, especially when metallic ion impurities (lead, iron, zinc) are present.

The effect of dose rate, total dose, selective bleaching and temperature will be related to optical absorption and emission measurements, laser tests, optical quality and magnetic resonance studies. Comparison against both a non-radiated and an undoped laser glass sample will be made.

INTRODUCTION

One of the important objectives in laser research is to improve the overall efficiency of the laser device. This can be tackled in numerous ways: material development (i.e. optimising concentration, temperature or lifetime), selective pumping techniques and special cavity configurations. The approach presented in this paper is directed towards increasing the optical pumping efficiency of neodymium doped laser glasses, since these are some of the most useful and practical laser media as yet realised.

The efficiency, i.e. laser radiation out-to-electrical energy supplied, of glass lasers has so far not exceeded a few percent. One of the reasons for this low efficiency has been the poor spectral match between the broad band laser pump lamps and the relatively few absorption bands found in glasses doped with trivalent rare earth ions. One way of increasing efficiency is to add another ionic species to the crystal which has broad absorption bands and which will transfer the energy they absorb to the rare earth ions (i.e. a sensitization or activation process). For example, the transfer of excitation radiation from Cr^{3+} to Nd^{3+} ions in various host media (including glass) has been observed (1). However, the inclusion of additional dopants is not always successful as the material quality and therefore, the optical homogeneity may deteriorate. Our approach for increasing the optical pump efficiency of Nd doped glass is to modify the intrinsic absorption characteristics and induce new absorption bands in the laser material by the formation of color centers under the influence of gamma rays and X-rays (2,3). Therefore, the objective of the work performed has been to examine whether or not this method is a possible mechanism for increasing the overall glass laser efficiency by transfering directly or indirectly some of the energy absorbed by optical excitation of the color centers to the Nd^{3+} system. In this manner a kind of percharged laser system is obtained which can be utilized upon optical pumping. This technique then represent a marriage of two established technologies: lasers, on the one hand, and photon irradiation of glasses, on the other. The usefulness of this approach has been explored to some extent with ruby (4,5,6). So far no definite conclusions have been obtained because the mechanisms for gamma ray excitation of fluorescence in solids is not well understood. In addition to improved optical pumpability, irradiation of a solid state laser system may have certain other practical merits in terms of:

1) direct pumping of a laser material by X-ray or gamma ray excitation

2) generation of giant laser pulses by saturable absorption of induced short lived color centers (7).

EXPERIMENTAL APPROACH

Our experimental program is directed towards studying to what extent the stimulated emission and spectroscopic characteristics of commercially available laser glasses are affected by energetic photon irradiation. As this program is not completed, only preliminary results are reported here so far.

The most convenient experimental approach for irradiating different configurations (i.e. rods and slabs) of laser glass has been to use a 4 kilocurie Co^{60} radioactive isotope source. This source emits gamma rays at a rate of 3.9×10^5 rads per hour with a photon energy of 1.1 - 1.3 MeV. Two glass types have been available to us so far:

1) Eastman Kodak ND 11 laser glass in slab form -- (a sodium silicate glass containing lanthanum)

2) Schott LG 55 silicate glass in rod form.

The former was exposed to different amounts of radiation varying from 10^5 up to 10^8 rads at room temperature before strong coloration effects appeared. The Schott glass required only 5×10^5 rads to produce considerable coloration. The irradiated glass was then examined under various experimental conditions.

Laser Tests

The most direct test to study the interaction of gamma rays and laser glasses is to compare the threshold energy of laser oscillations, near field and far field patterns, spiking behavior, energy and power output of an irradiated with a normal laser glass rod.

An experimental, medium energy pulsed laser facility with an external laser cavity arrangement and linear pump lamp configuration is now under construction, and will allow us to observe in detail any changes in the stimulated emission properties of photon irradiated glass systems. In the meantime, an assembly with a helical flash lamp was used to examine a Schott LG 55 laser glass rod. The threshold for the rod was found to be 110 joules. Irradiation with a dose of 3.9×10^5 rads increased the threshold beyond the maximum rating of the laser device (i.e. 175 joules) but heating normalised the threshold again. For example, the rod lased with an input energy of 155 joules after heating it in ½ hour intervals up to 250°C and reached its initial values of 110 joules again around 400°C. This is depicted in Fig. 1. On the same figure the laser output energy is displayed for an input energy of 175 joules. As one might expect from the threshold data, the energy

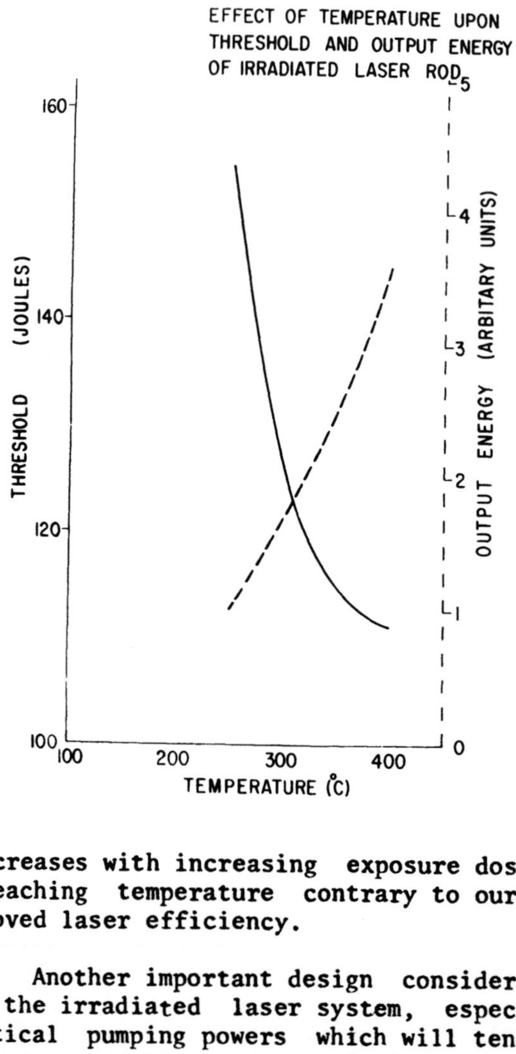

Fig. 1. Effect of temperature upon threshold and output energy of irradiated laser glass rod.

decreases with increasing exposure dose and increased again with bleaching temperature contrary to our desired objective of improved laser efficiency.

Another important design consideration will be the stability of the irradiated laser system, especially when exposed to high optical pumping powers which will tend to bleach induced color centers. However, for a single or even several laser pulses under favorable operating conditions (e.g. filtering) the intrinsic problem of color center fading may be minimised.

Optical Studies

Optical measurements are particularly useful in evaluating locally prepared laser glass samples in slab form, supplementing the previously described laser tests. Two effects have been studied:

Optical Absorption. A Beckman recording spectrophotometer was used to monitor transmittance and particularly spectral absorbance of the two available glass types (i.e. EK and Schott) in the wavelength range from 3500 and 9000 A° where most of the intrinsic absorption occurs for rare earth doped glasses.

The measured absorption data can be related to laser performance criteria from which threshold energy and other data may be predicted on the basis of a simple theoretical model in the absence of saturation effects.

$$P_{output}\ (4\ \text{level}) = \Delta N \frac{h\gamma}{\tau_{spont}}$$

where the number of excited atoms generated by optical pumping is given by

$$\Delta N = \frac{kV\ \tau_{obs}}{\gamma_p} \sum_{\Delta\lambda_p} P(\lambda,t)\ \alpha(\lambda)\ \eta(\lambda)\ \Delta\lambda_p\ e^{-\alpha(\lambda)z}$$

and where V	is the active volume
$\eta(\lambda)$	the quantum efficiency for the pump light (~ 0.65)
k	is the coupling efficiency of the light source to the medium
γ and γ_p	the emitting laser and pump frequency respectively.
h	is Planck's constant
τ_{obs}	$= \tau_{spont}\ \phi$ = radiative lifetime of metastable atoms
ϕ	quantum efficiency
$\Delta\lambda_p$	absorption bandwidth
$P(\lambda,t)$	normally incident pump light available per band from high pump power Xenon pump source, which can be characterized in terms of a black body radiator at (say) 10,000°K
$\alpha(\lambda)$	absorption coefficient, cm^{-1}
z	thickness of absorbing medium

A set of the absorption curves of ND 11 glass with the main absorption peak centered at 5800°A is shown in Fig. 2. The effect

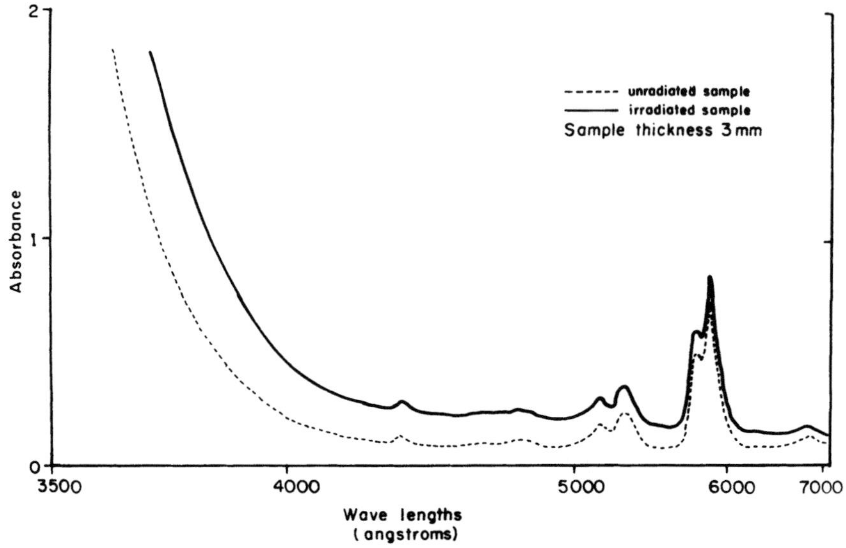

Fig. 2. Visible absorption spectra of neodymium laser glass (EK - ND 11)

of irradiation upon the absorption characteristics is apparent as Figure 3 shows the absorbance for both the irradiated and unirradiated samples. (Undoped ND 11 base glass irradiated under the same conditions will be investigated later). As one may anticipate the effective absorption crossection is increased. The net change in optical density may be obtained from these curves and plotted as function of photon energy or wavelength for low and high levels of irradiation.

Not only the laser threshold values can be predicted from the absorption data but also the number of color centers giving rise to induced absorption can be obtained using a well developed procedure due to Smakula (8).

If the centers which give rise to the induced absorption are sufficiently dispersed that they do not interact with one another, the area under the absorption curve is directly proportional to the concentration of the absorbing centers. Applying classical dispersion theory, one can write

$$Nf = \frac{9\,m\,c}{2\,e^2} \cdot \frac{n}{(n^2+2)^2} \quad \alpha_{max}\,\Delta\gamma$$

$$= 0.87 \times 10^{17} \cdot \frac{n}{(n^2+2)^2} \quad \alpha_{max}\ \Delta\gamma\ \text{(Gaussian)}$$

where N is the number of color centers per cm^3.

n the refractive index of glass at the wavelength of the absorption band m,e,c, have the usual meanings, the mass and charge of electron, and velocity of light respectively.

f is the oscillator strength of the optical transition, producing absorption ($\Sigma f = 1$)

α_{max} the absorption coefficient in cm^{-1} at the peak of the absorption and $\Delta\gamma$ the full width of the absorption band in eV measured at the half maximum points.

The validity of this approach could be checked using electron paramagnetic resonance techniques (9).

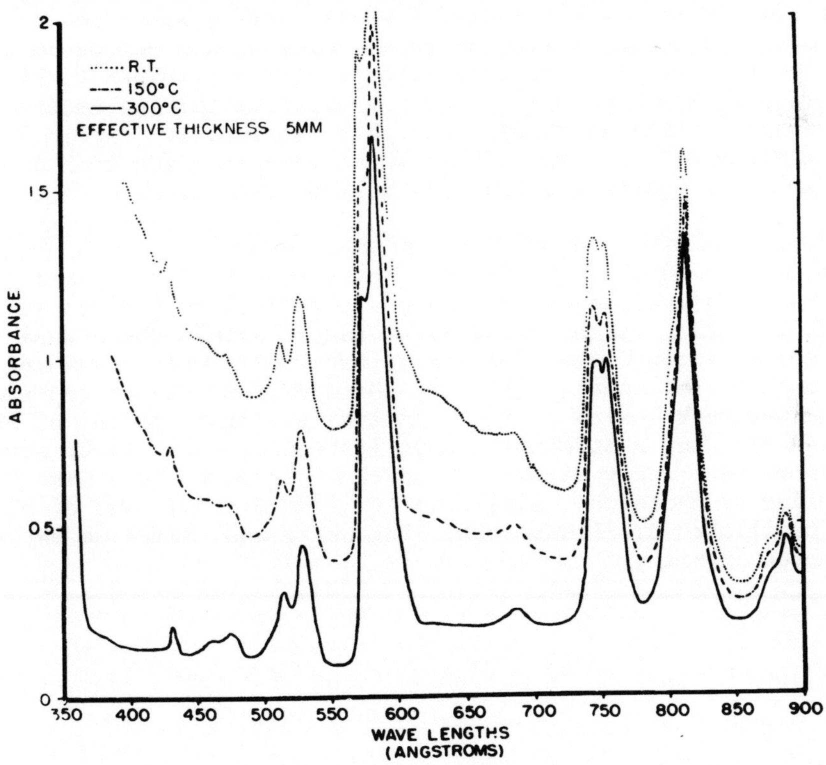

Fig. 3. Effect of bleaching temperature on spectral absorbance of irradiated laser glass rod.

In this manner we can establish a relationship between dose (i.e. the accumulated total exposure), change of optical density, number of color centers and laser operating conditions.

Irradiated glass can be bleached and returned to its normal state either by heating or by light excitation. Fig. 3 shows the thermal bleaching effect upon the absorption characteristics of ND 11 laser glass.

Fluorescence Measurements. Two alternative methods for measuring relative fluorescence intensity levels and lifetime are used. One method is a spectroscopic method analysing the spontaneous emission of an optically excited sample. Figure 4 shows the fluorescent spectrum obtained with an optimized Leitz prism monochromator system employing a PbS detector, lock-in amplifier and pen recorder display arrangement. The glass sample is optically excited with a 650 watt quartz tungsten iodine lamp. For laser action the main fluorescent line located at 1.06 microns is utilized. This line is traced out in detail as shown in Fig. 5, both for an unirradiated and irradiated ND 11 laser glass slab. So far no apparent increase in fluorescence emission and change in linewidth was observed for the irradiated specimen examined within the accuracy and reproducibility of the experimental parameters despite improved initial absorption characteristics. Figure 5 also shows a linewidth of about 350 A which results from both homogeneous and inhomogeneous broadening of the laser line.

A considerable disadvantage of this method is that the excitation level required to obtain a good fluorescent signal (good in terms of signal-to-noise ratio) is such as to produce considerable bleaching of the irradiated sample within the actual time of sweeping through the spectrum, especially when scanning the wavelength at low speed. This will be overcome with an experimental arrangement using a high intensity light pulse of short duration from say a strobotac unit instead of a CW light source. The fluorescent decay signal can now be picked up by a fast photo multiplier or phototube sensitive at 1.06 microns and displayed on an oscilloscope. Fluorescent intensity and lifetime may then be deduced directly.

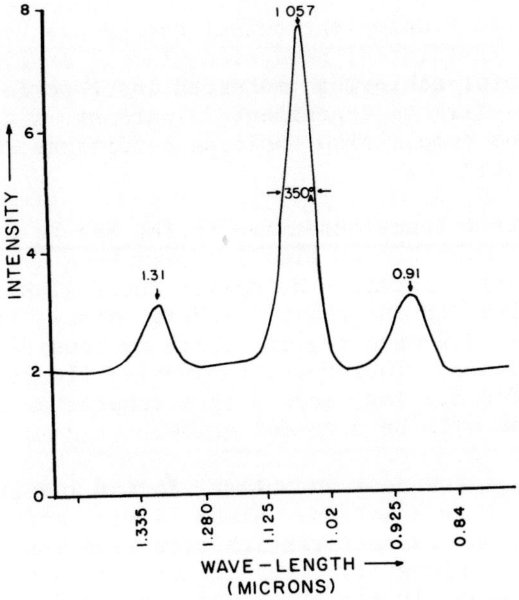

Fig. 4. Fluorescence spectrum of neodymium laser glass (fast scan).

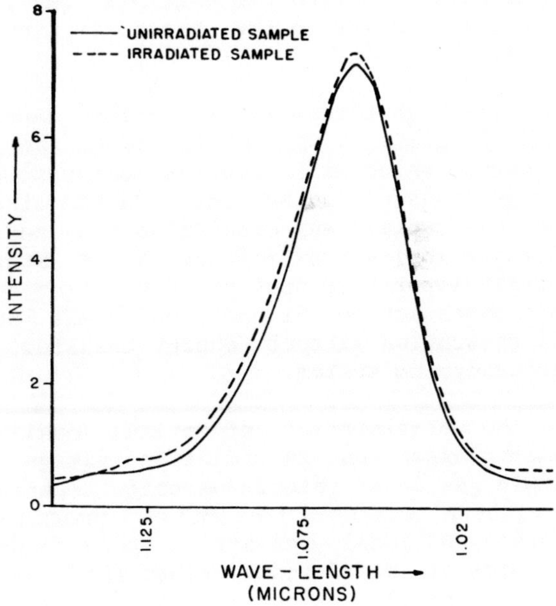

Fig. 5. Main fluorescence line of neodymium laser glass (slow scan).

CONCLUSION

Although no improvement in fluorescent output energy has been observed so far with gamma irradiated neodymium glass, several possibilities and approaches for achieving enhanced laser performance remain to be explored. Various experimental parameters are at our disposal: these include composition, melting conditions and controlled presence of impurities.

The decrease in laser efficiency observed so far may be due to unfavorable composition, impurities, lifetime quenching by in duced color centers and radiation damage. Neodymium doped glasses may be particularly sensitive to quenching effects since the spontaneous lifetime is in the 0.5 msec region which is considerably shorter than for ruby (3msec). Therefore, monitoring fluorescence lifetime as function of radiation dose with strobotac excitation as previously suggested will be a useful guide.

Favorable composition is the most important factor leading to enhanced laser operation by irradiation, since it will affect both fluorescence and absorption characteristics over wide ranges from one glass to another. Unfortunately, the actual composition of commercially available glasses is almost always a guarded company secret, so that a certain unknown factor will always exist. We plan to test numerous glass forming systems from which we hope to correlate certain compositional changes in glass and the effect of these changes on the spectral properties of neodymium laser glass.

The high threshold for color center formation observed for EK ND 11 glass is apparently due to the presence of a small amount of cerium which will suppress coloration and most likely inhibit energy transfer mechanisms. On the other hand the Schott glass does not contain any desolarization additives. In this case; the decrease in laser efficiency may be attributed to the presence of certain impurities such as lead, iron and zinc or ruptured covalent bonds between Si and O which all would tend to produce overall absorption without energy transfer from the color centers to the neodymium system.

No deterioration of optical quality of the two available laser glasses due to radiation damage has been observed using a simple gas laser interferometric technique (10). A visible He-Ne laser beam was used to compare quantitatively the optical homogeneity and local refractive index changes of irradiated glasses in terms of interference fringe shift and distortion. Furthermore, researchers (5) using irradiated ruby reported large increases in efficiency only when operating at very large optical pump energies, and this possibility will be checked for in our glass media as soon as the new laser facility now under consideration and capable of 2000 joules input is completed.

In brief, the possibility of improved laser performance induced by gamma irradiation may well be feasible despite our initial results for only two specific glass systems.

ACKNOWLEDGMENTS

The authors wish to thank Dr. J.H. Sanders for the use of the facilities at the Clarendon Laboratory and the UAR Atomic Energy Establishment for the use of its irradiation source.

REFERENCES

1. *R.C. Ohlmann and R. Mazelsky, Physics of Quantum Electronics, Puerto Rico 1965, p. 322 (McGraw Hill).*

2. *K.H. Sun and N.J. Kreidl, The Glass Industry, Oct., Nov. and Dec., 1952.*

3. *A.M. Bishay, J. Chem. UAR 6, 141 (1963).*

4. *W. Low, Bern Laser Conference, Oct. 1964.*

5. *W. Flowers, Proc. IEEE 51, 858 (1963).*

6. *V.R. Johnson, Proc. IEEE 52, 185 (1964).*

7. *E. Snitzer and R. F. Woodwock, 4th Int. Quantum Electronics Conf., Phoenix 1966.*

8. *A. Smakula, Z. Physick 59, 603 (1929).*

9. *W.P. Slichter, Non-Crystalline Solids, Frechette 1960, p. 199.*

10. *A.E. Blume and K.F. Tittel, Applied Optics 3, 527 (1964).*

DISCUSSIONS

E. DEEG (American University in Cairo, U.A.R.): *What kind of induced color centers would you want to have in order to allow you to decrease the threshold of a* Nd^{3+} *doped laser glass?*

F. TITTEL: *Radiation color centers giving rise to increased and induced absorption within the visible wavelength range are desirable in order to achieve maximum coupling of available pump light to the laser glass.*

MOSTAFA EL AZAB (American University in Cairo, U.A.R.): *Could you explain the mechanism responsible for energy transfer between the color center and the Nd^{3+} ions?*

F. TITTEL: *The mechanism leading to energy transfer between induced color centers and the Nd^{3+} ions is probably quite complicated and so far not well understood. Some possible mechanisms are the following:*

a) Gamma rays create radiation centers and defects in the glass. Some of these centers emit light over a wide wavelength range. This emitted light may then be reabsorbed by the rare earth ions.

b) Photo-electrons, released by photon irradiation can excite all the electronic levels by inelastic scattering processes.

c) Electronic recombination processes of trapped electrons and positive holes in the neighborhood of the Nd ions may also assist population inversion of the Nd^{3+} laser system.

R. WOODCOCK (American Optical Co., Research Division, Southbridge Mass., U.S.A.): *Was any direct or extrapolated measurement of the transmission at 1.06μ made before or after irradiation?*

F. TITTEL: *So far no absorption measurements at 1.06 microns have been carried out. Certainly, this would be an additional method to determine whether photon irradiation will enhance or deteriorate the intrinsic gain coefficient of Nd laser glass.*

THERMOLUMINESCENCE AND COLOR CENTERS IN SOME IRRADIATED VITREOUS AND CRYSTALLINE SILICATES, BORATES AND PHOSPHATES

C. Bettinali

C.N.E.N., Universitandi Roma

Rome, Italy.

The thermoluminescence and the absorption spectra of some glasses and their devitrification products in the systems Li_2O-SiO_2, Na_2O-B_2O_3, and Na_2O-P_2O_5 were studied. The activation energies and the thermoluminescence spectra were studied either in glasses or in devitrified systems from the glow-curves. The color centers were investigated by obtaining the absorption spectra and the ESR spectra. The thermoluminescence shows a different behavior in the different systems. In the silicate glasses, the devitrification modifies the intensity and the temperature of the low-temperature peak. The glow-peaks at 80 and 220°C are strongly increased by a 10^4 factor. In the phosphate glasses, the low temperature peak is modified as in the silicate glasses, but the high temperature thermoluminescence does not change in intensity from the vitreous to the crystalline form. The vitreous sodium borates show a similar behavior. The relation between the thermoluminescence and the color centers is different in the crystalline and vitreous products. The mechanisms of the thermoluminescence and the color centers formation are discussed.

INTRODUCTION

Diffusion of cations and anions far from their original position has been suggested as part of the mechanism explaining color center formation in glasses and in crystals (1,2,3,4,5). If one considers a solid of the same composition in glassy and in crystalline form, differences must arise in color center formation due to different mechanisms of diffusion. The diffusion coefficient D in glasses is related (4) to the distance r between two non-bridging oxygen atoms by the relation $D = 1/6\ r^2\ \Gamma$ where Γ is the jump frequency. A factor of ½ is usually introduced into this equation in order that it may be more realistically applied to alkali diffusion in glass. In the case of a crystalline body centered structure the symmetry factor is ¼ instead of ½ and so D becomes smaller. Furthermore, the values of r in glasses have a larger spread than in a crystal, and so more atoms having a lower r occur. It seems probable, if this point of view is correct, that the processes of color center formation involving cation diffusion are favored in glasses. In order to check this hypothesis, the color center formation in gamma irradiated $NaPO_3$, $Na_2B_4O_7$ and $Li_2Si_2O_5$ in glassy and crystalline form were studied.

EXPERIMENTAL

Materials were prepared from reagent grade chemicals in accord with the methods reported in the literature (6,7,8). Thermoluminescence measurements were made, using an apparatus described earlier (9). Optical absorption spectra were measured by a Beckman DK spectrometer, and ESR spectra by a Hilgher and Watt spectrometer. The irradiations were performed with a Co^{60} cell at a dose rate of 2×10^5 r/hr.

RESULTS AND DISCUSSION

The results obtained are given in Table I. In $NaPO_3$ the main features observed, going from glassy to crystalline compounds, are:

a) the low temperature thermoluminescence peak decreases in relative intensity (column c);

b) the spectral composition of thermoluminescent light is shifted in the $(NaPO_3)_3$ ring structure to higher energy (column e);

c) the 2.4 eV optical absorption band decreases in intensity in both crystalline compounds. A new well resolved 4.7 eV absorption band appears in ring $NaPO_3$ (column f);

d) electron spin resonance decreases in intensity (column i) and absorption curves are modified: in the $(NaPO_3)_n$ chain a three-line curve appears.

The most obvious explanation is that an unpaired electron reacts with two equally spaced P nuclei whose nuclear spin is 2½. This happens in centers of type III (see Table II). In $(NaPO_3)_3$, a very weak absorption appears. The position of absorption lines in this crystal seems to indicate that electron centers are modified, the g value being lower than that of a free electron. However, the very low value of absorption does not allow accurate measurements.

TABLE I

COMPOUND	THERMOLUMINESCENCE				OPTICAL ABSORPTION		ESR RESONANCE ABSORPTION	
a	b	c	d	e	f	g	h	i
	T_g C°	I a.u*	E-eV	mμ	eV	I a.u*	Derivative absorption	intensity a.u*
$NaPO_3$ (glass)	80	I	0.71	620	2.4	1	Two line symmetric	10
	220	5	0.7	620	3.5	1	g = 2.009	
$(NaPO_3)_n$ Chains	70	3	0.7	620	2.4	0.2	Three lines symmetric	2
	220	2	0.69	620	3.5	1	g = 2.009	
$(NaPO_3)_3$ ring	70	0.5	0.7	530	2.4	0.01	Multiline unsymmetric	0.2
	220	4.5	0.7	"	4.7	1	g = 1.910	
$Na_2B_4O_7$ (glass)	80	8	0.72	615	2.4	1	Isotropic singlet	10
	140	10	"	"	3	1	g = 2.014	
	260	10	"	"				
$Na_2B_4O_7$ tetragonal	80	2	0.72	380	2.4	1	Seven-line	
	220	10	0.72	"	3.4	1.2	mean g = 2.015	1.2
					4.67	1		
$Li_2Si_2O_5$ (glass)	80	1	0.69	630	4	5	Two asym. independent lines	
	220	3	0.69	"	2.8	3	g = 2.010	10
					1.9		g = 1.960	10
$Li_2Si_2O_5$ (crystal)	80	10^4	0.68	410	4	5	Two line-Line having	
					2.8	3	g = 2.010 unchang.	10
	220	$3x10^4$	0.69		1.9	6	g = 1.960 Two sym. lines	2

* a.u. arbitrary units.

TABLE II

		Suggested Color Centers Formation Reaction	Characteristics	Notes
	I	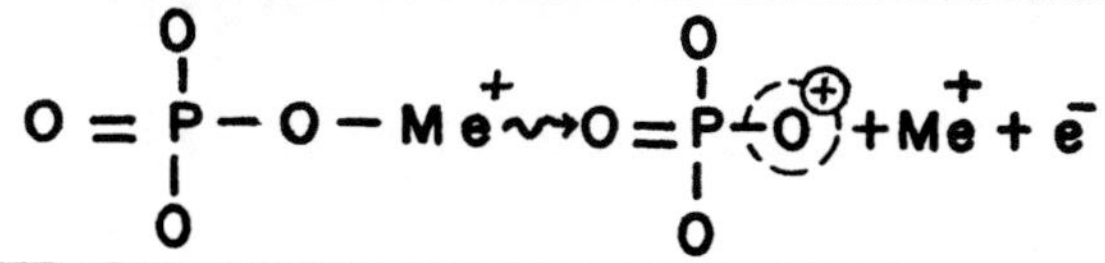	2.4 eV optical absorption ESR doublet g=2.008 T.L:T_g=220°;620mμ (absorption)	holes in glass diffusion (ref. 10, 11, 12).
$NaPO_3$	II	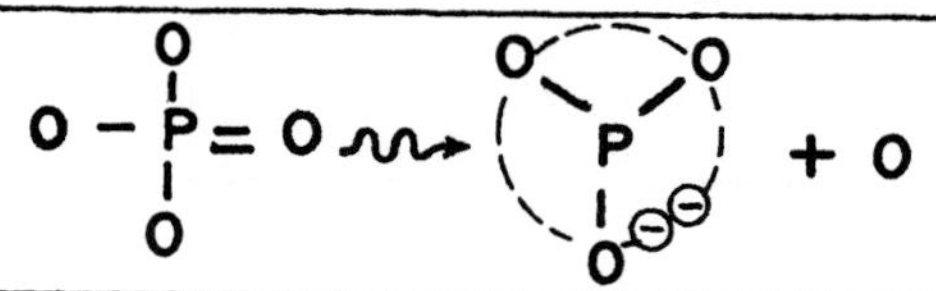	3.5 eV optical absorption no ESR T.L:0.7 eV (activation energy)	electron trap glass and cryst. no diffusion
	III	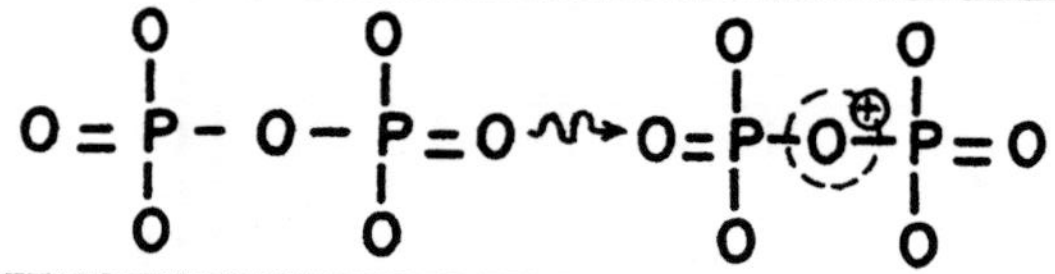	3.5 eV optical absorption ESR triplet g=2.007 T.L:T_g 220°;330mμ emission	holes crystal no diffusion
	I	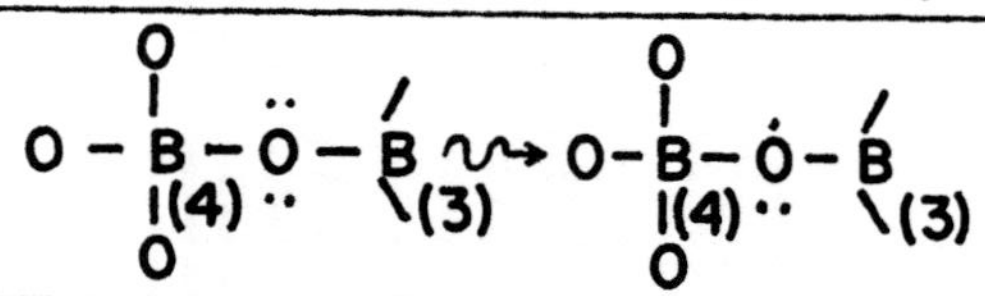	4 eV optical absorption ESR singlet g=2.010 T.L:615 mμ emission	holes in glasses diffusion (ref 13,14,15)
$Na_2B_4O_7$	II	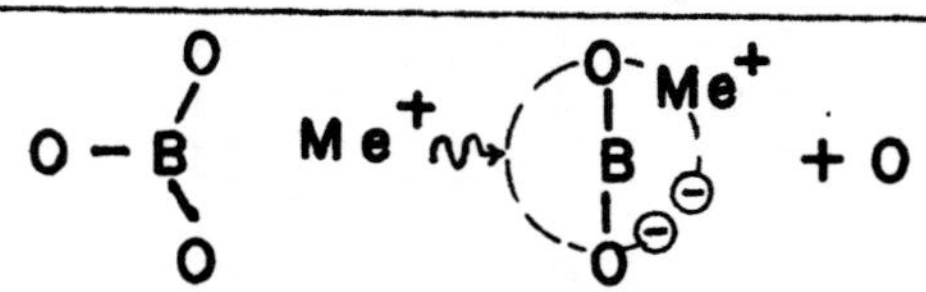	2.5 eV optical absorption no ESR T.L:0.7 eV activation energy	electron trap glass and cryst. no diffusion
	III	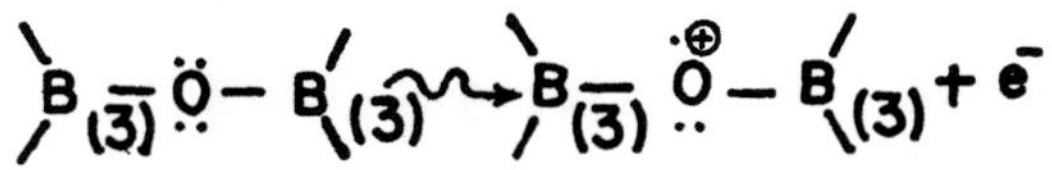	4 eV optical absorption ESR : seven-line T.L: 390 mμ emission	holes in crystal no diffusion

TABLE II (Cont'd)

$Li_2Si_2O_5$	I	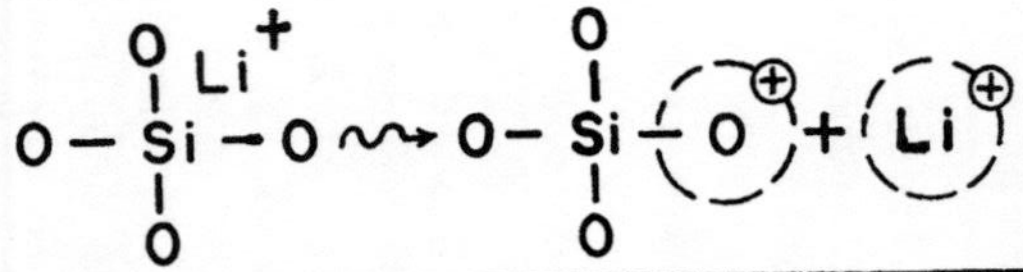	4 eV optical absorption ESR g=2010 T.L: 590 mμ emission	hole centers in glass diffusion (ref. 16)
	II	idem	2.8 eV absorption ESR g=1.960 T.L: 0.7 eV activation energy	electron trap in glasses diffusion
	III	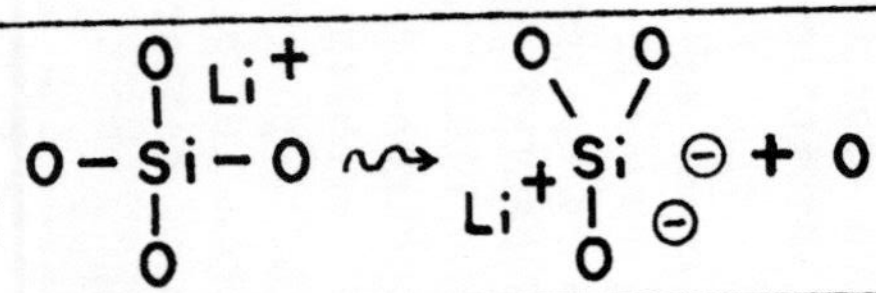	1.9 eV optical absorption no ESR T.L?	electron traps in glasses and crystal no diffusion
	IV	//	4 eV optical absorption g=1.960 unchanged T.L:420 mμ absorption	hole centers crystal
	V	//	4 eV optical absorption ESR g=1.960 2 symm. lines	electron center? crystal

T.L. = Thermoluminescence.

The main observations in the case of $Na_2B_4O_7$, going from glassy to crystalline compounds, are:

a) the low temperature thermoluminescence peak decreases in relative intensity, and the thermoluminescence spectrum is shifted to higher energies;

b) a well resolved peak appears at 4.67 eV in the absorption spectrum;

c) the ESR intensity decreases about 10 times and the isotropic singlet observed in the glass changes to a seven-line spectrum of $g = 2.015$.

In the case of $Li_2Si_2O_5$ going from glassy to crystalline structure:

a) thermoluminescence intensities increase by a factor of 10^4, and the spectral emission is shifted to higher energy;

b) the optical absorption at higher energy shows a well resolved maximum at 4.2 eV and a decrease in absorption at 5 eV;

c) the ESR resonance is not modified in the lines whose g value is lower than that of a free electron and corresponds to an electron center. An asymmetric absorption is developed.

Table II summarizes some types of color centers induced in irradiated glasses as suggested by different authors. Observing $NaPO_3$, it is clear that the 2.4 eV optical absorption and the ESR doublet are connected to a hole center produced by reactions involving Na diffusion (10). On the other hand, the 3.5 eV optical absorption is connected to an electron-trapping, non-paramagnetic center. Taking into account the results of Table I, it is clear that in crystalline sodium metaphosphate, centers of Type I are formed with less probability than in glass. On the other hand, the formation of electron trapping centers of Type II, involving oxygen diffusion, does not change. Hole centers are deeply modified and centers not requiring diffusion are formed in crystalline materials as Type III centers. In $NaPO_3$ the poor resolution of ESR signals does not allow suppositions on the nature of formed centers, but the change in thermoluminescence emission spectra supports the view of a change in hole centers. In glasses, the prevailing thermoluminescence mechanism seems to be that of electrons coming from a single level recombining with hole centers of different depth and capture cross section (9).

In borate glasses, a hole center produced by means of cation diffusion gives rise to the ESR singlet and to a 4 eV optical absorption. Change in crystalline materials gives rise to a new 4.67 eV optical absorption band and a seven-line spectrum. The observation that thermoluminescence spectral composition shifted to higher energies can support the hypothesis of a change in hole center structure. This seven-line ESR spectrum is that predicted for a center in which the unpaired electron reacts with two boron nuclei whose nuclear spin = 3/2. These centers are indicated as Type III centers and their formation does not require cation diffusion.

In $Li_2Si_2O_5$, electron trap centers do not seem to be modified from glass to crystalline compounds. Hole centers, on the other hand, are modified in structure but not in number. In fact, optical absorption in the high energy field of the absorption spectrum is modified but the intensities are of the same order. ESR shows only slight modifications in the symmetry of the absorption associated with electron centers. Furthermore, thermoluminescence increases in intensity and its spectrum is shifted to higher energies. These facts can be explained if we recall that alkali silicate glasses are believed to contain the alkali in some kind of interstitial position (10). Thus the color center formation does not require diffusion.

The observed increase in thermoluminescence emission can be explained by there being a greater efficiency of optical recombination in crystals than in glasses. This fact is compensated in other glasses by the reduced number of color centers formed.

CONCLUSIONS

The results obtained seem to indicate that electron trapping centers are less influenced than hole centers when glasses crystallize. This fact can be related to the mechanism of formation: electron trapping centers require cation diffusion, while hole centers require oxygen diffusion. The first process seems less probable in crystalline materials. This fact is in accord with some views on the mechanism of cation diffusion in glasses (4). In lithium silicate, lithium atoms seem to preexist as interstitials, and consequently no differences in electron trapping centers are observed.

The study of differences in color center formation in glasses and crystals of the same composition may be used as a means to check proposed models of color center formation.

REFERENCES

1. J.H. Schulman and W.D. Compton. "Color Centers in Solids" Pergamon Press 1962, pp. 209 - 237.

2. J.M. Stevels and Y. Volger. Philips Research Report 17 - 283, 1962.

3. A. Kats and J.M. Stevels. Philips Research Report II, 115, 1956.

4. R.J. Charles. J. Appl. Phys. 32, 1115, 1961.

5. C. Zenner. "Imperfections in Nearly Perfect Crystals". John Wiley and Son Inc., 1952, p. 215.

6. J.R. Van Wazer. "Phosphorus and Its Compounds". Inst. Pub. Inc. 1958, p. 668.

7. Ibid., p. 683.

8. Ibid.

9. C. Bettinali and G. Ferraresso. J. Chem. Phys. 44, 2262, 1966.

10. P. Beekenkamp, H.J.A. Van Diyk and J.M. Stevels, VII Cong. Int. du Verre 47, 3-11, 1965.

11. Yasuo Nakai. Bull. Chem. Soc. Japan 36 - 8, 1965.

12. J.W.H. Schreurs and R.F. Tucker. "Proc. of V Int. Conf. on Non-Crystalline Solids". Delft 1964, 116.

13. S. Lee, P.J. Bray. J. Chem. Phys. 40 - 2982, 1964.

14. P. Beekenkamp. "Color Centers in Borate, Phosphate and Borophosphate Glasses". Ph. D. Thesis, Technische Hogeschool Eindhoven 1965.

15. Yasuo Nakai. Zeit fur Phys. Chem. N.F. 44, 2, 1965.

16. J.S. Van Wieringen and A. Kats. Philips Research Report 12, 432, 1957.

THERMOLUMINESCENCE OF VITREOUS SODIUM

ALUMINO SILICATES AND GERMANATES

C. Bettinali, V. Gottardi, G. Ferraresso

Universita Cattolica di Roma

and Stazione Sperimentale del Vetro - Murano, Venezia

The correlation between color centers and thermoluminescence was studied in gamma-irradiated glasses of the Na_2O-Al_2O_3-SiO_2 *and* Na_2O-Al_2O_3-GeO_2 *systems.*

The perturbing effect of increasing Al content on the thermoluminescence activation energies was investigated. The activation energy values obtained from the glow-curves analysis agree with the values calculated, assuming, for the electron trap, a hydrogenoid model.

The absorption spectra of the samples containing Al/Na >1 and Al/Na<1 were also investigated and compared to the thermoluminescence spectra. The thermoluminescence between 20 and 300°C is related to electron-hole recombination.

INTRODUCTION

It is well known that thermoluminescence of irradiated solids results from different types of transitions: between atomic or molecular levels or crystal field levels.

Recombination of trapped electrons with positive hole centers, after irradiation, occurs through a different mechanism (1,2,3,4, 5).

In earlier work (6) on thermoluminescence of vitreous sodium silicates, it was found that thermoluminescence from 20 to 250°C results from a recombination of trapped electrons with positive hole centers of different capture cross sections, and that activation energies correspond to that of a hydrogen-level trap. In order to check if this mechanism is general in glassy systems, it was decided to study the thermoluminescence and the absorption spectra of two sodium aluminum silicate glasses containing Al/Na >1 and Al/Na<1, and the thermoluminescence of some sodium aluminum germanate glasses.

Introducing aluminum in the glass must give rise to a perturbation in both electron trap centers and positive hole centers; thus influencing the electronic field of Na and non-bridging oxygen atoms, respectively.

EXPERIMENTAL

The glasses were prepared by melting reagent grade materials in a platinum crucible. The percentage composition of the glasses is given in Table 1.

TABLE I

Percentage Composition of Glasses Studied

Glass No.	SiO_2%	GeO_2%	Al_2O_3%	Na_2O%
1	66.6	-	-	33.3
2	71.67	-	11.14	17.19
3	62.1	-	19.86	18.04
4	-	66.6	-	33.3
5	-	66.4	7.5	26.1
6	-	66.7	22.4	10.7

Glow curves, thermoluminescence and fluorescence spectra were obtained by the use of an apparatus described earlier by the authors (6,7); the emitted light was detected by means of an EMI 6255 B photomultiplier. The absorption spectra of the samples were measured by a Beckman DK spectrophotometer.

RESULTS

The thermoluminescence glow peak temperature produced at a heating rate of 0.2°C/min, the activation energies calculated from the glow curves (6) and the spectral distribution of thermoluminescence light are shown in Table II.

TABLE II

Thermoluminescence of Gamma Irradiated Glasses

No.	Glass	Temp. of TL peaks (0.2°C/sec)	Activation Energy (exper.) eV	Activation Energy (calcul.) eV	TL emission eV
1	Na_2O-SiO_2	85 130	0.8 0.8	0.74 0.74	2.95 2.40
2	$Na_2O-Al_2O_3-SiO_2$ (Al/Na<1)	85 145	0.99 1.00	0.96 0.96	2.95 2.40
3	$Na_2O-Al_2O_3-SiO_2$ (Al/Na > 1)	85 210	1.01 1.00	0.96 0.96	3.25 2.55
4	Na_2O-GeO_2	65 125	0.53 0.52	- -	broad band
5	$Na_2O-Al_2O_3-GeO_2$ (Al/Na<1)	60 123	0.76 0.77	- -	broad band
6	$Na_2O-Al_2O_3-GeO_2$ (Al/Na> 1)	62 185	0.75 0.78	- -	broad band

In the $Na_2O-Al_2O_3-SiO_2$ system it is evident that:

a) the activation energies are the same for both peaks. This fact supports the postulated thermoluminescence mechanism in which electrons come from a single level and recombine with traps of different capture cross section and different depth;

b) the presence of Al shifts the activation energies to higher values;

c) with the Al/Na ratio greater than 1, the thermoluminescence light spectrum is shifted to higher energies.

The intensity of thermoluminescence and the number of color centers were determined at different radiation doses for samples containing Al/Na > 1 and Al/Na<1, in order to find out if the difference in the intensity of thermoluminescence for an equal radiation dosage is related to a higher number of trapping centers, or to a greater luminous efficiency of the recombination process.

The intensity of thermoluminescence was measured by the glow peak area and the number of color centers was deduced from the increase of absorption in the wave length range from 330 to 700 mμ.

The induced optical absorption in the spectral range between 330 and 700 mμ is considered by the authors to be caused by trapped electron centers.

Figure 1 shows the results for two sodium aluminum silicate glasses in which Al/Na > 1 and Al/Na<1, respectively. The parallel between the increase in thermoluminescent light emission and the induced absorption is clear.

In germanate glasses the results are more ambiguous, but they are like those for aluminosilicate glasses. Activation energies are less than in sodium silicate glasses. The spectral emission is more widely distributed and no shift was clearly detected.

DISCUSSION OF RESULTS

Figure 1 shows that the electron traps which give rise to thermoluminescent light do not preexist but are formed during irradiation. A careful observation of the thermoluminescence vs. dose curve shows a sharp rise for low radiation dosage: this could be explained assuming a number of preexisting traps. Similar behavior has been reported in alkali halogenides (8). Furthermore, it is clear that the differences in thermoluminescence from samples of different composition can be related to differences in the induced optical absorption in these samples (9).

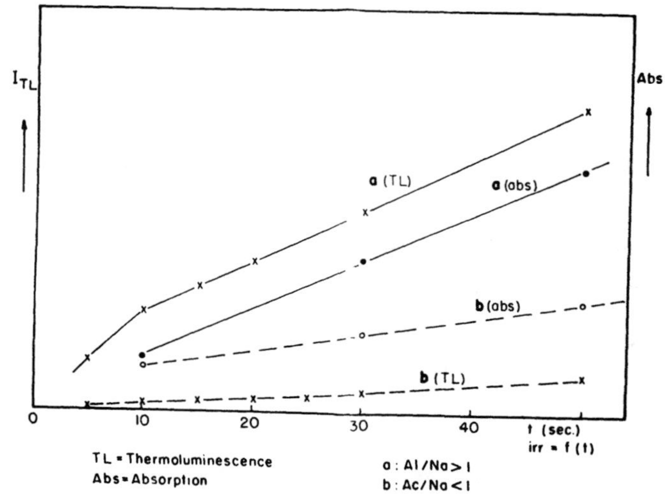

Fig. 1. Effect of increasing dose on intensity of thermoluminescence and induced absorption for two samples of different composition.

The results obtained seem to indicate that thermoluminescence comes from the recombination of the trapped electrons which give rise to optical absorption, as is the case in alkali halogenides (5). Based on this postulate, the activation energies were calculated assuming a hydrogen-like model for the trapped electrons (10). In such a model, the activation energy of an electron is given by the relation

$$E = R/K^2_{eff} \qquad (1)$$

where

$$R = 13.5\ eV;\ K_{eff} = 1/K + 5/16\ (1/K_o - 1/K) \qquad (2)$$

K = the dielectric static constant, K_o = the optical dielectric constant (equal to n^2), n = the refractive index. For vitreous aluminum sodium silicate, $K = 7.75$ and $n = 1.50$ (11). By introducing such values in the relation (2) and substituting the values so found in the relation (1) we obtain 0.68 eV.

This value is much lower than values found by experimental glow curves. This fact can be explained assuming a perturbation of the electrical field caused by aluminum. Such a perturbation may be thought, in the first approximation, to be provoked by some AlO^+ groups situated in substitutional positions with respect to sodium. On supposing this model, the relation (1) changes into

$$E = R/K^2_{eff} + e^2/r_{AB} \cdot K \qquad (3)$$

where e is the number of positive charges acting on the electrical field of the trap (in our case, 2) and r_{AB} is the distance between positive charges which give rise to electron trapping center.

In sodium silicate the distance Na-Na and Na-Si are about 3.8 A^o and the distance Na-O is 2.3 A^o (14). Relation (3) will give a value E = 0.906 which is in agreement with values obtained from thermoluminescence glow curves.

CONCLUSION

The mechanism of thermoluminescence in sodium aluminosilicate and sodium aluminogermanate glasses fits with the model proposed for the thermoluminescence of crystalline solids: an electron trapped in a single level hydrogen-like trap which recombines with positive hole centers having different capture cross sections. The introduction of Al gives rise to a perturbation in both electron capturing and positive hole centers according to the Al/Na ratio.

The behavior of sodium germanate glasses is like that of the corresponding silicate glasses, but the activation energies are lower.

REFERENCES

1. M. Magat, J. Chem. Phys. 1, 142 (1966).

2. A. Charlesby, R.H. Partridge, Proc. Roy. Soc. A 272, 312 (1964).

3. J.J. Hill, P. Schwed, J. Chem. Phys. 33, 652 (1954).

4. G. Bonfiglioli, P. Brovetto, C. Cortese, Phys. Rev. 114, 951 (1959).

5. A. Halperin, A.A. Braner, Phys. Rev. 117, 408 (1960).

6. C. Bettinali, G. Ferraresso, J. Chem. Phys. 44, 2262 (1966).

7. C. Bettinali, V. Gottardi, G. Ferraresso, International Conference "Giornate del Vetro", Roma 14-16 Ott. 1965, Published in Vetro e Silicati X, 1-12 (1966) (Eng.).

8. D.A. Wiegand, P.V. Mitchell, R. Smoluchowsky, Proc. International Conference on Color Centers and Crystals Luminescence, 8-12 Sept. 1960, Turin (Italy) Instituto Elettrotecnico Nazionale "G. Ferraris" Under ARDC - USAF.

9. V.V. Vargin, S.A. Stepanov, Dokl - Akad - Nauk. SSSR, 1962,147 3, 609-11 (Russian).

10. D. Curie, J. Phys. et le Radium 17, 16 (1956).

11. G.W. Morey, The Properties of Glass, 2nd. Edition Rheinolds, Publish. Corp. 477 (1954).

12. Ibid. p. 567.

THERMOLUMINESCENT GLASS ACTIVATED BY MANGANESE

J. Kocik and J. Nebrensky

State Glass Research Institute

Hradec Kralove, Czechoslovakia

The use of glass for the dosimetry of ionizing radiation can be substantially based on one of the three following phenomena:

a) radiocoloration - change of the spectral absorption
b) radiophotoluminescence effect
c) radiothermoluminescence effect.

The influence of different factors on the resulting thermoluminescent properties of the phosphate glasses activated by manganese was studied.

The following items are considered:

- *the content of various impurities in the glass.*
- *the concentration of the activator*
- *the chemical composition of base glass*
- *the effect of the Al_2O_3 - P_2O_5 substitution*
- *the effect of Li_2O content in the glass*

In the case of phosphate glasses activated by manganese it is possible to achieve promising properties for their use as thermoluminescent phosphors for dosimetry purposes. The glow curve is well - defined, with one maximum around 300°C, the dose response is nearly linear and the emission spectrum is also favorable.

INTRODUCTION

Dosimetry represents a very important branch of radiation physics and many research workers have paid great attention to its problems during the last 20 years. The appropriate dosimetry control plays an important role in the development of the industrial use of atomic energy. The new dosimetry materials which enable the reliable measuring of the ionizing radiation in a wide range of doses are therefore intensively studied.

Glass is becoming more and more significant as a material suitable for high energy radiation dosimetry. Its use is usually based on one of the three following phenomena:

a) radiocoloration
b) radiophotoluminescence effect
c) radiothermoluminescence effect.

In the case of radiocoloration measurements of radiation doses are based on the change of spectral absorption caused by irradiation. The practical evaluation is usually carried out from the changes in the spectral absorption at a certain wavelength.

The well-known brown coloration of silicate glasses irradiated for high doses of gamma radiation (10^5-10^8r) is due to color centers causing the change in absorption at certain wavelengths, which is proportional to the magnitude of radiation doses. However, the fading which appears even during the course of irradiation proves to be disadvantageous. The change in absorption of transition elements caused by radiation is often used in practice (e.g. the change in valency of chromium serves for measuring doses of 10^3 - 10^7 r at a wavelength of 550 mμ and in the case of phosphate glasses containing cobalt the intensive absorption band appears in the 350 - 450 mμ range enabling the measurement of high energy radiation in the dose range of 5 x 10^3 - 4x10^6 r (1,2).

Induced photochemical reactions associated with the change in spectral absorption can occur in glasses containing two or more polyvalent ions exposed to radiation. For example the combinations Mn - Fe, Mn - V, Mn - V - Fe (3), Bi - As, Mn - As (4) with respect to high level radiation dosimetry have been studied.

The use of silver - activated phosphate glasses developed for dosimetric purposes by Schulman (5) is based on the photoluminescence effect.

The new stable luminescence centers are created in glasses by high energy radiation. When excited by ultraviolet rays, the irradiated glass emits a secondary, orange fluorescence proportional to the exposure dose; its peak being near 620 mμ. The original dose range 10 - 600 r (6) in which measurements were possible is now considerably widened. For example, in the case of $LiPO_3$ containing glasses manufactured by Tokyo Shibaura Electric

Co, the measuring dose range of 10^{-2} - 10^{4} r is quoted in literature (7).

Radiothermoluminescence is a widespread phenomenon in which the emission of visible and eventually ultraviolet light is observed in irradiated solids when subjected to heating or to infrared rays.

The effect of heat causes the return of excited electrons from higher energy levels to their ground states while the energy is being released and the phosphor is emitting light or "glows".

Radiothermoluminescence seems to be very useful for precise dosimetry measurements of high energy radiation. Thermoluminescent LiF, CaF_2 and $CaSO_4$ crystals activated by manganese or samarium are widely used but their preparation is considerably expensive.

Glass appears to be a suitable material for thermoluminescent dosimetry and its chief advantages as compared with crystals are: low preparation costs, easy shaping, high mechanical strength, and adequate chemical durability.

The thermoluminescence of glasses in many different phase - systems has been studied intensively during the past ten years. A considerable number of papers deals with thermoluminescence of fused quartz, germanium, silicate and borate glasses (e.g. 8-12). Good results with respect to dosimetry have been achieved by using the aluminophosphate glasses activated by manganese (13 - 16). These glasses have also been successfully used by cosmonauts during their space flight (17 - 18). Their measuring range is quoted to be $2x10^{-2}$ to $2x10^{6}$r. For the measurement of radiation doses higher than 10^{4}r radiocoloration phenomenon instead of thermoluminescence is utilized.

EXPERIMENTAL

This paper deals with some of the results of studies on thermoluminescent phosphate glasses activated by manganese which were carried out at the State Glass Research Institute at Hradec Kralove and in which four component glasses of the P_2O_5 - Al_2O_3 - MgO-SrO systems were used. This glass system was chosen because it gives glasses suitable for dosimetry purposes and of a good chemical durability. The measurements of dosimetry properties were carried out in the research laboratories of Tesla Liberec and in the Nuclear Research Institute, Prague. During research, attention was paid not only to the properties important from the dosimetry point of view, but also to other significant factors such as preparation costs, technological aspects and practical usage of this glass.

Preparation of Samples

Laboratory melts of 700 - 1000 g were carried out in a gas heated tangential furnace. Fused silica crucibles and propeller stirrers were used for these melts. Melting was carried out at temperatures ranging from 1300°C to 1400°C (depending on the composition of glass). Homogenized glass was shaped by casting into rectangular plates and annealed. Samples for dosimetry measurements (size 15 x 15 x 2 mm) and for other physical and chemical tests were prepared. Special attention was paid to the purity of the raw materials, especially to their Fe_2O_3 content. The P_2O_5 used was in the form of polymeric crystalline H_3PO_4 containing less than 0.001 % Fe_2O_3 (in a few cases a liquid containing 85 % H_3PO_4 was used). Al_2O_3 was added as $Al(OH)_3$ and Li_2O, MgO, SrO and MnO in the form of carbonates. The Fe_2O_3 content in Sr, Li and Mn carbonates did not exceed 0.003 % and in $Al(OH)_3$ and $MgCO_3$ 0.008 %.

Measuring Equipment and Methods

The rate of crystallization and liquidus temperature were determined in a gradient tube furnace and the hydrolytical durability was evaluated from the evaporated extract of powdered glass (according to DIN 12 111). The dosimetric evaluation was carried out on the basis of registered glow curves.

The samples were heated at a uniform rate of 4-8°C/sec. The peak-height, peak temperature of the glow curve and the area below this curve (defined by the glow curve and the coordination axis) were estimated. The intensity of radiothermoluminescence was evaluated from peak - height (19) whereas the area below the glow curve served as a measure of the absorbed radiation dose. This area is proportional to the total amount of light energy emitted. In the case of a well defined glow curve with one maximum only, the peak height alone could be used for evaluation of radiation doses (19). Good reproducibility of the heating rate is a necessary condition for preventing peak - height fluctuations (20,21). The peak temperature of the glow - curve also gives information about the depth of trapped electrons energy level in eV.

Samples were irradiated by ^{60}Co, at a dose rate of 40 r per minute. The heating of samples to a temperature of 340°C within a period of 40-80 sec was carried out by indirect means using an electrically heated brass plate. The radiation was converted to an electrical current by a photoelectric multiplier type 61 Pk 412 with a SbCs cathode. The light from the sample impinges on the photocathode of a photoelectric multiplier through a light conductor made of quartz glass and provided by heat absorbing filters on both frontal surfaces. The integral sensibility of the photoelectric multiplier operated at the Voltage of 1050V was 12.1 A/Lm, the temperature of the source (W-bulb) being 2850°K.

The photomultiplier current was amplified and the sensibility of the Metra Vareg miliamperemeter reached $1x10^{-7}$ A per 1 cm of the curve height registered.

RESULTS AND DISCUSSION

Influence of Impurities in Glass on Its Radiothermoluminescent Properties

A series of glasses activated by 0.15 % MnO by weight with Fe_2O_3 contents in the range of 0.003 to 0.050 % by weight was melted in fused silica crucibles. It was found that the shapes and peak - heights of thermoluminescent glow curves are strongly influenced by Fe_2O_3 content in glass. If the Fe_2O_3 content is higher than 0.05 % by weight, radiothermoluminescence does not occur at all (see Table I).

TABLE I

The Influence of Fe_2O_3 Content in Glass on the Thermoluminescence Intensity

Fe_2O_3 content in glass (% by weight)	Thermoluminescence intensity I_{TL} (relative units)
0.003	100
0.005	90
0.010	60
0.025	35
0.050	-

The preliminary laboratory melts were carried out in porcelain crucibles. The value of I_{TL} of these glasses was about 30 - 50 % lower than of those melted in fused silica crucibles. This difference was caused by the fact, that the fused silica contained only 0.012 % Fe_2O_3 by weight while the Fe_2O_3 content in porcelain was about 100 times greater. Glass melted in fused silica crucibles contained approximately 0.003 % Fe_2O_3 while the Fe_2O_3 content of a glass prepared in porcelain crucibles was approximately 0.010 - 0.015 % Fe_2O_3 by weight.

It is obvious, that the Fe_2O_3 content in thermoluminescent glasses for dosimetry purposes should be as low as possible. For this reason appropriate attention must be paid to the raw materials, crucibles, stirrers and protection of the glass against possible contamination by Fe_2O_3 during its preparation process.

The negative influence of As_2O_3 on thermoluminescence of glass has also been found. If As_2O_3 in amounts of 0.7 - 1.0 % by weight was added to the melt, the thermoluminescence vanished completely. For this reason As_2O_3 cannot be used as a fining agent. The effect of NH_4Cl on radiothermoluminescence is also negative. The addition of 1 % NH_4Cl by weight to the batch caused a 50 % decrease of I_{TL}; simultaneously shifting the peak temperature from 280° to 400°C. This is in good agreement with the observation of Keirim, Marcus and co-workers (13) concerning the negative influence of ammonium salts.

Concentration of the Activator

In accordance with literature data (13,15) manganese was chosen as activator, and introduced into the batch as $MnCO_3$. Series of glass melts from the four component system studied with MnO content ranging from 0.01 to 0.50 % MnO by weight was carried out. It has been found that the optimal sensitivity can be achieved with the MnO content in the glass in the range of 0.05 - 0.15 % MnO by weight (see Table II).

Since the maximum of thermoluminescence intensity and its temperature was found to change only slightly when MnO concentration in glass varies from 0.05 to 0.15 % MnO by weight, the upper limit of MnO concentration range in glass (0.15 % MnO by weight) was chosen for this study for technological reasons.

TABLE II

The Influence of the Activator Concentration on the Thermoluminescence Intensity I_{TL} (Glass System: P_2O_5 - Al_2O_3 - MgO - SrO)

MnO Content in Glass (% by weight)	Thermoluminescence intensity I_{TL} (relative units)
0.01	60
0.05	98
0.10	100
0.15	100
0.25	91
0.50	65

Secondary Maximum on the Glow - Curve

It has been found that the chemical composition of glass of the P_2O_5 - Al_2O_3 - MnO - SrO system has a considerable effect on the shape of the glow - curve. The secondary maximum appears at temperatures from 100° to 120°C in glasses containing more than 10 mole % Al_2O_3 and to a certain extent also in glasses with the MgO content over 20 mole %. These secondary peaks are undesirable since they increase the fading possibility. The effect of the change in P_2O_5/Al_2O_3 ratio on the temperature and height of the secondary peak is illustrated in Figure 1. The molar composition of these glasses are shown in Table III. All these glasses were activated by 0.15 % MnO by weight.

In order to obtain a well - defined glow - curve without any pronounced secondary peak, the Al_2O_3 content in glass should not exceed 9 - 10 mole %. It has been found that the secondary maximum also appears if the MgO content exceeds 20 mole %.

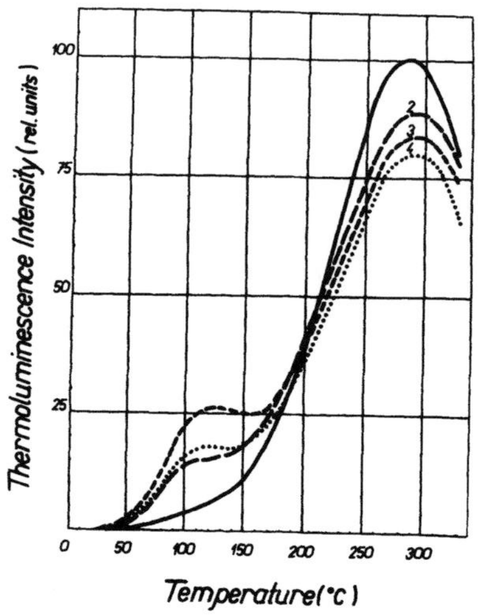

Fig. 1. Effect of the Al_2O_3 - P_2O_5 substitution on the secondary maximum of the glow - curve.

TABLE III

The Compositions in Mole % of a Series of Experimentally Melted Glasses

Glass No	P_2O_5 %	Al_2O_3 %	MgO %	SrO %
1	73.5	8.5	10	8
2	70.0	12.0	10	8
3	66.5	15.5	10	8
4	63.0	19.0	10	8

The Effect of Li_2O Content in Glass on the Temperature T_m of the Maximum on the Glow Curve

The temperature corresponding to the maximum on the glow curve T_m of the irradiated glass activated by manganese is influenced by the chemical composition of the base glass. The molar substitution MgO/SrO increases the peak temperature T_m from

280°C to 300 - 320°C. A similar effect can be achieved when reducing agents as C, Si, NH_4Cl are added to the batch. The lowering of the temperature T_m could be achieved by introducing a further component into the P_2O_5 - Al_2O_3 - MgO - SrO system. This lowering of the temperature may be advantageous if a more rapid and economical evaluation of the thermoluminescence is required. For this purpose lithium oxide seems to be the most effective component whereas silica is less suitable because of the necessity of using much higher melting temperatures.

For influence of the change in the Li_2O/P_2O_5 ratio on the glow - curve peak position see Table IV and Figure 2.

TABLE IV

The Effect of the Molar Substitution Li_2O / P_2O_5 on the Peak - Temperature T_m of the Glow Curve

Glass No	P_2O_5 %	Al_2O_3 %	MgO %	SrO %	Li_2O %	T_m
5	75	10	8	7	-	280
6	65	10	8	7	10	240
7	55	10	8	7	20	200

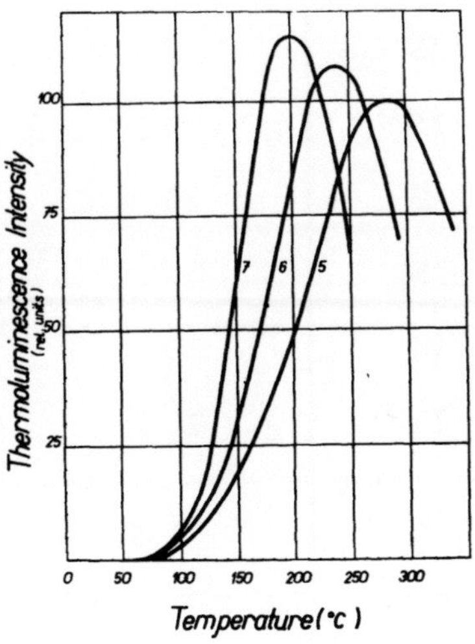

Fig. 2. The shift of the peak - temperature T_m of the glow curve caused by a successive Li_2O / P_2O_5 molar substitution.

The Effect of the Composition of the Base Glass on Its Properties

The effect of the composition of the base glass activated by manganese on its properties was studied in the P_2O_5 - Al_2O_3 - MgO-SrO system. Several series of melts were carried out. As an example a series of glasses (No 11 - No 19) is presented, the composition of which varies as follows:

P_2O_5	Al_2O_3	MgO	SrO
60 - 85 %	5 - 15 %	5 - 15 %	5 - 10 %

The compositions of this series of glass are given in Table V.

TABLE V

Molar Composition of Glasses in the P_2O_5 - Al_2O_3 - MgO - SrO System

Glass No	Molar Composition in Mole %			
	P_2O_5	Al_2O_3	MgO	SrO
11	85	5	5	5
12	75	15	5	5
13	75	5	15	5
14	65	15	15	5
15	80	5	5	10
16	70	15	5	10
17	70	5	15	10
18	60	15	15	10
19	72.5	10	10	7.5

Glasses of this series activated by 0.15 % MnO by weight were experimentally melted under the same conditions and the following properties were determined (Table VI).

TABLE VI

The Properties of Glasses in P_2O_5 - Al_2O_3 - MgO - SrO System

Glass No	h mg	d g/cm^3	α 10^7	Tg ^oC	T_L ^oC	A_{TL} rel.u.	I_{TL} rel.u.
11	30	2.680	75.5	575	1260	128	37
12	7	2.634	65.2	652	1265	136	34
13	98	2.612	79.1	525	1198	170	54
14	8	2.609	66.6	573	1250	179	43
15	131	2.725	83.6	512	1223	148	44
16	9	2.690	70.5	585	1290	161	40
17	151	2.666	83.4	495	1197	171	49
18	20	2.648	69.3	548	1132	174	43
19	23	2.664	73.4	600	1270	169	50

h = hydrolytical durability (weight of residue in mg after evaporation.
d = density
α = mean linear thermal expansion coefficient in the temperature region 20 - 300°C
Tg = Transition temperature in ^{o}C
T_L = liquidus temperature in ^{o}C
A_{TL} = the area defined by the glow curve and coordination axis
I_{TL} = thermoluminescence intensity, estimated from the height of glow - curve peak

The variations in the above mentioned properties have been evaluated by using statistical methods and the effect of each glass component on the individual properties of glass was expressed mathematically. The evaluation of individual factors enables the calculation of the properties from the chemical composition of the glass in the part of the four component system specified on Table V and even (by means of extrapolation) outside its boundaries if a lower accuracy of results is acceptable. The equations for the calculation of individual properties of glasses are presented in Table VII. The symbols m_x represent the mole % of the component x. The sign + or - denotes the positive or negative sense of the effect of the glass component and the magnitude of factors characterizes the intensity of this effect.

From the additive equations given on Table VII, it is possible to deduce the effects of individual glass components on its properties. The hydrolytical durability h is considerably improved by an increased Al_2O_3 content, the opposite being true for SrO. The density d of glass increases especially when the content of the heaviest component SrO is raised; on the other hand, the contribution of MgO to the density is the smallest.

TABLE VII

Equations for Calculation of Glass Properties From its Molar Composition

h	=	$0.607m_{P_2O_5}$	$-\ 8.563m_{Al_2O_3}$	$+\ 3.077m_{MgO}$	$+\ 9.007m_{SrO}$
d	=	$0.0266m_{P_2O_5}$	$+\ 0.0240m_{Al_2O_3}$	$+\ 0.0218m_{MgO}$	$+\ 0.0362m_{SrO}$
$\alpha.10^7$	=	$0.781m_{P_2O_5}$	$-\ 0.471m_{Al_2O_3}$	$+\ 0.871m_{MgO}$	$+\ 1.801m_{SrO}$
Tg	=	$6.104m_{P_2O_5}$	$+\ 12.384m_{Al_2O_3}$	$+\ 1.524m_{MgO}$	$-\ 3.136m_{SrO}$
T_L	=	$13.257m_{P_2O_5}$	$+\ 14.757m_{Al_2O_3}$	$+\ 6.757m_{MgO}$	$+\ 6.737m_{SrO}$
A_{TL}	=	$1.047m_{P_2O_5}$	$+\ 1.867m_{Al_2O_3}$	$+\ 4.067m_{MgO}$	$+\ 3.087m_{SrO}$
I_{TL}	=	$0.376m_{P_2O_5}$	$-\ 0.236m_{Al_2O_3}$	$+\ 1.262m_{MgO}$	$+\ 0.752m_{SrO}$

The expansion of glass is increased especially by an increasing SrO content, while an Al_2O_3 content has the opposite influence. The temperature Tg is increased by the addition of Al_2O_3 while it decreases as SrO content increases. The liquidus temperature, which is important for the preparation of glass, is strongly increased when the content of Al_2O_3 and P_2O_5 increases.

The area below the glow curve A_{TL} increases with increasing the concentration of divalent metal oxides MgO and SrO. The relative peak - height of the glow - curve I_{TL} increases as the MgO and SrO contents increase while it decreases as the Al_2O_3 content is increased.

All the glasses containing 15 mole % Al_2O_3 exhibit a distinct secondary maximum on the glow curve.

On the basis of the analysis of the effects of the individual components on the properties important for the preparation and usage of phosphate glasses, the conclusions enabling the preparation of the corrected series of melts in the P_2O_5 - Al_2O_3 - MgO - SrO system could be drawn.

Some Properties of Thermoluminescent Glass

As an example, some properties significant for dosimetry purposes of a thermoluminescent glass of the P_2O_5 - Al_2O_3 - MgO - SrO system activated by manganese will now be discussed.

Glass No 50 of the molar composition 68 % P_2O_5, 9 % Al_2O_3, 18 % MgO, 5 % SrO is activated by 0.15 % MnO (by weight). It exhibits a well - defined glow curve having one maximum at 280°C only (see Fig. 3).

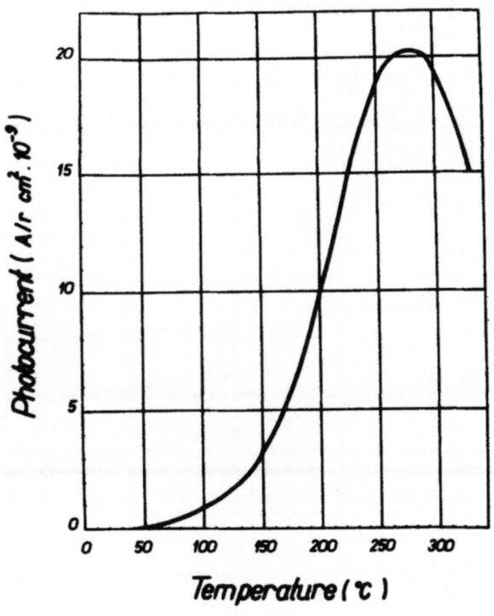

Fig. 3. The typical glow - curve of glass No 50 irradiated by the dose of 20r.

This glass exhibits linear response of thermoluminescence over a range of doses from 1 up to 10,000 r. The glass plates 15 x 15 x 2-4 mm can be used for radiation measurements down to 0.1r, and, the sensitivity of this sample can be further increased by using a higher heating rate. In this case, the glass sample is placed directly on the heating plate, the surface temperature of which is 340°C. When white coating of the surface adjacent to the heating plate was used a further increase in photoelectric current of 60 - 80% was achieved. The sensitivity to the radiation dose could also be increased by using a sample of larger size. The use of larger samples is limited by the fact that they require a slower heating rate due to the cracking danger.

We can conclude that the glass No. 50 is suitable for the measurement of ionizing radiation doses ranging from 10^{-2} to 10^4r.

The water and atmosphere resistance of the glass No. 50 is fully satisfactory. Since crystallization occurs above 750°C there is no danger of devitrification at the working temperatures.

The reproducibility of the properties of glass No 50 is satisfactory and the differences in thermoluminescence intensity found for laboratory melts of 700 g do not exceed ± 6%. In the case of larger melts better reproducibility can be naturally expected. Fading can be practically neglected if the samples are used up to the temperature of 50°C, but at temperatures above 100°C the glass exhibits a distinct signal fading as can be seen from Figure 4.

The emission spectrum of glass No. 50 recorded after the exposure to 1.4 x 10^4r exhibits one peak at 580 mμ, the half width of the absorption curve being 60 - 70 mμ. If the radiation doses are higher than 10^5r of gamma rays, the glass becomes blue (Phenomenon of radiocoloration).

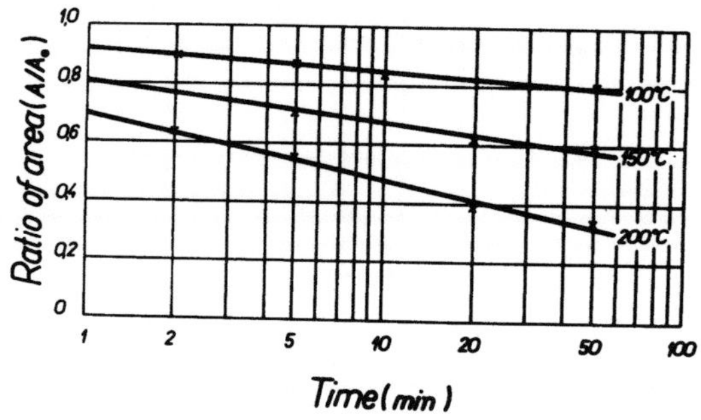

Fig. 4. The fading of thermoluminescent glass No. 50 irradiated to 100r. (Expressed as ratio A/Ao of glow - curve area vs time).

The absorption spectrum of glass No 50 irradiated by various doses of γ-radiation (of the order 10^5 and 10^6 r) is shown in Figure 5. Heating this radiocolored glass leads to a distinct fading.

A somewhat high energy dependence of phosphate glasses activated by manganese represents a certain disadvantage when compared with e.g. LiF crystals.

As seen from Figure 6 the energy dependency curve reaches its maximum in the region of 40 keV and is about six times higher as compared with the influence of ^{60}Co gamma rays. This energy dependency can be reduced by suitable metal filters.

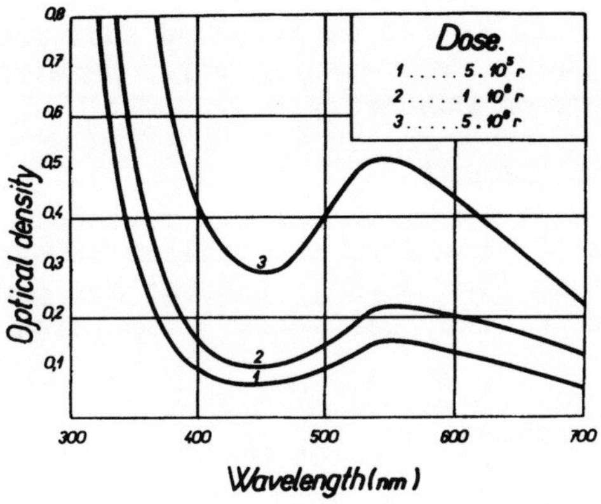

Fig. 5. Absorption curves for the glass No 50 irradiated with γ-rays of the order 10^5 and 10^6r (Thickness 2 mm).

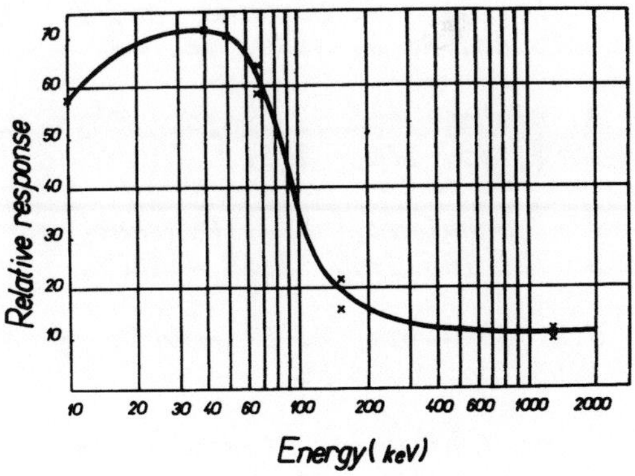

Fig. 6. Wavelength dependence of the glass No 50 irradiated by 100 r doses in the energy region of 10 - 1250 keV.

CONCLUSION

Glasses of the four component system P_2O_5 - Al_2O_3 - MgO - SrO activated by manganese are suitable for measurements of the ionizing radiation in the dose range of 10^{-2} - 10^4 r. High purity of raw materials is a basic condition for achieving high sensitivity. Special attention must be paid to Fe_2O_3 content in glass (max. 0.003 %). The presence of a secondary peak, the temperature T_m of the main peak and the shape of the glow curve can be considerably influenced by changing the chemical composition of the glass. This composition affects naturally the chemical and physical properties of glass which are important for its preparation and use. The energy dependence represents a certain disadvantage which can be compensated by the use of suitable filters.

On the basis of available data it can be presumed that phosphate glasses activated by manganese represent a very promising material for use in thermoluminescent dosimetry.

ACKNOWLEDGMENT

The authors wish to thank Z. Spurny from the Nuclear Research Institute, Prague, and Z. Vitamvas from Tesla, Liberec for their advice and for the measurement of the dosimetry properties of glass samples.

REFERENCES

1. *N.J. Kreidl and G.E. Blair. Nucleonics Vol. 14, No. 1, 1956, pp. 56 - 60*

2. *N.J. Kreidl and G.E. Blair. Nucleonics Vol. 14, No 3, 1956, pp. 82 - 83*

3. *J. Paymal, M. Bonnaud and P. Le Clerc. J. Amer. Ceram. Soc. Vol. 43, No 8, 1960, pp. 430 - 436.*

4. *A. Bishay and S. Arafa. Phys. and Chem. Glass Vol. 6, No 4, 1965, pp. 134 - 142.*

5. *J.H. Schulman, R.J. Ginther* et *al., J. Appl. Phys. Vol. 22, No 12, 1951, pp. 1479 - 1487.*

6. *J.H. Schulman, W. Shurclift* et *al. Nucleonics Vol. 11, No 10, 1953, pp. 52 - 56.*

7. *Tokyo Shibaura Electric Co. Ltd. "Toshiba Fluoro - glass Dosimeters Type Fg D - 3B".*

8. *E. Lell. J. Amer. Ceram. Soc. Vol. 43, No 8, 1960, pp. 442 - 426.*

9. *S. Cohen and M. Priqueler. C. R. Acad. Sci. Paris 257 (7), 1963, pp. 1469 - 1472.*

10. *V. Garino - Canina and S. Cohen. J. Amer. Ceram. Soc. Vol. 43 No 8, 1960, pp. 415 - 416.*

11. *T. Kikuchi. J. Phys. Soc. Japan, Vol. 13, No 5, 1958, pp. 526-531.*

12. *C. Bettinali, G. Ferraresso and G. Bonetti. Vetro e Silicati Vol. 9, No 54, 1965 pp. 5 - 8.*

13. *I.B. Keirim - Markus, Z.M. Syrickaja and V.V. Jakubik. Steklo Vol. 14, No 2, 1961, pp. 77 - 89 in Russian.*

14. *J.A. Bocvar, I.B. Keirim - Markus* et *al. Russian patent No. 154379, 1961.*

15. *I.B. Keirim - Markus, V.V. Jakubik* et *al. Steklo Vol. 16, No 2, 1963, pp. 75 - 84 in Russian.*

16. *I.A. Bocvar, A.A. Vasiljeva* et *al. Atomnaja Energija Vol. 15, No 1, 1963, pp. 48 - 52 in Russian.*

17. *I.B. Keirim - Markus and I.A. Bocvar. Kom. Issledovanija Vol. 1, 1963, p. 179 in Russian.*

18. *I.A. Bocvar and I.B. Keirim - Markus. Kosm. Issledovanija Vol. 2, 1964, p. 304 in Russian.*

19. *F. Urbach. Sitz. Akad Wiss., Wien, Abt. IIa, 139, 1930, pp. 363 - 372.*

20. *Z. Spurny. Jaderna Energie Vol 7, No 6, 1961, pp 205-206 in Czech.*

21. *Z. Spurny. Atomic Energy Review - Proc. IAEA, Vol. 3. No 2, 1965, pp. 61 - 115.*

PROBLEMS IN THE IRRADIATION OF POLYMERS

A. Charlesby

Royal Military College of Science

Shrivenham, England

The study of radiation effects in polymers has wide remifications, ranging from the basic phenomena of interaction of radiation and matter, radiation physics and chemistry, to industrial applications on the one hand, and radiobiology on the other. Early observations on crosslinking and degradation of simple polymers have been subjected to further and more detailed study using such techniques as electron spin resonance, optical absorption, pulse radiolysis and thermoluminescent spectroscopy to determine the intermediate processes. The influence of crystal structure, of molecular mobility in solution, and of additives present in very small concentration, have all been analysed. A long sequence of events appear to be involved, from the initial absorption of energy to the final physical and chemical change.

INTRODUCTION

For far too long polymer science has been considered as a somewhat exotic subject, a dubious offspring of organic chemists with which honest physicists should not be associated. This attitude is now changing, as it is being realised that many of the interesting problems with which it is concerned are analogous to those dealt with for example in metal physics. The inclusion of this subject in the present important conference on the solid state is therefore a happy recognition of this fact, though I must admit that it does raise several questions as to the meaning of the term "solid state".

In the long-past age when I was a young student, the solid state was still defined for me in terms of a regular crystallographic structure; materials which showed no such regularity (as shown by X-rays for example) could, when necessary, be classified as supercooled liquids and hence relegated to a remote region of science from which awkward questions could be readily avoided. In recent years this addiction to crystallinity has lessened as it became obvious that many of the most interesting patterns of behavior were precisely due to defects in crystallographic perfection. Where are we to place those polymers which contain little or no crystallinity, and yet have many physical properties similar to those of crystalline polymers? And what of polymers in which a single molecule can be situated partly in a crystalline, and partly in an amorphous (liquid?) environment? We can envisage polymeric materials which are essentially solid in one direction, liquid in another. And how do we classify a material, consisting of a single giant molecule, such as a perfect diamond, a piece of bakelite, or a rubber car tyre - is this a solid, a liquid or a gas? These may be amusing philological questions. Perhaps we can simply accept the pragmatic concept which I trust is in the mind of our Chairman - *a solid is as a solid does.*

These considerations are directly relevant to the subject matter of my talk - radiation effects in polymers - since the physical properties of polymers (e.g. crystallinity or flow, elastic properties) are closely related to the geometrical arrangement of the individual molecules (often in a way which can be determined on purely theoretical grounds), and exposure to high energy radiation is a very effective method of modifying this arrangement. We can therefore make use of radiation in several ways:

(a) as a means of modifying and studying polymer properties in a relatively simple and controllable way,

(b) as a technique for investigating the basic interaction of high energy radiation with matter,

(c) for the improvement of existing polymers, or the production of new ones,

(d) for the guidance it can afford in the vastly more complicated subject of radiobiology, it being accepted that for this purpose biological molecules are essentially complex polymeric systems.

RADIATION-INDUCED CHANGES IN POLYMERS

It may be useful at this stage to summarise the major conclusions reached from past work.

(i) Both temporary and permanent changes can occur. The former are evidenced by such phenomena as thermoluminescence and electrical conductivity; the latter by crosslinking and degradation resulting in modified mechanical and chemical properties.

(ii) These changes are ultimately due to excitation and ionisation, and not to any significant extent to a reaction of radiation with the nuclei. Excitation and ionisation lead to the formation of reactive groups - either ions or radicals, and it is these which result in the observed changes.

(iii) Although the incident energy is usually assumed to be absorbed almost at random, the changes it produces are highly selective. Thus some form of so-called "energy transfer" (reactivity transfer is a more accurate description) must be envisaged.

(iv) These changes can be further modified by changes in configuration, temperature and by the presence of small amounts of additives (this is one basis for "radiation protection").

(v) Compared with many solid state radiation-induced reactions those in polymers often occur at low energy inputs. Doses and dose-rate of a few rads or rads/min can produce appreciable conductivity and thermoluminescence. For permanent chemical and physical changes only a minute fraction of the bonds need be modified. Thus viscous liquids can be converted to rubber-like solids for an expenditure of only 30 ev per molecular weight of one million, and a single 1 MeV electron can permanently change some 35,000 molecules - a total molecular weight in the region of 10^{10}. This provides a permanent and highly amplified indication of the passage of radiation.

CONFIGURATION OF POLYMERIC MATERIALS

Before developing the implications of these bald statements, it may be useful to present a main classification of polymeric materials. These are constituted by an array of monomer units (either of one or several types) bound together by homopolar bonds into one-dimensional, two-dimensional or three-dimensional arrays to form single giant molecules, usually well above 10^5 in molecular weight. If the regularity of chemical structure and of geometrical orientation are sufficient, a crystal lattice may be formed. Examples of such regular structures are fibres (one-dimensional), graphite (two-dimensional) and diamond (three-dimensional).

If the molecules are linear and flexible, the polymer behaves as a very viscous fluid. If they are bound together very closely a hard material results, while if they are crosslinked in this way but lightly, rubber-like properties are obtained. The search for relationships between mechanical properties and geometric arrangement in such systems can be fascinating to the mathematically minded. For example a theoretical relation exists between Young's Modulus E and the average molecular weight between crosslinks M_c in a rubber-like system, in a simple form

$$E = 3\,\rho\,RT/M_c$$

where ρ is the polymer density, R, T have their usual meaning.

A more empirical relation relates volume v, external pressure p and temperature T.

$$(p + p_0)\ (v - v_0) \propto RT$$

where p_0 is the polymer density, and v_0 represents an irreducible volume. It is not necessary to emphasize the analogies which these relationships present with those derived in the kinetic theory of gases. The basic assumption is that the molecular chains can take up a most probable configuration, subject to the constraints imposed by their ends (represented by crosslinks). The further inhibitions imposed by the finite volume of each element of chain, not allowed for in the simple theory, has called for much mathematical ingenuity. Furthermore, as the density of crosslinks increases with radiation dose (and hence M_c diminishes), the chains are no longer able to take up this most probable configuration due to the interaction of their terminating crosslinks, and departure from this form of elasticity is observed. Ultimately a brittle form of irregular three-dimensional structure, reminiscent of a glass can be achieved. Since radiation is a simple and quantitative method of increasing the crosslink density and hence altering M_c we are able to study the changes in mechanical properties of a variety of materials, ranging from rubber-like to glass-like, in terms of the size of the loops and circuits formed by these chains.

Recent attempts to produce new geometrical arrangements have been successful; we may mention ladder-like polymers, graft copolymers and block copolymers

```
        B
        B
  - AAAAAAAAAA -        and        - AAAABBBBAAAAABBBBA -
           B
           B
           B
           B
```

(where A and B are two different monomer units combined in the same molecular chain);

isotactic (parallel) and syndiotactic (alternating) as well as atactic (random) polymers

- AAAAAAA - - AVAVAVVAVA - - AAVAAVVA -

depending on the configuration of successive (identical) monomer units along the chain. As yet comparatively little of a fundamental character is known relating these different arrangements to the resulting mechanical properties.

RADIATION EFFECTS IN ORGANIC SOLIDS: FATE OF THE ELECTRON

Much of our thinking in radiation matters is colored by our knowledge of what occurs in gases. Information on what happens in the condensed phase is less precise. For example how does ionisation in an organic solid or liquid compare with that of the identical molecular in the gas phase? There is some - rather limited - evidence that in simple hydrocarbons the ionisation potential is reduced by some ten percent. We can also assume that an ion is formed when the ejected electron reaches a distance from its parent ion such that its thermal energy exceeds the potential energy of electrostatic attraction to its parent ion. Everything therefore depends on the rate at which it loses energy by collision. If this is sufficiently rapid the electron will not escape the electrostatic field, but return to form a highly excited molecule. We may therefore distinguish between ionised and excited molecular states, with a possible very different ratio to that in gases. Here we can call on polymer science, where very different chemical reactions occur depending on whether initiation results from an ionic or a radical species - the radical being an uncharged species with an unpaired electron, as results from excitation. Present evidence, by no means conclusive, indicates a G value of about 0.3 for ionic reactions, 3 for radical reactions in any simple hydrocarbons. Thus most electrons are recaptured. The G value represents the number of species formed per 100 ev absorbed in the system. It must be emphasized that these values are estimates only, and in simple hydrocarbons can vary from one system to another by 50%, or more in special cases. The assumptions on which they are based are in any case not universally accepted.

The role of the electron can be investigated further in several ways - by the use of electron traps with a resultant change in chemical product; by investigation of thermoluminescence in irradiated polymers and by the induced conductivity.

THERMOLUMINESCENCE

When polymers are exposed to gamma radiation at liquid nitrogen temperature, and then allowed to warm up slowly, light is emitted. The glow curve, which is a plot of light emitted against time and temperature, may show several peaks depending on polymer, crystallinity, etc. In polyethylene, irradiated in the absence of oxygen, three main peaks are observed, denoted alpha, beta and gamma in order of rising temperature. The alpha peak originates in the crystalline component, the gamma peak in the amorphous component, while the beta probably arises near the crystal surface. Thus a measurement of the alpha / gamma ratio provides a simple method of indicating the degree of crystallinity for the same polymer, following different thermal treatments.

It might have been supposed that this light emission is related to the reaction of radicals, trapped in the polymer, and able to interact as the temperature is raised (chemoluminescence). In fact measurements in a range of polymers show that the temperatures at which these emission peaks occur correspond primarily to transitions in mobility of the polymeric chains, e.g. increased flexibility. This technique is now being used to compare the effect of temperature on mechanical properties of related polymers.

Measurement of the spectrum of each glow peak shows that this consists of the fluorescent and phosphorescent spectra of the polymer - or rather of some impurities in the polymer, since some can be removed or transferred very readily. In polyethylene many of these impurities are (surprisingly) polycyclic aromatics - particularly a simple derivative of phenanthrene.

One more piece of evidence is provided by the effect of oxygen which when present in polyethylene during irradiation removes the beta and gamma peaks, but not the alpha peak. This oxygen must be present as dissolved oxygen, and not as combined oxygen as in carbonyl, when it is ineffective. Since oxygen cannot penetrate the crystallites, its lack of effect on the alpha peak is readily understood.

We are therefore led to envisage the following process -- similar in many respects to that previously studied in inorganic crystalline materials. The electron ejected by ionisation is trapped on the polymer structure - and when oxygen is available preferably by oxygen molecules with their high electron affinity. When the temperature is raised and new forms of chain mobility permitted, these electrons are released and react with luminescent centers. These centers which do not appear to be the base polymer itself in the case of polyethylene, are excited to the triplet or the singlet state and can decay either by a radiationless transition, or by a phosphorescent triplet to ground state decay, or by a fluorescent type singlet decay to ground state.

These observations still leave open a number of questions including the nature of the electron trap, the remarkable prominence of certain aromatic impurities as luminescent centers, and the difference in behaviour of the crystalline and amorphous regions. The intensity of light emitted is also very small, down by many orders on the number of photons which might be expected to result from electrons initially formed with G values of about 0.3.

Somewhat similar glow curves can be observed from irradiated DNA and its constituents, but their full interpretation has not been achieved, due in part to the difficulty of obtaining pure specimens. Taken together with information derived from ESR work, they should provide useful information on the mechanism by which energy is transferred to the most sensitive sites within the DNA molecule.

RADIATION-INDUCED CONDUCTIVITY

The minute electrical currents flowing through thin polymer films in an electrical field can be greatly increased by exposure to radiation. In spite of the considerable amount of experimental work available, it cannot be said that any general agreement has been reached on the cause of this conduction - whether it arises from electrons, holes or even H ions.

The phenomenon can be outlined as follows:

When a thin polymer film is exposed to radiation, a current i flows through it which is proportional to the applied field at low field values. The variation with radiation intensity I is less simple:

$$i = KI^{\Delta}$$

where the exponent Δ can vary from 0.5 to 1. (This behavior may be compared with a changeover from second-order to first-order kinetics). After the onset of radiation, the current i increases to a steady value which may be attained only after an interval of seconds or minutes. When radiation ceases, the decay in current is protracted, often over some hours. To explain this behavior Fowler has calculated a distribution of electron traps, which release their electrons at varying times. However, further measurements lead to different distributions; no definite conclusions can be drawn from these measurements.

FRACTURE OF CHEMICAL BONDS - RADICAL MOBILITY

When an organic molecule is excited, either directly by radiation, or indirectly by recapture of an electron, a homopolar bond may be broken, and the molecule becomes severed into two uncharged groups which are still reactive due to the presence of an unpaired electron. Such radicals may be small - as for example an H atom - or large, e.g. a radical polymer chain. The fracture pattern in the condensed phase can be very different from that observed in similar compounds irradiated in the gas phase; in the former case ionic reactions are presumed to be less frequent (recapture of electrons); at the same time there is less opportunity for larger fragments to escape from their environment (cage effect). The net result is that in spite of the assumed random absorption of energy, the nature of the radicals produced is fairly selective. In solid polymers many of these radicals can live for long periods, and their nature and subsequent reactions can be traced by such techniques as electron spin resonance and product analysis. In practice it is often convenient to irradiate the polymer specimen at liquid nitrogen temperature, and examine it periodically after various times of annealing.

Hydrogen gas is one of the most frequent gases produced during the irradiation of simple linear polymers such as polyethylene. This is to be expected from the large number of C-H bonds present, and from the small size of the H atom, which enables it to escape from its cage. In ESR studies this simplest of radicals can rarely be seen.

These simple observations raise major questions.

(1) What is the relationship between the point at which energy is absorbed, and the location of the resultant radical?

(2) How do these radicals react with each other and ultimately disappear?

(3) How do the individual H atoms interact to give the evolved H_2 gas?

To explain the selectivity of radical formation, several possibilities are available. The first, energy transfer, involves the movement of excitation energy within the molecule, or between adjacent molecules, and its interpretation involves quantum theory; it has been widely investigated in other branches of science. A second possibility, less closely studied, involves ionisation, followed by motion of the hole within the chain to a more stable configuration, with a final recapture of the electron at this new site to give a highly excited system, which then fractures to produce the observed radicals.

For a radical to disappear, it must react with another radical. If the polymer radicals are formed in close proximity this combination will occur very readily, and we cannot therefore account for the long lives often observed. Considerable evidence is now accumulating that these radicals have a limited degree of mobility which enables them to move round through the crystalline or amorphous regions in the polymer, where they eventually pair up and disappear. In polyethylene there are both crystalline and amorphous regions, and radiation produces radicals in both. By stretching the sample, the crystalline regions can be oriented, and the ESR patterns of the radicals imbedded therein then show asymmetry. By studying the rate of decay of this asymmetry, and of the total radical population, one can show that the radicals trapped in the crystallines take longer to migrate and disappear in larger crystals, and that they do this by a bimolecular reaction. We may therefore envisage a slow motion into the amorphous regions where by virtue of greater chain mobility, conditions for linking two molecules by the interaction of two polymer radicals are far more favorable.

Where in the course of its travels, a radical encounters a trap, such as an unsaturated group (- CH = CH -) it reacts to form a new type of radical of much greater stability, which can last for many days. This is readily shown with ESR techniques.

Further experimental evidence favoring the radical rather than the ionic explanation of crosslinking is now coming to light. We expect the fundamental radiation-induced reaction in polyethylene and in simple paraffins to be basically similar, so that an irradiated mixture of polyethylene and hexane should show polymer-polymer, polymer-hexane and hexane-hexane links. Only the first of these will be effective in forming an insoluble polymer network. If polyethylene swollen in hexane is irradiated, the gel fraction is greatly reduced, although the hexane can only penetrate the amorphous regions. Present measurements may be summarised as follows:

(i) Low Density Polyethylene (about 45% amorphous); crosslinking reduced by a factor of two when swollen with 10% of hexane.

(ii) High Density Polyethylene (about 15% amorphous); crosslinking reduced by a factor of four when swollen with 5% of hexane.

Radicals may be presumed to form about equally readily in the crystalline and amorphous regions. Thus in high density polyethylene, where 85% of the total radicals are initially present in the crystallites, the four-fold reduction in crosslinking requires the loss of most of these radicals, as well as the 15% initially present in the amorphous region. This loss can only occur by the motion of radicals into the amorphous region where they react with hexane. The short lifetime of ions at room temperature would be quite inadequate to allow for a similar transfer.

Further work is continuing along these lines.

FATE OF H ATOM

Surprisingly little attention has been paid to the fate of the H atoms. When first ejected from their parent molecule, they have considerable kinetic energy and can very readily abstract a second H atom from a neighboring molecule

$$\_\_\_\ CH_2\ \_\_\_ \longrightarrow \_\_\_\ \dot{C}H\ \_\_\_ + H\ (\text{energetic})$$

$$H\ (\text{energetic}) + \_\_\_\ CH_2\ \_\_\_ \longrightarrow \_\_\_\ \dot{C}H\ \_\_\_ + H_2$$

leaving a second polymer radical in the vicinity of the first. Combination of these two radicals then occurs to give a crosslink. When abstraction does not take place in the first few encounters, the H atom becomes thermalised, and can then suffer many encounters before abstracting to form a second radical. By using H scavengers as additives to the polymer it becomes possible to prevent such abstraction from the base material, therefore reducing the net effect of radiation. This represents one form of radiation protection, and the amount of additive needed can be small, since the thermalised H atom has such a long path during which it can react.

Certain sulphydryl compounds used in radiobiology are particularly effective in this respect, since they can readily collect, store and release H atoms. They can therefore both prevent the second radical being formed, and restore an H atom at the first radical site. In solids the lack of mobility prevents the process from being fully effective below about 1% of additive; in polymer solutions a few parts per million of additive can be remarkably effective. Further work along these lines show many interesting analogies between the behavior of simple polymers and the more complex biological macromolecules.

EFFECT OF STATE

It may be appropriate at this stage to mention the importance of the state of the sample on chemical changes induced by radiation. Since polymers cannot be obtained in the gaseous state, it is perhaps more convenient to compare gas yields (as an example of such changes) for a range of simple paraffins, which can be irradiated as gases, liquids and solids. Polyethylene can be considered as the ultimate member of the series. G values for H_2 production show no significant trend when going from ethane (two C atoms per molecule) to hexadecane (with 16). Production of methane from CH_3 and H shows a fall with increasing chain length, as is to be expected if CH_3 is formed by scission of the terminal C-C bond. Again low density and high density polyethylene (i.e. low and high percentage of crystallinity) show the same rate of radical formation, although these radicals disappear far more readily in the former material, where they can join up more rapidly.

Although these yields depend but little on physical state, they do show a marked temperature dependence. At high temperatures, the increased amplitude of vibration favors permanent dissociation of a bond in the excited state.

SOLID STATE POLYMERISATION

So far we have been largely concerned with radiation mechanisms, and the reactions by which reactive groups, such as radicals, are produced. A different problem arises in crystalline and in amorphous monomer solids, where (surprisingly) radiation can initiate chain reactions. Here we can perhaps ignore radiation mechanisms, and discuss mobility of larger molecules within the matrix, radiation being considered merely as a very convenient initiator of a chain reaction, which subsequently propagates throughout the material. The effect of single ionisations and excitations is greatly enhanced by the considerable length of the resultant polymer chain.

Many monomers can form good single crystals, or when cooled more rapidly, polycrystalline or glassy solids. Irradiation of single crystals of Trioxane for example initiates polymerisation, and a good single crystal of polymer is formed. The outstanding feature of this reaction is that it requires some degree of mobility in each successive monomer unit, to allow it to link up with its neighbor, and to do this without disrupting the crystal lattice. In the liquid phase the reaction is slow or non-existent. Thus monomer orientation assists the reaction.

In other monomer crystals such as acrylamide the reaction leads at first to polymer showing some degree of order (birefringence) which disappears on warming. In this material polymerisation is not aided by the regular spacings of the lattice, and polymer is formed preferentially in the neighborhood of crystal imperfections. Since the polymerisation reaction proceeds slowly, it can be followed e.g. by electron microscopy of the surface of irradiated monomer crystals. The original polymer nuclei within the crystal grow to globular structures of diameter 300-400Å, distributed along the directions of stress or imperfection in the lattice. The use of this method of decorating latice defects, deserves further attention. Examination under polarised light reveals a polymerisation wave, travelling slowly through the crystal lattice; ahead of this wave new nucleation centers, previously trapped, can be seen to grow as the stresses resulting from the main reaction reach them.

Many other phenomena relating to mobility in monomeric solids can be investigated using this powerful new technique. The influence of performed polymer, of temperature and of pressure have revealed some surprising features. (The growth of unoriented polymer can be inhibited temporarily by the application of pressure to the crystal). Time and space do not permit a fuller description, particularly as no general pattern has yet emerged to describe the behavior of the large number of monomeric solids which can be studied.

PRACTICAL APPLICATIONS

Exposure to radiation offers an almost unique method of modifying polymers in the solid state. It is now widely used for the crosslinking of polyethylene which it converts from a thermoplastic to a rubberlike material at temperatures above 115 - 135°C. Applications are electrical fittings, cables, shrinkable film, etc. The production of certain grafted material is also being envisaged as is the direct formation of polymeric solids by solid state polymerisation.

The original obstacle of radiation costs has largely disappeared and the major task is now the search for novel applications of the many new types of materials in which these processes can be used. One obvious example is cotton, which unfortunately degrades (main chain scission) on irradiation. If means could be discovered for converting the radicals from the degrading to the crosslinking type an entirely new range of fibres would result.

CONCLUSIONS

In presenting this paper I have tried to avoid spending too much time on certain branches of radiation studies which, while important to our general understanding, are not specifically applicable to the solid state. I have necessarily had to skim very lightly over the surface of topics, emphasizing those which still merit fuller attention, and can only hope that this does not convey the impression that most of the subject is but dimly understood. Each step forward has opened up new vistas and new territories worthy of exploration, and confirmed that as a subject, radiation treatment of polymers has ramifications in very diverse fields.

Gas Yields for Aliphatic Hydrocarbons in the Gaseous, Liquid and Solid State

	$H(CH_2)_nH$: Fast Electrons or Gamma Radiation			
	State	n	$G(H_2)$	$G(CH_4)$
Ethane	Gas	2	5.6	0.56
Propane	Gas	3	3.8	1.25
Pentane	Liquid	5	4.2	0.4
Hexane	Liquid	6	4.0 5.0 4.9	0.12 0.15 0.4
Heptane	Liquid	7	4.7	0.09
Hexadecane	Solid	16	4.8	0.04

MECHANISM OF THE TRANSFORMATION OF CRYSTALLINE QUARTZ INTO AMORPHOUS SILICA BY NEUTRON IRRADIATION

R. Comes, M. Lambert and A. Guinier

Université de Paris

FRANCE

A single crystal of quartz becomes amorphous after a strong irradiation (dose > 10^{20} n/cm^2), its structure being very similar to that of vitreous silica. The structure of intermediate stages, corresponding to increasing neutron doses, was studied in order to follow the transformations.

In the first stage (< 2×10^{19} n/cm^2), the defects are homogeneously dispersed through the crystalline lattice giving a normal pattern diffraction, but the lattice parameters are increased and the macroscopic dimensions of the crystal vary in the same direction.

When the dose of neutrons increases, the structure of the quartz is no longer homogeneous: islands of highly perturbed matter are imbedded in a still crystalline matrix. The proportion of the two phases may be evaluated by the measurement of the intensities of the crystalline reflections. The discrepancies between the variations of lattice parameters and macroscopic dimensions are quantitatively explained.

The disordered phase for intermediate stages is not quite isotropic: it keeps some relation with the parent crystal structure. This is apparent in the diffraction pattern which is not yet the simple ring pattern of the amorphous silica. On the other hand, the crystalline matrix is somewhat disturbed. The diffraction lines are broadened and an easily detectable characteristic diffuse scattering appears. This scattering is localized on the (101) and (100) planes of the reciprocal lattice and seems similar to the scattering observed in unirradiated quartz at a temperature *very near the α ⇄ β transformation point. It can be interpreted by very numerous micro Dauphine twins. Above 6 x 10^{19} n/cm^2, the crystal has a hexagonal symmetry. This explains the disappearance of the α ⇄ β transformation by neutron irradiation.*

INTRODUCTION

Fast neutron irradiation transforms a crystal of α quartz, when the dose is higher than 10^{20} n/cm^2, into an amorphous substance (1) which is similar to ordinary vitreous silica (2,3). This amorphisation of the α quartz takes place progressively with variation of many of the most important physical properties: variation of the lattice parameters and macroscopic change of length (1,2,4); variation of the optical constants (2); variation of the elastic constants and change in the transition α ⇄ β (4).

The authors tried to elucidate the mechanism of this transformation by studying the structure of the irradiated quartz by means of X-ray methods (diffraction as well as diffuse scattering). Two schemes appear a priori possible: in the intermediate stage, amorphous domains might be formed, which progressively grow inside the crystal line matrix, or the entire crystal might be disturbed nearly homogeneously, passing from the crystal line to a more and more paracrystalline structure in which the parameters of the disorder increase continuously. The first aim of this work was to decide between these two possibilities.

The irradiated samples are the same as those which were previously used by Mayer (4); thus it was possible to correlate directly our X-ray measurements with the previous measurements of macroscopic change of dimensions.

RESULTS AND DISCUSSION

Existence of two Phases in the Intermediate Stage

Figure 1 shows the relative variation with irradiation dose of the lattice parameters *a* and *c*, together with the increase of macroscopic lengths in the directions parallel and normal to the optical axis. These data are in agreement with already known results: up to the dose of 3.10^{19} n/cm^2, there is complete agreement between the measurements of macroscopic lengths and of the lattice cell. Above this dose, the increase of the lattice parameters is larger than the increase of the macroscopic length in both directions, the difference increasing with increasing dose. Thus, for doses larger than 3.10^{19}, independent measurements show that:

$$\frac{\Delta a}{a} > \frac{\Delta \ell_a}{\ell_a} \tag{1}$$

$$\frac{\Delta c}{c} > \frac{\Delta \ell_c}{\ell_c} \quad ,$$

ℓ_a being a length of the specimen along $\vec{a}$, and ℓ_c along $\vec{c}$.

This is a rather surprising result, because it is in contradiction to the behavior of many other irradiated crystals. For instance, the measured density of heavily irradiated LiF is less than the one calculated from the lattice parameter: this is explained by the presence of holes inside the crystals (5).

The situation must be different in irradiated quartz. Let us consider a block of unirradiated quartz of volume V, composed of, say, 100 unit cells. After irradiation, each cell has a volume of $v_c + \Delta v_c$, for a given dose ($>3.10^{19}$). Δv_c is related to the variation of the parameters a and c by the equation:

$$\frac{\Delta v_c}{v_c} = 2\,\frac{\Delta a}{a} + \frac{\Delta c}{c} \tag{2}$$

But the 100 cells would have a volume $100\ (v_c + \Delta v_c)$, which is larger than the observed volume $V + \Delta V$. Thus we are necessarily led to the conclusion that the number of crystalline cells in the irradiated sample is less than 100. That means that there must be another phase besides the crystallized phase which has been slightly modified by the irradiation. If x is the proportion of the remaining quartz after irradiation, we can write the following equation:

$$V + \Delta V = 100x(v_c + \Delta v_c) + 100(1 - x)(v_c + \Delta v_a) \tag{3}$$

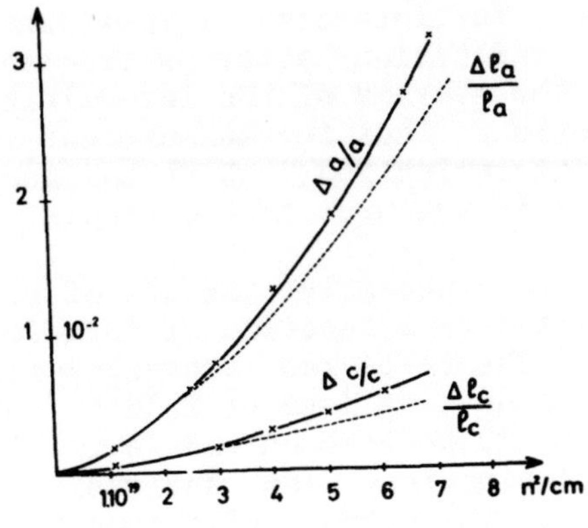

Fig. 1. Parameters and macroscopic lengths variation as a function of irradiation dose.

The total volume $V + \Delta V$ is the sum of the volume of the crystallized phase, i.e. 100 x cells of volume $v_c + \Delta v_c$, and the volume of the second phase corresponding to the initial volume of $100(1-x)v_c$ and which we suppose modified in the proportion $\Delta v_a/v_c$.

But the experimental data so far reported give only one equation (3) which contains two unknowns, the proportion of the remaining crystalline part, x, and the parameter Δv_a which depends on the nature of the second phase not yet defined. A first series of experiments was conducted to measure the parameter x.

Measurement of the Proportion of the Crystalline Phase

The part of the sample crystallized as quartz is responsible for a series of characteristic X-ray reflections. If several samples of the same shape are compared under the same experimental conditions, the ratio of the reflected intensities for the same reticular planes may give the amount of quartz crystal in the different samples. The slight changes in lattice parameters do not appreciably change the intensity of the reflection. Thus it is possible to take as the approximate proportion of crystal quartz the ratio

$$\frac{I^{hK\ell}_{irr.}}{I^{hK\ell}_{o}}$$

of the intensities of the crystalline reflections of indices $hK\ell$ given respectively by the irradiated and unirradiated samples. But many precautions are necessary to avoid too large errors.

All the samples are lamellae of the same thickness and the same orientation (parallel to (100) planes). They are placed on the diffractometer, the counter being fixed behind a widely opened window to receive the reflected beam. The crystal is rotated through the position of reflection. The intensity I represents the integrated intensity during the reflection after subtraction of the background which is taken as the average of the intensities recorded before and after the reflection. Thus the measurement is independent of the angular width of reflection, which may vary with irradiation, and also of the mosaic textures of the sample.

Measurements were made with three successive orders of reflection: 110, 220 and 330. The results as a function of irradiation doses are given in Figure 2. The reflected intensity for (110) increases first with irradiation up to a dose of 2.10^{19}n/cm^2, this phenomenon being smaller for the 220 reflection and invisible for the 330: it is an effect of extinction. The unirradiated quartz is a perfect enough crystal, so the strong reflections are

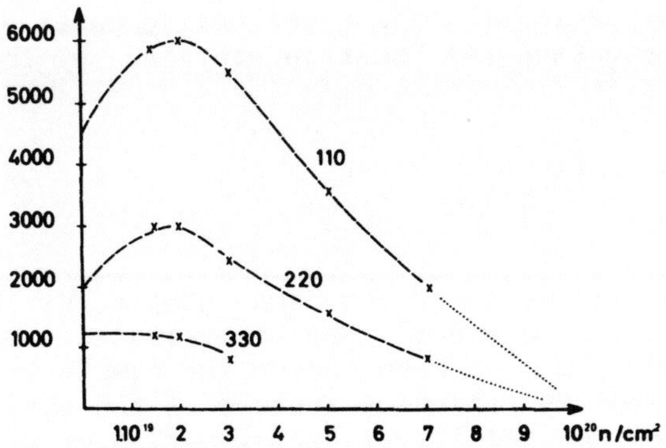

Fig. 2. Variation of the intensity of the reflections on (110) planes with increasing radiation dose.

markedly reduced by primary extinction. Even after a small irradiation dose, the perfection of the quartz crystal is damaged and the extinction vanishes, and, besides, there is no extinction for the high order reflections.

Thus we can conclude that, up to $2.10^{19}n/cm^2$, only the quartz phase is present. For higher doses, the reflected intensity decreases, but in a proportion which depends on the indices, which is not in agreement with the formula:

$$x = \frac{I^{hK\ell}_{irr.}}{I^{hK\ell}_{o}} \qquad (4)$$

On the other hand, for the high doses, measurements are no longer possible with the 330 reflection, because it is too weak to be distinguished from the continuous background. This can be explained by a distortion of the crystalline part similar to the disorder caused by the thermal agitation of a crystal. It is known that in the latter case the intensity of the Bragg reflections are reduced by the Debye-Waller factor:

$$\exp\left(-\,2\,B'\,\frac{\sin^2\theta}{\lambda^2}\right) ,$$

where θ is the Bragg angle and λ the wave length of the radiation used for the diffraction measurements. If the irradiation produces displacements of the atoms isotropically and distributed homogeneously in the crystal without destroying the long range order, it may be admitted that the Bragg reflections will be affected by an artificial temperature factor with a coefficient B' depending on the irradiation dose. Thus the intensity of a reflection in the irradiated sample is reduced for two reasons, the decrease of the crystalline fraction and the decrease of the reflecting power of the remaining crystal. So equation (4) must be replaced by the following

$$I_{irr.}^{hK\ell} = I_o^{hK\ell} \exp\left(-2\,B' \frac{\sin^2\theta}{\lambda^2}\right) \tag{5}$$

If the reflected intensities are measured for each sample for at least two lines, both coefficient x and B' may be evaluated. On account of the extinction effects, we have chosen the sample irradiated at a dose of 10^{19} n/cm^2 as reference instead of the unirradiated sample. Table I gives the results of determination of the amount of crystallized phase, x, as a function of the irradiation dose.

TABLE I

The Proportion of Crystal Phase as a Function of Irradiation Dose

dose (n/cm^2)	proportion of α-quartz, x
3.10^{19}	90%
5.10^{19}	60%
7.10^{19}	35%

The values obtained for the disorder factor B' will be discussed in the next paragraph.

By extrapolating the values of x for higher doses, it is predicted that the crystalline part would vanish for a dose of 10^{20}, and in fact it is observed that for this dose the irradiated quartz is completely amorphous.

The values of x cannot be considered as accurate, because of the difficulty of measuring the intensities and because of the approximations used to establish formula (5). Thus the error in x if of the order of at least 10%. Nevertheless two phases can clearly be distinguished at the intermediate stage: the crystalline quartz with slightly modified parameters and some local atomic disorder in its structures; and regions of which the composition is also SiO_2, or certainly very near its stoichiometric ratio, but which do not give a crystalline pattern, so that they can be identified with amorphous silica.

This idea has been confirmed in different ways. The successive values of x given in Table I, may be substituted in Equation (3). Since $V = 100\ v_c$,

$$\frac{\Delta V}{V} = x\,\frac{\Delta v_c}{v_c} + (1 - x)\,\frac{\Delta v_a}{v_c} \qquad (6)$$

From the experimental values of the relative variation of the macroscopic volume and of the fraction x of crystal quartz, we can calculate the dilatation $\Delta v_a/v_c$ of the amorphous phase relative to the original crystal. The results for two irradiation doses are as follows:

TABLE II

Dilatation of the Disordered Phase as a Function of Irradiation Dose

dose n/cm^2	$\frac{\Delta V}{V}$	$\frac{\Delta v_c}{v_c}$	$\frac{\Delta v_a}{v_c}$
5.10^{19}	3.6%	4.2%	2.6%
7.10^{19}	5.8%	7.0%	5.1%

The values found for $\Delta v_a/v_c$ seem very reasonable; they show that, as expected, the non-crystalline regions have a density lower than that of the original quartz crystal but higher than that of the vitreous silica obtained after a long irradiation ($\Delta v_a/v_c$ = 14%). The structure of these domains varies with the irradiation and it approaches the final structure of amorphous silica.

The most important feature of the disordered phase appears in the diffraction patterns. These patterns were obtained with a stationary sample receiving a monochromatic beam (CuKα radiation) (Fig. 3). Thus a weak anomalous scattering could be recorded besides the ordinary Bragg reflection spots. For low doses of irradiation, the patterns are similar to those given by unirradiated

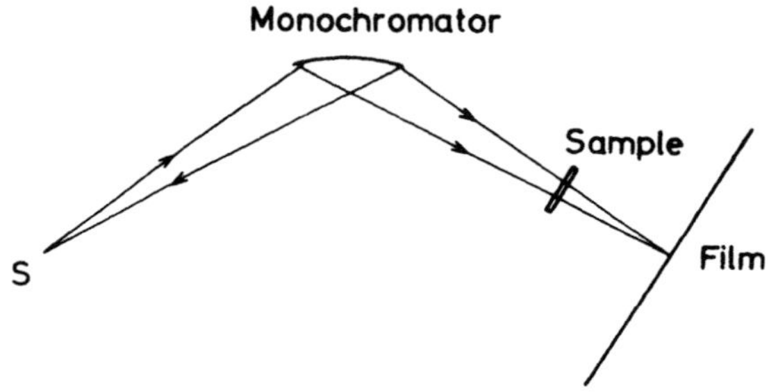

Fig. 3. Schematic Diagram of the camera used to study the diffuse scattering.

quartz. However, in the intermediate stage ($\approx 10^{19}$ n/cm^2), the crystalline pattern shows some modifications which will be described in detail later on, and, in addition, there appears a region of scattering around the center of pattern, which resembles the diffuse ring characteristic of vitreous silica. When the dose increases, the intensity of this diffuse ring increases and the distribution of scattering is somewhat modified. For a sample which has received a dose of 7.7×10^{19} n/cm^2, and has been annealed after irradiation at 1250°C, the separation between the crystalline pattern and the amorphous pattern is very clear (Figure 4). For very high doses ($> 10^{20}$ n/cm^2), the crystalline pattern disappears completely and thus there only remains the amorphous ring, which is almost identical to the ring given by pure vitreous silica.

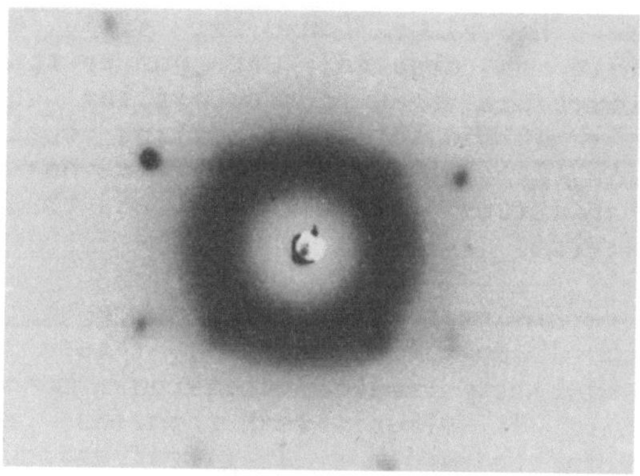

Fig. 4. Diffuse scattering of a sample irradiated at a dose of 7.7×10^{19} n/cm^2 and annealed at 1250°C. Orientation as in Fig. 5 a).

The existence of two phases and the progressive development of the amorphous phase during the irradiation process seems well established. However, the crystalline phase is not the normal perfect quartz phase, and the amorphous phase at the beginning is not identical to the vitreous silica. A discussion follows of what it has so far been possible to determine about the atomic structures of both phases.

Structure of the Crystalline Part

In these domains, the structure can be described as a first approximation as α-quartz, but some significant differences are observed. We have already noticed that the average lattice has a slightly different unit cell. That may be explained by the presence of point defects, vacancies, interstitials and broken bonds isolated and dispersed at random in the lattice. These disorder centers decrease the average packing density of the atoms, and furthermore destroy the rigorous periodicity of the perfect crystal. The homologous atoms are not situated exactly at the nodes of a perfect lattice, nevertheless a long-range order is preserved. The first effect on the diffraction pattern is the decrease in the intensities of the high-order reflections which has already been noticed. The coefficient B' of formula (5) is related to the quadratic average $\overline{\Delta X}$ of the atomic displacements: $B' = 8\pi^2\overline{\Delta X^2}/3$. Table III gives the values of B' and $\overline{\Delta X}$ deduced from the diffraction intensity measurements.

TABLE III

Atomic Disorder Factors in the Crystalline Part

dose n/cm^2	B' (Å^2)	ΔX (Å)
3.10^{19}	0.64	0.15
5.10^{19}	0.95	0.19
$7.2\ 10^{19}$	1.30	0.22

Since the intensities for the irradiated samples are referred to the normal quartz, the values of ΔX given in Table III do not include the disorder due to thermal agitation. It is interesting to note that the factor B' for the normal quartz is equal to .70 $Å^2$ at room temperature; it is then of the same order of magnitude as the factor found in quartz irradiated at moderate doses. Thus it is understandable that the quartz lattice may accept these static disorders without losing its characteristic long-range order.

The defects of periodicity produce another effect in diffraction patterns. They give rise to diffuse scatterings outside the directions of Bragg reflections. Figures 4, 5, 6 and 7 give some examples of scattering patterns. Whereas the decrease of the intensity of the Bragg reflections gives only a measure of the average atomic displacements, the repartition of the anomalous scattering gives information about the nature of these displacements.

In the patterns of quartz irradiated with doses higher than 3.10^{19} n/cm^2, the regions of diffuse scattering form broad streaks which generally join neighboring diffraction spots. This distribution of scattering is similar to that which is produced by the thermal agitation in a normal quartz crystal (Fig.8). However, there are two important differences: the intensity of the diffuse streaks in the irradiated crystals is about four times as strong, and their intensity is independent of the temperature, whereas the thermal streaks are roughly proportional to temperature. We have verified that at liquid nitrogen temperature the disorder streaks are still visible and the thermal streaks vanish.

It should be noted that α-quartz has a trigonal symmetry around its $\vec{c}$ axis and this symmetry is apparent in the diffraction pattern and the thermal diffuse patterns (Fig. 5): but in strongly irradiated quartz the patterns have a hexagonal symmetry (Fig. 6) for the diffraction spots as well as for the diffuse streaks.

In order to locate the regions of strong scattering in the reciprocal space, a series of patterns was obtained for the same sample with different orientations. Thus, it could be shown that the scattering is concentrated along two sets of planes in the reciprocal lattice, the (100) and (101) planes. In a given pattern, the diffuse streaks correspond to the intersections of these two sets of planes with the Ewald sphere (6). Figure 7 shows that the streaks are actually observed along the predicted lines; their intensities vary over a large range between the different planes of scattering.

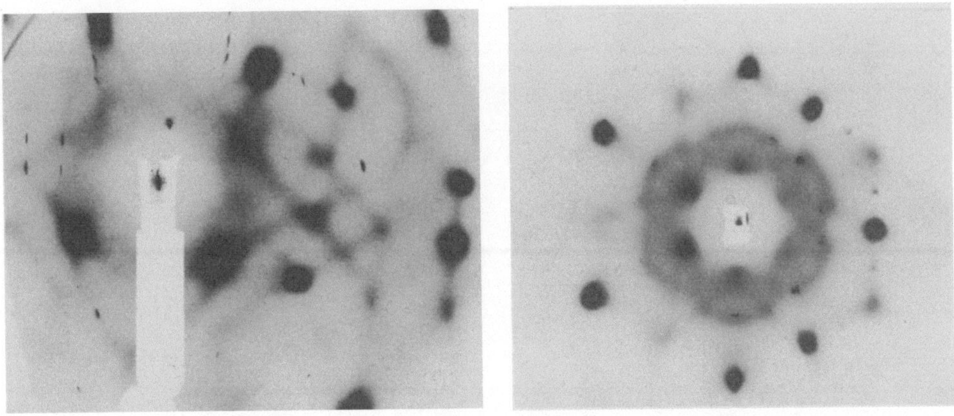

(a) (b)

Fig. 5. Diffuse scattering from a sample irradiated at the dose of $3.10^{19}n/cm^2$; orientation of the quartz crystal: pattern a) axis $\vec{c}$ vertical, axis $\vec{a}^x$ of the reciprocal lattice parallel to the X-ray beam; b) axis $\vec{c}$ parallel to the X-ray beam.

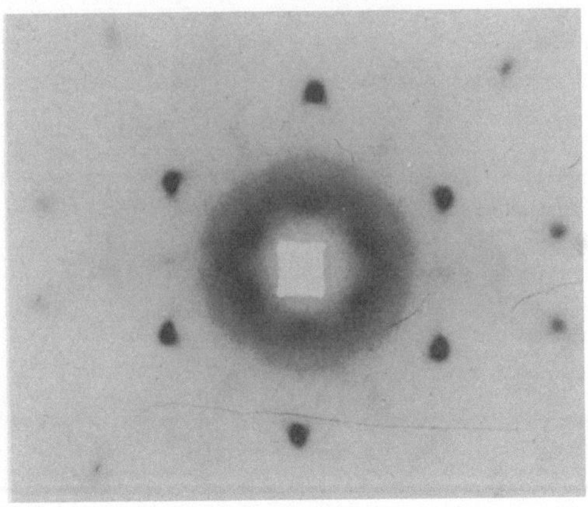

Fig. 6. Diffuse scattering of a sample irradiated at a dose of $5.10^{19}n/cm^2$; orientation as in Figure 5b.

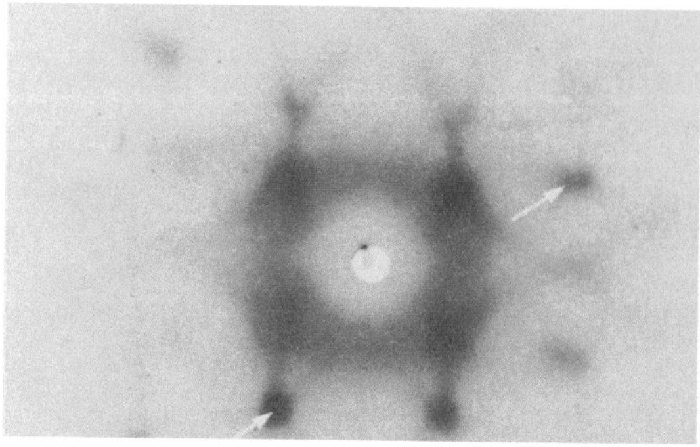

Fig. 7a. Diffuse scattering of a sample irradiated at 5.10^{19}n/cm^2.

Fig. 7b. Schematic diagram of the traces of the (100) and (110) planes calculated from their intersections with the Ewald sphere (same scale as in pattern a).

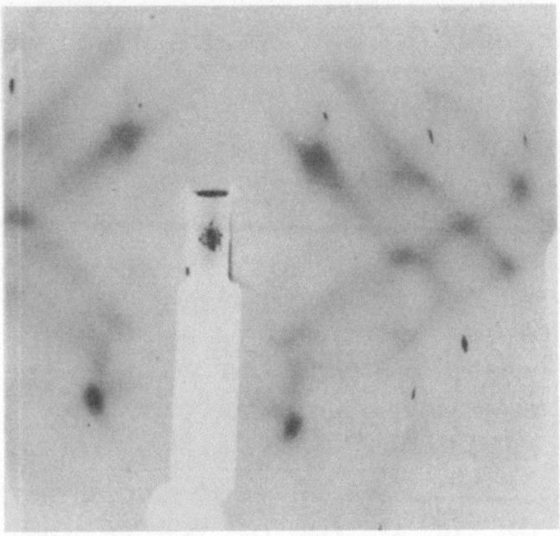

(a)

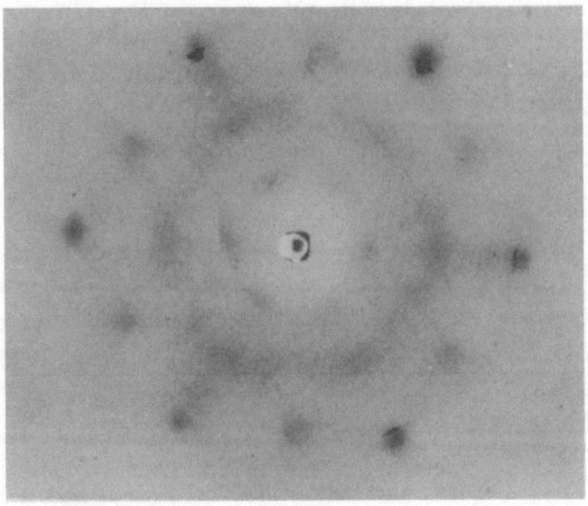

(b)

Fig. 8. Diffuse scattering by an unirradiated quartz crystal. Same orientations of the crystal as for Figure 5a) and b).

From the observed distribution of scattering, some general conclusions about the nature of the imperfections in irradiated quartz can be drawn. A careful investigation of the region of low angle scattering did not show any noticeable scattering. This proves that the disorder is due to the displacements of atoms, and that these displacements are small in comparison to the cell unit dimensions; in other words there are no large local density fluctuations; there are no clusters of vacancies or interstitials as in irradiated ionic crystals (LiF).

Because the scattering is not uniformly distributed in reciprocal space, there must be some correlation between the atomic disorders in neighboring cells. In particular, the localization of the scattering along reticular planes (100) and (110) means that in the disordered regions, chains of atoms parallel to [100] or [110] axes move together, so that the periodicity is preserved along the chain and disturbed between the chain and its neighbors. From the thickness of the region of scattering along the reticular plane, it is possible to deduce the order of magnitude of the lengths of the displaced chains: the number of cells in the chain is roughly equal to the ratio of the interreticular distance of the planes to their thickness. So we arrive, in this case, at a length of 40 A°.

It is not possible to find the structure of a distorted crystal directly from the diffuse scattering, even if it is quantitatively determined. We have to imagine models satisfying, of course, the general conditions stated above and then compare their calculated effects on diffraction patterns with the observations. The following speculative hypotheses have been tried.

Quartz is formed by an ensemble of SiO_4 tetrahedra linked together by their apices. A simple assemblage is realised in β-quartz, stable above 574°C; the tetrahedra are regular and the symmetry is hexagonal. But in α-quartz the tetrahedra are slightly distorted and their arrangement has only a trigonal symmetry. Starting from one β-quartz crystal, it is possible by small displacements of atoms to build two α-quartz crystals which are transformed into each other by a rotation of 180° around the $\vec{c}$ axis. Figure 9 represents the projection of the two α-structures on the plane (001). The tetrahedra of oxygen atoms only have been drawn, and the silicons are indicated only as S_1 and S_2. The oxygen atoms are situated in B sites in the β-phase and in A_1 or A_2 sites in the α-crystals. For the silicon atom, the β-position is located at the middle of a [100] axis and in the α-crystals it is displaced along the $\vec{a}$ direction either in position $\frac{1}{2} - \varepsilon$ or in position $\frac{1}{2} + \varepsilon$.

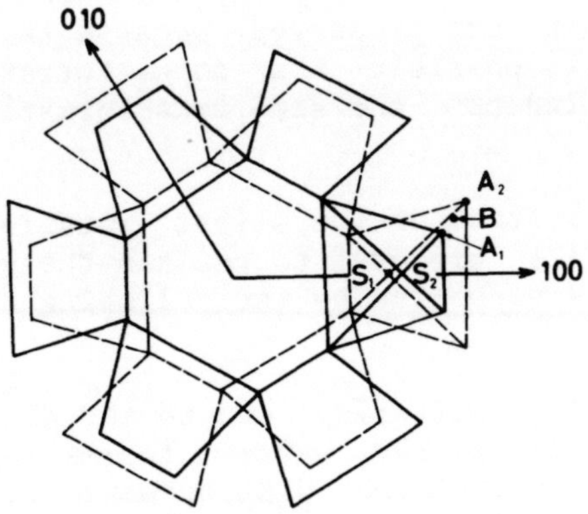

Fig. 9. Schematic diagram of the structure of α_1 and α_2 quartz formed from a β crystal.

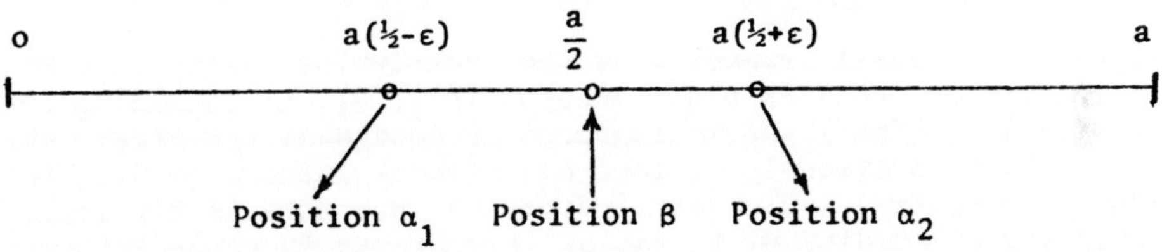

It is very common to find large crystals of quartz divided in domains all having the same $\vec{a}$ and $\vec{c}$ axes, but with structures α_1 and α_2 respectively. These are called Dauphine twins. Generally the twinned domains have a macroscopic size and are visible after suitable etching. R.A. Young (8), studying the intensity of a series of reflections of different planes at a temperature just below the transition $\alpha \rightleftarrows \beta$, suggested that within a few degrees the α-quartz is formed of microtwins of the Dauphine type. Since at 574°C in the β phase the middle position of the atoms corresponds to the minimum of energy, at a few degrees below this temperature only a very small energy is required to pass from one form of α-quartz to the other. In this transition state, Arnold (7) has observed diffuse scattering patterns of quartz which are very similar to the patterns observed in this work with irradiated crystals.

Therefore, it is suggested that the disorder produced by neutron irradiation is due to the passage of small groups of SiO_4 tetrahedra from their original α form to the opposite one. These groups have the shape of small chains of tetrahedra in the directions [100] or [110], these shapes being such that the interfacial energy with the matrix is the smallest. In the neighborhood of the track of an atom knocked by a fast neutron, the atoms receive an impulsion and are displaced, but they very rapidly find a new equilibrium. The SiO_4 tetrahedra are preserved because the Si-O bonds are very strong. But it is possible that an α_1-tetrahedron takes the α_2-configuration and induces the same transformation in the neighboring tetrahedra.

This mechanism is a special case of the slight rotations of the tetrahedra by which Primak (2) proposed to explain the structure of irradiated quartz. It is supported by the following series of observations:

a) the similarity between the scattering due to the disorder in irradiated crystals and that due to the natural thermal agitation in normal crystals suggests that the displacements during radiation damage are those which are excited by thermal energy: and it is known that raising the temperature is sufficient to displace the atoms from the α to the β position, passing by a mixture of α_1 and α_2.

b) the trigonal symmetry of the scattering disappears when the irradiation dose is high enough (Fig. 6, corresponding to a dose of 5.10^{19} n/cm^2) and is replaced by hexagonal symmetry: this fact has been previously noticed by several authors (4,9). This hexagonal symmetry would result from the presence in the irradiated sample of domains of α_1 and α_2 structures in equivalent proportions: that does not mean that the irradiation changes the α structure into a β structure, which is really hexagonal.

c) the piezoelectric constant d_{11}, which is connected to the trigonal symmetry of α-quartz, decreases with the irradiation dose and vanishes at 6.5×10^{19} n/cm^2 (4), although for this dose the proportion of crystal is still 35%.

d) the anomaly of specific heat (4) occurring at the transition $\alpha \rightleftarrows \beta$ diminishes for irradiated crystals and disappears for high doses. Thus the irradiated quartz seems to be nearer to the β form, which is in agreement with the model of the microtwins.

e) according to Zachariasen (10), natural samples of quartz may contain micro-Dauphine twins, even if they appear untwinned by ordinary tests. This proves that the formation of a microtwin is easy.

It then seems justifiable to make use of the hypothesis of microtwins to explain the disorder revealed by the X-ray diffuse scattering of irradiated quartz. In order to solve the structure problem of these irradiated crystals, a precise model must be found, the scattering of which is in quantitative agreement with experiment. Such a success has not yet been reached, but encouraging results have been obtained with some models and trials are currently being made in order to improve them.

Structure of the Amorphous Part

The combined measurements of macroscopic density, lattice parameters and fraction of crystalline part give two data: the proportion of the amorphous phase and the dilatation of the domains becoming amorphous. It is interesting to note that in spite of the crystalline part, it is not necessary to assume that the amorphous regions are compressed (11), which would be difficult to understand in the general evolution of the irradiated quartz. Actually, according to Table II there is a volume expansion of a few percent. But from these figures it is not possible to deduce directly the density of the amorphous phase, because some atoms may have been transferred to the surrounding crystal leaving vacancies inside the amorphous domain and forming interstitials inside the crystal. It must be remembered that the structure of quartz is not compact and thus may easily contain interstitials. However, the absence of any small angle scattering shows that the fluctuations of density remain relatively weak. Therefore, we can conclude from the continuous increase of the dilatation of the amorphous regions with increasing dose, that their density decreases to reach, for very high doses, the density of totally amorphous quartz (14%). A confirmation of the evolution of the structure is found in the diffraction patterns.

As already mentioned, the existence of the amorphous part is proved by the appearance of a "ring" in the pattern characteristic of amorphous substances. Unfortunately, the scattering of the crystalline parts formed by broad diffuse streaks is superposed upon this ring, and it would be impossible accurately to separate the two effects; thus only qualitative observations are feasible. Nevertheless, there is no doubt that the ring changes with irradiation dose. For small doses, it has no circular symmetry, but exhibits intensity reinforcements which have the symmetry of the crystal. This is particularly apparent in Figure 6, where the hexagonal axis is parallel to the X-ray beam: the diffuse ring has clearly a six-fold symmetry. This symmetry disappears in Figure 4, for a higher dose. At the beginning, in the disordered domains, the structure is not quite isotropic but is still somewhat governed by the symmetry of the crystal in which they are imbedded.

No information is given by the X-ray patterns about the dimensions of the amorphous domains: in the diffraction figure of a body with irregular structure, it is not possible to separate the effects of the irregularity of the structure and of the size of the diffraction body. It is likely that the domains are still small when the action of the surrounding crystal produces an effect of anisotropy and that they grow larger for increased irradiation doses: the disorder is then larger and the structure becomes isotropic and independent of the remaining crystalline parts.

It is interesting to note that a not completely amorphous structure has been already postulated to describe the state of the silica in the superficial layers of fine particles of crushed quartz crystals (12).

CONCLUSION

It seems well established that before becoming completely amorphous, irradiated quartz consists of two phases, the crystal phase, and an amorphous phase the proportion of which grows with the dose of irradiation up to the total disappearance of the crystal form. However, two important corrections must be made to this too-simple description. The crystal has no more a perfect lattice: it has still a well defined long-range order, but contains many local perturbations. On the other hand, the amorphous part at the beginning is not completely disordered and keeps some remnants of the order of the crystal from which it is formed. Furthermore, there must exist a border region of transition between the two phases, the structure of which passes gradually from the ordered to the disordered state.

What is permanent and nearly unchanged during the whole process of radiation damage is the building unit of silica, that is, the tetrahedron SiO_4, and what is modified is the mutual arrangement of these tetrahedra. The existence of many varieties of crystallized silica as well as of the vitreous silica shows that the bonding of tetrahedra is rather flexible and the passage from one configuration to the other requires little energy, whereas the SiO_4 bonds are very difficult to break. Such a breakage occurs only in the center of the highly disordered domain at the end of the track of a high energy knock-on atom. The SiO_4 tetrahedra surrounding the broken bonds are displaced and this may be the incipient amorphous domain. This domain grows because the tetrahedra at the border are in a relatively unstable position and may rotate under the influence of fast particles passing in their vicinity. At larger distances, the lattice is only agitated by the passage of the particle but generally returns to its original structure. Nevertheless some permanent damage may be left; that is, the disorder observed in the crystal for which the micro-Dauphine twins seem to play an important role. When the irradiation is continued the disordered domains progressively extend to the whole sample. This process is essentially heterogeneous. The disorder starts from the particular points where the primary damage has been by chance very efficient.

The characteristics of the radiation damage in quartz are related to the flexibility of the arrangement of the SiO_4 units. The individual SiO_4 units, on the other hand, have a very stable structure. The situation is quite different in an ionic crystal like LiF, for instance. These ions can only form a crystal which can afford only a limited number of point defects, vacancies or interstitials. If the point defects are too numerous, they are expelled from the lattice to form clusters (clusters of vacancies, i.e. cavities, or clusters of lithium atoms (5)), and the crystal remains nearly undistorted. There are no non-crystalline domains of the composition LiF.

REFERENCES

1. *M.C. Wittels, Phil. Mag. 2, 1445 (1957).*

2. *W. Primak, Phys. Rev. 110, 1240 (1958).*

3. *I. Simon, J. Am. Ceram. Soc. 40, 150 (1956).*

4. *G. Mayer and M. Lecomte, J. Phys. Rad. 21, 846 (1960).*

5. *M. Lambert, These Paris (1958).*

6. *A. Guinier, Theorie et Technique de la Radiocristallographie, Dunod Paris (1964), p. 623.*

7. *H. Arnold, Z. fur Krist. 121, 145 (1965).*

8. *R.A. Young, Personal communication.*

9. *E.V. Kolonstova, L.A. Ignateva, Kristallografiya 4, 821 (1959).*

10. *Zachariasen, Acta Cryst. (Inter) 18, 710-716 (1965).*

11. *P.G. Klemens, Phil. Mag. 1, 938 (1956).*

12. *A. Liberti and G.D. Francesco, Chem. Ind. 1983 (Dec. 1963).*

DISCUSSIONS

CHARLES GREENE (Alfred University, U.S.A.): *Do the α_1 and α_2 regions envisioned in irradiated quartz correspond to right and left quartz?*

A. GUINIER: *No it is another kind of twinning.*

A. BISHAY (American University in Cairo, U.A.R.): *In our irradiation studies of Libyan Desert silica glass we noted that after about 10^{17} nvt the X-ray pattern, infrared and density changed in the direction of a regular structure. Increased dose* *resulted in a change in the opposite direction. Do you have an explanation for this experimental observation?*

A. GUINIER: *No, I think that such a phenomenon is not observable with ordinary amorphous silica.*

J.D. MACKENZIE (Rensselaer Polytechnic Institute, U.S.A.): *Assuming a total random distribution of SiO_4 tetrahedra, then one can have a density value for the amorphous phase which is much higher or lower than about 2.22 gm/cc. Why is it we always get this saturation value? Does it mean that the random network is not "really" random?*

A. GUINIER: *It seems very likely that the lattice is not random; this is shown by the model which allows only a partial freedom to the bonding between SiO_4 tetrahedra.*

H. PEISL (T.H. Darmstadt, Germany): *Is any thermal activation necessary for this transformation, or would you get the same transformation at very low temperatures?*

A. GUINIER: *Only samples irradiated at the reactor temperature have been studied. Low temperature irradiation would very likely bring interesting information.*

M. RAFIK SOLIMAN (National Research Center, U.A.R.): *You mentioned that in the case of ionic crystals such as LiF, the defects produced by irradiation move outside the lattice leaving only small islands of lattice imperfections. In other words it is hard to damage the crystallinity in the case of ionic crystals. Is this a general statement and is it because of strong bonding?*

A. GUINIER: *Yes, it is a general statement. It is not possible to damage ionic crystals as also is the case of metallic bonding, but it is possible to damage a covalent bonding.*

G.J. DIENES (Brookhaven National Laboratory, U.S.A.): *Since in your picture the extensive agitation around the path of the irradiating particle promotes the transformation, would you expect that even heavy electron irradiation would not produce these effects?*

A. GUINIER: *We did not study this kind of irradiation.*

E. LELL (Bausch and Lomb. Inc., U.S.A.): *Would you care to comment on Weissman's picture of radiation damage in quartz and his interpretation of islands with hexagonal symmetry as displaced Si-atom arranged in hexagonal fashion?*

A. GUINIER: *The model of Weissman was supported by the observation of small angle X-ray scattering. But, in spite of very careful trials we never observed significant scattering at small angles. Therefore, we think that the fluctuations of electronic density are smaller than required by the Weissman model.*

P. LUCASSON (University of Paris, FRANCE): *There should be a possibility of checking the transformation mechanism you are suggesting; by using high energy charged particles. Since the specific energy losses are lower the higher the energy of the particle, it should even be possible to determine a transformation threshold energy, related to the energy losses per unit volume.*

A. CHARLESBY (Royal Military College of Science, Shrivenham, ENGLAND): *We can derive some valuable data on interaction of radiation with solids, since these experiments give almost an instantaneous frozen picture of damage. Most atoms will escape the primary collision, so from these data we can deduce the number of atoms involved indirectly in each collision. However, we would expect an exponential decrease in the residual crystallinity, and not a linear drop. Furthermore, the size of each remaining crystalline domain should decrease with further radiation. If there is no X-ray diffraction evidence for such very small crystallites, it must be concluded that below a certain size they are no longer stable.*

S.W. BARBER (Owens - Illinois Incorporated, U.S.A.): *In my paper in this conference, I expect to display a distribution function which I think is related to the disorder in vitreous silica. It was broadened by a very mild exposure to fast neutrons ($\sim 10^{15}$ neutrons). This is contrary to remarks from the floor suggesting that order may be improved by such irradiation.*

CHANNELING AND RELATED EFFECTS IN THE PASSAGE OF CHARGED PARTICLES THROUGH CRYSTAL LATTICES

C. Erginsoy

Brookhaven National Laboratory

Upton, New York, U.S.A.

Anisotropic effects, often referred to as "channeling", have been observed recently in electronic, atomic and nuclear interactions when a beam of charged particles is transmitted through a single crystal. Related "blocking" effects have also been demonstrated when charged particles are emitted from a single crystal source or emerge from a single crystal after undergoing large-angle Rutherford scattering. This paper discusses the theoretical model of axial and planar effects and compares the predictions of the model with experimental results. For the purpose of this comparison the main emphasis is on experiments concerning the energy loss and large-angle scattering yields of energetic protons (3-11 MeV) in silicon single crystals. The problem of the minimum energy loss is discussed in detail. Possible uses of the anisotropic effects in solid state research are outlined.

INTRODUCTION

Interactions between energetic ions and solids give rise to a variety of physical phenomena such as secondary electron emission, sputtering, radiation damage, excitation of inner electron shells or a nuclear reaction. For all such phenomena the probability for the basic event to occur is high only when the ion comes sufficiently close to an atom or nucleus. In an amorphous solid the impact parameters of successive collisions suffered by the ion are *randomly* distributed and the total yield of a reaction over a given pathlength is independent of the direction along which the ion moves. However, in a crystalline solid atoms are regularly arranged in rows and in planes. Under certain conditions the impact parameters of successive collisions may become *correlated* and their distribution is not random. Hence, the yield of the same reaction for a given pathlength of the ion may strongly depend on direction. In this paper we discuss some aspects of the theory of such directional effects.

"CHANNELING" AND "BLOCKING" EXPERIMENTS

Directional effects were originally predicted for heavy ions in the keV range in machine calculations (1,2) and were experimentally observed in ion penetration studies (3,4,5). The observation of directional effects for fast charged particles is more recent than those associated with heavy slow ions.

There are basically two types of experiments in which directional effects can be observed. In one type a narrowly collimated beam of particles enters a single crystal parallel to some low-index crystal axis or plane. This is often called a "channeling" experiment because it maximizes the fraction of particles going into the "channels" between the atom rows and planes. The distribution of successive impact parameters is heavily weighted towards the largest possible values. Correlation effects, when they exist, tend to keep this heavily weighted distribution from changing into a random one.

In the other type of experiment the distribution of impact parameters is weighted towards small values. This is achieved either by looking at charged particles emitted directly from lattice sites or after particles are scattered to large angles by lattice atoms. If a collimated beam of particles is used the angle of incidence with respect to an atom row or atomic plane can be made sufficiently large so as to minimize "channeling" and put the maximum fraction of particles into the dense rows or planes, that is, to maximize small impact parameter encounters at the start. Again the effects of correlation, when they exist, tend to keep this non-random distribution of impact parameters from changing into a random one. We refer to this as a "blocking" experiment because the trajectories are effectively blocked by the high atomic density in the direction of rows or planes and particles are deflected away from such directions.

Admittedly this distinction between the two opposing types is arbitrary from the theoretical viewpoint but experimental studies of directional effects fit conveniently into one or the other type. Table I shows the variety of particles, energies and crystals for which one or the other type of experiment has been reported in the literature.

The first theoretical analysis of "channeling" experiments with charged particles in terms of a model is due to Nelson and

TABLE I

Experimental Studies on Directional Effects with Charged Particles

A. Experiments with Collimated Beams ("Channeling" Experiments)

Particle	Energy	Crystal	Reaction	Reference
Protons	50-75 keV	Au	Transmission and reflection	5
Protons	2.1 MeV	Si	Energy loss	6
α particles from external source	∿5.5 MeV	Si	" "	7
Protons	3 MeV	Si	" "	8
Protons	2.8 MeV	Cu	$Cu^{65}(p,n)Zn^{65}$	9
Protons	∿400 keV	Al,Si	$Al^{27}(p,\gamma)Si^{28}$ and $Si^{29}(p,\gamma)P^{30}$	10
Protons	75-110 keV	Al,Cu	K-shell and L-shell x-rays	11
Protons Deuterons	4.25-7.75 MeV 7.63 MeV	Ge	Energy Loss	12
Br^{79} and I^{127}	20-80 MeV	Au	" "	13,14,45
Protons	400 keV	Ta	Rutherford Scattering	15,16
α particles	40 MeV	Si	Energy Loss	17
Electrons	10 KeV	Cu	K-shell X-rays	18
	5 MeV	Si	Energy Loss	19,20
Protons	2.5 MeV	Cu	Energy Loss, Temperature, rad. damage effects	21
Protons	∿1.38 MeV	Al	$Al^{27}(p,\gamma)Si^{28}$	22

B. Experiments with Internally Generated or Scattered Particles ("Blocking" Experiments)

α particles from decay of Rn^{222}	5.5-7.68 MeV	W	Emission from lattice sites	23,24
Electrons from Internal Conversion	44.8-75.8 keV	Ta	" " " "	24,25
Protons	4-7 MeV	Si	Rutherford-scattering from lattice atoms	26
Protons	3 MeV	W	" " " "	27
Protons	500 keV	Mo	" " " "	28
Protons	3-6 MeV	W	" " " "	29

Thompson (5). A detailed theory of directional effects has been given by Lindhard (30) on the basis of axial (or "string") effects. Planar effects, first investigated in detail in energy loss studies (8,17,19,20), have been treated briefly by the author (31).

AVERAGE POTENTIALS FOR ATOM ROWS AND PLANES

The trajectory of a particle in the classical sense is always governed by deflections due to atomic fields even when the energy loss is primarily due to inelastic encounters. The atomic fields are the screened Coulomb fields of the nuclei.

In a "channeling" type of experiment in which the angle ψ to an atomic row is very small the trajectory of the particle can be projected on a plane perpendicular to the atomic row and the motion of the particle becomes equivalent to the motion of a particle of momentum $Mv\psi$ or energy $E\psi^2$ in an average potential due to many atoms. This average potential $\overline{V}(\rho)$ at a distance ρ from the row can be calculated if the potential V(r) between the particle and an individual atom at a separation r is known. For an atomic row with spacing d $\overline{V}(\rho)$ is given by

$$\overline{V}(\rho) = \frac{1}{d} \int_{-d/2}^{+d/2} dx\; V(\sqrt{\rho^2+x^2}) = \frac{2}{d} \int_0^{\infty} dx\; V(\sqrt{\rho^2+x^2}) \qquad (1)$$

As the particle trajectory approaches the atomic row and moves away, it comes to a distance of closest approach ρ_{min} given by the classical relation

$$1 - \frac{p^2}{\rho_{min}} - \frac{\overline{V}(\rho_{min})}{E\Psi^2} = 0 \qquad (2)$$

where p is the impact parameter of the "collision" that takes place in the transverse plane upon which the trajectory has been projected.

As an extreme case (giving the lowest value of ρ_{min} for a given ψ) we may put p=0. This occurs when the trajectory is in the same plane as the row. Eq. (2) then becomes

$$\overline{V}(\rho_{min}) = E\Psi^2 \qquad (3)$$

or

$$\Psi = \left[\frac{\overline{V}(\rho_{min})}{E}\right]^{\frac{1}{2}} \qquad (4)$$

This relation was first used by Nelson and Thompson (5) and also by Lindhard (30) in defining the limiting incidence angle if the minimum distance of closest approach ρ_{min} for stable channeling is known. Lindhard (29) has suggested that ρ_{min} may be of the order of a_{TF}, the Thomas-Fermi screening radius of the atom.

For an atomic plane there is only one angle Ψ between the trajectory and the plane and Eq. (4) is to be used directly, with $\overline{V}(\rho)$ as the planar average potential. For $\overline{V}(\rho)$ in a planar channel we have

$$\overline{V}(\rho) = \frac{2\pi}{A}\int_0^\infty x\, V(\sqrt{x^2+\rho^2})\, dx \qquad (5)$$

where 1/A is the atomic density in the plane.

As far as V(r) is concerned several types of screened Coulomb potentials have been proposed (32, 33, 34). These are shown in Figure 1 for the case of a proton and a silicon atom. We can use the analytical approximation to the Thomas-Fermi potential due to Molière (33). This potential is of the form

$$V(r) = \frac{Z_1Z_2e^2}{r}\left[0.1\exp(-6r/a_{TF}) + 0.55\exp(-1.2r/a_{TF}) + 0.35\exp(-0.3r/a_{TF})\right] \qquad (6)$$

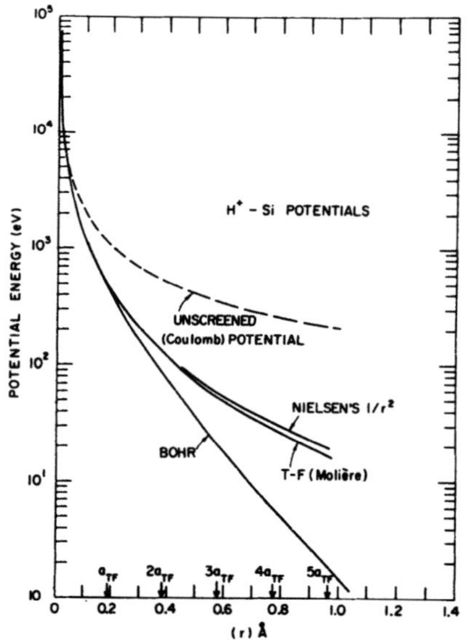

Fig. 1. Interaction potentials between a proton and a silicon atom at different separations. The distance a_{TF} = 0.1943 A° is the Thomas-Fermi radius of the atom.

where $a_{TF} = 0.8853\ a_0/Z_2^{1/3}$ is the Thomas-Fermi screening radius of the atom.

The row average potential (31) then becomes

$$\overline{V}(\rho) = (2Z_1Z_2e^2/d)\left[0.1\ K_0\ (6\rho/a_{TF}) + 0.55\ K_0\ (1.2\rho/a_{TF}) + 0.35\ K_0\ (0.3\rho/a_{TF})\right] \qquad (7)$$

where $K_0(x)$ is a zero-order modified Bessel function of the second kind (35). Figure 2 shows the average row potential and the corresponding incidence angles for protons of different energies and the [110] row of silicon.

The planar average potential for the Molière potential is of the form

$$\overline{V}(\rho) = (2\pi Z_1Z_2e^2a_{TF}/A)\left[(0.1/6)\exp(-6\rho/a_{TF}) + (0.55/1.2)\exp(-1.2\rho/a_{TF}) + (0.35/0.3)\exp(-0.3\rho/a_{TF})\right] \qquad (8)$$

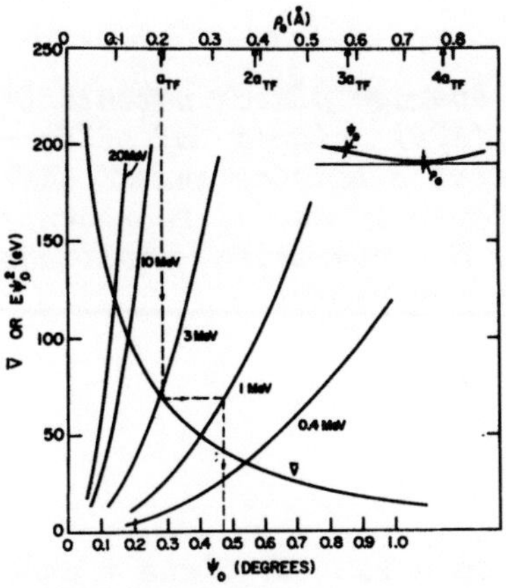

Fig. 2. Average row potential for a proton colliding with the [110] atomic row in silicon. To an incidence angle Ψ_o corresponds a distance of closest approach given by ρ_o. An example for obtaining Ψ_o for $\rho_o = a_{TF}$ in the case of 1 MeV protons is shown by the dotted lines.

When thermal motion is taken into account, each term in the potential which has the general form

$$\overline{V}_o(\rho) = (C/A)\ \exp\ (-\rho/\alpha) \tag{9}$$

becomes

$$\overline{V}(\rho) = \overline{V}_o(\rho)\ \exp\ (u_\perp{}^2/2\alpha^2)\left\{1 - \tfrac{1}{2}\left[1 - \Phi\left(\frac{\rho}{\sqrt{2u_\perp^2}} - \frac{u_\perp}{\sqrt{2\alpha^2}}\right)\right] + \frac{e^{2\rho/\alpha}}{2}\left[1 - \Phi\left(\frac{\rho}{\sqrt{2u_\perp^2}} + \frac{u_\perp}{\sqrt{2\alpha^2}}\right)\right]\right\} \tag{10}$$

where Φ (x) = erf (x) is the error function and $u_\perp^2$ is the transverse component of the square vibration amplitude.

As ρ becomes large we find that $\overline{V}$ (ρ) tends to the limit (31)

$$\overline{V}\ (\rho) = \overline{V}_o\ (\rho)\ \exp\ (u_\perp^2/2\alpha^2), \tag{11}$$

indicating that $\overline{V}$ (ρ) increases slightly with increasing temperature. However, for a small ρ the average potential (and therefore the critical incidence angle Ψ) decreases with increasing temperature. Such a decrease has been observed experimentally (29).

Figure 3 shows the potential for a proton in the (110) planar channel of silicon. It can be seen that the potential decreases substantially below its static lattice value for small ρ.

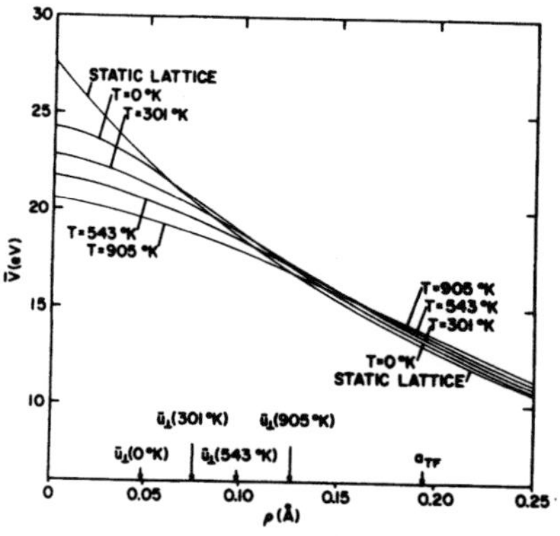

Fig. 3. Average planar potentials for the (110) planes of silicon for the static lattice and at different temperatures. $\overline{u_\perp}$ represents the R. M. S. vibration amplitude normal to the plane.

The distance of closest approach to a row or plane - and, therefore, the maximum incidence angle - is not the same for all physical phenomena. For Rutherford scattering, for instance, $\rho=u_\perp$ may be appropriate while for energy loss $\rho=a_{TF}$ may be more meaningful.

In Fig. 4 and Fig. 5 we compare the calculated Ψ's with those measured recently by L.C. Feldman for the scattering of protons to 40° as they pass through a 5μ-thick silicon crystal. The agreement is reasonable and shows the predicted $1/\sqrt{E}$ dependence of the critical angle.

MULTIPLE SCATTERING OF PARTICLES INTO CHANNELS

It has been shown (20) that planar channeling effects exist independently of any axial channeling in a lattice. There is evidence to suggest that quite apart from this, planar channeling effects are of importance even when the narrowly collimated beam of particles is incident along a crystal axis.

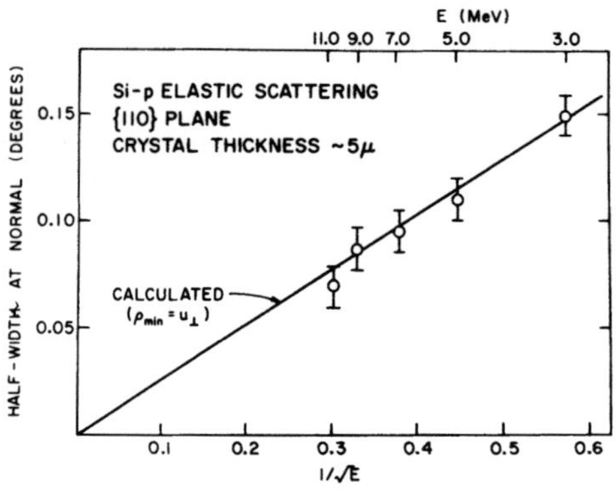

Fig. 4. Experimental and calculated angular widths for Rutherford scattering for protons in the (110) planar channels of silicon (Reference 39).

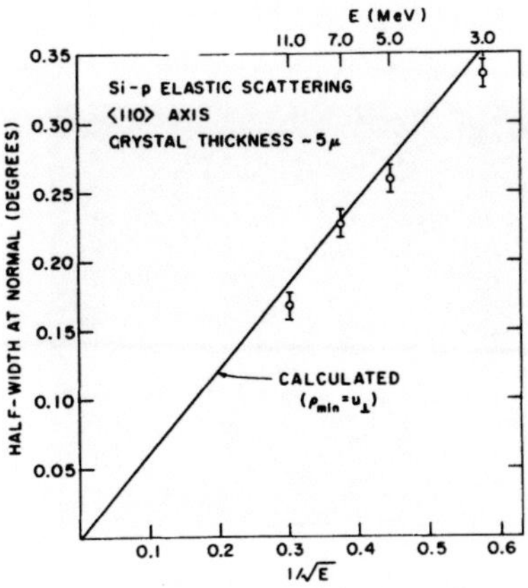

Fig. 5. Experimental and calculated angular widths for Rutherford scattering of protons in the [110] axial channels of silicon (Reference 39).

Figure 6 was obtained by Schiffer and Holland (36) by photographing the fluorescence produced by transmitted protons on a quartz plate. The size of the spot is (bottom left) an indication of the angular divergence of the beam. When the beam is incident parallel to the [111] axis (top left) one sees, in addition to the bright spot exactly along the incident direction, a star-pattern which indicates the presence of particles moving in the three (110) planar channels which cross each other along the [111] direction.

The pattern at the top right and bottom right has been obtained by keeping the beam direction fixed and tilting the crystal by 2° upwards and downwards respectively. Even for such a large angle of incidence the star pattern is clearly visible.

These results are a pictorial proof that particles are fed into planes by multiple scattering inside the crystal even when incidence is axial. We have also other evidence from the analysis of the energy spectra of transmitted particles (19).

The solid angle for particles that would be directly transmitted in the same axial channel is negligibly small, being of the order of $\pi\theta^2$ where $\theta = w/t$, w is the width of a channel (say $3A^o$) and t is crystal thickness. It follows that all axially channeled particles must "wander" from one axial channel to another and, in the process, acquire some transverse momentum that causes the beam to broaden. This broadening accounts for the fact that the bright spot at the center of the star-pattern in Figure 5 is larger than the spot due to the raw incident beam.

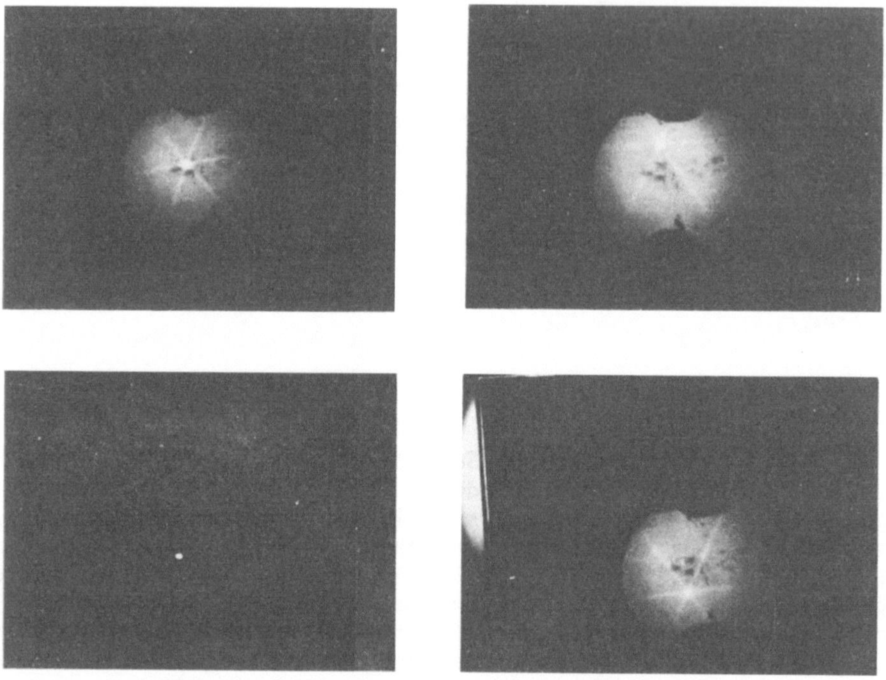

Fig. 6. Demonstration of the role of multiple scattering in feeding particles into channels. The star pattern (for 3.5 MeV protons passing through 2 mil silicon in the [111] direction) does not disappear with 2° tilt of the crystal. The bright spot in the bottom left square shows the small angular divergence of the incident beam (Reference 36).

Particles wandering from one axial channel to another without changing their azimuthal angle must necessarily sample electron densities that are characteristic of the different planar channels rather than the substantially lower electron density along a narrow axial channel. It was found that the limiting energy loss for incidence along an axis is equal to the energy loss for the incidence of the same beam along the most open of the contributing planar channels (37,38).

This result has led the author to attempt an estimate of the spectrum of transmitted particles when incidence is parallel to an axis from the measured spectra when the particle is incident parallel to the different contributing planes. Figure 7 shows this

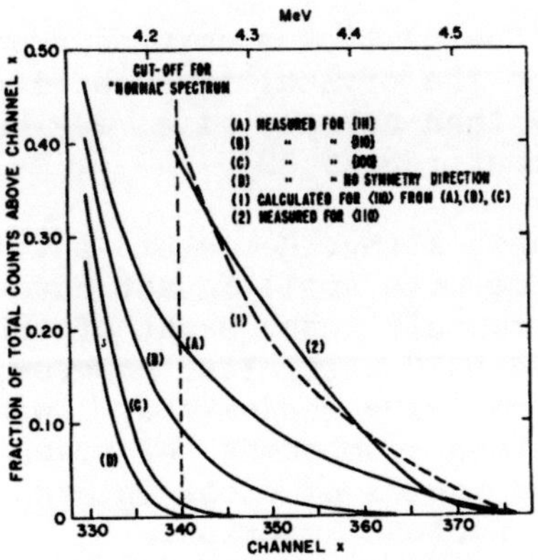

Fig. 7. Calculated and measured values of the fraction of particles losing less energy than normal (Reference 20).

comparison for the [110] axis of silicon and the contributing planes of the (111), (110) and (100) types. The shape of the spectra are shown by the fraction of total particles that are counted in the different electronic channels of the multichannel analyzer.

The basic idea is that the probability $p_{axis}(E)$ that a transmitted particle incident along the axis will have an energy larger than E in emergence can be expressed in terms of the corresponding p_{pl}'s (referring to the contributing planes).

For the [110] axis we take two (111) planes, one (110) plane and one (100) plane (others do not contribute for the thick crystal). Therefore,

$$1 - p_{[110]}(E) = \left[1-p_{(111)}(E)\right]^2 \times \left[1-p_{(110)}(E)\right] \times \left[1-p_{(100)}(E)\right] \qquad (12)$$

The agreement between the calculated spectrum (shown by the curve marked (1) in Figure 7 and the actual spectrum measured for incidence parallel to the [110] axis (curve marked (2)) is reasonable, considering that we have assumed equal probability of diffusion into the different planar channels and neglected those trajectories that wander from one planar channel to another.

The problem of the diffusion of the trajectories in the transverse plane is very interesting and bears more detailed study.

CORRELATED MOTION OF PARTICLES WITH SMALL IMPACT PARAMETERS

In a "blocking" experiment some of the particles emitted from lattice sites can be scattered throughout the crystal in such directions that they may lose less energy than normal, i.e. may be considered as channeled to a certain extent (26,29,39).

Also, a "channeling" experiment cannot altogether avoid putting some of the particles into the lattice with small impact parameters. Many of these particles will rapidly lose sight of the crystal and fall into the category of "normal" particles. However, correlation effects keep some of them from being randomized. Such particles are responsible for the high-loss component of transmitted spectra (8, 19) and for increased yields of close interactions such as Rutherford scattering or a nuclear reaction.

It is necessary to inquire into the nature of the correlation that maintains the distribution of impact parameters from changing into a random one over a finite thickness of the crystal. Lindhard (30) has stated compensation rules according to which, if there is no energy change, the yield of a given reaction averaged over all incidence angles must be equal to the yield for the amorphous solid. In other words, if there are some directions which give a low energy loss or a decreased reaction yield, there must necessarily be other directions that give a high energy loss or high yield. The statement of such a compensation rule does not answer the question regarding the nature of the mechanism that must be responsible for the correlation.

Computer calculations into this problem are currently being carried out by L. C. Feldman and the author. For particles being emitted from lattice sites lying in a plane one has to calculate, for all directions, the most probable angle of emergence from the plane and the distribution of emergence angles. The emergence direction is determined by the several small-angle collisions that the particle makes. The case is similar to that of plural scattering, where the number of scattering events is not so high that the theory of multiple scattering applies, but also this number is not one as it might be supposed in the absence of thermal vibrations.

CHANNELED PARTICLES: THE PROBLEM OF LEAST ENERGY LOSS AND STRAGGLING

One of the important questions to which a theory of channeling is required to find an answer is that of the least energy loss of the channeled particles. It has been suggested by Lindhard (30, 40) that, on the average, one half of the stopping power of a fast particle in a system of electrons is due to close collisions (which are minimized by channeling) and one half to resonance-type distant interactions (which cannot be avoided by channeling). Thus the stopping power of a channeled particle could be reduced by one half at the most.

This prediction by the equipartition rule was believed (8) to be confirmed by early experiments on the energy loss of protons in silicon but later experiments (12, 41) in germanium showed significant disagreements.

We have looked at the energy loss of well-channeled particles on a different basis (42). As far as core electrons in silicon are concerned, these are in the atomic K- and L- shells. One can estimate the maximum distance from nuclei at which excitation of such shells are still possible by establishing an adiabaticity criterion

$$\rho_{max} = \frac{\hbar v}{\Delta E} \tag{13}$$

where v is the velocity of the particle and ΔE is the binding energy of the electron in its shell.

For a K-shell electron in silicon ΔE = 1845 eV and ρ_{max} for a 3 MeV proton becomes as small as 0.05A°. For well-channeled particles in the (111) planar channel experimental ρ's are of the order of 1A°. Therefore, K-shell excitation can be neglected.

For L-shell electrons ΔE is between 100 eV and 150 eV. Taking the value of 100 eV one finds that ρ_{max} is now about 1.5A°. Therefore, L-shell excitation cannot be altogether neglected. However, energy loss scales in the ratio

$$\ln \frac{\rho_{max}}{\rho_{exp}} \Big/ \ln \frac{2mv^2}{\Delta E}$$

where $\rho_{exp} \simeq 1.0$ A° for the (111) planes. It follows that the energy loss due to L-shell electrons can be reduced by more than 90% by channeling. We shall neglect this loss in what follows.

Energy loss to valence electrons alone can be written in the form

$$\left(-\frac{dE}{dx}\right)_{val} = \frac{4\pi Z^2 e^4 L}{mv^2} \tag{14}$$

where

$$L = Z_{val} \ln \frac{v}{v_F} + Z_{loc} \left[\ln \frac{2mvv_F}{\hbar\omega_p} \right] \tag{15}$$

The first term on the right hand side of equation 15 corresponds to the contribution of collective (plasma) excitations which use the total density of valence electrons (Z_{val} = 4 in silicon). The second term gives the contribution of single-particle excitations. This is due to the local density of electrons in the channels. This division of L into two contributions is possible in terms of the random-phase approximation of Bohm and Pines (43). In the above equation v is the velocity of the particle, v_F the Fermi velocity of the free electron gas describing the valence electron, $\hbar\omega_p$ the plasmon energy (16.6 eV in silicon), Z_{loc} is the local density of electrons in the channels divided by the atomic density N.

Knowing the crystal thickness ΔX from the measurement of normal energy loss, we can write

$$\Delta x = \int_{E_C}^{E_I} \frac{dE}{(-dE/dx)_{val}} \tag{16}$$

where E_I is incident energy and E_C is the energy of channeled particles. We obtain Z_{loc} in the different planar channels by integrating Eq. 16 with the use of Eq. 14 and Eq. 15. For the (111) planar channels of silicon and for a proton beam of 3.0 MeV. $Z_{loc\,(111)}$ = 3.8 by this method.

Independently from the measured values of energy peaks the straggling of the channeling peak, too, can give information regarding Z_{loc}. Since collective excitations do not cause any appreciable straggling in energy loss we can write

$$\frac{\Omega_N^2 - \Omega_I^2}{\Omega_C^2 - \Omega_I^2} = \frac{Z_{total}}{Z_{local}} \tag{17}$$

where Ω_N^2, and Ω_C^2 are the square of the widths of energy straggling due to normal energy loss and channeled loss and Ω_I^2 is the square of the energy width of the incident beam.

Z_{total} = 14 in silicon and we have the measured values

Ω_N = 29.71 keV

Ω_I = 14.97 keV

Ω_C = 20.18 keV

for E_I = 3.00 MeV

This gives $Z_{loc\ (111)} = 3.9$, which agrees well with the $Z_{loc\ (111)} = 3.8$ estimated by the essentially independent measurement of the energy peaks.

USE OF DIRECTIONAL EFFECTS IN SOLID STATE PHYSICS

The study of directional effects has now progressed beyond the stage of mere demonstration and it is natural to seek methods of utilizing the effects in research into basic processes. Lindhard (30) has discussed some of these methods. Various uses have already been found for directional (channeling and blocking) effects.

Location of impurities and defects: Substitutional impurities or defects can be distinguished from interstitials by whether interactions with a beam of particles is lowered or increased by channeling. Already such determinations have been made (23, 24, 44).

Mapping of valence electron densities in channels: As discussed in this paper the energy loss of well-channeled particles can, at low energies, give good information on the density of valence electrons in semiconductors, insulators or metals.

Range studies: Channeling minimizes the fraction of particles suffering elastic collisions with atoms and thus allows penetration and range studies to extend to such low energies where normal stopping power due to elastic collisions would mask any inelastic stopping power (4).

Radiation damage: It has been shown that directional effects are sensitive to radiation damage (5, 21). Thus, some parameter (such as the fraction of particles that are well-channeled) may be used to detect radiation damage. Also, by the use of channeled beams it may become practical to use protons in radiation damage studies.

Estimate of interatomic potentials: As discussed in this paper the angular widths of directional effects depend sensitively on the average potentials of rows or planes. Recently additional information on these average potentials has been obtained by the structure in the energy loss spectra (45). Although an estimate of actual interatomic potentials at different separations from the row or planar average potentials is not entirely unambiguous, this method may provide a new means of investigating the important problem of interaction potentials.

Crystallography: The use of the blocking patterns (26, 28) may help in crystallographic studies in identifying the various planar and axial symmetries in complicated structures.

Particle detectors: Already in the applied field of particle detectors (17, 46) an improvement of detectors has been achieved by manufacturing them in orientations that minimize channeling effects. On the other hand silicon detectors for fission fragment detection can be improved considerably by maximizing channeling (47).

Study of short nuclear lifetimes: The possibility of estimating very short nuclear lifetimes ($<10^{-17}$ sec) by contrasting channeling and blocking effects have been suggested (26, 27) but no results have so far been reported.

ACKNOWLEDGMENT

I am grateful to my colleagues B.R. Appleton and W.M. Gibson of Bell Telephone Laboratories, L.C. Feldman of Rutgers University and H.E. Wegner of Brookhaven National Laboratory for continuous discussions on all aspects of channeling and blocking effects. Some of their experimental work discussed here have not so far been published. Stimulating discussions with J.A. Davies, G. Leibfreid, J. Lindhard, M.W. Thompson and G.H. Vineyard are gratefully acknowledged.

This work was supported by the U.S. Atomic Energy Commission.

REFERENCES

1. *M.T. Robinson and O.S. Oen, Appl. Phys. Letters 2, 30 (1963).*

2. *M.T. Robinson and O.S. Oen, Phys. Rev. 132, 2385 (1963).*

3. *G.R. Piercy, F. Brown, J.A. Davies and M. McCargo, Phys. Rev. Letters 10, 399 (1963).*

4. *E. V. Kornelsen, F. Brown, J. A. Davies, B. Domeij and G. R. Piercy, Phys. Rev. 136, A849 (1964) and references therein.*

5. *R. S. Nelson and M. W. Thompson, Phil. Mag. 8, 1677 (1963); ibid. 9, 1069 (1964).*

6. *G. Dearnaley, IEEE Trans. Nucl. Sci. NS-11 No 3, p. 243 (1964).*

7. *T.C. Madden and W.M. Gibson, IEEE Trans Nucl. Sci. NS-11, No 3 p. 254 (1964).*

8. C. Erginsoy, H. E. Wegner and W. M. Gibson, Phys. Rev. Letters 13, 530 (1964).

9. M.W. Thompson, Phys. Rev. Letters 13, 756 (1964).

10. E. Bøgh, J. A. Davies and K. D. Nielsen, Phys. Letters 12, 129 (1965).

11. W. Brandt, J.M. Khan, D.L. Potter, R.D. Worley and H.P. Smith, Jr., Phys. Rev. Letters 14, 42 (1965).

12. A.R. Sattler and G. Dearnaley, Phys. Rev. Letters 15, 59 (1965)

13. S. Datz, T.S. Noggle and C.D. Moak, Phys. Rev. Letters 15, 254 (1965).

14. S. Datz, T.S. Noggle and C. D. Moak, Proceedings of the Aarhus Conference on Electromagnetic Isotope Separators and their Application, Nucl. Inst. Methods 38, 221 (1965).

15. E. Bøgh and E. Uggerhøj, Phys. Letters 17, 116 (1965).

16. E. Bøgh and E. Uggerhøj, Proceedings of the Aarhus Conference on Electromagnetic Isotope Separators and Their Applications, Nucl. Inst. Methods 38, 216 (1965).

17. H. E. Wegner, C. Erginsoy and W. M. Gibson, IEEE Trans. Nucl. Sci. NS-12 No. 1 p. 240 (1965).

18. F. Bell and R. Sizmann, Phys. Letters 19, 171 (1965).

19. W.M. Gibson, C. Erginsoy, H.E. Wegner and B.R. Appleton, Phys. Rev. Letters 15, 357 (1965).

20. B.R. Appleton, C. Erginsoy, H.E. Wegner and W.M. Gibson, Phys. Letters 19, 185 (1965).

21. B. W. Farmery, R. S. Nelson, R. Sizmann and M. W. Thompson, Proceedings of the Aarhus Conference on Electromagnetic Isotope Separations and Their Applications, Nuc. Inst. Methods 38 231 (1965).

22. J.U. Anderson, J. A. Davies, and K. O. Nielsen, Proceedings of the Aarhus Conference on Electromagnetic Isotope Separations and Their Applications, Nucl. Inst. Methods, 38, 207 (1965).

23. B. Domeij and K. Bjorkvist, Phys. Letters 14, 127 (1965).

24. *B. Domeij, Proceedings of the Aarhus Conference on Electromagnetic Isotope Separators and Their Applications, Nucl. Inst. Methods 38, 207 (1965).*

25. *G. Astner, I. Bergstrom, B. Domeij, L. Erikson and A. Persson, Phys. Letters 14, 308 (1965).*

26. *D. S. Gemmell and R. E. Holland, Phys. Rev. Letters 14, 945 (1965).*

27. *A.F. Tulinov, Soviet Physics-Doklady, 10, 463 (1965).*

28. *A.F. Tulinov, B.G. Akhmetova, A.A. Puzanov and A.A. Bednyakov, JETP Letters 2, 30 (1965).*

29. *A.F. Tulinov, V. S. Kulikauskas and M. M. Malov, Phys. Letters 18, 304 (1965).*

30. *J. Lindhard, Phys. Letters 12, 126 (1964) Mat. Fys. Medd. Daw. Vid. Selsk. 34, No. 14 (1965).*

31. *C. Erginsoy, Phys. Rev. Letters 15, 360 (1965).*

32. *N. Bohr, Mat. Fys. Medd. Dan. Vid. Sels. 18, No. 8 (1948).*

33. *G. Molière, Zeits. f. Naturforschung 2a, 133 (1947).*

34. *K. O. Nielsen, Electromagnetically Enriched Isotopes and Mass Spectrometry, Academic Press Inc., New York, 1965, p. 68.*

35. *G. N. Watson, Theory of Bessel Functions, Cambridge University Press, Cambridge 1958, 2nd ed., p. 698.*

36. *J. P. Schiffer and R. E. Holland, Bull. Am. Phys. Soc. 10, 54, (1965) and private communications.*

37. *B. R. Appleton, PH.D. Thesis presented to Rutgers, the State University, May 1966.*

38. *B.R. Appleton, C. Erginsoy, and W.M. Gibson, (to be published).*

39. *L.C. Feldman, (private communication).*

40. *J. Lindhard and A. Winther, Mat. Fsy. Medd. Dan. Selsk. 34, No. 4 (1964).*

41. *B. R. Appleton, M. Altman, L. C. Feldman, E. J. Ludwig, W. M. Gibson and C. Erginsoy, Bull. Am. Phys. Soc. 11, 176 (1966).*

42. *C. Erginsoy, B. R. Appleton and W. M. Gibson, Bull. Am. Phys. Soc. 11, 176 (1966).*

43. *See, for example, D. Pines, Elementary Excitation in Solids, W.A. Benjamin Inc., 1963 p. 148.*

44. *J.A. Davies, Bull. Am. Phys. Soc. 11, 230 (1966).*

45. *H. O. Lutz, S. Datz, C. D. Moak and T. S. Noggle, Phys. Rev. Letters 17, 285 (1966).*

46. *W.M. Gibson, IEEE Trans. Nucl. Sci. NS-14 (1966).*

47. *C. D. Moak, J. W. T. Dabbs and W.W. Walker, Bull. Am. Phys., Soc. 11, 101 (1966).*

THE STRING EFFECT AND ITS USE IN SOLID STATE INVESTIGATIONS

E. Bogh

Institute of Physics, University of Aarhus

Aarhus C, DENMARK

This paper deals with the influence of crystal lattice periodicity on the motion of energetic, heavy charged particles penetrating a crystal.

A short summary of Lindhard's theory and of the experimental information obtained in measurements of Rutherford scattering yields in single crystals is given. In the scattering experiments, a very marked decrease (between a factor of 10 and a factor of 100) in scattering yield is found whenever the particle beam is aligned within a certain critical angle of a major crystal axis. The critical angles obtained from the experiments are all in good agreement with theoretical estimates. A strong reduction in multiple scattering of the aligned beam was also observed. The reversibility of particle trajectories in single crystals predicted in the theory is experimentally demonstrated for 500 keV protons moving through a W crystal.

The experiments indicated that valuable information about the solid state parameters of the scattering crystal can be obtained. This is illustrated in (i) detection of a disordered surface layer on W, (ii) a preliminary study of the distribution of radiation damage introduced into W and Si by heavy ion bombardment, and of the dependence of this damage on the direction along which the particles are injected into the crystal, and (iii) determination of the sites of impurity atoms in a crystal lattice. In the present case, the position of Tm atoms in a Si crystal was determined.

INTRODUCTION

Several experimental investigations of various phenomena associated with the penetration of fast charged particles (in the keV-MeV energy region) through a solid target have demonstrated the importance of considering the crystal structure of the target. In particular, an orientation dependence has been observed in measurements of (i) sputtering (1) and (ii) ranges of heavy ions (2). An even more pronounced dependence has been observed in measurements of (iii) the yield of processes like nuclear reactions, Rutherford scattering, and characteristic X-ray production (3-5), and (iv) emission of charged particles from radioactive nuclei embedded in single crystals (6).

In the low-energy region (i.e. the experiments (i) and (ii) listed above), a theoretical treatment has been given in terms of "channeling" as introduced by Oen and Robinson (7) and by Nelson and Thompson (8). In the high-energy region ((iii) and (iv) above), the theoretical treatment is based on the "string approximation" as introduced by Lindhard (9).

It should be mentioned that the term "channeling" is often used to describe directional effects in general - also in cases where the description is based on the string approximation.

It turns out that directional effects provide a tool for solid state investigations, mainly because the string approximation makes it possible to distinguish between lattice points and interstitial points in a single crystal.

THE STRING EFFECT

Theoretical Estimates

According to theory (9), a parallel beam of energetic heavy charged particles will, on passing through a crystal surface, split up into two parts: the random beam and the aligned beam. The part to which a particle belongs is determined by the asymptotic angle ψ between the particle's path and a close-packed crystal axis; a string. If this angle is smaller than a certain critical angle ψ_c, the particle belongs to the aligned beam; if it is larger than ψ_c, it belongs to the random beam. ψ_c is estimated to be

$$\psi_c = \psi_1 = \left(\frac{2Z_1Z_2e^2}{E \cdot d}\right)^{\frac{1}{2}} \qquad \text{if } \psi_1 < \frac{a}{d}$$

$$\psi_c = \psi_2 = (\psi_1 \, a/d)^{\frac{1}{2}} \qquad \text{if } \psi_1 > \frac{a}{d} \,. \tag{1}$$

Z_1 and Z_2 are the atomic numbers of the particle and the target atoms, respectively, e is the electronic charge, E the particle energy, d the distance between neighboring atoms on the crystal axis in question, and, a is the Thomas-Fermi screening distance.

If the particle beam is well-aligned with the axis before it enters the crystal, the random beam is estimated to contain $100\pi a^2 Nd$% of the total particle flux, where N is the number of target atoms per cm^3.

For the particles in the random beam, the penetration phenomena are expected to be as in a target of randomly distributed atoms, whereas the particles in the aligned beam are predicted never to come closer than $\sim$ a from the target atoms. Transitions between random and aligned beams are in a first approximation prohibited in the ideal case, i.e. the case of a perfect crystal without lattice vibrations.

In a more detailed theoretical treatment, both the passage of the particles through the crystal surface and the thermal vibrations of the atoms should be taken into account. At the comparatively low depths below the crystal surface where multiple scattering can be disregarded and slowing-down is small, the transverse energy, $E_\perp = E \cdot \psi^2 + U(r)$ is conserved in the planes perpendicular to the axis and half-way between atoms; U(r) is the average ion-atom potential at a distance r from the axis. In this case, the probability of hitting the center of the crystal atoms may be calculated, by simple numerical computations, as a function of the angle between the external particle beam and the crystal axis. Such computations have been performed by Andersen (10) who for U(r) inserted the standard potential proposed in (9).

As a rule-of-thumb, theoretical estimates imply that the width of the dip in the curve: probability of hitting a crystal atom versus angle between external particle beam and crystal axis, taken as the width at half the minimum probability, is closely equal to $2\psi_1$ when the upper equation in (1) applies. The minimum probability is estimated to be $\sim 100\pi(a^2 + \rho^2)N \cdot d$% of the probability of hitting an atom in the corresponding amorphous target. ρ^2 is the average square amplitude of atomic vibrations.

Summary of Experimental Investigation

Since only the particles in the random beam are able to take part in reactions requiring impact parameters smaller than a, such reactions may be used as "detectors" for investigating the theoretical predictions. During the past two years, investigations of this type have been performed at the Institute of Physics in Aarhus. The reactions used have been (p,γ) resonance reactions (3) and wide-angle Rutherford scattering(4). Similar studies have been reported from other laboratories (5,11,12).

The results obtained in (3) and (4) have shown a good general agreement with Lindhard's theory. A very marked decrease in reaction yield is found whenever a particle beam is aligned with both low and high indexed crystal directions. The angular widths of the dips in reaction yield as functions of Z_1, Z_2, d, and E are in accord with formula (1). In many cases, the observed minimum yields are very close to the predicted value and are found to vary very slowly as a function of the depth beneath the crystal surface at which the reaction occurs. This indicates that the multiple scattering of the aligned beam is much reduced compared to normal multiple scattering. For instance, it is found that about 93% of a 3 MeV proton beam originally aligned with a [100] axis in a W crystal is still aligned to within $\pm 1^o$ at a depth of $\sim 10\mu$. In an amorphous W target, the same angular spread would be obtained from normal multiple scattering after penetration of approx. 0.1μ.

An example of a "string dip" is given in Fig. 1. Curve 1a is the yield from a W crystal as a function of the angle between a proton beam and the [100] axis. Curve 1b is the yield as a function of the angle between the direction of the outgoing protons and the [100] axis for a fixed angle between crystal and incoming protons. In both cases, the energy of the incoming protons was 500 keV $\pm$ 0.5 keV. The scattering zone was, as will be described in the next section, defined to be 0-1500 A^o below the crystal surface. The angular resolution was: Curve 1a:0.1^o; curve 1b:0.3^o.

Apart from showing general features of a string dip, Figure 1 also illustrates two other theoretical predictions: (i) the rule of angular compensation which states that the dip in yield occurring around a string should be approximately compensated for by a rise over normal yield for angles somewhat larger than the critical angle. (ii) the rule of reversibility which states that in cases where slowing-down can be neglected, the probability of hitting a nucleus in a crystal by an external beam of particles as a function of the angle of incidence (curve 1a) equals the angular distribution outside the crystal of particles emitted from atomic nuclei in the crystal (curve 1b). See also reference (6) and (11).

APPLICATIONS IN SOLID STATE PHYSICS

Lattice imperfections influence the ratio between the number of particles in the random and in the aligned beam. This suggests the possibility of using reaction yield measurements in the study of crystal defects.

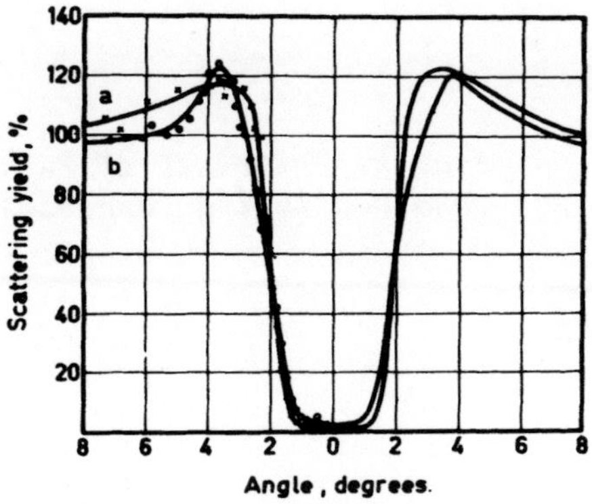

Fig. 1. Scattering of 500 keV protons in W. Scattering zone: 0-1500A° beneath surface. Curve a: yield versus angle between incident proton beam direction and [100] axis. Tilting plane not coinciding with any low-index plane. Curve b: Yield versus angle between emission direction and [100] axis. Incoming beam not aligned with any low-index plane or axis.

Principle of Scattering Experiment

Reactions used in such studies should be of a type allowing the reaction to be confined to a well-defined thin zone at a chosen depth beneath the surface of a thick target. Resonance reactions (e.g. (p,γ)-resonances) and Rutherford scattering are good examples. For a general application, the Rutherford scattering has proved to have several advantages over e.g. a nuclear reaction (4), although in special cases, a specific nuclear reaction might be a better choice. In the following, only Rutherford scattering is considered.

The principle of a scattering experiment is illustrated in Figure 2. A monoenergetic particle beam (particle mass M_1 and energy E_1) enters the target (atomic mass M_2) at an angle θ_1 with respect to the surface normal. At a perpendicular distance R beneath the surface, a particle is elastically scattered through an angle θ_L and emerges from the target with energy E_2 and at an angle θ_2 with respect to the surface normal. E_2 is given by

$$E_1 - E_2 = \int_{R/\cos\theta_1}^{0} \left(\frac{dE}{dx}\right) dx + (1-\alpha)\left\{E_1 - \int_{R/\cos\theta_1}^{0} \left(\frac{dE}{dx}\right) dx\right\} + \int_{0}^{R/\cos\theta_2} \left(\frac{dE}{dx}\right) dx, \qquad (2)$$

where

$$\alpha^{1/2} = \frac{M_1\cos\theta_L}{M_1 + M_2} + \left[\left(\frac{M_1\cos\theta_L}{M_1 + M_2}\right)^2 + \frac{M_2 - M_1}{M_2 + M_1}\right]^{1/2}$$

dE/dx, which is a function of E, is the rate of particle energy loss along the path x.

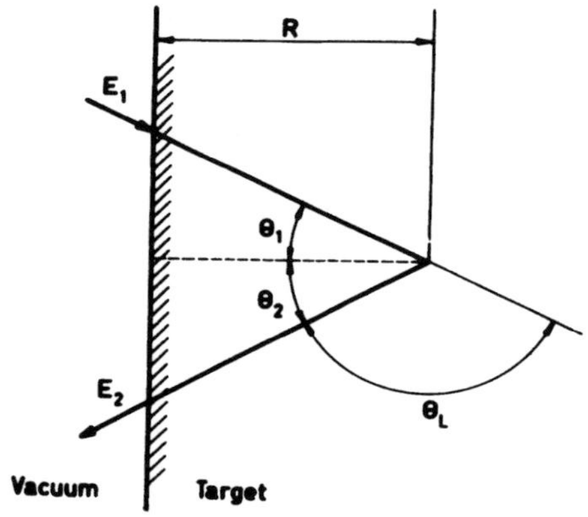

Fig. 2. Principle of Rutherford scattering experiment.

The first and the third term in (2) is the particle energy loss along the path lengths $R/\cos\theta_1$ and $R/\cos\theta_2$, respectively. The second term is the energy loss in the elastic scattering. Formula (2) relates the energy of a particle emerging from the target to the depth in which the particle has been scattered, i.e. by performing an energy analysis of particles scattered through a given angle θ_L, the scattering process may be confined to a thin zone at a chosen depth R. Also, in the case of a target containing impurities, having an atomic mass which differs from that of the host material, it is possible in the energy spectrum to distinguish between protons scattered on the atoms of the host material and protons scattered on the impurity atoms.

The thickness ΔR of the scattering zone is, neglecting straggling in the particle energy loss, determined by the energy resolution ΔE_2 of the spectrometer used in the energy analysis. In the first approximation, ΔR is given by

$$\Delta R = \frac{\Delta E_2}{(dE/dx)} \left[\frac{1}{\cos\theta_1} + \frac{1}{\cos\theta_2} \right]^{-1} \quad (3)$$

where the value valid for the motion of the particle through the thin zone should be inserted for dE/dx.

In single crystals, the orientation dependence of dE/dx (13) implies a slight uncertainty in R and ΔR as calculated from (2) and (3). However, if only one of the path lengths, $R/\cos\theta_1$ or $R/\cos\theta_2$, coincides with a crystal axis, the uncertainty amounts to maximum 25%, and it can be made small by choosing the path length along the axis small compared to the other.

Experimental Apparatus

The preliminary experiments that will be presented here were performed at the 600 keV heavy ion accelerator in Aarhus.

The experimental set-up is shown in Figure 3. The combination of the tilt ψ of the ball A and the rotation θ of the crystal holder B enables any direction of the crystal lying within a 20° cone around the normal of the crystal, to be aligned with the particle beam. The accuracy of the angle readings is $\pm 0.05^o$.

The pressure in the scattering chamber is $\sim 3 \times 10^{-6}$ mmHg, and an effective liquid-air cooling trap prevents carbon contamination of the crystal surface. The beam current is monitored by the solid state detector.

In all cases presented below, a 500 keV $\pm$ 0.5 keV proton beam was used. The proton current was $\sim 10^{-8}A^o$, the beam cross section 0.8 mm^2, and the angular spread of the beam less than $\pm 0.1^o$. The scattering angle was 90°, the energy resolution of the magnetic analyser $2^o/oo$. The over-all energy resolution, including stability of the accelerator, was $\sim 3^o/oo$.

RESULTS AND DISCUSSION

Figure 4 shows the energy spectrum of protons scattered in a W single crystal. The depth resolution ΔR as calculated from (3) is $\sim$ 50 A^o. Curve 4a is the spectrum measured with the protons entering an electropolished crystal along the (111) axis. Curve 4b shows the spectrum after $\sim$ 2 atomic layers of the crystal anodic oxidation. Curve 4c is measured after $\sim$ 10 atomic layers have been converted into oxide on the surface. Curve 4d is the spectrum measured with the protons entering the crystal along a direction far from any prominent plane or axis.

Figure 5 shows the energy spectrum of protons scattered in a W single crystal. $\Delta R \sim 50\ A^o$. Curve 5a is the same as curve 4a. Curve 5b and 5c show the spectrum after the crystal has been bombarded with $\sim 10^{15}$ 200 keV N ions per cm^2 and 10^{18} 200 keV N ions per cm^2, respectively. The N ions were injected along the [111] axis. Curves 5b' and 5c' correspond to 5b and 5c, respectively, but the N ions have been injected far from any prominent crystal axis or plane. Curve 5d is the same as curve 4d.

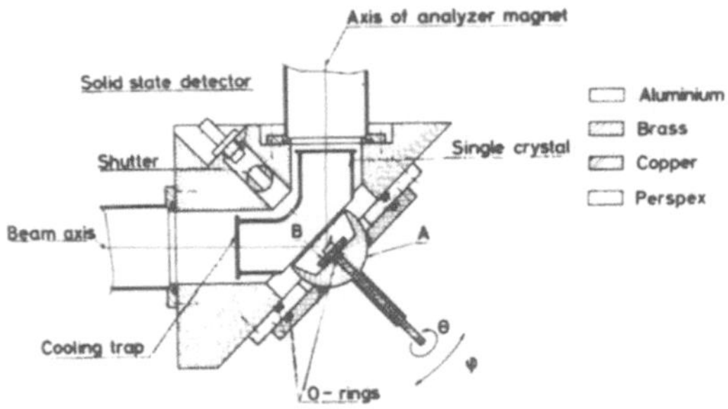

Fig. 3. Schematic drawing of the experimental assembly.

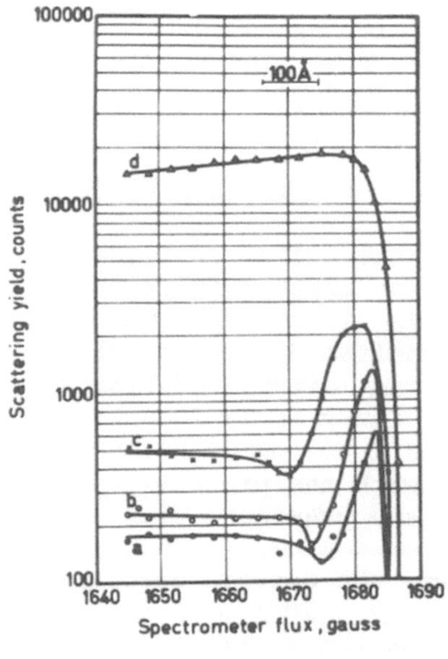

Fig. 4. Energy spectrum of 500 keV protons scattered 90° in W. Curves a, b, and c: Proton beam aligned with [11$\bar{1}$] axis. Curve d: Proton beam not aligned with any major axis or plane. Curve a: Electropolished surface. Curve b: Surface covered with ∿ 2 atomic layers of oxide. Curve c: Surface covered with ∿ 10 atomic layers of oxide.

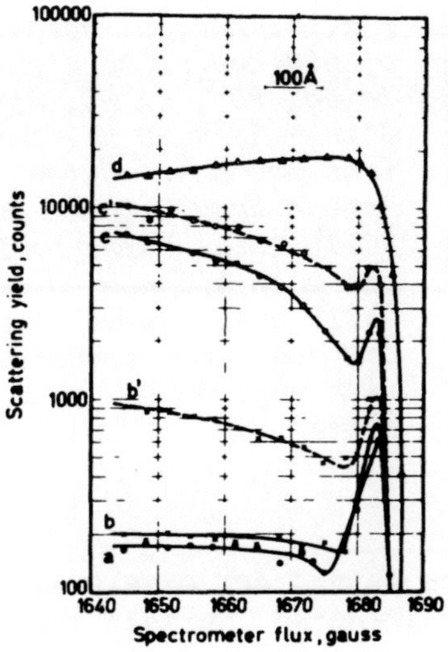

Fig. 5. Energy spectrum of 500 keV protons scattered 90° in W. Curves a, b, b', c, and c': Proton beam aligned with [111] axis. Curve d: Proton beam not aligned with any major axis or plane. Curve a: Perfect crystal. Curve b: After 10^{15} 200 keV N ions per cm^2 along [111] axis. Curve b': As in curve b, but N ions not aligned. Curve c: After 10^{18} 200 keV N ions per cm^2 along [111] axis. Curve c': As in curve c, but N ions not aligned.

Finally, the energy spectrum of protons scattered in a Si crystal doped with Tm^{169} is shown in Fig.6. The Tm was injected at room temperature with an energy of 70 keV by means of an electromagnetic isotope separator. The ions were injected within 5° of a [011] axis. The total dose was $\sim 10^{15}$ ions per cm^2, giving concentrations of approximately 0.5 atomic per cent in the first $\sim$ 300 A° of the crystal. Curves 6a and 6b are the spectra measured with the protons entering the crystal along an [011] axis and a [111] axis, respectively. Curve 6c is the spectrum measured with the protons entering the crystal far from any prominent crystal axis or plane. The spectrum measured with the incoming protons aligned with a [100] axis was found to coincide with curve 6b.

It should be noted here that surface atoms can be hit by the particle beam even in the case of a perfect crystal. Therefore, the yield of scattering events confined to the very first atomic layers will always be the "normal yield", i.e. the yield measured in an amorphous target. The peak in curves 4a and 5a occurring at $\sim$ 1680 gauss shows this. The width of the peak is determined by the depth resolution $\Delta R \sim 50$A°. The total peak area corresponds to normal yield from $\sim$ 1 atomic layer which indicates that the crystal has an essentially perfect surface, where the first atomic layer is effectively "shadowing" the atoms in the following layers. It may also be concluded that the amplitude of the lattice vibrations with wavelengths of the order of the distance between

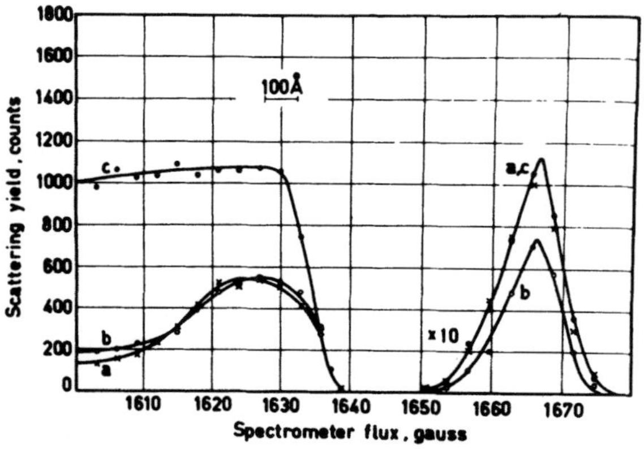

Fig. 6. Energy spectrum of 500 keV protons scattered 90° in Si doped with ∿ 10^{15} 70 keV Tm^{169} ions per cm^2. Curve a: Proton beam aligned with [011] axis. Curve b: Proton beam aligned with [111] axis. Curve c: Proton beam not coinciding with any major axis or plane.

nearest neighbors must be small. Curves 4b and 4c merely illustrate the sensitivity in detection of disordered surface layers. The vacuum in the scattering chamber used in this experiment is too poor to allow more quantitative surface studies. For instance, the effect of a thin, chemisorbed layer of e.g. nitrogen or oxygen on the spectra in Figure 4 should be considered; provided it does not change the positions of the W atoms, and the contaminated surface will erroneously appear perfect. In fact, the electropolished crystal used to measure curve 4a very probably has some chemisorbed oxygen on the surface. This specific sensitivity to disorder may, of course, in some cases be an advantage. The shadowing effect of an adsorbed monolayer might be used to give information of the position of the adsorbed atoms relative to the crystal surface atoms.

If, for some reason (damage, thermal vibrations, etc.), a crystal atom is moved more than ∿ a away from its normal lattice position, it may be hit both by the aligned beam and by the random beam and, consequently, the yield will rise over that from the perfect crystal. This is illustrated in Figure 5. The effect of damage introduced in the W crystal by the N ions is clearly seen. Since well-known scattering formulas (Rutherford scattering and multiple scattering) relate the rise in yield measured at a certain depth to the concentration of interstitials, it is in principle possible to deduce the distribution of interstitials. In the present connection, however, it shall only be mentioned that the steady increase in yield as a function of depth in curves 5b, 5b', 5c, and 5c' indicates that the crystal is rather homogeneously damaged over the depth shown in the figure. The difference between curves 5b and 5b' indicates a strong dependence of the damage on the direction along which the N ions are injected. A similar effect has been seen in electron microscope studies by

Noggle and Oen (14). In the aligned case, the N ions are, by the steering effect of the crystal lattice, kept away from the crystal atoms and consequently the damage is reduced. As the crystal is damaged, the steering effect of the lattice becomes less efficient and the difference in damage caused by an aligned beam and a non-aligned beam gradually disappears as seen by comparing the difference between curves 5b and 5b' with the difference between curves 5c and 5c'.

If impurity atoms are substitutionally embedded in a crystal lattice, the yield of a reaction between an impurity atom and a particle beam will show the same orientation dependence as the yield of a reaction between the particle and an atom of the host lattice. If the impurity atom is interstitially embedded in the lattice, the yield will show an orientation dependence which differs from that of the reaction between particles and atoms of the host lattice. Based on these considerations, it may be concluded from the curves in Figure 6 (where the peak occurring at $\sim$ 1665 gauss, according to formula (2) is the yield of protons scattered on Tm) that the Tm ions injected into the Si are, at least partly, placed outside the [011] axis, but on the [100] and [111] axis. This suggests the interstitial positions shown in Fig. 7. Figure 6 also indicates that the injection of Tm ions has caused a considerable damage of the crystal in the first 400-500A° beneath the surface. This damage prevents a more accurate estimate of the number of Tm atoms in the suggested positions.

The author is indebted to J. Lindhard and K.O. Nielsen, for fruitful discussions, and to P.B. Kjaer, Ths. Norgaard and J.B. Rasmussen who helped with the measurements.

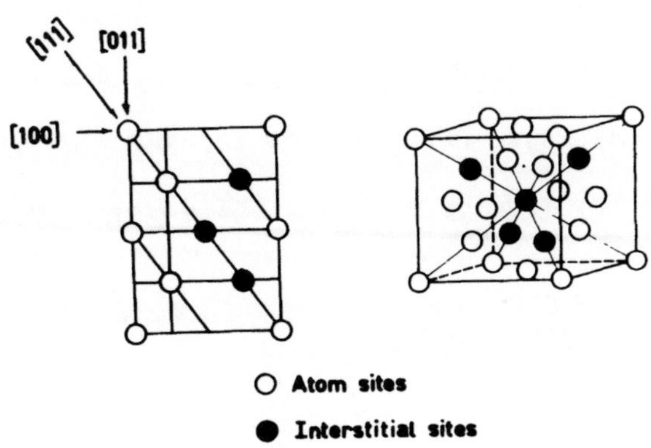

Fig. 7. Positions of Tm atoms in Si as obtained from Figure 6.

REFERENCES

1. *G.K. Wenner, Advances in Electronics 7, 239 (1955).*

2. *G.R. Piercy, F. Brown, J.A. Davies, and M. McCargo, Phys. Rev. Letters 10, 399 (1963).*

3. *E. Bogh, J.A. Davies, and K.O. Nielsen, Physics Letters 12, 129 (1964).*
   J.U. Andersen, J.A. Davies, and K.O. Nielsen, Nucl. Instr. Meth. 38, 209 (1965).

4. *E. Bogh and E. Uggerhoj, Physics Letters 17, 116 (1965).*

   E. Bogh and E. Uggerhoj, Nucl. Instr. Meth. 38, 216 (1965). The measurements reported in this reference have later been extended to:
   50 keV - 10 MeV protons and α-particles,
   100 keV - 1 MeV Li and B ions,
   scattered in Si and W crystals.

5. *J.M. Khan, D.L. Potter, and R.D. Worley, Phys.Rev. 148, 413 (1966).*

6. *B. Domeij, Nucl. Instr. Meth. 38, 207 (1965).*

7. *O.S. Oen and M.T. Robinson, Appl. Phys. Letters 2, 30 (1963).*

8. *R.S. Nelson and M.W. Thompson, Phil. Mag. 8, 1677 (1963).*

9. *J. Lindhard, Mat. Fys. Medd. Dan. Vid. Selsk. 34 No.14 (1965).*

10. *J.U. Andersen, unpublished thesis.*

11. *A.F. Tulinov, Soviet Physics, DOKLADY 10, 463 (1965).*

12. *M.W. Thompson, Phys. Rev. Letters 13, 756 (1964).*

13. *C.Erginsoy, Proceedings of Cairo Solid State Conference American University in Cairo, 1966.*

14. *T.S. Noggle and O.S. Oen, Private Communication and Bull. Am. Phys. Soc. 11, 177 (1966).*

SOLID STATE REACTION WITH ACCELERATED IONS

T. Andersen and G. Sorensen

University of Aarhus

DENMARK

An electromagnetic isotope separator and a 600 keV ion accelerator have been used for a study of the chemical fate of energetic, singly charged ^{32}P ions implanted in pure and doped crystals of the alkali chlorides. Energetic ^{32}P ions implanted in alkali chlorides may be able to simulate ^{32}P recoil atoms produced by the $^{35}Cl(n,\alpha)^{32}P$ reaction and give new information about the chemical effects following nuclear transformations.

The chemical distributions of the ^{32}P atoms was analyzed by dissolving the crystals in water containing hypophosphite, phosphite and phosphate carriers. The ^{32}P atoms are distributed between the three fractions.

The chemical distribution depends on the target preparation method, the energy of the accelerated ions, the dose implanted in the target and the irradiation temperature. The effects of different types of defects were investigated. No influence on the chemical distribution of the ^{32}P atoms was observed by enriching the sodium chloride crystals with sodium, exposing the sodium chloride crystals to a total dose of 2 x 10^7 rad of 10 MeV electrons or by introduction of small amounts of divalent cations before the $^{32}P^+$ exposure. Sodium chloride crystals doped with Cu^+, Ca^+, or Ag^+, using the ion implantation technique, exhibited marked temperature dependent effects on the chemical valence state of subsequent implanted $^{32}P^+$ ions.

INTRODUCTION

During the last few years, ion accelerators have been used rather extensively in solid state research. Accelerators based on the electromagnetic isotope separation principle produce energetic ion beams of well-separated radioactive isotopes as well as beams of stable isotopes. The application of radioactive ion beams makes it possible to perform implantation studies with doses of 10^{10} - 10^{15} ions/cm^2 in crystalline matter. The radioactive tracer method allows a detailed investigation of some of the processes occurring when an energetic ion interacts with crystalline matter.

The present paper reports on a preliminary investigation of the chemical fate of singly charged energetic phosphorous ions implanted in pure and doped crystals of alkali chlorides.

Energetic beams of radioactive $^{32}P^+$ ions ranging from 5 keV to 500 keV were obtained from the Aarhus isotope separator (1) and the Aarhus heavy ion accelerator (2). The isotope separator is shown in Figure 1. The ions are extracted from an ion source of the magnetic type with end extraction. The extracted ions are accelerated and focussed by a lens system after which follows the deflection chamber placed in a gap of 55 mm, 1.50 m radius, 90° magnet. On leaving the magnet, the resolved isotopic beams proceed to the target chamber. Beams in the μA region can be obtained of most stable isotopes. The energy of the ion beams can be varied from 5 keV to 70 keV for singly charged ions. Figure 2 shows the Aarhus heavy ion accelerator with an energy range from about 70 keV to 550 keV for singly charged ions.

It is established that energetic ions penetrating a solid slow down by elastic and inelastic energy loss processes resulting in radiation damage of the target crystal. Radiation damage gives rise to different types of centers, and in radiation unstable systems, even decomposition processes may occur. A detailed study on the decomposition of nitrates by light energetic ions has been published by Logan and Moore (3).

Radioactive atoms from nuclear reactions recoil as ions in crystalline matter. While a considerable number of investigations (4) concerning the chemical effects of nuclear transformations in ionic solids have been carried out in "hot atom" chemistry, the present knowledge about the chemical fate of energetic ions implanted in ionic solids is rather limited. As shown in Figure 3, the two types of processes seem identical. An investigation of the alkali halide systems seems appropriate from a solid state point of view. Radioactive beams of singly charged ^{32}P ions have been used allowing a comparison between the results obtained and the published data (5-9) for the chemical effects of the $^{35}Cl(n,\alpha)^{32}P$ process in the alkali chlorides.

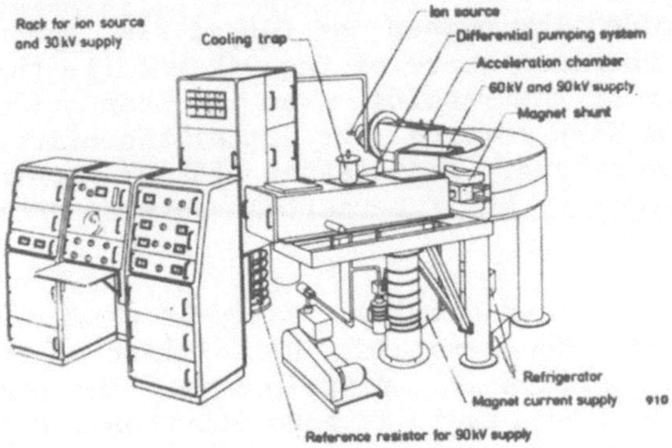

Fig. 1. The isotope separator at the Institute of Physics, University of Aarhus, Denmark.

Fig. 2. The 600 keV heavy ion accelerator at the Institute of Physics, University of Aarhus, Denmark.

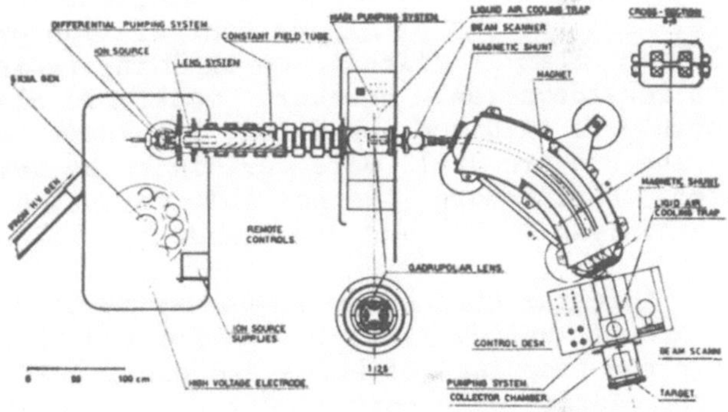

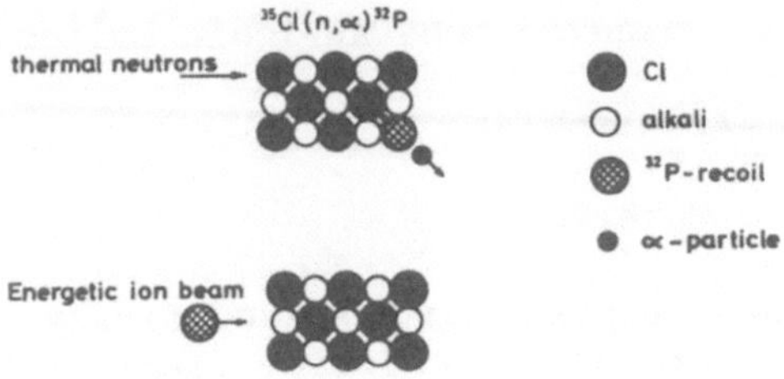

Fig. 3. Comparison of the experimental technique with "hot atom" chemistry.

The ^{32}P recoil atoms formed by the $^{35}Cl(n,\alpha)^{32}P$ reaction in the reactor have a recoil energy of $\sim$100 keV. Concurrent with the formation of ^{32}P recoils in the reactor, production of F and V centers takes place due to γ radiation. These centers, exceeding by several orders of magnitude the number of formed ^{32}P recoil atoms, influence the final chemical state of the ^{32}P recoils thus effecting the reproducibility of the results. The ion beam method for the study of recoil chemistry removes the disturbing effects of ionizing radiation in the reactor and creates cleaner experimental conditions for the study of the interaction between energetic ions and solids.

At the present time, a detailed chemical analysis cannot be carried out without dissolving the crystals. Dissolution of the crystals destroys the whole solid state distribution, and only results expressed as yields of ions stable in aquo solution can be obtained. The phosphorus atoms are present in solutions as P^{+5} (phosphate), P^{+3} (phosphite), or P^{+1} (hypophosphite). However, the chemical distribution in a solution does not give direct information about the chemical state of the ^{32}P atoms in the crystals. Using Lidiard's criterion (10), the presence of a P-atom with valence state +2 would be possible in an alkali halide lattice, but not one of higher charge. The presence of a P^{+5} ion in solution seems difficult to explain taking Lidiard's criterion into consideration. However, monatomic elemental phosphorus can probably be oxidized by water, at least to P^{+3}. Species such as P^{+1} and P^{+2} are still more certain to be oxidized upon solution in water. Furthermore, the possible formation of P-Cl-compounds able to hydrolyse to P^{+5} and P^{+3} cannot be excluded.

Since the phosphorus atoms may be considered as impurities in the alkali halide lattice, their oxidation and reduction may reflect the trapping of holes and electrons, and the changes might be correlated with corresponding change in color center concentration. Thus investigation of the influence of different types of point defect or dislocation on the chemical distribution of ^{32}P atoms may be able to shed some light on the mechanism determining the chemical distribution of implanted ions in ionic solids.

EXPERIMENTAL

Material

The alkali halide targets exposed to beams of singly charged ^{32}P ions were single crystals of sodium chloride (semi-elements), deposits of A.R. sodium chloride, or potassium chloride prepared by vacuum evaporation. Alkali chlorides doped with divalent impurity atoms were prepared either by melting A.R. grade chemicals together in a crucible, crushing the crystals to a fine powder and transferring the mixed crystal to an aluminum backing by an electrophoretic technique, or by ion implantation of the impurity using high-energy ion beams. The electrophoretic technique consists of the dispersing of the crystals in acetone and then applying a D.C. potential of about 400 volts between the aluminum plates. According to the nature of the crystals, a fine deposit is formed on either the positive or the negative plate.

Irradiations

Neutron irradiated phosphorus (red modification) was loaded into a furnace, and vapors of phosphorus were evaporated into the ion source, ionized, and electromagnetically separated. The beam of stable ^{31}P ranging from 10 to 40 μA was measured in a Faraday cup, yielding a relative measure for the number of radioactive ^{32}P ions implanted in the target. Twelve alkali halide targets were placed in a rotating wheel in the collection chamber and exposed in vacuum, normally at room temperature. Facilities for effecting exposures at -100°C by connecting the rotating wheel to a liquid-air freezing trap were available.

The ions hitting the target are mainly $^{32}P^+$ ions, but O_2^+ from rest gasses in the ion source may also be present. However, the amount of O_2^+ is small when red phosphorus is used as charge material. Contamination from ^{31}P varies from time to time, but was normally kept below 0.5%. This corresponds with the specific activities normally used for exposures of a target of 4x6 mm with a dose of 10^{12} ions divided equally between stable ^{31}P and radioactive ^{32}P. This means that the interactions taking place in the crystal are reflected by the radioactive ions.

Analytical Procedures

The exposed alkali halide systems were analyzed for the distribution of ^{32}P atoms among the possible chemical forms produced when the alkali halides are dissolved in water. Since the concentration of implanted phosphorus ions is too small to be responsible for the formation of compounds containing more than one phosphorus atom, the possible chemical forms existing in the solution of the target are hypophosphite (P^{+1}), phosphite (P^{+3}), and phosphate (P^{+5}). The exposed alkali halides were dissolved in a solution containing carriers of phosphate, phosphite, and hypophosphite ions. First, all phosphorus present as phosphate was precipitated as magnesium-ammonium phosphate. The resulting precipitate was filtered off, and the phosphite was oxidized with iodine (1 h) in a neutral bicarbonate medium and precipitated as phosphate. The precipitates were dissolved and transferred to a liquid counting tube, Frieseke and Hoepfner FH2 44, and the beta disintegration rate was determined.

This was compared to counting rates of standards containing all phosphorus species, yielding the chemical distribution of the implanted phosphorus ions. All results are based on double determinations, and the accuracy of the method is 1-2%.

RESULTS AND DISCUSSIONS

Comparison With "Hot Atom" Studies of $^{35}Cl(n,\alpha)^{32}P$

If the ^{35}Cl ion was undergoing a (n,α) reaction leading to ^{32}P without perturbation of orbital electrons, the primary charge of ^{32}P would be -3. However, both the α-emission and the slowing down of the recoil may lead to ionization. The ^{32}P may then be raised to P^{-2}, P^{-1}, P^0, or even high oxidation states. It should, however, be pointed out that the most important factors determining the final chemical state of the ^{32}P recoil in the crystal are the surroundings in which the phosphorus atom finds itself after having dissipated its recoil energy, and the interaction with radiation-induced defects and thermal annealing effects before analysis of the crystal.

Table I shows a comparison between the chemical distributions obtained with ^{32}P, originating from $^{35}Cl(n,\alpha)^{32}P$ and when a 70 keV $^{32}P^{+1}$ ion is implanted in potassium chloride. These data indicate that the primary charge of the recoiling or implanted ^{32}P ions does not influence the final charge of the stopped atom. The results are in agreement with the general assumption that charge equilibrium is readily obtained after very few collisions between the energetic particle and the target material. Thus the primary charge state should not influence the final chemical state.

TABLE I

Comparison of Results Obtained With 70 keV $^{32}P^+$ ions in KCl and Recoils of 100 keV ^{32}P Ions From $^{35}Cl(n,\alpha)^{32}P$ in KCl(8).

	$^{35}Cl(n,\alpha)^{32}P$%	70 keV P^+ in KCl %
P^{+5}	∿ 43	46
P^{+3}	∿ 27	20
P^{+1}	∿ 30	34

The quantitative comparison of the results may indicate that the present technique is suitable to obtain further information about the chemical effects associated with nuclear tranformations. These results for A.R. potassium chloride are included for the sake of comparison; the major part of the experimental data, however, concerns single crystals of sodium chloride.

Effects of Target Preparation

When targets of sodium chloride prepared in different ways are exposed to the same ion dose, differences in the appropriate chemical distribution occur. Table II shows the variations in the chemical distributions due to the target preparation technique. The targets have received identical doses of ^{32}P ions. For targets prepared by vacuum evaporation, an increase in the P^{+5} fraction is observed. This might be due to differences in the densities and in impurity contents leading to different concentration of radiation-produced centers. The oxygen possibly included in evaporated targets may also have a marked effect.

Experimental work was performed with all three types of target and the same general features were obtained. Further data reported here refer only to single crystals of sodium chloride and electrophoretic deposited sodium chloride.

TABLE II

Variations in Chemical Distributions due to Target Preparation Technique

	P^{+5}	P^{+3}	P^{+1}
NaCl single crystals	26.7	16.3	57.0
NaCl (crushed) Electrophoresis	31.0	18.0	51.0
NaCl evaporated from single crystals	37.8	16.7	45.5

Effect of Dose and Dose Rate

The final charge state of the accelerated ion is not independent of the total dose of ions implanted in the alkali halide. But it seems to be independent of the speed at which the ion beam reaches the target. In Figure 4 is shown the chemical distribution of ^{32}P as a function of the total dose in μcoulombs received on a 4 x 6 mm area of a sodium chloride single crystal (10 x 10 x 1 mm).

Even though the variations are rather small, the present data show that the chemical valence state of the slowed down particle mainly depends on the interaction between the particle at the end of its track and defects produced by the particle itself. To some extent it also depends on the influence of charge carriers produced by other particles. Thus targets used for comparison were given the same ion dose.

Some experiments were performed with different dose rates. but they did not show variations under the conditions normally present in the isotope separator.

Effects of Ion Energy and Irradiation Temperature

Figure 5 shows the chemical distribution of ^{32}P; varying the ion energy from 40 to 500 keV. A marked energy dependence is observed, yielding more oxidized species when the energy is increased. This is in agreement with observations on the energy dependence of decomposition processes by Logan and Moore (3) for ion bombardment and with previous data by the authors (11) on K_2CrO_4 and K_2SO_4. The observed energy dependence may not represent a real energy effect but rather indicate that increased energy deposit per unit volume will initiate annealing reactions able to change the chemical distribution. The de-excitation process of the accelerated ion, consisting of an energy depending mixture of electronic and nuclear stopping, may also be responsible for the effect, the two types of stopping leading to different crystal defects.

Some experiments in which the target temperature was decreased from 25°C to -100°C, were also performed. At this low temperature, some of the radiation-induced centers may not be reactive. While increasing ion energy normally leads to oxidation processes, the opposite effect is observed at -100°C. As shown in Figure 6, the exposures at -100°C tend to yield more reduced species with increasing energy. Several hypothesis may explain the observed data; for instance, a larger number of collisions of the electronic type along the path yielding more free electrons. However, more experimental data are needed to clear up this temperature dependence.

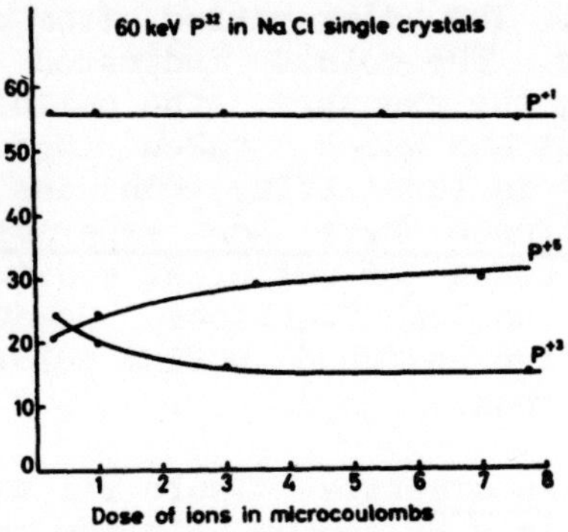

Fig. 4. Chemical distribution of 60 keV singly charged ^{32}P implanted in single crystals of NaCl as a function of ion dose.

Fig. 5. Chemical distribution of singly charged ^{32}P in single crystals of NaCl as a function of ion energy.

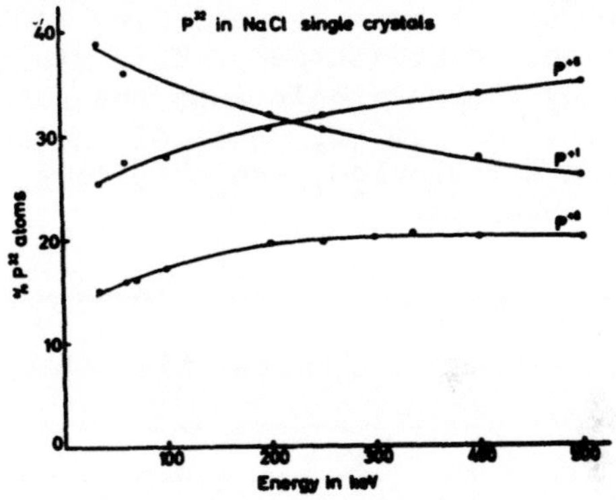

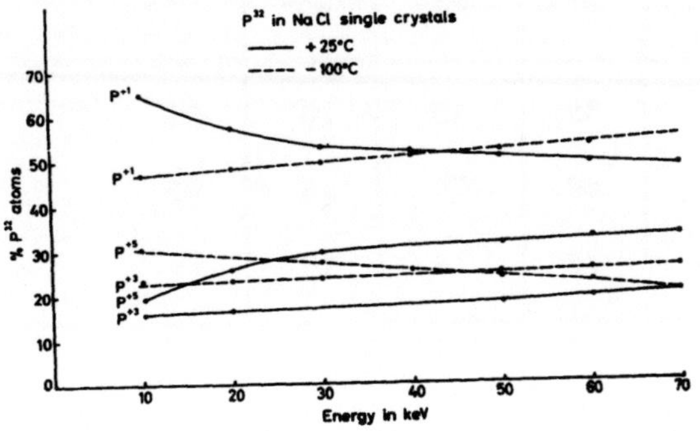

Fig. 6. Comparison of the chemical distribution of ^{32}P in single crystals of NaCl at -100°C and +25°C, varying the energy.

Effects of Different Types of Centers

Sodium-enriched crystals. Single crystals of sodium chloride were heated in sodium vapor to produce a surplus of sodium atoms which results in F center formation. The color varied from dark bluish purple to brown or pale blue. The colors indicated the presence of a colloidal band. During the exposure, the colloidal sodium may provide a supply of electrons which, taken together with the increased density of defects in the lattice, enhances the probability of stabilizing the P^{+1} valence state. Some experiments (12) have demonstrated a marked influence of colloidal sodium in sodium chloride on recoil atoms from nuclear reactions. However, no reducing effect was found when sodium-enriched sodium chloride was exposed to a beam of 60 keV $^{32}P^{+}$ ions.

Electron bombarded sodium chloride crystals. Table III shows the variations in chemical distributions in sodium chloride crystals, pre-irradiated with 10 MeV electrons to doses of 20 and 100 Mrad and then exposed to either neutrons or 60 keV ^{32}P ions.

Figure 7 shows the chemical distribution of ^{32}P implanted in single crystals of sodium chloride pre-irradiated with different doses of 10 MeV electrons. The exposure of ^{32}P was made at 25°C and -100°C. At room temperature, only very high doses influence the chemical distribution. This might be due to oxidative centers created by the energetic ions and subsequent coalescence with the F-centers. At -100°C, some of the centers produced by the energetic ion are blocked, and therefore, a decrease in oxidized species is observed.

TABLE III

Variations in Chemical Distributions Due to Pre-Irradiation

	$^{35}Cl(n,\alpha)^{32}P$			60 keV ^{32}P in NaCl		
Dose	P^{+5}	P^{+3}	P^{+1}	P^{+5}	P^{+3}	P^{+1}
0 Mrad	50	15	35	30	16	54
20 Mrad	19	12	69	30	13	57
100 Mrad	8	14	78	15	10	75

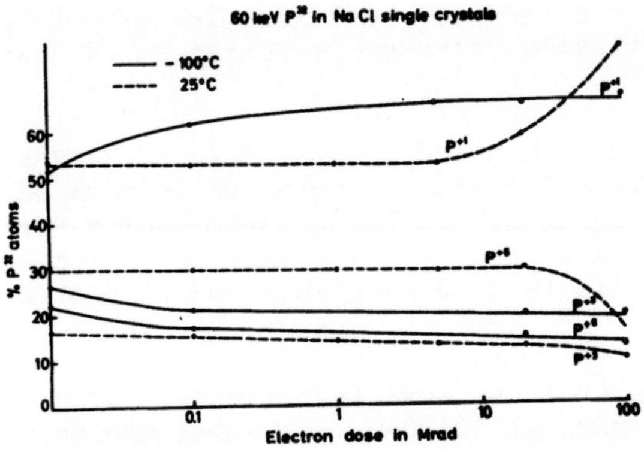

Fig. 7. Chemical distribution of 60 keV ^{32}P in single crystals of NaCl pre-irradiated with different electron doses.

The present data show a considerable difference in the pre-irradiation conditions necessary to obtain the same reducing effect. However, while the concentration of P-atoms in the (n,α) experiment is homogeneous in the crystals, the relatively high local concentration of implanted P-atoms may explain both the missing reducing effect in the sodium-doped crystals and the missing reduction at 20 Mrad. If the ratio of the centers available for electron supply to the number of P-atoms is considerably different for the two experiments, the present data indicate that 1) either F-centers provide a potential supply of electrons for capture by the phosphorus, or 2) perhaps more important, the provision of vacancies suitable for occupation of anionic species, i.e. species leading to formation of P^{+1} upon solution, can explain the results obtained.

Alkali Halides Doped With Divalent Ions by Melting

Sodium chloride was melted with chlorides of calcium, cesium, lead, and nickel in concentrations from 0.01 to 1 Mole%. The variations in the chemical distributions were negligible when such doped NaCl targets were exposed to a 60 keV ^{32}P beam. However, crystals prepared by the same technique show an increase in reduced species of phosphorus when the ^{32}P recoil originates from $^{35}Cl(n,\alpha)^{32}P$.

Alkali Halides Doped with Ion Implantation

Single crystals of sodium chloride were exposed to beams of stable isotopes (eg: Cu^+, Ca^+, or, Ag^+) before exposure to energetic ^{32}P ions. The dose of Cu-ions varied from 8×10^{13} to 5×10^{15} ions/cm^2.

The heavy ion induced defects have a considerable influence on the chemical properties of the ^{32}P atoms. Single crystals of sodium chloride were pretreated with stable beams of 60 keV Cu^+, Ca^+, and Ag^+. Figure 8 shows the chemical distribution of ^{32}P in Cu-doped sodium chloride crystals at room temperature and at -100°C.

There is a marked difference between the two sets of curves. While the data at -100°C reflect the same general pattern as observed for the dose dependence at 25°C, the presence of a "reducing maximum" at 25°C may be due to a center formation connected to the copper ions. Similar effects have been observed with other implanted ions.

The crystals doped with Cu-ions were investigated by u.v. spectroscopy. An absorption band at ∿2300A° was superimposed on a continuum. No absorption characteristic of the F-center was observed. It has been reported that the Cu^+ ion doped in sodium chloride by suitable melting technique absorb at 2500A°.

CONCLUDING REMARKS

The present work only considers a few aspects of the interaction occurring when accelerated ions penetrate ionic solids. The reactions occurring are very complex, involving all types of center formation, electron-hole production, and disorder in the crystalline matter. The final charge of the accelerated ions is, in some ways, reflected by the chemical distribution investigated in the present work.

The present data may be considered only preliminary. At the present time, we have only a very rough picture of the process determining the chemical fate of implanted ions. Further studies are needed; exposures at different temperatures and the use of ion-doped targets seem to give the best prospect of obtaining information at the present time. A combination of chemical analysis with u.v. spectroscopy, thermoluminescence, and E.S.R. may be a possible way forward.

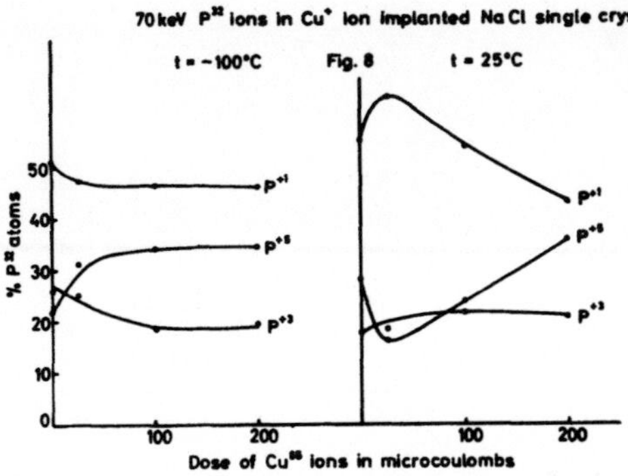

Fig. 8. Chemical distributions of 70 keV ^{32}P in single crystals of NaCl pre-irradiated with Cu^+ ions at different doses.

REFERENCES

1. *K.O. Nielsen, and V. Toft, to be published.*
2. *E. Bogh, P. Dahl, H.E. Jorgensen, and K.O. Nielsen, Report, University of Aarhus (1965), to be published.*
3. *S. Logan, and W. Moore, J. Phys. Chem. 67, 1042-48 (1963).*
4. *See: Proceedings of Symposium in Chemical Effects of Nuclear Transformations, IAEA (Prague) 1960, (Vienna) 1961, and (Vienna) 1965.*
5. *G. Harbottle, Ann. Rev. Nucl. Sci. 15, 89-124 (1965).*
6. *J.S. Butterworth, and I.G. Campbell, Nature 196, 982-83 (1962)*
7. *J.S. Butterworth, and I.G. Campbell, Trans. Faraday Soc. 59, 2618-24 (1963).*
8. *J. Cifka, J. Inorg. Nucl. Chem. 26, 683-91 (1964).*
9. *J. Cifka, Proceedings of Symposium on Chemical Effects of Nuclear Transformations, IAEA (Vienna) 1965, Vol. II, 71-80.*
10. *A.B. Lidiard, Handbuch der Physik 20, 246-349 (1957).*
11. *T. Andersen, and G. Sørensen, Trans. Faraday Soc., in press, and Nucl. Instr. Meth. 38, 204-06 (1965).*
12. *A.G. Maddock, and R.M. Mirsky, Proceedings of Symposium on Chemical Effects of Nuclear Transformations, IAEA (Vienna) 1965, Vol. II, 41-64.*

ELECTRON AND POSITIVE HOLE REACTIONS IN KCl - TlCl

C.J. Delbecq, A.K. Ghosh, and P.H. Yuster

Argonne National Laboratory

Argonne, Illinois, U.S.A.

During γ-ray irradiation at 77°K, Cl_2^- and thallium atoms (Tl^o) are formed in KCl-TlCl crystals. On warming the irradiated crystals, two glow peaks are observed. The low temperature glow peak at 208°K is caused by the diffusion of Cl_2^- and the subsequent electron-hole annihilation involving nearby pairs of Cl_2^- and Tl^o. In addition to diffusing into the neighborhood of Tl^o, a Cl_2^- can diffuse into the neighborhood of a Tl^+ and become trapped to form Tl^{++}. The high temperature glow peak at 300°K results when electrons are thermally released from thallium atoms into the conduction band and recombination of these electrons and Tl^{++} takes place. Spectral analysis of the low temperature glow peak shows that it consists of two emission bands with peaks at 410 and 298 mμ while the high temperature glow peak consists only of the 298 mμ band. The processes which give rise to these emissions are discussed. The optical absorption of Tl^o and Tl^{++} has been measured. The absorption spectrum of Tl^o consists of very weak bands at 1500 and 1260 mμ, a strong band at 380 mμ and weak bands at 250, 300, and 640 mμ. Tl^{++} has strong absorption bands at 220, 262, 294, and 364 mμ.

INTRODUCTION

Considerable work has been reported on the optical absorption and luminescence properties of alkali halide crystals containing thallous ion impurity (1-7). The effects of x-ray or ultraviolet irradiation on these crystals has also been investigated (8-12). Delbecq, Smaller, and Yuster (13,14), while studying the optical spectrum of Cl_2^-, reported that after x-ray or γ-ray irradiation at liquid nitrogen temperature, KCl-TlCl crystals showed thermoluminescence glow peaks at 208°K and 300°K. The first of these peaks was associated with the disappearance of Cl_2^- molecule ions, as shown by ESR measurements. The present work (15) using optical absorption, ESR, and thermoluminescence techniques was undertaken in order to study the products formed during x-ray or γ-ray irradiation and the subsequent changes which produce thermoluminescence. From the results obtained in the present investigation, it is concluded that the two thermal glow peaks in KCl-TlCl result from recombination of electrons and holes.

EXPERIMENTAL RESULTS

Optical Absorption Spectra

Figure 1 shows the optical absorption spectrum, at 77°K, of a potassium chloride crystal containing 1.28×10^{-2} mole per cent thallous chloride; the well-known A, B, and C absorption bands of Tl^+ are observed at 247, 209, and 196 mμ, respectively.

Figures 2A and 2B show the induced absorption spectra, measured at 77°K, of a KCl-TlCl crystal which has been exposed to γ-ray radiation and given subsequent treatments. Note that because the long wavelength transitions are very weak the ordinate in Figure 2B has been expanded by a factor of 100. Curve (a) shows the net absorption after a 15 min. exposure to Co^{60} γ-rays at 77°K and subsequent optical bleaching to remove F' centers. The spectrum, curve (a), is complicated and consists of bands with peaks at 367 mμ (Cl_2^- centers), 540 mμ (F centers), and Tl^0 bands with peaks at ∿300, 380, 650, 1260 and 1500 mμ. In addition, a weak 262 mμ band is observable indicating the formation of a low concentration of Tl^{++} ions (Fig. 4). Curve (b) in Figure 2 shows the induced absorption after warming to 233°K for 5 min. This treatment eliminates Cl_2^-; it has been shown that Cl_2^- in KCl becomes thermally unstable near 173°K (14). In KCl-TlCl, Cl_2^- has a maximum rate of decay and a half-life of about 2 min. at 208°K(13,14).

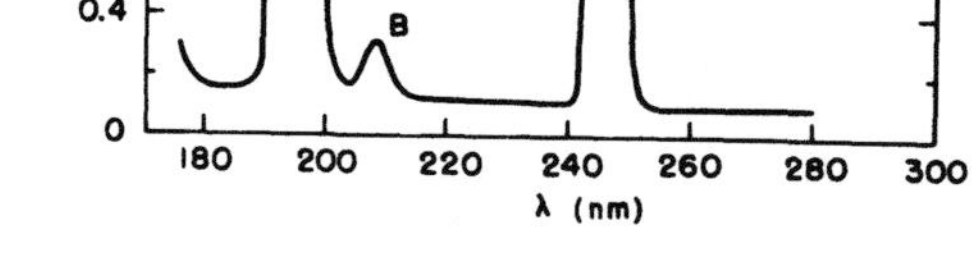

Fig. 1. Absorption spectrum, at 77°K, showing the A, B, and C bands of Tl^+ in a 0.21 mm thick KCl-TlCl crystal containing 1.28×10^{-2} mole per cent Tl^+.

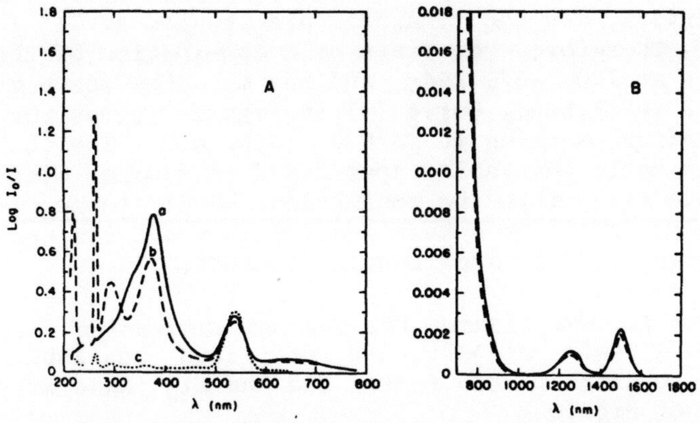

Fig. 2, A and B. Induced absorption of a KCl-TlCl crystal at 77°K after
(a) a 15 min. exposure to Co^{60} γ-rays at 77°K,
(b) warming the crystal to 233°K for 5 min., and
(c) warming the crystal to 320°K for 5 min.

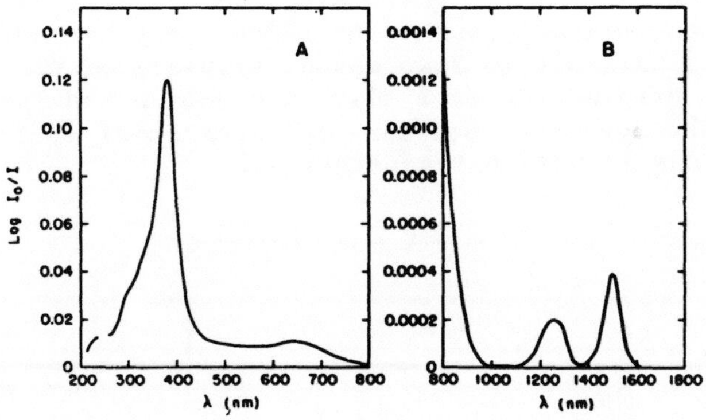

Fig. 3. Absorption spectrum of Tl^0 in a KCl-TlCl crystal at 77°K.

Similar to the case of KCl-AgCl (16), a fraction of the Cl_2^- is annihilated by radiative recombination with electrons from Tl^o, a very small fraction is annihilated by recombination with electrons from F centers, and the remainder are trapped at Tl^+ to form Tl^{++}. Curve (b), therefore, consists of a combination of the Tl^{++} bands with peaks at 220, 262, 294, and 364 mμ, the above-mentioned Tl^o bands, and the F band. Curve (c) in Figure 2 shows the induced absorption after warming to 320°K for 5 min. The thallium atoms become thermally unstable near 288°K releasing electrons at an appreciable rate into the conduction band; these electrons then recombine with Tl^{++} or a very small number are trapped at negative ion vacancies to form additional F centers.

Figure 3, which shows the optical absorption spectrum of Tl^o in KCl, with peaks at ∿300, 380, 640, 1260 and 1500 mμ, was obtained by subtracting the F band and the Cl_2^- absorption band from curve (a) of Figure 2.

The absorption spectrum of Tl^{++}, with peaks at 220, 262, 294, and 364 mμ, is shown in Fig. 4. The Tl^{++} spectrum has been determined in two ways. (1) The spectrum indicated by curve (a) was obtained by subtracting the Tl^o absorption, Fig. 3, from curve (b) of Figure 2. (2) The spectrum indicated by curve (b) was obtained from a KCl crystal containing two added impurities, Tl^+ and NO_3^-. A KCl crystal grown from a melt to which KNO_3 has been added contains NO_3^-, NO_2^- and other decomposition products, some of which can readily trap electrons. The absorption bands of none of the species originating from NO_3^- overlap strongly with the 262, 294, or 364 mμ bands of Tl^{++}. Therefore, in a KCl-TlCl-KNO_3 crystal which has been γ-rayed, warmed to 230°K, and bleached to get rid of Tl^o, it is possible to trap essentially all of the electrons at the species originating from NO_3^- and obtain the Tl^{++} absorption spectrum. The agreement between the spectra of Tl^{++} obtained by the two methods is very satisfactory.

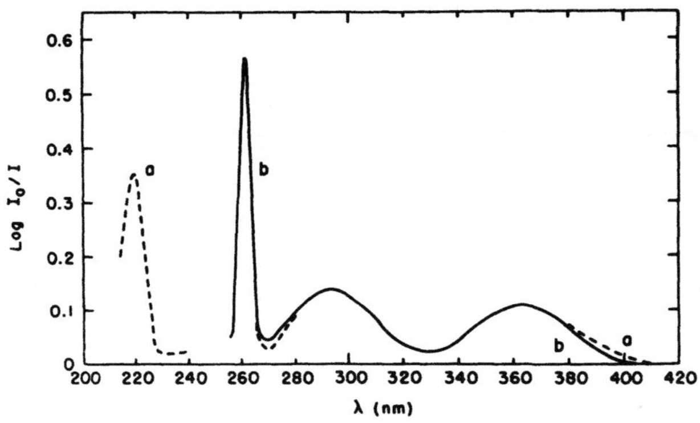

Fig. 4. Absorption spectrum of Tl^{++} in a KCl-TlCl crystal at 77°K

Luminescence

Figure 5 shows a plot of the luminescence intensity of a KCl-TlCl crystal, which had been exposed to Co^{60} γ-rays at 77°K, as it is warmed. Two emission bands are observed, one with its peak at 410 mμ and the other at 298 mμ. The relative intensities of these two bands vary with the temperature. The afterglow, at 77°K and up to 170°K, shows only the 410 mμ emission band. The low temperature glow peak is composed of both the 410 mμ and 298 mμ emission bands, while the high temperature glow peak consists of only the 298 mμ band. The low temperature glow peak occurs at about 208°K (obtained by adding the two emission components) and the high temperature glow peak occurs at about 300°K. During the low temperature glow peak Cl_2^- disappears, there is some decrease in the Tl^o concentration and Tl^{++} centers are formed. During the high temperature glow peak the Tl^o and Tl^{++} centers disappear.

Figure 6 shows emission spectra of a KCl-TlCl crystal after exposure to Co^{60} γ-rays at 77°K. Curve (a), taken at 77°K, shows the afterglow luminescence spectrum. Curves (b), (c) and (d) represent spectra taken at the low temperature side (184°K), maximum (208°K), and the high temperature side (219°K) of the glow peak, respectively. The relative intensities of the 410 and 298 mμ emission bands in the region of the low temperature glow peak depend reversibly on the temperature; e.g., after having warmed the crystal to 219°K, cooling the crystal to 207°K again gives an emission spectrum in which the two bands have the same ratio as shown in curve (c). The 410 mμ emission band broadens and shifts only a small amount between 77° and 219°K. The 298 mμ emission band, which is the sole component of the high temperature glow peak, also broadens and shifts little between 200° and 330°K. It is believed that the 410 mμ emission arises from the annihilation of an electron and hole of a nearby pair of Tl^o and Cl_2^-. This process in KCl-TlCl is similar to the one reported for KCl-AgCl(17). The origin of the 298 mμ emission will be discussed later in this paper. Figure 7 shows a plot of the ratio of the intensities of the 410 mμ band to the 298 mμ band versus 1/T. The points fall on a good straight line which indicates an activation energy of 0.31 eV.

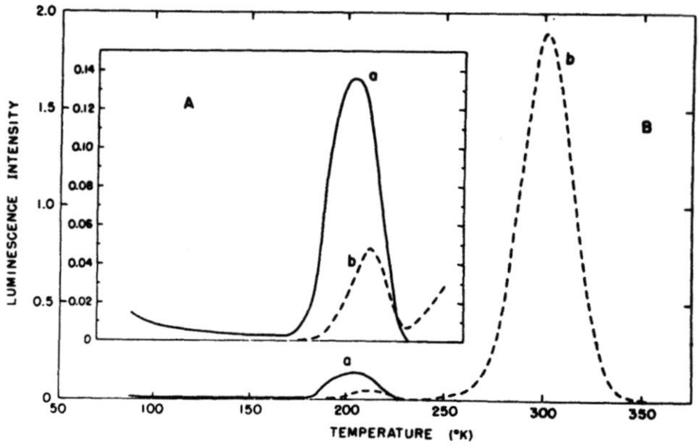

Fig. 5. Glow peaks of a KCl-TlCl crystal which had been exposed to Co^{60} γ-rays at 77°K. The ordinate indicates the relative intensities of the (a) 410 mμ and (b) 298 mμ emission bands. The ordinate for part A is expanded by a factor of ten.

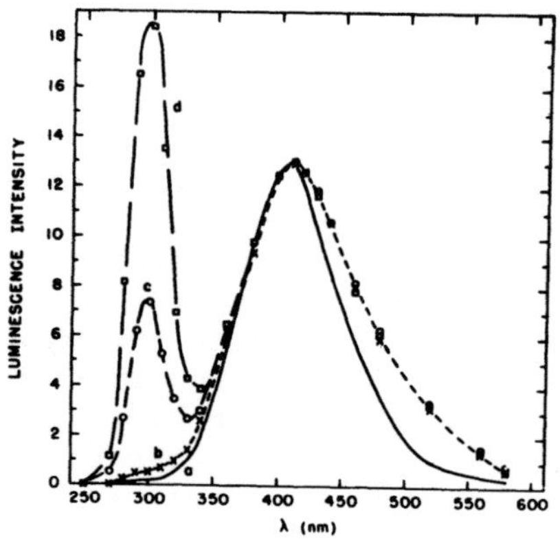

Fig. 6. Emission spectra--taken at (a) 77°K, (b) 184°K, (c) 208°K and (d) 219°K---characteristic of a KCl-TlCl crystal after exposure to Co^{60} γ-rays at 77°K.

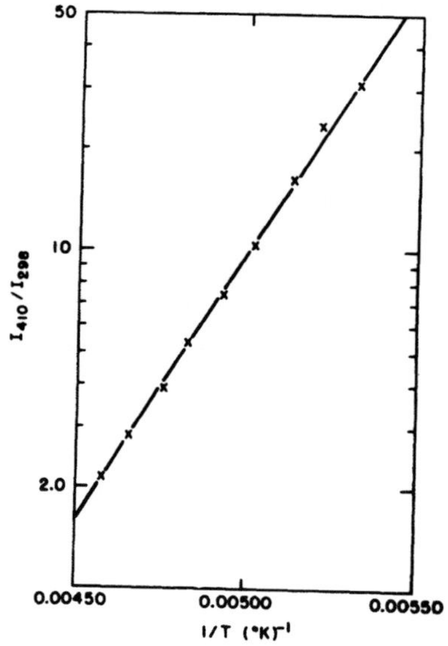

Fig. 7. A plot of the logarithm of the ratio of the intensities of the 410 mμ band to the 298 mμ band versus 1/T.

DISCUSSION OF RESULTS

Formation of Tl^{o} and Tl^{++}

Previous investigations (13,14,15,16,18) have shown that x-irradiation of KCl-TlCl, KCl-AgCl, and KCl-$PbCl_2$ crystals at 77°K results in the very rapid formation of Cl_2^-. It was proposed that the ions Tl^+, Ag^+, and Pb^{++} in KCl act as very efficient electron traps, thus decreasing the recombination of electrons and holes and allowing a rapid buildup of Cl_2^-. In the system KCl-AgCl it was observed that after x-irradiation at 77°K both Cl_2^- and isolated silver atoms, Ag^o, were present in the crystal. Ag^o results from the capture of an electron by Ag^+ while Cl_2^- results from the capture of a positive hole, Cl^o, by one of the Cl^- ions in the crystal. The optical and ESR absorption of Ag^o as well as Cl_2^- have been identified. It was shown through disorientation experiments that Cl_2^- begins to undergo a translational motion in the crystal starting at 170°K. As the temperature is raised the rate of diffusion of Cl_2^- increases and results in its disappearance. Upon warming a crystal containing Ag^o and Cl_2^- a glow peak is observed at about 208°K which corresponds to the maximum in the rate of disappearance of Cl_2^- and Ag^o. This glow peak, with a maximum in the emission spectrum at 550 mμ, arises from the recombination of electrons from Ag^o with Cl_2^- which have migrated into the neighborhood. However, as shown in the ESR and optical absorption spectra, after all the Cl_2^- have disappeared it is observed that 50-90% of the Ag^o remain and that a new hole center, Ag^{++}, has formed. These Ag^{++} result from the retrapping of positive holes at Ag^+ ions. Upon warming further a second glow peak, with a maximum in the emission spectrum at 275 mμ, is observed at about 100°C. In this temperature range the Ag^o are thermally unstable and release electrons into the conduction band; the luminescence arises from the recombination of these electrons with Ag^{++}. When this process is completed the Ag^o and all but a small fraction of the Ag^{++} have disappeared.

Although the ESR absorptions of Tl^o and Tl^{++} have not as yet been observed, it is firmly believed that these species are formed in KCl-TlCl crystals in complete analogy with the formation of Ag^o and Ag^{++} in KCl-AgCl crystals. Upon warming a KCl-TlCl crystal, there is a glow peak associated with the disappearance of Cl_2^- which arises from recombination of electrons from Tl^o with Cl_2^- ions that have migrated into the neighborhood. Between 50 and 90 per cent of the Cl_2^- ions are retrapped at Tl^+ ions to form Tl^{++}; the formation of the 220, 262, 294, and 364 mμ bands of Tl^{++} coincides with the disappearance of Cl_2^-. A second glow peak occurs at 300°K, consisting only of the 298 mμ emission band. Experiments have shown that this glow peak results from the thermal instability of Tl^o and the capture of the freed electron by Tl^{++}.

This suggested mechanism for the second glow peak can be confirmed through the use of a KCl-TlCl crystal containing a second electron trap which leads to the formation of an electron excess center having a higher thermal stability than Tl^o, e.g., Ag^o. After x-irradiation at 77°K a KCl-TlCl-AgCl crystal contains Ag^o, Tl^o and Cl_2^-. After warming to 233°K the crystal contains Ag^o, Ag^{++}, Tl^o and Tl^{++}. Further warming to 320°K removes Tl^o completely, reduces Tl^{++} and Ag^{++} somewhat, and increases the concentration of Ag^o slightly. Thus it is the thermal instability of Tl^o, not Tl^{++}, which causes the glow peak at 300°K. On raising the temperature of the crystal to 373°K, where Ag^o is unstable, emissions characteristic of recombination of electrons and Tl^{++} and electrons and Ag^{++} are observed. Thus, independent of the source of electrons, 298 mμ emission results from the recombination of electrons and Tl^{++}. The rise in the F-band during the high temperature glow peak is also in agreement with electrons being thermally released into the conduction band by Tl^o.

In addition to the analogous behavior of KCl-AgCl and KCl-TlCl crystals described above, the decrease of the A band of Tl^+ upon γ-irradiation at 77°K and warming to 233°K also supports the proposal that electrons and holes are trapped by Tl^+. The complete recovery of the A band of Tl^+, on warming the irradiated crystal to 320°K, is also consistent with the mechanism proposed for the high temperature glow peak.

Absorption Spectra of Tl^o and Tl^{++}

In the ultraviolet and visible region of the spectrum the absorption of Tl^o consists of a strong rather sharp absorption band with peak at 380 mμ (3.26 eV) and weaker bands at 250 (4.96 eV), 300 (4.13 eV), and 640 mμ (1.92 eV). Excitation in any of these bands results in freeing an electron into the conduction band even at temperatures as low as 5°K. In absorption, the lowest energy permitted transition of the free thallium atom is from the $6s^26p^1$, $^2P_{1/2}$ ground state to the $6s^27s^1$, $^2S_{1/2}$ excited state, with an energy separation of 3.28 eV (19). It is possible that the strong resonance-like absorption at 380 mμ is to be associated with a transition from the $^2P_{1/2}$ ground state to a $^2S_{1/2}$ resonant level in the conduction band and that the remaining absorption results from transitions from the ground state of Tl^o to conduction band states. The weak Tl^o bands at 1260 (0.98 eV) and 1500 mμ (0.83 eV) are believed to arise from the forbidden $^2P_{1/2} \rightarrow {}^2P_{3/2}$ transition in the free atom. M. H. Cohen has suggested to us that the splitting of the $^2P_{3/2}$ level of Tl^o, giving two absorption bands, arises from the dynamical Jahn-Teller effect.

The Tl^{++} absorption spectrum shows peaks at 220, 262, 294, and 364 mμ. Excitation of the three shorter wavelength transitions, at 77° or 5°K, causes the Tl^{++} ion to decompose and form Cl_2^-. This decomposition can be clearly seen in a crystal originally containing only Tl^0 and Tl^{++} by examining the ESR and optical absorption spectra and also by observing the regeneration of the low temperature glow peak. The evidence is strong that the species we have indicated as Tl^{++} is in fact a complex ion. The ESR absorption spectrum shows that Ag^{++} is a complex ion in which a Jahn-Teller distortion has taken place producing a square planar configuration. In this configuration a positive hole is shared principally by an Ag^+ and the surrounding four halide ions (16) and is less strongly coupled to the two remaining halide ions. It is believed that a somewhat analogous complex ion forms involving a positive hole, a Tl^+ ion and the neighboring Cl^- ions. In the case of Tl^{++}, however, no Jahn-Teller distortion would be expected if the ground state of Tl^{++} is ^2S; such a state would be produced if Tl^+ ($6s^2$) lost one of its 6s electrons to become Tl^{++} (6s).

Luminescence Processes

The 298 mμ emission band, observed during the low and high temperature glow peaks, has the same shape and energy as that observed upon excitation of the A band of Tl^+ and, therefore, is considered to be a characteristic emission of the Tl^+ ion. This emission is generally thought to result from the transition of the thallous ion from the $^3P_1$ excited state to the $^1S_0$ ground state.

The 410 mμ emission, observed during the afterglow and low temperature glow peak, is not a characteristic emission of the Tl^+ ion. It is believed to result from tunneling between a Tl^0 and a nearby Cl_2^- in which the 6p electron from Tl^0 and the trapped hole, Cl_2^-, recombine radiatively. The 410 mμ emission during the afterglow and during the low temperature glow peak are both believed to arise from this process. At 77°K the concentration of nearby tunneling pairs decreases with time resulting in a decay of the afterglow. However, upon warming the crystal, Cl_2^- starts to diffuse near 170°K thus increasing the number of tunneling pairs and, therefore, increasing the intensity of the 410 mμ emission. As the temperature is further raised the mobility of Cl_2^- increases and the rate of production of nearby pairs increases until a maximum is obtained at the glow peak. The probability of the tunneling process, giving rise to the 410 mμ emission, is believed to be essentially temperature independent. The process giving rise to the 298 mμ emission during the low temperature glow peak, however, has a thermally activated step as indicated by Figs. 5, 6, and 7.

The ratio $\left(\frac{I_{410}}{I_{298}}\right)$ observed in the low temperature glow peak is a constant at a given temperature independent of the concentration of Tl^+ of the length of time the crystal is kept at the given temperature. It is, therefore, believed that the 298 mμ emission observed during the first glow peak arises again from a tunneling process, but in this case a thermal activation is required. The process involves annihilation of one of the $6s^2$ electrons of Tl^0 and the hole of Cl_2^-, and results in a Tl^+ in an excited state (6s6p, $^3P_1$) from which 298 mμ emission occurs.

In the case of the high temperature glow peak, the luminescence process involves the capture of an electron from the conduction band by a Tl^{++}. The fact that in this case the emission appears to be identical to the 298 mμ emission characteristic of the thallous ion suggests that the emitting state is the $^3P_1$ state of Tl^+.

REFERENCES

1. *R. Hilsch, Z. Physik 44, 860 (1927).*
2. *F. Seitz, J. Chem. Phys. 6, 150 (1938).*
3. *F.E. Williams, J. Chem. Phys. 19, 457 (1951).*
4. *R.S. Knox and D.L. Dexter, Phys. Rev. 104, 1245 (1956).*
5. *R.S. Knox, Phys. Rev. 115, 1095 (1959).*
6. *R. Edgerton and K. Teegarden, Phys. Rev. 129, 169 (1963).*
7. *A. Fukuda, Science of Light 13, 64 (1964). This reference contains a very extensive bibliography of work done on alkali halide crystals containing Tl+ impurity.*
8. *V. Bunger and W. Flechsig, Z. Physik 69, 637 (1931).*
9. *P.D. Johnson and F.E. Williams, J. Chem. Phys. 21, 125 (1953).*
10. *H.N. Hersh, J. Chem. Phys. 31, 909 (1959).*
11. *R.D. Murray and F.J. Keller, Phys. Rev. 137, A942 (1965).*
12. *W.B. Hadley and R.G. Kaufman, J. Chem. Phys. 44, 1311 (1966).*
13. *C.J. Delbecq, B. Smaller and P.H. Yuster, Color Center Symposium at Argonne National Laboratory (1956).*
14. *C.J. Delbecq, B. Smaller and P.H. Yuster, Phys. Rev. 111. 1235 (1958).*
15. *A preliminary report on this work was given--A.K. Ghosh, C.J. Delbecq, and P. H. Yuster, Bull. Am. Phys. Soc. II, 7, No.9*

(December 1962).

16. *C.J. Delbecq, W. Hayes, M.C.M. O'Brien and P.H. Yuster, Proc. Roy. Soc. 271, 243 (1963).*

17. *C.J. Delbecq, A.K. Ghosh and P. Yuster, Bull. Am. Phys. Soc. II, 9, No. 6 (1964).*

18. *C.J. Delbecq, B. Smaller, and P.H. Yuster, Phys. Rev. 121, 1043 (1961).*

19. *Atomic Energy Levels, National Bureau of Standards, Circular 467 (1958).*

This paper is based on work performed under the auspices of the U.S. Atomic Energy Commission.

THE NATURE OF LUMINESCENCE INDUCED BY PLASTIC DEFORMATION AND LIGHT ILLUMINATION OF γ-RAYED KCl CRYSTALS

A. S. Krugloff, Physics Institute, Obninsk, USSR

I. A. El-Shanshoury, Atomic Energy Est., Cairo, UAR

Luminescence of γ-rayed KCl crystals induced by deformation and illumination with light in the F and M absorption bands was studied. The intensity of light-induced luminescence was found to increase after loading the crystals above the elastic limit. Results confirming the dichroism of color centers responsible for light-induced luminescence were obtained. The crystals bleached with light in the M band recover their ability to emit luminescence after a period of time. The nature of both deformation and light-induced luminescence is discussed. The interaction between color centers and edge dislocations was proposed to be a general factor affecting the luminescence process. The presence of V-type centers within the dislocation core is considered.

INTRODUCTION

The instantaneous spikes of light during deformation of irradiated KCl and NaCl crystals were observed by Ueta et al.(1) The luminescence was supposed to be a result of excitation of F centers by heat released during the movement of dislocations. A more detailed study of luminescence (2) initiated by deformation showed that the process of light emission is rather complex and can hardly be explained by the mechanism suggested by Ueta et al. The interaction of color centers with edge dislocations was assumed to be responsible for the process of deformation induced luminescence.

A further development of the methods used to study the process of luminescence made it possible to observe a new phenomenon named light-induced luminescence. It has been found that an irradiated crystal emits light in the visible region of the spectrum under the effect of illumination with light of certain spectra. Light in the range of the M or F absorption bands of KCl was discovered to induce luminescence in irradiated crystals, as well as in those deformed after irradiation.

The present work is considered to be a summary of data concerning the properties of deformation and light-induced luminescence. The conclusions are supposed to be preliminary and require further study of the phenomenon.

EXPERIMENTAL PROCEDURE

The KCl crystals of known impurities (Table 1) were irradiated by a Co^{60} γ-source at a room temperature of 30°C at a dose rate of 4.23×10^5 r/hr.

Element	Mg	Pb	Al	Cu	Ca	Sr	Ba
wt.%	3×10^{-4}	1×10^{-4}	1×10^{-3}	1.5×10^{-4}	1×10^{-3}	Nil	Nil

Luminescence induced by deformation and light illumination was recorded with the use of a system consisting of a photomultiplier, cathode follower and chart recorder. The photomultiplier ФЭУ-36 type was operating at a cathode voltage not less than 1600 volts. In addition to the facilities employed for studying deformation induced luminescence (2), a system consisting of an illuminator and filters was designed to record light-induced luminescence. Shielding filters absorbing light used for the illumination were placed in front of the window of the photomultiplier.

Two methods of crystal illumination were employed to study the properties of color centers responsible for the luminescence phenomenon. The first method consists of continuous illumination of a crystal, while the second one depends on the application of light pulses of durations ranging from 1/50 up to 1 sec. The basic difference between the two methods is that as a result of continuous illumination a number of centers responsible for luminescence vanishes due to bleaching, whereas in the case of pulse illumination only a negligible part of these centers is believed to be bleached.

The system finally employed makes it possible to study the simultaneous effect of deformation and light illumination upon the intensity and development of crystal luminescence.

RESULTS

Properties of Luminescence

Deformation induced luminescence. The luminescence of a crystal during deformation, as a function of load, qualitatively follows the stress - strain curve with the intensity of luminescense corresponding to strain. Fig. 1 shows a typical curve obtained during deformation of an irradiated crystal at a loading rate of 0.007 kg/mm^2 sec. The curve may be subdivided into four characteristic parts. The first part represents the course of crystal luminescence on loading below the elastic limit. The second part is characterized by a sharp increase in the luminescence intensity caused by the initial and further development of plastic deformation. A point of transition from the region of the curve, which is believed to correspond to elastic deformation, to that of a rapid increase in luminescence intensity was taken as the "Luminescence yield stress" (2). The third part of the curve corresponds to luminescence of the crystal during slip band formation. It is characterized by sharp light spikes of a rather short duration. The fourth section of the curve shows the behavior of luminescence after the end of loading.

Pre-irradiation treatment of crystals throws light on the feature of the luminescence - load curve. Figs. 1 and 2 show the luminescence - load curves for the crystals, differently treated before irradiation. One crystal was deformed up to 0.7 kg/mm^2 (Fig. 1), while the second (Fig.2) was deformed up to the same stress and then annealed at 400°C for 0.5 hr. As is seen from the curves, the luminescent intensity of the first crystal is much higher than that of the second. In the annealed crystal a regular slip band formation starts right after the luminescence yield stress is achieved. However, the amplitudes of light spikes are lower than those of the first crystal. Although both crystals were loaded at the same stress, the amount of light emitted (area under the curve) is higher for the first crystal deformed before irradiation.

It may be reasonable to assume that the deformed but not annealed crystal would emit more light because of a higher number of color centers formed during irradiation (3, 4, 5). However, the experience shows (2) that the increase in F- center density, approximately by three times, shows practically no effect on the amount of light emitted and the shape of the luminescence-load curve. In this case, the crystal which was deformed but not annealed before irradiation, shows a density of F centers (after irradiation) about 70% higher than that of the annealed crystal.

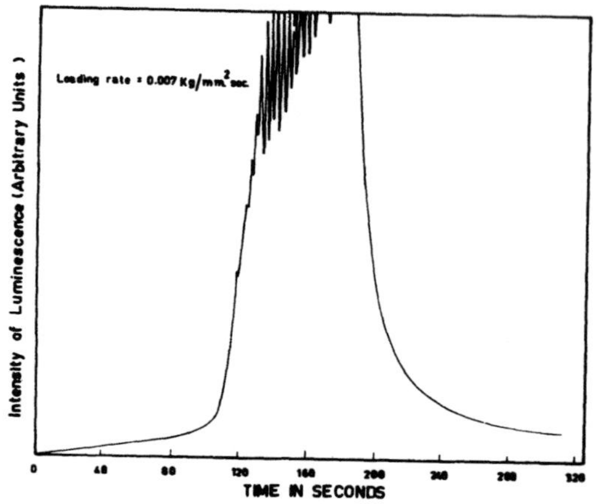

Fig. 1. The luminescence-load curve for a crystal deformed before irradiation at a stress of 0.7 kg/mm^2.

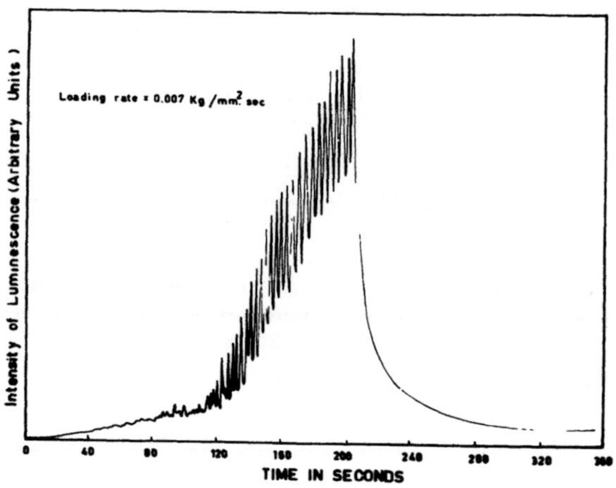

Fig. 2. The luminescence-load curve for a crystal stressed up to 0.7 kg/mm^2 and annealed at 400°C for 0.5 hour.

An increasing ability of the crystal to emit luminescence under the effect of red light may be easily observed in the experiment on crystal deformation with pulse illumination. Fig. 3 shows the luminescence-load curve obtained through a blue shielding filter. The sharp peaks, marked with letter P, are the emission of light by the crystal under and after 1/25 sec red light pulse illumination. The amplitude of each peak characterizes the crystal ability to emit light at the corresponding moment under the effect of red light illumination. Figure 4 shows the course of this ability during deformation and after the end of loading. It should be noted that this ability begins increasing only after the luminescence-yield stress is overcome, i.e. when plastic deformation starts.

The above observation suggests that on deformation of an irradiated crystal the emission of light is not a simple process. During deformation, at least two processes take place simultaneously--emission of light due to deformation, and accumulation of some product which is able to emit light under red light excitation. Both processes are to some extent interrelated. From Fig. 4 it is seen that accumulation of a red light sensitive product in the crystal is proportional to the stress applied.

Luminescence induced by M and F band light. According to the transmission spectrum of the filter, red light used in experiments is found to cover the M absorption band of KCl crystal, as well as some R bands. This light is assumed to excite mostly the centers characterized with the M absorption band and called " M band light ".

The ability of irradiated crystals to emit light under the effect of M and F band light illumination seems to be concerned with the deformation suffered by the crystals. Although irradiated crystals show the ability to emit light under illumination, the intensity of light-induced luminescence sharply increases after deformation of irradiated crystals.

Luminescence induced by M band light is experimentally found to be more powerful than that initiated by F band light of the same intensity. A special experiment was conducted to compare the ability of crystals to emit light under the effect of F and M band light illumination. The spikes of luminescence induced by F and M band light pulses, were recorded respectively after different treatments of the crystal. This included bleaching of irradiated crystals by F and M band light, either before or after load appli-

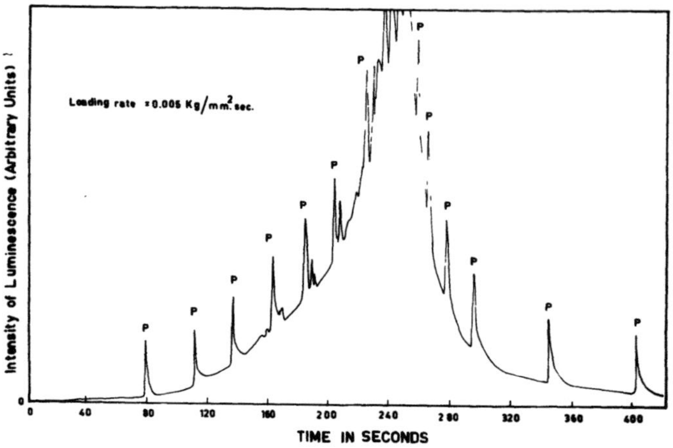

Fig. 3. The luminescence-load curve and luminescence spikes (letter P) caused by pulses of M-band light during crystal deformation.

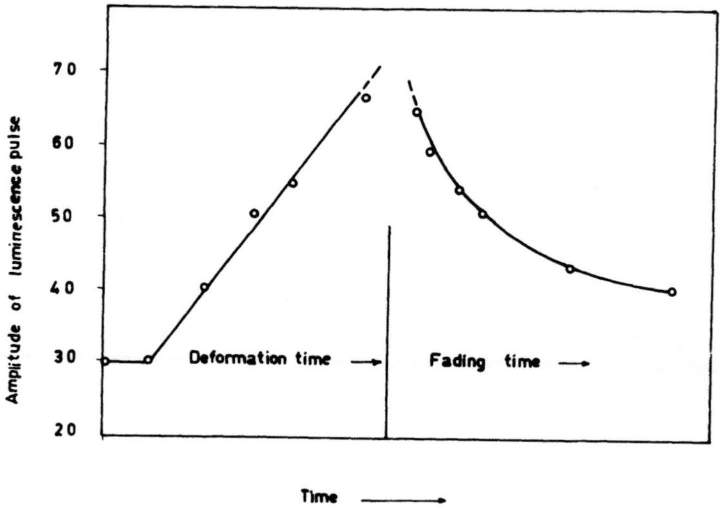

Fig. 4. The curve of the luminescent peak intensities induced by a 1/10 sec. M-band light pulse applied during crystal deformation.

cation. The results are given in Fig. 5. The curves marked with "F" correspond to luminescence spikes induced by a 1/10 sec. F band light pulse, whereas the curves marked with "M" are those induced by the M band light pulse of the same duration. The intensity of illuminating F band light was 1.5 times higher than that of the M band.

Curves F_1 and M_1 are recorded for the untreated irradiated crystal, while F_2 and M_2 after bleaching of the same crystal by F band light for 3 min. Curves F_3 and M_3 are obtained after bleaching by M band light for 1.5 min. After that the crystal was deformed at a stress of 1.35 kg/mm^2 and curves F_4 and M_4 were recorded. The procedure of bleaching was, then, repeated. Curves F_5 and M_5 show the effect of the second bleaching by F-band light for 3 min., whereas curves F_6 and M_6 correspond to the M band light bleaching which lasted for 2 min. As is seen from the curves, the bleaching by F band light depresses the ability of the crystal to glow under F and M band light illumination. A continued bleaching of the crystal with M band light practically deprives the crystal of the ability to emit light. A conclusion of dichroism based on the results of this experiment may prove to be very important in the understanding of the nature of light-induced luminescence as a whole. One may assume that irradiated crystals contain the centers with two electron energy levels coinciding with the F and M absorption bands. Moreover, these centers are responsible for luminescence induced by applying light of these two bands. The excitation efficiency of M band light is higher than that of the F band light.

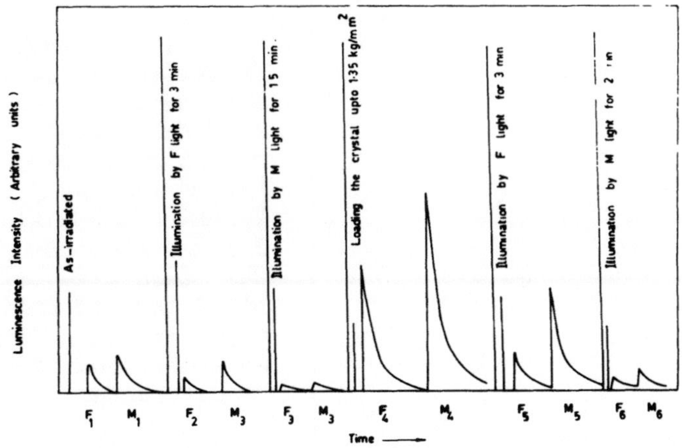

Fig. 5. The luminescence initiated by pulse illumination with light in the F (marked F) and M (marked M) bands depending on the crystal treatment after irradiation.

One of the characteristics of luminescence is its fading at the end of loading. In the case of deformation induced luminescence, the shape of the luminescence decay curve characterizes stability of the centers responsible for the emission of light. In the process of crystal luminescence induced by light, the rate of decay in luminescence intensity characterizes quantitatively and qualitatively the process associated with light excitation.

Fig. 6 shows the luminescence decay of crystals after loading up to 1.2 kg/mm^2 and 2.4 kg/mm^2 (curves 1 and 3, respectively). Curves 2 and 4 show the course of luminescence under the effect of M band light application after the same deformation as for curves 1 and 3. It is seen that a relative rate of luminescence decay after the end of loading increases with the stress at which the crystal has been deformed. Furthermore, the increase in the rate of deformation results in slowing down of the decay process of luminescence induced by light applied after the deformation. The luminescence decay curve, both for the deformation and M band light application at the initial moment of decay is well described by the following equation:

$$\frac{Io}{I} = 1 + Bt$$

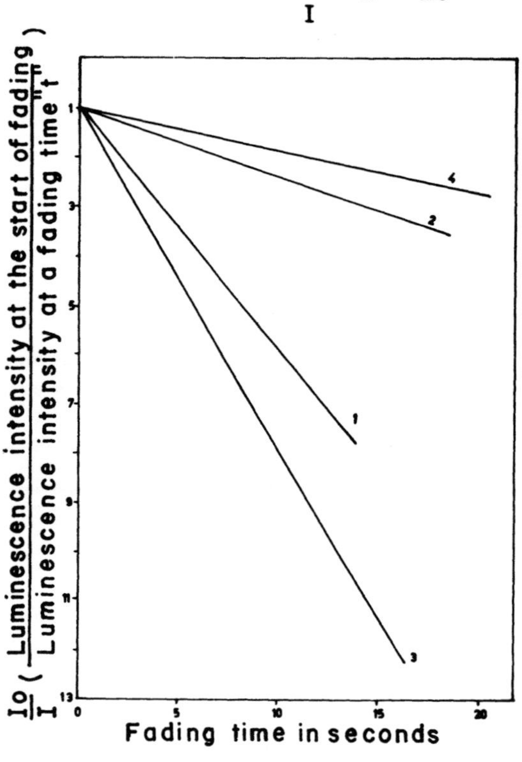

Fig. 6. The change in the intensity of deformation-induced luminescence after loading (decay) the crystal up to a stress of 1.2 kg/mm^2 (curve 1), and 2.4 kg/mm^2 (curve 3). Curves 2 and 4 show the change in the intensity of M-band light induced luminescence after crystal deformation up to 1.2 and 2.4 kg/mm^2, respectively.

where I = luminescence intensity at time t,
Io = luminescence intensity at the initial moment of decay,
B = constant, and t = fading time.

At longer fading times, the shape of the curve does not obey this relation and follows more closely the exponential law. When the crystal is illuminated by F band light the decay of luminescent intensity begins to follow the exponential law much earlier than at M band light illumination.

Recovery process. The luminescence which has faded completely after the deformation of a crystal and its subsequent illumination was found to appear after a certain period of time. This phenomenon is defined as a recovery effect. The rate of recovery depends on the treatment of the crystal. It has been found that the recovery is faster for irradiated crystals that are deformed at high stresses. The recovery of light induced luminescence may be observed provided that the irradiated crystal is completely bleached by light. Fig. 7 shows the recovery of light induced luminescence after the crystal has been deformed at a stress of 1.2 kg/mm^2 and bleached, by M band light till complete luminescence fading. The process of recovery proceeds at a higher rate during the first 40 min, and then continues more slowly. Fig. 8 shows the curves of the M band light induced luminescence of an irradiated crystal: (1) before deformation, (2) after deformation and (3) 80 minutes after complete fading of luminescence caused by

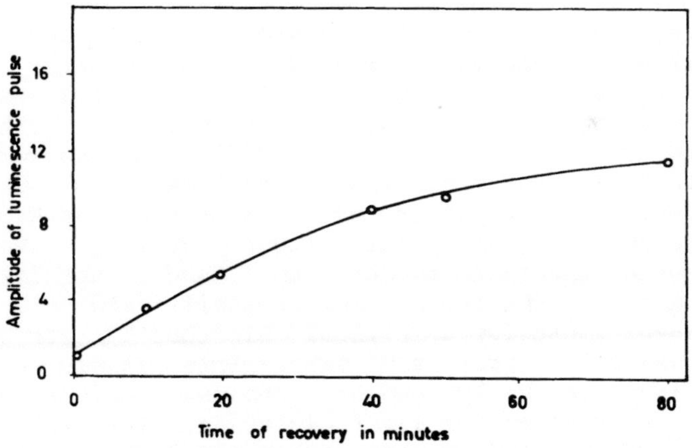

Fig. 7. The recovery of crystal ability to emit light-induced luminescence. The crystal was deformed and bleached with light in the M band.

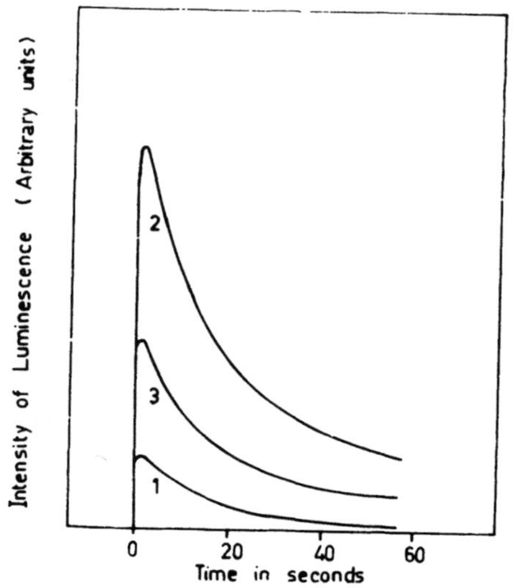

Fig. 8. The curves of luminescence intensity obtained with M-band light applied to crystals. Curve 1 for a non-deformed irradiated crystal. Curve 2 for a crystal deformed after irradiation. Curve 3 obtained 80 minutes after the end of M-band light bleaching of the deformed crystal.

a previous M band illumination under the effect of red light excitation. Fig. 9 shows the luminescence curves of an irradiated crystal excited by M band light. Curves 1 corresponds to luminescence induced by the primary application of light, and curve 2 shows the luminescence of the same crystal, an hour later in the dark. In the case of a non-deformed crystal, the recovery of light induced luminescence goes much faster and the number of centers is almost completely restored. Unfortunately, all the experiments were carried out at a room temperature of 30°C and temperature development of the recovery process was not possible to examine. Nevertheless, some data on the rate of color center excitation in a crystal are obtained by bleaching the crystals with light of a high intensity. Fig. 10 (curves 1 and 2) shows the remaining luminescence in 30 seconds after the end of bleaching. Curves 1' and 2' correspond to the amplitude of a luminescence spike caused by a 1/10 sec pulse of M band light (recovery curves). For a crystal bleached with sunlight through an F band transmission filter (curves 1 and 1'), the luminescent intensity and rate of recovery is less than those for a crystal bleached by direct sunlight (curves 2 and 2'). From many experiments it can be summarized that the higher the background of a residual luminescence induced either by deformation or light bleaching, the faster is the recovery rate of the centers responsible for light induced luminescence.

Spectrum of luminescence. It was not possible in this work to determine the exact spectrum of luminescence, although with the aid of band filters some peculiarities of the spectrum distribu-

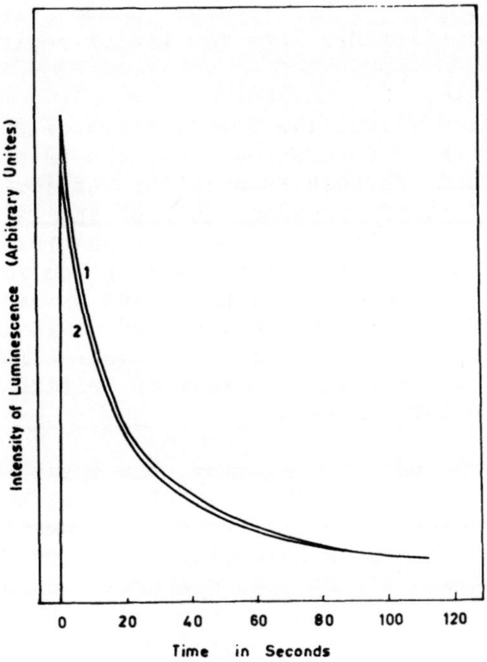

Fig. 9. The recovery effect for the non-deformed irradiated crystal. The luminescent intensity induced by light in the M-band at the first illumination (curve 1), and that at the second illumination one hour later (curve 2).

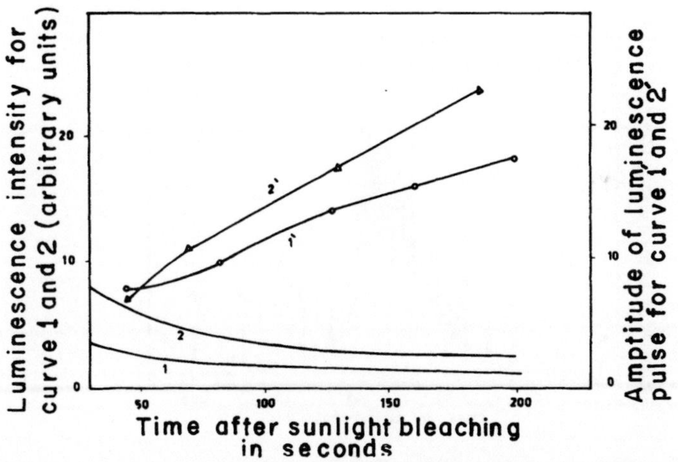

Fig. 10. The growth of light-induced luminescence ability as a function of residual luminescence intensity observed after sunlight exposure. Curve 1 shows the residual luminescence intensity after a 30 sec. sunlight exposure through the F band transmission filter. Curve 2 shows the residual intensity of the crystal after a 30 sec. exposure in direct sunlight. Curves 1' and 2' show the change in M-band light induced luminescence ability of a crystal for the cases 1 and 2.

tion were qualitatively studied. The spectrum of deformation induced luminescence is monochromatic, for it is either distributed, at least in the visible range, or has several bands. The intense part of luminescent light stretches into the violet region of the spectrum.

With the use of a filter which allows the transmission of the spectral region corresponding to the F absorption band, the intensity of luminescent light recorded through this filter was found to be extremely low as compared to that recorded through the filter transmitting the remaining part of the spectrum and absorbing light in the F band. This may suggest that luminescence during deformation contains a negligible amount of light with a wavelength corresponding to the crystal's F absorption band. It is likely that the process of electron re-capture by the negative ion-vacancies is negligibly small as compared to that of electron interaction with other defects in the crystal.

A remarkable result is obtained by comparing the spectrum characteristic of deformation induced luminescence with that induced by the M band light excitation. Fig. 11 shows the luminescence during deformation recorded through a shielding filter absorbing light in the F band, but transmitting the spectral region

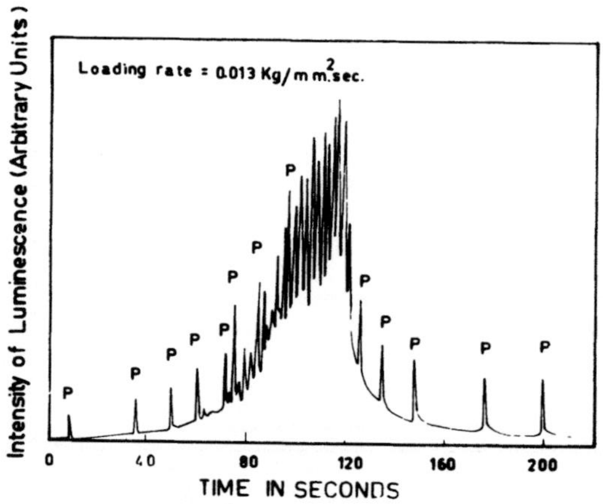

Fig. 11. The luminescence-load curve recorded with the shielding filter before the window of the photomultiplier. The filter transmits only the violet range of the spectrum. Letter P marks the luminescence spikes induced by pulse illumination of a crystal with light in the M band at the moment of loading.

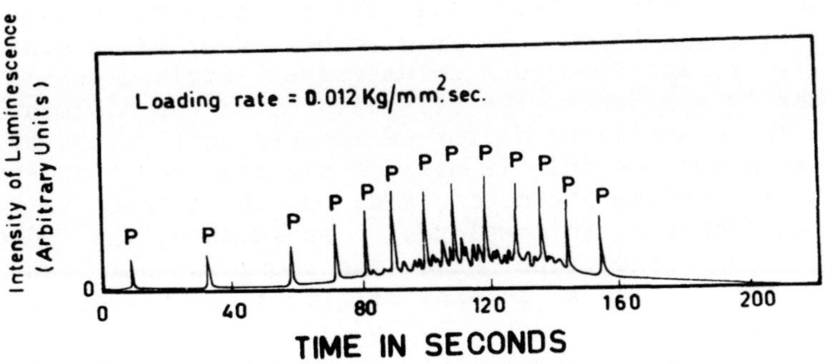

Fig. 12. The luminescence-load curve recorded with the shielding filter before the window of the photomultiplier. The filter transmits only the light corresponding to the F absorption band. Letter P marks the luminescence spikes induced by pulse illumination of a crystal with light in the M band at the moment of loading.

stretching to the violet range. The crystal was at the same time illuminated by 1/10 sec M band light pulses. The peaks in the curve (mark "p") correspond to the spikes of light-pulse induced luminescence. Fig. 12 shows the luminescence of an analogous crystal under the same conditions, when the filter was substituted by one transmitting the F band light only. The difference in the spectrum of deformation and light induced luminescence is evident from the two curves. Since the curve (Fig. 12) was recorded through the F band transmission filter, one may conclude that unlike the deformation induced luminescence, that caused by M band illumination, contains a substantial amount of light with a wavelength corresponding to the F absorption band. Thus the processes of deformation and M band light induced luminescence are somewhat different by their nature.

DISCUSSION AND CONCLUSIONS

The experimental results suggest that luminescence appearing as a result of a plastic flow of irradiated crystals and the excitation by light with a definite wavelength depends on the number of dislocations in the crystal. This suggestion may be logical, since edge dislocations in irradiated crystals are surrounded by regions with high concentration of color centers, most likely, F centers (3, 4, 5). In the foregoing work (2), the results obtained support this fact and moreover, indicate that the F centers located in the vicinity of dislocations are referred to a hard-bleached type.

One can say that the dislocation and color centers surrounding it stabilize each other. This should cause the hardening of crystal after irradiation. It is also believed that a dislocation, or its parts, escape from such color center regions when a sufficient stress is applied to the crystal. After the dislocation has moved out, the stabilizing factor of its presence stops affecting the centers in the region. It may, on the one hand, lead to a partial excitation of the F centers, and, on the other, to coagulation of these centers in complexes. Presumably, the release of electrons to the conduction band and their further capture by hole-type centers leads to the origin of luminescence during deformation.

The rather long time of luminescence after the end of deformation indicates that the process of electron release takes place not only during the application of excitation energy to the crystal through the stress, but also after it. Such post-luminescence may be conveniently considered together with that induced by light excitation, though light-induced luminescence is observed also for nondeformed irradiated crystals. In the case of light-induced luminescence the number of dislocations present in the crystal before irradiation is found to affect the intensity of glow. This may lead to the assumption that the centers involved in the process of light-induced luminescence are also associated with edge dislocation.

The authors suggest that the process of luminescence should be considered on the basis of an assumption that there is a great deal of interaction between color centers responsible for this luminescence and edge dislocations. Moreover, the assumption is that the edge dislocation in irradiated crystals is a peculiar place of the lattice containing both electron and hole-type color centers of high concentration.

A fair amount of experiments has been carried out to prove the existence of two types of F centers (3, 4, 6, 7, 8). However, few are concerned with the study of the properties of V-type centers. A well known fact is that the V_3 centers are extremely stable. When a colored crystal is bleached with light in the F band, the electrons released in this process are trapped by cluster-type vacancies, resulting in the formation of such centers as M, R and N, rather than combining with positive holes of V_3 centers. The authors assume that the stability of V-type centers is caused to a great extent by their position in the lattice, rather than by forces of electrostatic interaction with electrons. Experiments on deformation-induced luminescence may, to some extent, be interpreted in favor of such a model. During deformation of a crystal containing no easy-bleached F centers the luminescent intensity does not, in fact, differ from that of the crystal which has suffered no preliminary bleaching. This indicates that mostly hard-to-bleach F centers participate in the process of luminescence. Furthermore, the deformation seems to result in the activation of these centers and contributes to the appearance of such conditions under which the electrons can readily be trapped by V centers. This is believed to cause a crystal glow with a wavelength range in the violet region of the spectrum.

Let us consider the process of luminescence as a whole. As is assumed, the stability of some zones with high concentration of color centers decreases with plastic deformation. This leads probably to the interaction of the F and M center electrons with the holes of V centers in these zones. The process of center coagulation may also take place at the same time. Before deformation, the number of centers capable of participating in the process of light induced luminescence was controlled by the stabilizing action of dislocations. Because the dislocations escape from the zones during deformation, a larger number of centers which are able to be ionized upon light application appears. The number of such centers gradually decreases with time after the end of loading. The remaining centers show a great deal of stability and can be partially recovered after light bleaching. It is early to speculate about the mechanism of recovery process until more experimental data are available.

The study of deformation and light-induced luminescence may, to a great extent, be helpful in investigations of the nature of dislocation interaction with color centers. The effect of impurities on luminescence remains unclear.

The foregoing conclusions should be considered preliminary and disputable. Further experimental support and comparison with data on the properties of color centers is needed.

ACKNOWLEDGMENT

The authors wish to thank Anatoly Ouzoykin for his help in translating the manuscript.

REFERENCES

1. M. Ueta, M. Kojima, M. Hirai, J. Phys. Soc. Japan 17, 1668 (1962).

2. A. S. Krugloff, I. A. El-Shansoury and M. K. Matta, to be published in J. Phys. Soc. Japan.

3. F. Seitz, Rev. Modern Phys. 26, 7 (1954).

4. R. B. Gordon and A. S. Nowick, Phys. Rev. 101, 977 (1956).

5. A. S. Nowick, Phys. Rev. 111, 16 (1958).

6. J. J. Oberly, Phys. Rev. 84, 1257 (1951).

7. H. Rabin, Phys. Rev. 116, 1381 (1959).

8. S. C. Jain and P. C. Mehendru, Phys. Rev. 140, A957 (1965).

DISCUSSIONS

C. DELBECQ (Argonne National Laboratory, Argonne, Illinois, U.S.A.): *With the red light excitation are you exciting electron-excess centers or hole centers?*

A. S. KRUGLOFF: *We have used F-band and M-band light with the purpose to induce the luminescence. So, we believe that the centers responsible for luminescence are electron type centers.*

G.C. KUCZYNSKI (University of Notre Dame, Indiana, U.S.A.). *Do you believe that the recovery of luminescence may be due to the reformation of the color centers around dislocations?*

A.S. KRUGLOFF: *As is shown in the paper, the rate of recovery depends a great deal on the level of residual luminescence induced by sunlight bleaching. This may indicate that the recovery process is to some extent related to electron - vacancy interaction. However, it is early to make any speculations about the mechanism of recovery until additional kinetic results are obtained.*

CHARLES GREENE (Alfred University, New York, U.S.A.): *Are there any irregularities in the mechanical yielding of the crystals under steadily increasing stress which perhaps correspond to the spikes in light output?*

A.S. KRUGLOFF: *We did not measure the strain during deformation. The process of spike formation is described in detail in a previous paper (See reference 2).*

R.A. WEEKS (Oak Ridge National Laboratory, Tennesse, U.S.A.): *Could the pulses be due to release of dislocations at pinning points?*

A.S. KRUGLOFF: *The fact that the number of luminescent spikes is not high, and the luminescence at the moment when the spike appears is quite powerful, supports the idea of luminescence appearing at the slip-band formation rather than that proposed in the question. It would be more reasonable to say that luminescent spikes can occur at the moment of formation of new pinning points when moving fresh dislocations are crossing the color zones of old ones.*

J.E. WERTZ (University of Minnesota, Minneapolis, U.S.A.): *Will there not be a tendency of color centers stabilized at dislocations to follow the latter as they are caused by stresses to move?*

A.S. KRUGLOFF: *The velocity of dislocations during motion is known to be high, especially at the stresses when slip band formation occurs. The suggestion that the dislocation line escapes from color zones seems, therefore, to be reasonable.*

A. GUINIER (University of Paris, FRANCE): *Could you obtain images of the crystal showing the repartition of the sources of luminescent light inside the crystal?*

A.S. KRUGLOFF: *We have found that the crystal bleached up to the complete vanishing of easy-bleached F centers is able to emit the deformation induced luminescence of intensity just a little lower than that for the one without bleaching. Since we have found that hard-bleached color centers are mostly present around edge dislocations (see reference 2) one can say that centers responsible for stress - induced luminescence are located near dislocations.*

C. BETTINALI (New Instituto Chimico, Universita, Roma, Italy) *Were the crystals annealed under vacuum?*

A.S. KRUGLOFF: *All crystals were annealed in* N_2 *atmosphere.*

A. CHARLESBY (Royal Military College of Science, Shrivenham, ENGLAND): *We have observed a series of sharp peaks in light emision from a crystal of trioxane, irradiated at liquid nitrogen temperature, and allowed to warm up. This is related to polymerisation and not to the presence of an ionic lattice. It may, however, be linked to the formation and release of stresses within the crystal.*

INVESTIGATION OF SILVER CHLORIDE CRYSTALS

FOR DETECTION OF CHARGED PARTICLES

G. Haase and E. Schopper

University of Frankfurt

GERMANY

As is known from the investigations by Childs, it is possible, under certain conditions, to observe microscopically the tracks of charged heavy particles in silverchloride single crystals by decoration following the method of Schockley.

Observing and recording of particle-tracks in silverchloride crystals may be of particular interest in the field of radiation and nuclear physics. Also, the investigation of track-formation by particles in those crystals (combined with electronmicroscopic methods), provides information on the mechanism of the interaction of the particles with the crystal lattice.

The technique of track-decoration is described. A special technique of cleaving permits investigation of the structural properties of the crystals by electronmicroscopic methods, combined with the treatment of etching or decoration, and studies of the effects of radiation on a microscopic scale.

INTRODUCTION

In 1960 Childs and Slifkin (1,2) showed that lattice dislocations as well as tracks of charged particles in single crystals of silverchloride could be observed microscopically throughout the interior of relatively large crystals by a method of decoration combining the techniques of Hedges and Mitchell (3) and of Haynes and Schockley (4). A short report was given by Childs and Slifkin (5) in 1965 summarizing their latest results.

The study of particle tracks in silverchloride crystals may

also be of interest in some fields other than that of radiation detection; as an example we may mention the Cl (n,α) reaction for monitoring fast neutrons; the investigation of track formation by charged particles in those crystals should also provide information on the mechanism of the interaction of the particles with the crystal lattice on a microscopic scale.

Examples of tracks in silverchloride crystals are given in Figures 1 and 2.

These promising results encouraged several groups to employ and to investigate this technique. It was found, however, that the results were not clearly reproducible and were dependent on a number of conditions which have still to be cleared. The results of Childs, on the other hand, have shown that the technique works in principle.

One may ask for the reasons of the special interest in AgCl crystals, since the results by Walker et al (8) with particle recording using other solid systems as mica, glasses and plastics were quite successful. The interest in silver halides is based on their unique properties of combining the behavior of semiconductors, namely the transportation of electrons and holes, with properties resulting from their ionic structure, and the high mobility of silverions and holes (at room temperature) which helps the amplification of the primary effects of the particles. This is the real reason, too, for the unique role of silverhalides in photographic and ionographic systems of high sensitivity.

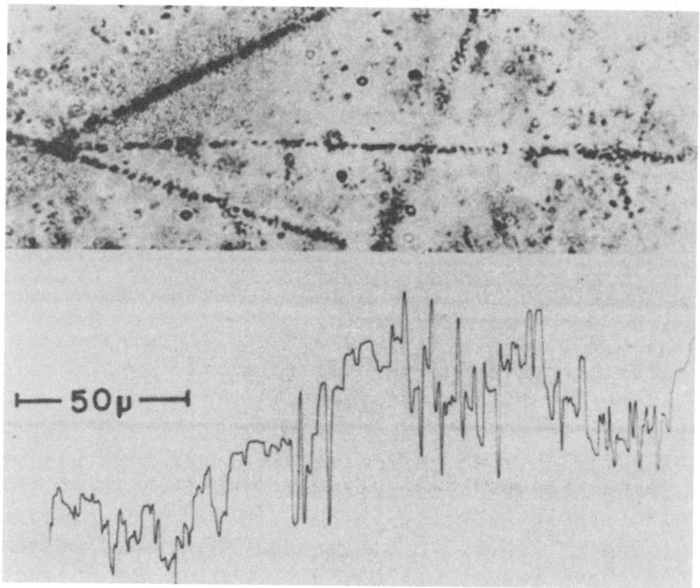

Fig. 1. Tracks of a nuclear disintegration and microdensitograph of the horizontal prong, showing dependence on energy loss along the path (after Childs (6)).

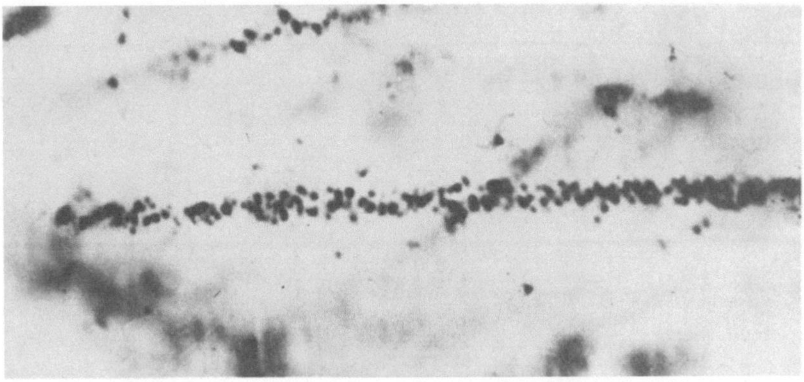

Fig. 2. Part of a track of a heavy ion, showing the general feature of tracks in silverchloride (after Schmitt (7)).

RESULTS AND DISCUSSION

Two factors have to be considered when studying the conditions for decoration of particle tracks in the crystal.

The Procedure of Decoration

Following the suggestions of Childs, an ultraviolet light-pulse of about 5μ sec of halfwidth filtered between 350-400 mμ was used. Conductivity electrons and holes were thus produced in a layer of about ten microns at one surface of the rectangular crystal, usually 1 by 1 cm^2 and of a thickness of 5-10 mm. The electrons are swept through the crystal by a pulsed electric field of usually 4 kV/cm; and may be trapped at disturbances of the crystal, e.g. along the path of a particle, hence leading to a migration of silverions into the track. The device, used by the authors is shown in Figure 3. The crystal is put between two insulating foils of milar or thin quartz plates into the field of the condensor. A stroboscopic device (AEG, Germany) with a Xenon-lamp (Osram XE 15 Watt) triggers with a repetition rate variable between 30-1000/sec the shut-down of the external electric field within about 1μ sec, and the lamp with a variable time-shift from 1-20μ sec, monitored by a multiplier. On account of the high ionic dark-conductivity of the crystal one generally uses the internal space charge field for sweeping the electrons through the crystal. This internal field is built up under the influence of the external field, and decays after shut-down of the external voltage with a relaxation time of about 100μ sec, depending on the conductivity. This important parameter was checked by a device, shown in Figure 4.

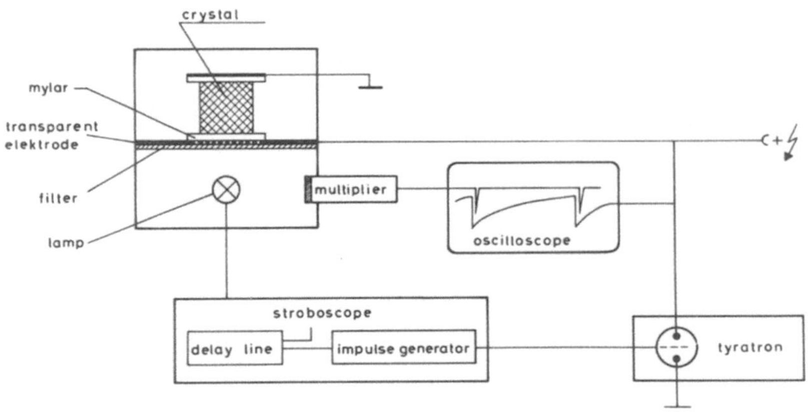

Fig. 3. Device for decoration of particle tracks in AgCl-crystals.

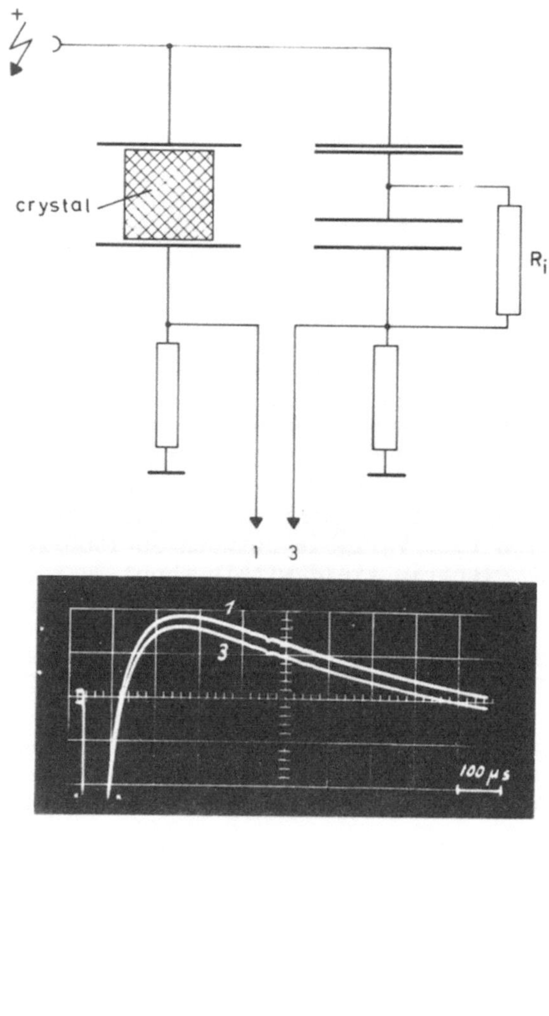

Fig. 4. Device for measuring the ionic conductivity of AgCl - crystals.

Simulating the crystal by a variable capacitor-resistor circuit, it was possible to adjust the circuit in Figure 4 to obtain equal voltages at points 1 and 3. This method permits the determination as well as the relaxation of the space charge field (Fig. 5).

Structural and Chemical Properties of the Crystals

These properties still require a number of fundamental investigations. However, some preliminary results may be given in this report.

It is reasonable to expect an influence of the structure of the crystal on the formation and decorability of tracks, as already mentioned by Childs. As an example, two specimens of crystals (prepared by Harshaw Chemical Company, USA), were compared by the authors:

a) a (real) pure crystal; real means as pure as available with some impurities in or below the ppm-region.

b) a Pb-doped crystal with 7 ppm of lead.

Lead was chosen because of its possible influence on the real structure of the crystal and on the formation of the latent image of the tracks. As expected, real structures differently decorated were observed. However, it was not possible to decorate any tracks of α-particles in both types of crystals so long as this was attempted after irradiation.

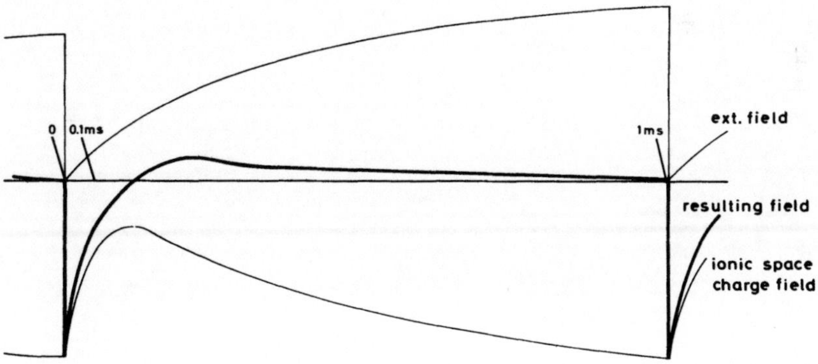

Fig. 5. Time-dependence of the external and of the space charge field in AgCl-crystals.

It was expected, as in the case of former experiments with "good crystals" by Childs, that the latent image would be stable with time. The situation changed, however, when the decoration took place simulataneously with the irradiation of the crystals.

The results, found in July 1966, are shown in Figure 6 for α-particles of ThC of energies of 6 and 8.8 MeV respectively in the Pb-doped crystal. The number of tracks recorded corresponds to the number of α-particles used; the tracks' lengths correspond to the different energies. Observation of the tracks suggests dependence of width upon the energy loss dE/dx by ionization. As also tracks crossing dislocation walls were clearly recorded, we may conclude that, under the conditions of the experiment, dislocations do not seriously influence the formation and decoration of the tracks.

The instability of the radiation effects with time is remarkably stronger in the pure crystal and recognizable at a certain diffuseness of the decorated tracks. Similar effects of fading have been observed by Tuerpe and Knorr (9) with self-decorated fission tracks in AgCl crystals.

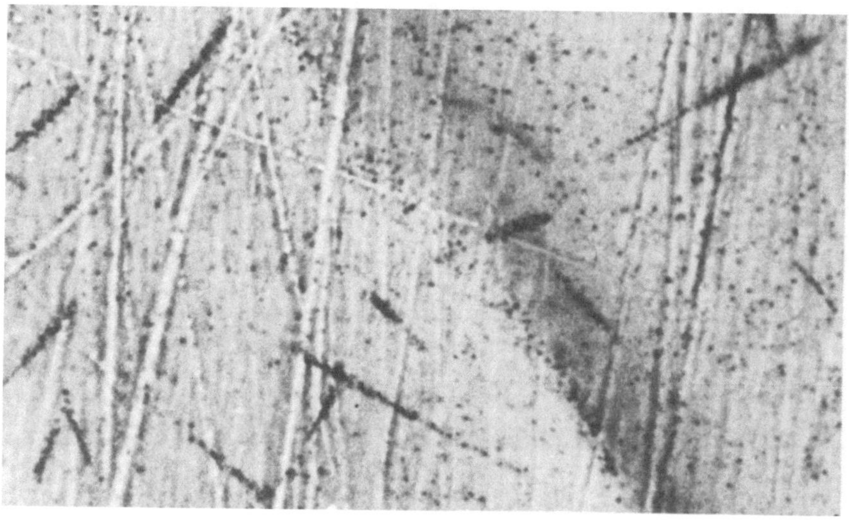

Fig. 6. Microphotograph of α-particle tracks from ThC, recorded in a Pb-doped AgCl-crystal by simultaneous irradiation and decoration.

a) random angle of incidence (M = 800).

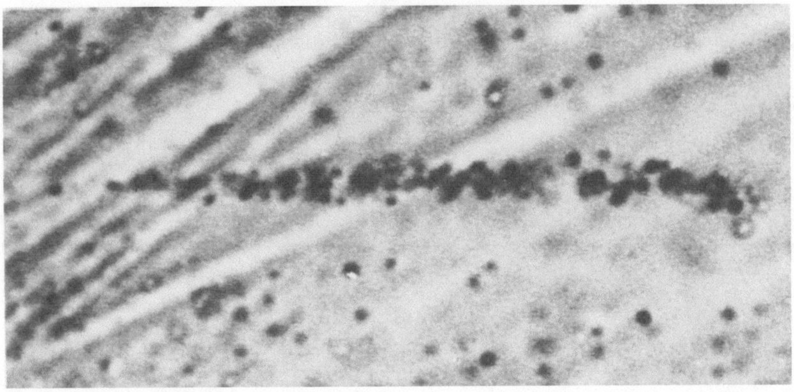

Fig. 6. Microphotograph of α-particle tracks from ThC. recorded in a Pb-doped AgCl-crystal by simultaneous irradiation and decoration.

b) track parallel to the surface at higher magnification (M = 2,100).

When studying radiation damage in general, some lattice effects should be taken into consideration (10). From the point of view of particle-detection the observations of the authors lead to the suggestion that it might be possible to use AgCl-crystals either with stable latent image of charged particles accumulating tracks, or, with highly unstable latent tracks, the recording of which can be triggered by choice. This also involves the question of temperature effects.

The decorated tracks have also proved unstable in those crystals, much less in the Pb-doped than in the pure crystal. Figure 7 gives a comparison of α-particle tracks in the Pb-doped crystal, immediately after and 3 days after decoration; some of the tracks are reduced to diffuse clusters.

Further investigations should be planned to find out the conditions for the stability of latent tracks. This requires extremely pure material, zone refining and doping, and special techniques of growing crystals.

Electronmicroscopic Techniques

Electronmicroscopy is considered a very useful tool for the investigation of the response of the crystals to charged particles This method requires well defined surfaces. A technique of cleaving single crystals of AgCl at liquid hydrogen temperature was

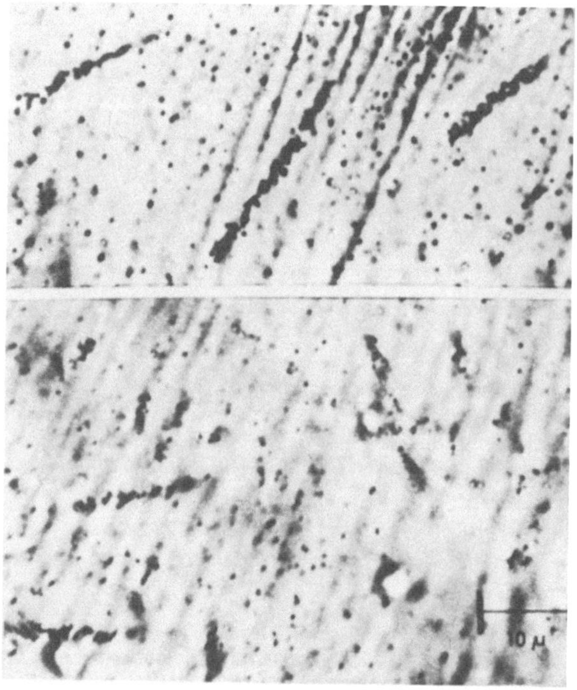

Fig. 7. Fading effects on decorated α-particle tracks in AgCl-crystals.
a) upper part: recent tracks
b) lower part: faded tracks after 3 days following decoration.

studied by G. Haase et al (11), as shown in Figure 8. Preceding experiments on x-irradiated cleaved surfaces of KCl-crystals (before the cleavage technique of AgCl was available) had given encouraging results. As the AgCl-experiments with cleaved surfaces are still in progress, an example of the response of a polished surface of a AgCl-crystal on irradiation with α-particles is given in Figure 9.

The number of elevations visible in Figure 9b corresponds roughly to the density of incoming α-particles; secondary effects of the interaction of the α-particles with the lattice at the surface of the crystal should be taken in consideration.

ACKNOWLEDGMENT

This work was supported by the German "Bundesministerium fur Wissenschaftliche Forschung". (St. Sch. 504).

We appreciate scientific contribution by Dr. F. Granzer, Dr. K. Breuer, W. Saul and F. Zorrgiebel.

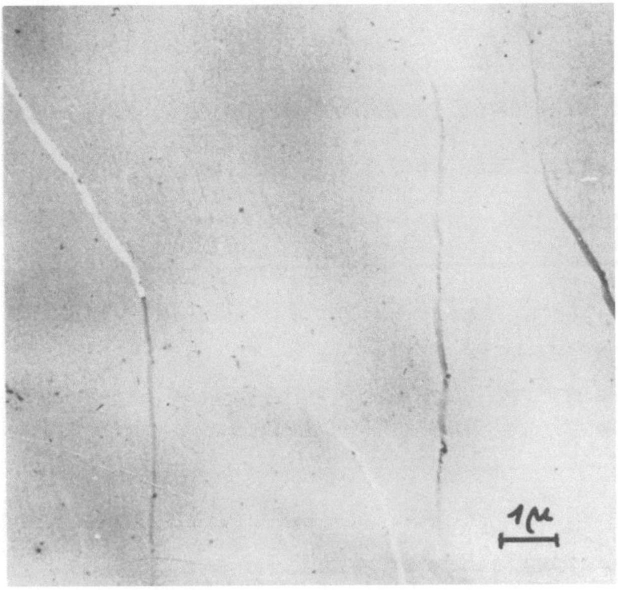

Fig. 8. Carbon replica of a nonetched (100) cleavage-plane of a AgCl-single crystal. (M = 5 000).

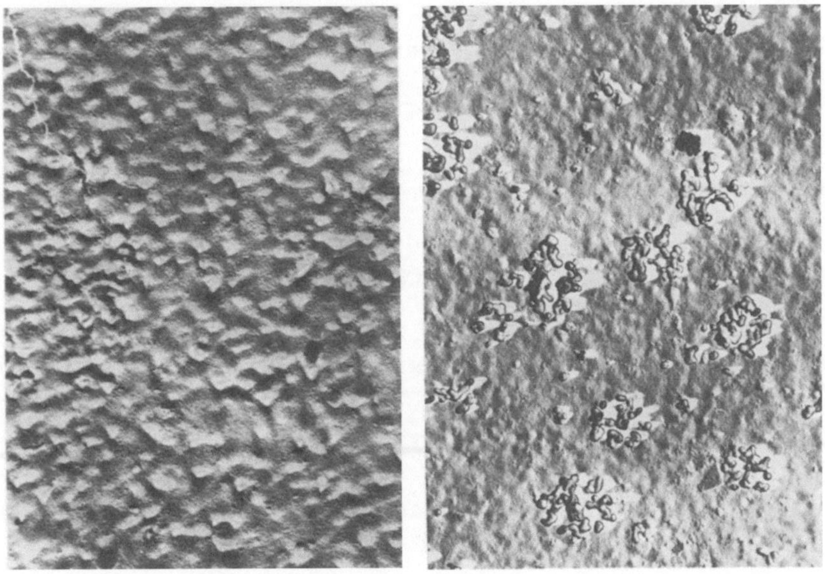

(a) (b)

Fig. 9. Carbon replica of a polished surface of a AgCl-single crystal after α-irradiation. a) nonirradiated surface (M=25000) b) α-irradiated surface (M = 15000).

REFERENCES

1. *C. Childs and L. Slifkin, Phys. Rev. Letters 5, 502, 1960.*

2. *C. Childs and L. Slifkin, Bull. Amer. Phys. Soc. 6, 52, 1961.*

3. *J.M. Hedges and J.W. Mitchell, Phil. Mag. 44, 223, 1953a.*

4. *J.R. Haynes and W. Shockley, Conference on Strength of Solids p. 151, Edited by Physical Society; London 1948.*

5. *C. Childs and L. Slifkin, Brit. J. Appl. Phys. 16, 771-777, 1965.*

6. *C.Childs and L. Slifkin, Proc. of the Vth Internat. Conference on Nuclear Photography, Geneva (1964) Vol. I, 3; Editor W.O. Lock, Geneva, CERN.*

7. *R.Schmitt, Symposium on "Recording of Tracks of Charged Particles in Crystals" (Editor P. Cuer and R. Schmitt, Strassburg, Centre de Rech. Nucl. 1963).*

8. *R.L. Fleischer, P.B. Price, R.M. Walker, Ann. Rev. Nucl. Sci. 15, 1-28, 1965.*

9. *D.R.Tuerpe and T.G.Knorr, Journ. Appl. Physics 33, 3257, 1962.*

10. *W. Von Der Osten, H.Peisl and W. Waidelich, Z. fur Physik 193, 400, 1966.*

11. *F. Granzer, K.G. Guetlich, G. Haase and W. Saul, Naturwissenschaften 52, 558-59, 1965.*

DISCUSSION

H. MOTZ (Oxford University, ENGLAND): *Can you explain the advantages of silver halide crystal tracks as compared to tracks in emulsion?*

SCHOPPER: *Higher homogeneity compared to the mixture of gelatin and silver halide in the case of the emulsion; dry development avoiding shrinkage; large available volumes and higher density; supposedly insensitivity against γ-ray background.*

V. STORIZKHO (Atomic Energy Establishment, U.A.R.): *Is it possible to obtain three dimensional spectrum by this technique?*

SCHOPPER: *In principle yes; good micrographs will require specially adapted optics with regard to the refractive index of*

AgCl.

H. PEISL (I. Physikalisches Institut, Darmstadt, GERMANY): *Could the tracks be stabilized by lowering the temperature?*

SCHOPPER: *This is one aspect for further experiments which have to deal with temperature effects.*

A THERMODYNAMIC TREATMENT OF RADIATION-INDUCED PHASE-TRANSFORMATIONS AND ITS APPLICATION TO SOME PEROVSKITES

O. Hauser and M. Schenk

The Academy of Sciences of the G.D.R.

Dresden, Germany

Samples of $BaTiO_3$, $PbTiO_3$, and $PbZrO_3$ were irradiated in the reactor for doses up to about 14×10^{18} n_f/cm^2. The paper discusses some aspects which may explain why certain polymorphic substances show radiation - induced phase - transformation, whilst others do not. It is shown that this type of transformation can be expected only if other defects cause a shifting of the phase transition temperature towards the temperature of measurement. In this case the defect-induced change of the enthalpy is $\delta H<0$ which is the case for $BaTiO_3$ and $PbZrO_3$.

INTRODUCTION

There are some polymorphic solids which can be converted from one crystalline modification into another by means of particle irradiation. However, it is not yet clear why certain polymorphic substances show radiation-induced phase-transformations whilst others do not. This paper will discuss some new aspects which may contribute to a better understanding of the problem of radiation-induced phase-transformations.

For this purpose let us take a solid which appears in the form of two different crystalline modifications. At a temperature $T = T_c$ the room temperature phase will transform into the high temperature modification. Both crystallized states can be characterized by the Gibbs' equation:

$$H_i(T) = E_i(T) - TS_i(T) + pV_i(T) \tag{1}$$

(i = 1: room temperature phase, i = 2: high temperature phase);

H = free enthalpy; E = internal energy; S = entropy; p = pressure; and V = volume.

For $T = T_c$ (T_c stands for the transition temperature of the perfect crystal) we have

$$H_1(T_c) = H_2(T_c) \tag{2}$$

If defects are created in the crystal by means of radiation the enthalpy will change accordingly. This change will be δH. In this way, we have at the transition point: (instead of equation 2):

$$H_1(T) + \delta H_1(T) = H_2(T) + \delta H_2(T) \tag{3}$$

and by applying

$$\Delta H = H_2 - H_1 = \Delta E - T\Delta S + p\Delta V \tag{4}$$

and $$\delta H = \delta H_2 - \delta H_1 = \delta E - T\delta S + p\delta V \tag{5}$$

Equation (3) can be transformed into

$$\Delta H + \delta H = 0. \tag{6}$$

In general, $\delta H_2 \neq \delta H_1$ which means that the transformation takes place at a temperature of $T \neq T_c$.

Figure 1 shows a schematic representation of the free enthalpy of the two modifications as a function of the temperature. T_c represents the transition point.

At temperatures lower than T_c : $\Delta H > 0$.
At temperatures higher than T_c : $\Delta H < 0$.

If the transition of the crystal disturbed by defects takes place at a higher temperature than T_c, ΔH will subsequently be negative, and $\delta H > 0$. (7)

On the other hand, if as a result of the induced defects transition takes place at a lower temperature than T_c, then $\Delta H > 0$ and thus $\delta H < 0$. (8)

The observation of a so-called radiation-induced phase-transformation of a modification which is stable at room temperature to a high temperature modification can therefore be attributed to a shifting of the transition temperature to a point below the temperature of measurement. Such a shift of the transition point can only be measured if annealing of the radiation-induced defects does not take place before the transition temperature is reached.

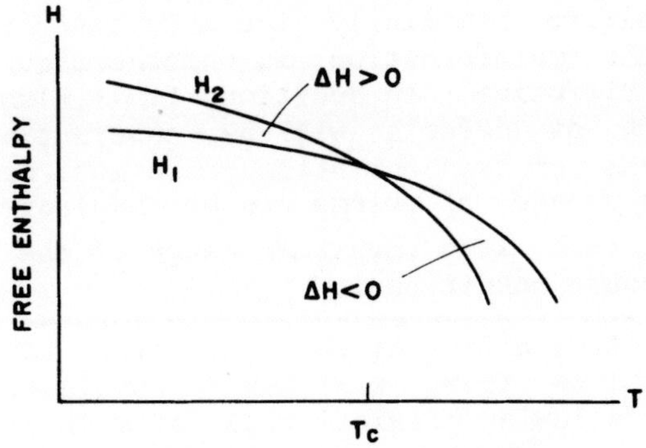

Fig. 1. Schematic representation of the free enthalpy for two modifications as a function of temperature.

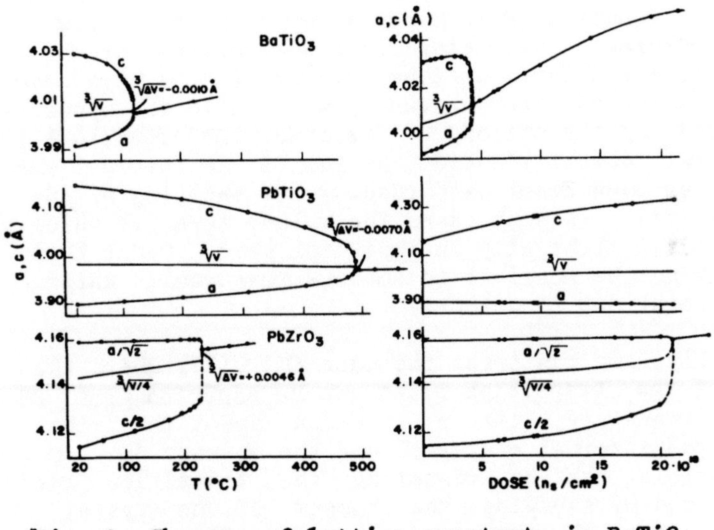

Fig. 2. Changes of lattice constants in $BaTiO_3$, $PbTiO_3$ and $PbZrO_3$.

Although equations (6), (7) and (8) are generally applicable, it is intended that for simplicity the only cases to be considered are those where the transformation is endothermic, reversible and does not involve diffusion. In addition, it is also expected that the concentration of defects will be substantially maintained during the process of transformation. In most cases the energy generated by the change of volume can be neglected owing to the fact that it is much less than the change of the internal energy under normal pressure conditions.

All possibilities affecting the transition temperature whilst neglecting the volume term, will now be considered. In case of $T<T_c$ we have the following possibilities for maintaining equations (2), (6) and (8):

$$
\begin{array}{llll}
\text{a)} & \delta E > 0 & \delta S > 0 & \text{if} \quad \delta E < T\delta S \\
\text{b)} & \delta E < 0 & \delta S > 0 & \\
\text{c)} & \delta E < 0 & \delta S < 0 & \text{if} \; |\delta E| > T|\delta S| \\
\text{d)} & \delta E > 0 & \delta S < 0 &
\end{array}
\qquad (9)
$$

Equation 9 (d) seems to be quite impossible, since it might only occur if $\delta E + T|\delta S| < p|\delta V|$ with $\delta V<0$.

Even if one takes into consideration that the change of entropy decreases with decreasing temperature; i.e. $\Delta S(T)<\Delta S(T_c)$, none of the other possibilities (a) to (c) can be excluded from the very outset. Equation (9b), however, permits the largest shift to lower temperatures. This means that the increase of the internal energy due to the defects will be smaller in the high temperature phase than in the room temperature phase, while the opposite effect is manifested by the entropy. The other two equations (9a) and (9c) show that defect-induced changes of the internal energy and entropy oppose each other with respect to shifting of the transition temperature. In such cases the volume term $p\delta V$ which has been neglected so far might play an important part. Under these conditions it may not be possible to observe measurable shifting of the transition point due to defects.

Similarly, we can treat the case $\delta H > 0$(7), where the largest shifting of the transition point can be expected when $\delta E > 0$ and $\delta S<0$. Unfortunately, little is known about the defect-induced changes of the internal energy δE and the entropy δS. However, it may be possible that by measuring the transition energy $T\Delta S + T\delta S \simeq \Delta E + \delta E$ and by studying the changes of the crystal structure one might estimate the change of the defect-induced internal energy δE. This is beyond the scope of the present paper. When dealing with the following experimental results we will restrict our discussion to the shifting of the transition point.

RESULTS & DISCUSSION

Three samples of the perovskite-type were irradiated in the reactor up to a total dose of about 2 x 10^{19} $n_f cm^2$.

$BaTiO_3$ and $PbTiO_3$ were in their tetragonal and ferroelectric modifications under the temperature of irradiation. $PbZrO_3$ was orthorhombic, quasi-tetragonal and antiferroelectric.

Figure No.2 shows the changes of the lattice constants as a function of increasing temperature (left side) and as a function of the neutron dose (right side). To simplify matters, the lattice parameters of $PbZrO_3$ were taken for the quasi-teragonal form.

$BaTiO_3$ was transformed to the cubic modification after a dose of about 4.2 x 10^{18} n_f/cm^2 and $PbZrO_3$ after a dose of about 21 x 10^{18} n_f/cm^2. However,$PbTiO_3$ behaves in a quite different way: The lattice constants do not approach each other, but move in opposite directions.

For understanding this different behavior we have measured the transition temperatures of the irradiated but not transformed samples. A dielectric method was used for $BaTiO_3$ while differential thermoanalysis was used for $PbZrO_3$ and $PbTiO_3$.

Figure 3 shows the shift in the Curie point of $BaTiO_3$ due to neutron bombardment. At low doses, the shifting of the transition point (tetragonal $\longleftrightarrow$ cubic) is small and increases exponentially after high doses of irradiation. At the position of the arrow the radiation-induced phase-transformation takes place; i.e. the transition temperature has shifted below room temperature.

Figure 4 shows the results of DTA measurements for $PbZrO_3$ and $PbTiO_3$ both unirradiated and irradiated for a dose of about 14×10^{18} n_f/cm^2. The sharp peaks belong to the unirradiated substances and agree with literature. The broadened and shifted peaks are attributed to the irradiated materials. The radiation-induced shifting of the transition points reaches about -50°C in the case of $PbZrO_3$ and about +115°C in the case of $PbTiO_3$.

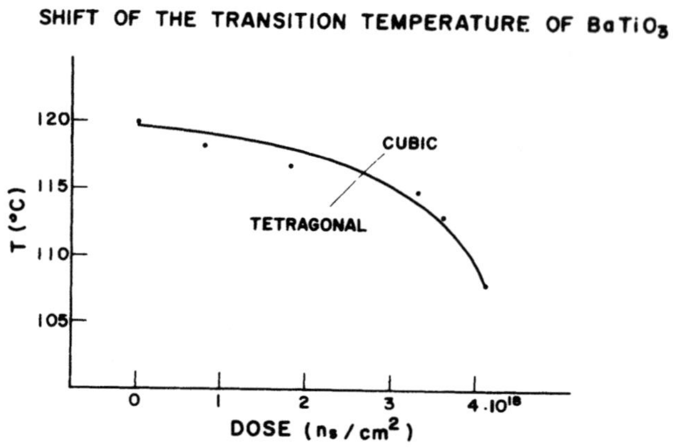

Fig. 3. Shift of the transition temperature for $BaTiO_3$.

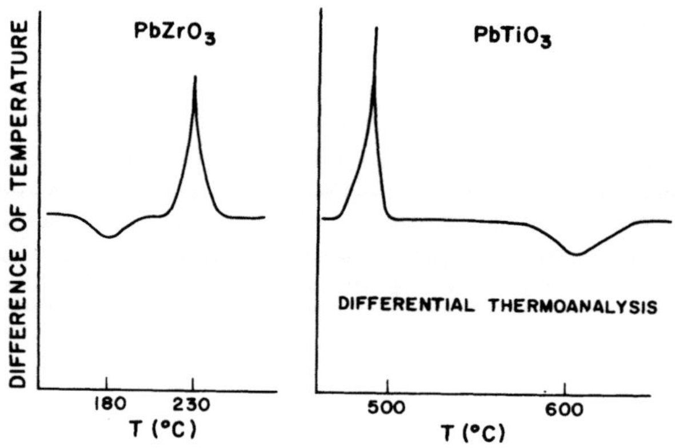

Fig. 4. Differential thermoanalysis for $PbZrO_3$ and $PbTiO_3$.

CONCLUSION

If one compares our initial thoughts with the experimental results, it becomes obvious that radiation-induced phase-transformation can be expected only if the defects cause a shifting of the phase transition temperature towards the temperature of measurement. This is demonstrated by our experiments with $BaTiO_3$ and $PbZrO_3$. In this case the defect-induced change of the enthalpy is $\delta H<0$. However, a shifting of the phase transition point towards higher temperatures was observed for $PbTiO_3$. Thus in this case $\delta H > 0$ which may explain the difficulty in obtaining radiation-induced phase-transformation to the high temperature modification at room temperature.

A NEUTRON DIFFRACTION STUDY ON MAGNETOPLUMBITE $PbFe_{12}O_{19}$

R. Amin, Y. Abbas, M. Fayek,

A. Murasik*, and I. Hamouda

Atomic Energy Establishment, Cairo, U.A.R.

A neutron diffraction study was carried out on a single crystal specimen of $PbFe_{12}O_{19}$. It was possible to obtain precise oxygen parameters on account of comparable neutron scattering lengths of atoms involved. Examination of magnetic reflections confirmed a ferrimagnetic structure with the spin alignment along the c axis.

INTRODUCTION

The neutron diffraction study on magnetoplumbite was undertaken in the hope of obtaining more accurate information on positional parameters of oxygen atoms than the data reported from earlier X-ray studies (1).

X-ray analysis is rather insensitive to the positions of light atoms in presence of heavy elements, causing considerable inaccuracy in the determination of their coordinates. However, the reflections of neutrons by oxygen atoms in $PbFe_{12}O_{19}$ can be observed with sufficient accuracy, since the neutron scattering powers of atoms involved are of the same order of magnitude (Pb,Fe : 0.96 x 10^{-12} cm, O; 0.577 x 10^{-12} cm).

Furthermore, an additional advantage of the neutron diffraction technique for magnetic materials arises from the possibility of detecting unpaired electron spins through their interaction with the magnetic moment of the neutron. Thus, by studying the magnetic Bragg scattering in $PbFe_{12}O_{19}$, additional data on its magnetic properties could be obtained.

**Instytut Badan Jadrowych, Swierk, Poland.*

Structure Information

According to Adelskold's(1) structure determination, $PbFe_{12}O_{19}$ crystallizes in the space group $P6_3/mmc$ (No 194) with the atomic positions given in Table I

The hexagonal unit cell of $PbFe_{12}O_{19}$ with the lattice dimensions a_o = 5.88 A° and c_o = 23.2 A° contains two molecules per cell.

The magnetoplumbite structure can be described as a succession along the c axis of spinel blocks with four oxygen layers denoted as a block S and S', which are "sandwiched" by the R and R' layers lying on the mirror plane and containing one lead, one iron and three oxygen atoms per unit cell.

Layers R' and S' can be obtained from layers R and S respectively by rotation over 180° around the c axis. Detailed description of this structure is given in references (2) and (3).

TABLE I

Atomic Positions For $PbFe_{12}O_{19}$

No	Atom Position	X		Z	
		Adelskold's Results	Present work	Adelskold's Results	Present work
2	Pb in 2(d)				
2	Fe_1 in 2(a)				
2	Fe_2 in 2(b)				
4	Fe_3 in 4(f)			0.028	0.029
4	Fe_4 in 4(f)			0.188	0.178
12	Fe_5 in 12(k)	0.167	0.172	-0.108	-0.109
4	O_1 in 4(e)			0.150	0.151
4	O_2 in 4(f)			-0.050	-0.041
6	O_3 in 6(h)	0.186	0.167		
12	O_4 in 12(k)	0.167	0.162	0.050	0.058
12	O_5 in 12(k)	0.500	0.500	0.150	0.146

EXPERIMENTAL

The neutron diffraction measurements were carried out on a single crystal specimen of magnetoplumbite kindly supplied by Dr. A. Braginski, Polfer, Poland.

From a carefully selected piece of a flat crystal of 1 mm thickness, the intensities of 13 basal reflections up to $2\Theta = 90^{\circ}$, were measured. The higher angles were not available due to the geometrical limitation of the spectrometer.

The neutron diffraction data were recorded on a simple two-axis spectrometer which was designed and constructed in this laboratory (4), and installed in front of a horizontal channel of the UAR-R-1 reactor.

The integrated intensities were read from the mechanical register and corrected for the background and second order contamination. To improve accuracy of measurements all intensities were measured three times and averaged. As the absorption cross sections of the atoms involved are small, no correction was applied.

In addition to the small size of the crystal, to minimize extinction errors, the crystal was immersed repeatedly in liquid nitrogen before data were collected. No correction for extinction was introduced.

RESULTS AND DISCUSSION

Redertimination of z Parameters

As it may be seen from the table of positional parameters there are seven independent z parameters and four of them, namely those assigned to the oxygen atoms, have idealized values.

Having this in mind at an earlier stage of structure calculations a test was made to find out whether X-ray results reported by Adelskold can fit the present neutron diffraction data. This was found not to be the case, giving the discrepancy factor $R \sim 40\%$. This discrepancy was considered as due to the improper oxygen location and it was therefore decided to keep the positional parameters z for iron atoms as determined by Adelskold varying only those assigned to oxygen atoms.

After several attempts during which a great number of arrangements were checked, the reliability factor R decreased significantly and agreement between F^2 calculated and F^2 observed became more reasonable but still insufficient.

TABLE II

Neutron Diffraction Data for $PbFe_{12}O_{19}$

h k l	F^2nuc.	F^2magn.	F^2tot.	F^2obs.
1 0 0	1.8	72.2	74.0	73.9
2 0 0	14.8	2.2	17.0	26.2
3 0 0	70.4	14.6	85.0	76.6
4 0 0	12.9	0.3	13.2	13.3
5 0 0	6.7	3.1	9.8	9.1
0 0 2	7.6	0	7.6	9.7
0 0 4	27.7	0	27.7	20.1
0 0 6	14.1	0	14.1	27.0
0 0 8	107.9	0	107.9	99.1
0 0 10	166.4	0	166.4	154.3
0 0 12	3.2	0	3.2	1.6
0 0 14	220.5	0	220.5	218.2
0 0 16	44.2	0	44.2	42.9
0 0 18	18.8	0	18.8	42.0
0 0 20	409.3	0	409.3	400.6
0 0 22	130.4	0	130.4	129.3
0 0 24	164.1	0	164.1	163.8
0 0 26	8.4	0	8.4	14.0

Calculations to examine whether changes in iron locations could improve the agreement with experimental data were also made. In fact it was soon found that displacement of Fe_4 strongly affects mutual relations of calculated and observed structure factors and after few tests, as a result of displacement of Fe_4 and small adjustment of Fe_3 and Fe_5, it was arrived to the arrangement given in Table I.

After applying an average temperature factor B=0.96 A^{o2}, the reliability factor R was found to be 0.066. A list of observed and calculated F^2 values is given in Table II.

Magnetic Structure of Magnetoplumbite

In absence of any known magnetic results concerning this compound we based our structural considerations on its apparent isomorphy with $BaFe_{12}O_{19}$.

The resultant magnetic moment in $BaFe_{12}O_{19}$ arises from the sum of moments of seven octahedral iron ions and one iron ion with fivefold coordination in the layer containing barium, reduced by the magnetic moments of two octahedral and two tetrahedral iron ions.

Thus, assuming S = 5/2 and complete quenching of orbital moment, the resultant magnetic moment per formula unit should be equal to 20 Bohr magnetons. In fact, measurements on a polycrystalline sample of $BaFe_{12}O_{19}$ at liquid hydrogen temperature and in fields up to 26,000 Oersted made by Stuijts confirmed this value.

If the above considerations are justified in the case of magnetoplumbite, then for room temperature measurements we can expect a value of about 14 Bohr magnetons (3) (H.B.G. Casimir).

Figure 1a shows the neutron diffraction pattern of h00 reflections made at room temperature. There appears five peaks; three of which have apparent magnetic contribution.

By combining these data with those obtained for the basal plane (Fig. 1b) it was easy to identify the ordering scheme, in spite of a relatively small number of observed reflections (further measurements are planned). Since the magnetic contribution was not detected for basal reflections, the spin alignment is taken to be along the c axis. Calculations of nuclear and magnetic structure factors with the new positional parameters x, given in Table I, and the spin arrangement for $BaFe_{12}O_{19}$ gave a reasonable agreement with experimental data. As a resultant magnetic moment the value 14.2 Bohr magnetons has been found. Results are presented in Table II.

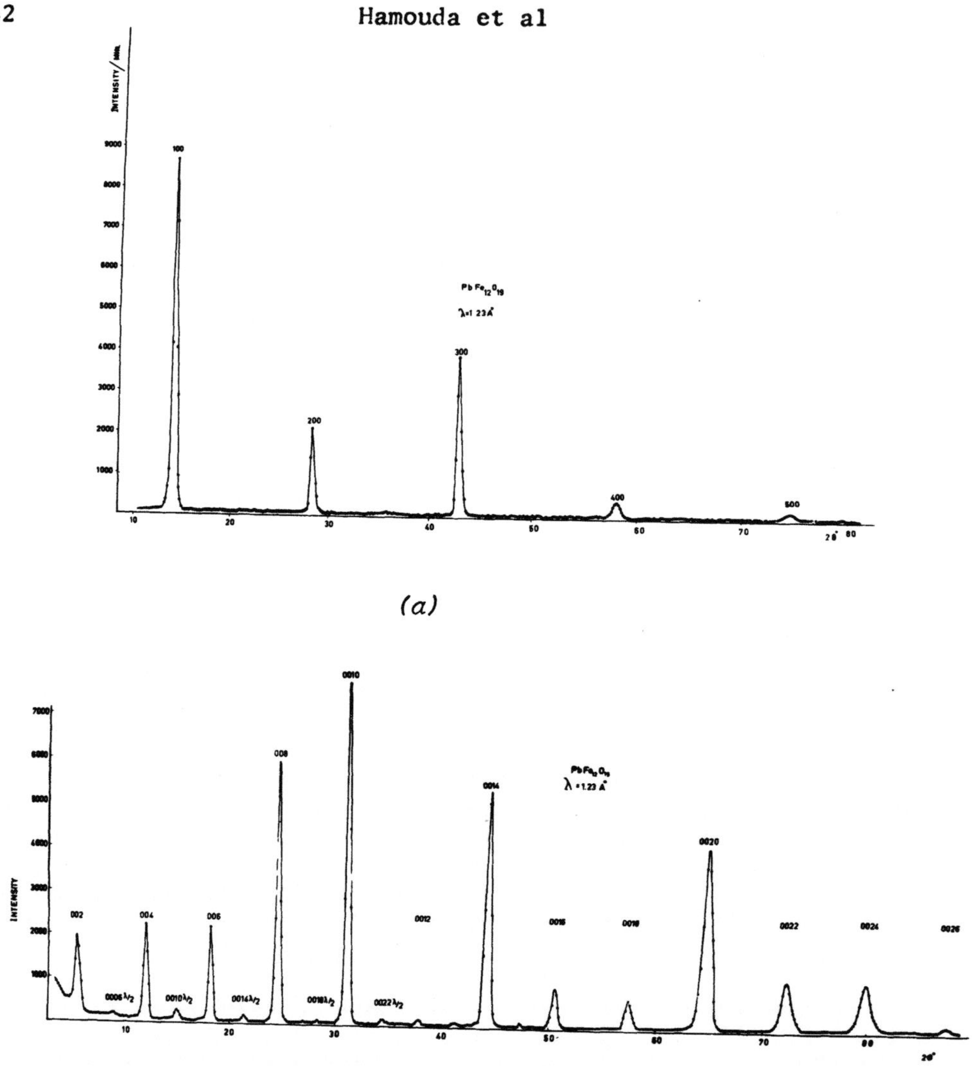

(a)

(b)

Fig. 1. Neutron diffraction patterns of $PbFe_{12}O_{19}$ obtained at room temperature for h00 and 00l reflections.

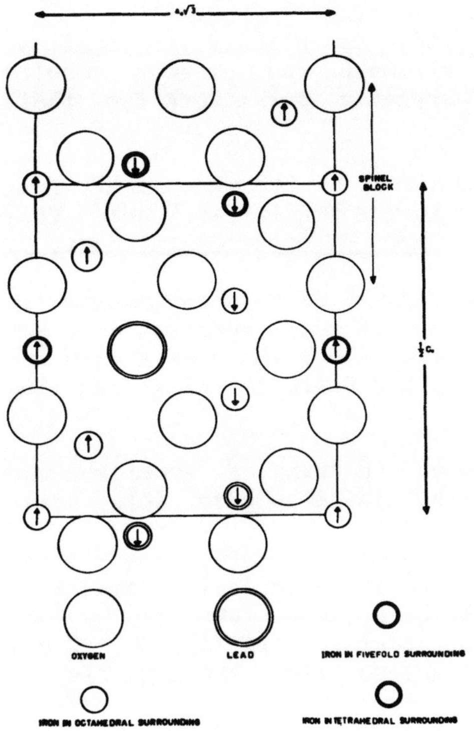

Fig. 2. Cross-section of $PbFe_{12}O_{19}$ structure with c axis vertical. Only atoms in mirror plane containing c axis are shown.

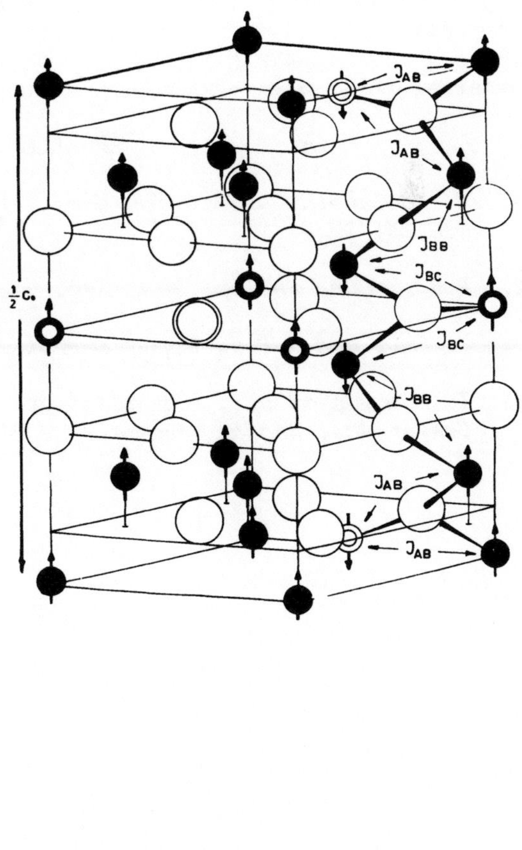

Fig. 3. Perspective drawing of $PbFe_{12}O_{19}$ structure, with indication of principal exchange interactions. The arrows indicate the spin orientations.

CONCLUSIONS

The neutron diffraction study led, essentially, to confirmation of Adelskold's structure, although the present results revealed significant departure of oxygen z parameters from their idealized values.

However, it should be emphasized that as far as x coordinates are concerned, their determination in the present work needs further refinement due to the limited number of experimental data.

The main feature of magnetic structure of magnetoplumbite, is its uniaxial character. The magnetic unit cell consists of a set of ferromagnetic layers which alternate along the c axis with individual moments oriented parallel to the direction of alternation (Fig. 2).

This ordering may be interpreted in terms of superexchange interactions which occur between ions in tetrahedral (A), octahedral (B), and five-fold surrounding (C).

Let us denote by A-B, B-B, and B-C the principal interactions which take place between ions in A, B, C, positions and by J_{AB}, J_{BB}, J_{BC} the corresponding exchange integrals. If J_{AB}, J_{BB}, J_{BC} are negative then all interactions attempt to align the magnetic moments of these ions antiparallel.

Thus assuming that B-B interaction is relatively small in comparison with A-B, we have along the c axis a chain of interactions ... B-A-B-B-C-B-B-A ... which create + - + - + - + -... ordering (Fig. 3).

REFERENCES

1. *V. Adelskold, Arkiv for Kemi, Mineralogi O. Geologi. Bd 12A. N:O 29, 1938.*

2. *P.B. Braun, Philips Res. Rep. 12, 491, 1957.*

3. *J. Smit and H.P.J. Wijn, Ferrites, John Wiley Sons, 1959.*

4. *I. Hamouda et al., Proceedings of the Fifth Arab Scientific Conference, Bagdad, March 27 - 31, 1966 (In print).*

THE INFLUENCE OF RADIATION ON METALLURGICAL PROCESSES

G.J. Dienes

Brookhaven National Laboratory

Upton, L.I., New York, U.S.A.

Irradiation of solids with energetic particles results in the production of defects in the solid. The basic aspects of this process will be reviewed with emphasis on the displacement process and hence the production of vacancies and interstitials. The connection between defect concentration and diffusion processes will be discussed in some detail. Since by means of irradiation one can produce defect concentrations higher than those present in thermodynamic equilibrium, one expects that radiation will enhance diffusion. The phenomenon of radiation enhanced diffusion has been observed and investigated systematically in several metal systems. Radiation enhanced diffusion can be of considerable metallurgical importance since it can lead to equilibrium states which are normally inaccessible because of slowness of the pertinent diffusion rates. Under favorable conditions, one can also obtain a great deal of basic information from such experiments.

In addition to enhancing diffusion processes, radiation induced defects can influence nucleation and precipitation processes and also interact with impurity atoms. Some systematic information concerning these processes has become available in recent years and the pertinent experiments will be described.

INTRODUCTION

Energetic particles impinging on a solid displace the atoms from their normal lattice sites. Heavy particles (neutrons, protons, etc.,) can impart a considerable fraction of their energy to a struck atom, while the energy transfer is small in the case of electrons (1,2). The energy transmitted to the struck atom, the primary knock-on, largely determines (3) the subsequent amount of damage and the distribution of damage.

Irradiation with electrons or gamma-rays results in the production of vacancies and interstitials and replacement collisions. The situation for such low energy collisions is illustrated in Figure 1 (4). In this figure, atom A was given 40 eV energy in the direction indicated. It struck atom B and pushed it off its lattice site, which in turn displaced atom C. Atom A now occupies the site of B, B that of C, and C became an interstitial at location D (in the split configuration). The empty site at A is, of course, the vacancy. Thus, in this simplest form of damage, two replacement collisions have occurred and an interstitial and a vacancy have been produced.

Irradiation with high energy neutrons is a much more complex process. The primary knock-ons are much more energetic and can create many other displacements resulting in a cascade of displaced atoms. Near the end of its path, a knock-on collides with almost every atom in its path and loses a lot of energy in a short distance, both thermally and by displacements. Hence regions of rapid energy loss are referred to as thermal spikes and displacement spikes. The nature of a thermal spike is illustrated in Figure 2, where the temperature transients are shown for a spherical heat pulse. It is to be noted that very high temperatures may be achieved but only for a very short time. Appreciable thermal diffusion of atoms is not expected. The initial configuration of a displacement spike, as suggested by Brinkman (5), is illustrated in Figure 3. This is a highly idealized situation but it shows the highly disturbed region which may occur. In a short time many of the atoms will relax (although probably not to the original sites), many will be driven away leaving voids behind, and parts of the spike may collapse to form dislocations.

Fission fragments are highly energetic and much of their energy is imparted to the electrons. Because of the high energies, thermal spike effects become important in addition to atomic displacements and displacement spikes. Fission fragments and other heavy charged particles produce typical tracks in thin films and in certain types of bulk materials (6). Clearly the agitation near the path of these particles is very high indeed.

The different radiation induced defects play various important roles in influencing rate processes in solids. With electron or gamma-ray irradiation two major effects are expected. Irradiation produces defects in excess of the thermodynamic concentration, and diffusion rates are, therefore, expected to be enhanced. On the other hand, solute atoms, particularly small interstitial ones, may become trapped by radiation induced defects, effectively

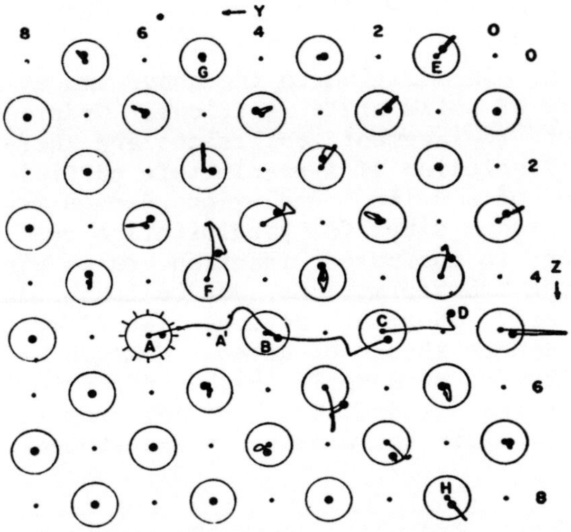

Fig. 1. Atomic paths, calculated dynamically, produced by a 40 eV knock-on in copper. The primary collision was at lattice position A. (Reference 4).

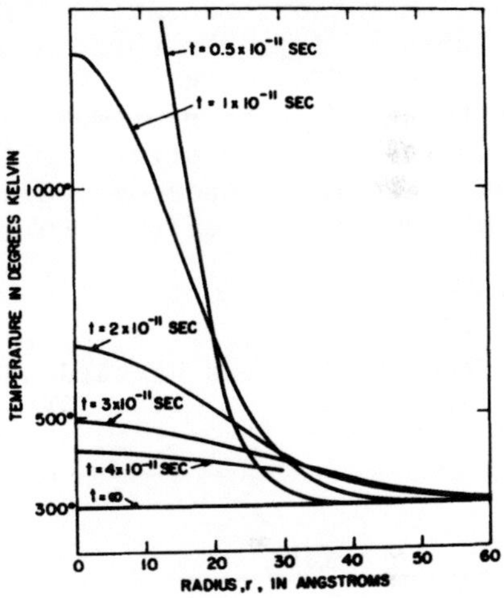

Fig. 2. Temperature distribution in a typical thermal spike at various times. The assumed heat pulse corresponds to Q = 1000 eV and the thermal diffusivity was taken as .001 cm² sec⁻¹. (Reference 1).

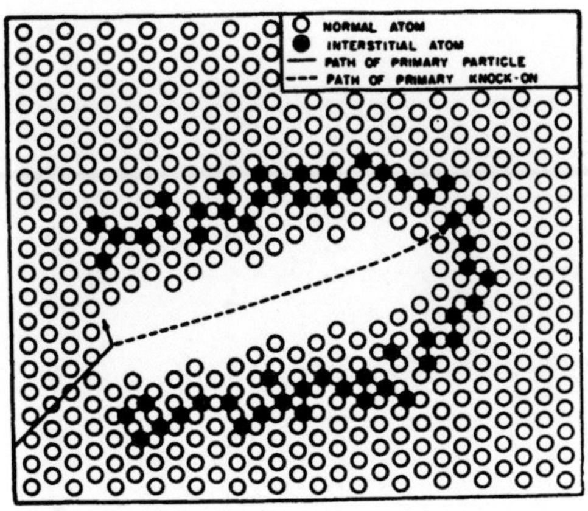

Fig. 3. Initial configuration of a displacement spike. (Reference 5).

decreasing their mobility.

Heavy particle irradiation, in addition to the above two effects, may influence a variety of solid state reactions. Ordered alloys may be disordered by both replacement collisions and thermal and displacement spikes. Resolution of a precipitate particle may occur in the neighborhood of a spike. Radiation damage regions may also serve as nucleation sites for precipitation reactions or phase changes leading to increased reaction rates via enhanced nucleation. Shear type transformations can be inhibited by interfering with the necessary shearing motion. Two effects peculiar to fission fragment damage should be noted. Homogenization of a segregated alloy into a metastable high temperature phase has been observed. Rare gas production, arising from the fission process, can drastically alter the dimensions and strength of the crystal.

The various radiation effects discussed above are summarized in Table I (7). In this paper attention will be focussed on enhanced diffusion and defect trapping, in metals and non-metals. The theory will be reviewed and some representative experiments discussed, but no attempt will be made at complete coverage. For that, and for a thorough discussion of the other topics in Table I, the reader is referred to the literature (6-11).

TABLE I

Effect of Various Bombarding Particles on Solid State Processes (After Damask (7)).

Effect	Light Particles	Heavy Part.	Fission Fragm.
Destruction of Order		x	x
Break-up of Precipitates		x	x
Enhanced Nucleation		x	x
Retarded Shear Transformation		x	x
Enhanced Diffusion	x	x	x
Defect Trapping (Retarded Diffusion)	x	x	x
Homogenization			x
Rare Gas Effects			x

RADIATION ENHANCED DIFFUSION

Basic Considerations

The basic idea of diffusion enhancement by radiation is very simple. Diffusion occurs by means of defect motion, in most cases by the mechanism of vacancy migration. Let a sample be irradiated at a temperature at which the defects are mobile. In this situation defects are produced at a constant rate, but they are also moving around, trying to get out of the solid, since they are present at a concentration higher than that for thermodynamic equilibrium. The result is that a steady state concentration of defects is set up in the crystal over and above the equilibrium concentration. The diffusion coefficient, since it is proportional to the defect concentration, is therefore enhanced by radiation.

The excess steady state defect concentration built up in the crystal depends on the mode of annihilation of the defects. Lomer (12) first derived the expression for enhanced diffusion for the case where vacancies are annihilated by interstitials and at fixed sinks. Dienes and Damask (13) explored the limits of the two annihilation effects and showed that in each case a different flux and temperature dependence of enhanced diffusion was to be expected. The simplest case is that of vacancies annealing at fixed sinks (often referred to as Case 1 for linear anneal). The time rate of change of the excess vacancies V is given by (with a completely analogous equation for interstitials):

$$\frac{dV}{dt} = K - K_v V \tag{1}$$

where K = rate of vacancy production, proportional to flux

$K_v = \alpha \nu_v \lambda^2$ = rate of vacancy escape at sinks

α = sink concentration (due to dislocations, surfaces, etc.)

ν_v = vacancy jump frequency

λ = jump distance.

Let V_o be the thermal concentration of vacancies. The diffusion coefficient, D, under thermal conditions alone is then $D = V_o \nu_v \lambda^2$. The diffusion coefficient under irradiation conditions, D', is proportional to the sum of the thermal and irradiation produced vacancies, i.e. $D' = (V+Vo)\nu_v\lambda^2$. After an initial build-up time a steady state condition is reached, which is given by

$$\frac{dV}{dt} = \dot{K} - K_v V = 0 \tag{2}$$

Thus,

$$V = K/K_v = K/\alpha\nu_v\lambda^2$$

and

$$D' = D + \frac{K}{\alpha} \tag{3}$$

The diffusion enhancement is

$$D' - D = K/\alpha \tag{4}$$

which is independent of the temperature, varies linearly with the flux, and is inversely proportional to the sink concentration.

It should be noted that the steady state approximation is applicable only in the temperature range where the vacancies are mobile and when the experimental diffusion time is long compared to the build-up time. It should also be noted that α in these equations is expressed as the sink concentration in atomic fraction (Ref. 8, p. 81).

If the sink concentration is low, and/or the defect concentration is high, vacancy-interstitial annihilation can no longer be neglected. When this effect, which obeys bimolecular kinetics for random recombination, is included a different solution for the diffusion coefficient results (12,13) (often referred to as Case 3). Since the interstitial is usually considered to be the more mobile defect, the recombination term is controlled by the interstitial jump frequency, ν_i. The rate equations now are:

$$\frac{dV}{dt} = K - K_v V - \nu_i (V+V_o)i$$

$$\frac{di}{dt} = K - K_i i - \nu_i (V+V_o)i \tag{5}$$

where K_i is the escape rate of the interstitials and is equal to $\alpha_i\nu_i\lambda^2$, where α_i is the concentration of interstitial sinks and may not be the same as that of the vacancy sinks. The steady state solution is given by

$$v = -\tfrac{1}{2}(\alpha\lambda^2+V_o)+\tfrac{1}{2}\left[(\alpha\lambda^2+V_o)^2+4K/\nu_v\right]^{\frac{1}{2}}$$

$$i = \frac{1}{2\nu_i}\left\{-(\alpha\lambda^2\nu_v+\nu_v V_o)+\tfrac{1}{2}\left[(\alpha\lambda^2\nu_v+\nu_v V_o)^2+4K\nu_v\right]^{\frac{1}{2}}\right\} \tag{6}$$

From Equation (6) $i\nu_i = V\nu_v$ and therefore $\Delta D_v = \Delta D_i$. At low temperature, where V_o is small and therefore the enhancement is large,

ΔD is dominated by

$$\nu_v (K/\nu_v)^{\frac{1}{2}} = (K\nu_v)^{\frac{1}{2}}, \qquad (7)$$

which leads to a square root dependence on the flux and an effective activation energy of $\frac{1}{2}E_m^V$, where E_m^V is the activation energy for vacancy motion.

It should be noted that neither in the linear case (Case 1) nor in the case of the combined linear and bimolecular anneal (Case 3) does one obtain any direct information as to which defect is responsible for diffusion. This is because the equations are completely symmetrical, in the steady state solution, with respect to vacancies and interstitials. Thus, from enhanced diffusion per se one cannot distinguish the two types of defect, although one may draw some conclusions from the value of a derived activation energy.

One can give a rather simple argument to indicate that the vacancy mechanism is likely to dominate over the interstitialcy mechanism, at least in the ordering alloys that have been studied in some detail. It is known experimentally, from the Kirkendall effect and other experiments, that vacancies are responsible for A-B interchanges, which are the only diffusion steps that can order an alloy. The ratio of A-B interchanges to sub-lattice interchanges (A-A or B-B) is not known, but the fact that A-B interchanges occur at significant rates shows that formation and migration energies for A and B type vacancies cannot be very different. There are no similar experiments to indicate that any interstitialcy ordering occurs. Theory indicates that the interstitial formation energies are high (4-5 eV vs. about 1 eV for vacancies) and, therefore, the difference in formation energies for A and B type interstitials is also apt to be high. If it takes more energy to form, say, a B interstitial than an A interstitial, an A is not expected to interchange with a B because of the energy requirement. Thus B interstitials would interchange with A lattice sites by means of an interstitialcy motion (pushing an A into an interstitial position and a B now occupying the lattice site) but the A interstitials would anneal out by migrating in the A sublattice only. Thus, while interstitialcy enhanced diffusion is conceivable, and has not been ignored theoretically, it has not been demonstrated experimentally and, therefore, it has generally been assumed, and with good justification, that the radiation enhanced diffusion experiments can be properly interpreted in terms of the vacancy mechanism.

A number of experiments are in agreement with the general outlines of the simple theory presented here. However, as shown in later sections, deviations have also been observed which are as yet unexplained.

Experiments on Alloys

Quantitative experiments (13) were first performed on Cu-Zn 30% after earlier experiments suggested enhanced diffusion in this alloy (14,15). The general procedure was to quench the alloy from about 300°C and then anneal isothermally at lower temperatures until equilibrium resistivity was achieved. In this way the equilibrium resistivity as a function of temperature was obtained as illustrated in Figure 4. From the time, at each temperature, required to achieve equilibrium the activation energy was determined, which compared favorably with that obtained by diffusion experiments (16). Samples were brought to equilibrium corresponding to some elevated temperature, quenched and then irradiated (17,18). The final point or equilibrium value of resistivity obtained by electron irradiation is plotted as a solid triangle at 50°C in Figure 4. This value was achieved in about 5 hrs. of irradiation. If one calculates the time required to achieve this state under thermal annealing alone, using the measured activation energy, it is about 30,000 yrs. The other two points were obtained by gamma irradiation, which will be discussed later, and it is seen that low temperature equilibrium conditions can be readily achieved in alloys with irradiation.

Ordering curves (by resistivity decay) were obtained at a series of temperatures in the Brookhaven reactor and the reciprocal of the characteristic decay time, τ, ($1/\tau = 12\ D/\lambda^2$) was evaluated. These results are illustrated in Figure 5 where it is seen that in the low temperature region the diffusion coefficient is temperature independent and is of the order of 10^{-20} cm^2/sec (13). The dashed line is obtained from the simple theory of Equ. (4) derived above. The flux dependence was shown to be linear by an other series of experiments (18). These were performed by placing samples at different distances from an unshielded reactor which was a prototype in the nuclear rocket development program. The complete resistivity decay curve could not be determined in this experiment, but a series of normalizations to the Brookhaven reactor results were carried out, and the results are shown in Figure 6. The final normalization in the presentation of the data was not performed, i.e., in Figure 6 the two straight lines from the two different tests on KIWI-A and B were not shifted, but it is readily seen that they would form one straight line, and the data seem to fit such a line over two orders of magnitude of flux. Thus, both temperature and flux dependence of enhanced diffusion of α-brass under neutron irradiation were shown to correspond to the prediction of linear annealing of the excess vacancies described by Equation (4).

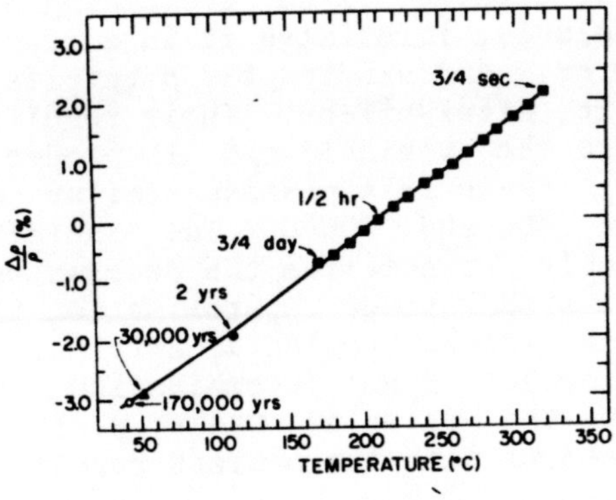

Fig.4. Equilibrium resistivity of α-brass at liquid nitrogen temperature, ■-thermal data, ▲-electron irradiation, ● and ○-gamma irradiation. Times indicate time needed to achieve thermal equilibrium thermally. (Reference 16).

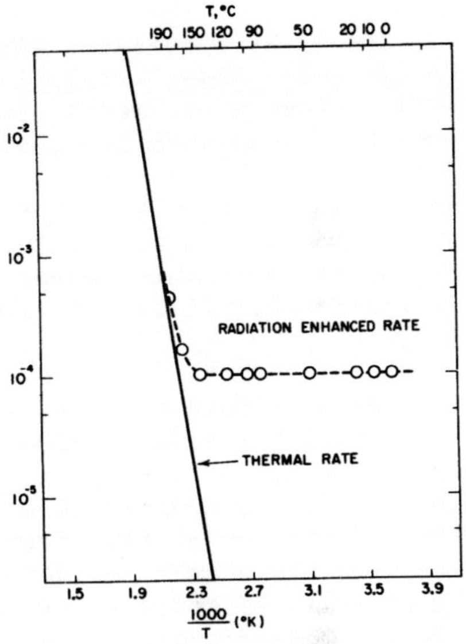

Fig.5. Rate of ordering for α-brass. Dashed line - theoretical curve from Equ. (4). Solid line - thermal rate. Circles - experimental points obtained by reactor irradiation. (Reference 13).

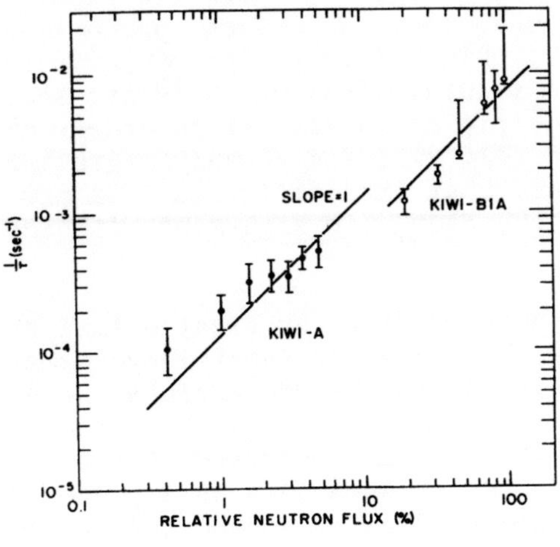

Fig.6. Enhanced ordering rate of α-brass vs. neutron flux from two different reactor experiments. (Reference 18).

Similar results have been obtained by Dietz and Balthesen in the iron-silicon system (19). They quenched iron containing 3% silicon from 1000°C, where Si is soluble, irradiated it in a reactor at a series of lower temperatures, and studied the precipitation rate by measuring the magnetic after-effect. Their results are illustrated in Figure 7, where the variable n_0D is proportional to the $1/\tau$ used in Figure 5; there is a plateau similar to that observed in α-brass between 300 and 400°C. The rapidly decreasing curve below 300°C probably arises from the decreasing mobility of the vacancies where the steady-state approximation is not valid. The slight slope in the plateau region is attributed by the authors to a small contribution from recombination of vacancies and interstitials. From the value of their plateau region, $K/\alpha \approx 10^{-20}$, which is in agreement with the α-brass results.

The first successful attempt to measure diffusion coefficients in the bulk under irradiation was made by Bierman and Keitkamp (20), whose experiments indicated an increase in the self diffusion of lead when the surface is subjected to α-bombardment. The diffusion coefficients measured were of the order of 10^{-18} - 10^{-19} and were detected by using Pb^{212} as the tracer. This experiment was followed by a temperature dependence measurement of the diffusion of Ni^{63} into copper under α-bombardment. A temperature independence, similar to that in the alloy systems, was observed (21).

As mentioned already, at high defect concentrations vacancy-interstitial recombination is to be taken into account. Under these conditions Equations (6) and (7) should be applicable. Arndt and Hines (22) studied accelerated diffusion in α-brass, but under bombardment by 30 keV neon ions. They used a 10% Zn alloy and measured the amount of dezincification at different temperatures by optical reflectivity of the surface. Figure 8 shows the results of the measurements. They estimated that the defect production rate was about 4.5×10^{-2} per sec., which is very high compared to the estimated production rate in a reactor of about 10^{-10} per sec. The slope of the curve of Figure 8 yields an activation energy of 0.4 eV which, via Equations (6) and (7), yields a vacancy migration energy of 0.8 eV. They also showed that the diffusion was proportional to the square root of the flux, in agreement with the theory.

Copper-aluminum alloys have been studied in great detail by the group at Oak Ridge. This alloy also exhibits short-range order and has the decided advantage that changes in the diffuse X-ray scattering with increased ordering are measurable. The increase of diffuse scattering with irradiation was reported first by Houska and Averbach (23), who made the measurements on a powdered Cu - Al (14.5 at.%) alloy furnace cooled from 570°C and then Borie

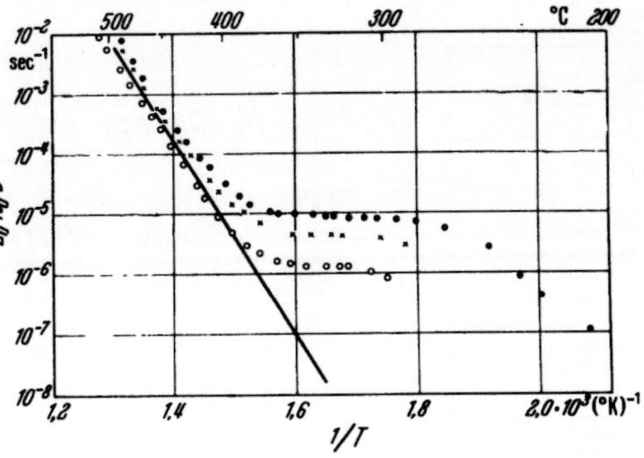

Fig. 7. Rate of diffusion vs 1/T for reactor irradiation of Fe-Si. Solid line - no irradiation. ○-12 cm from edge of reactor. ●, X - 4 cm from edge of reactor. (Reference 19).

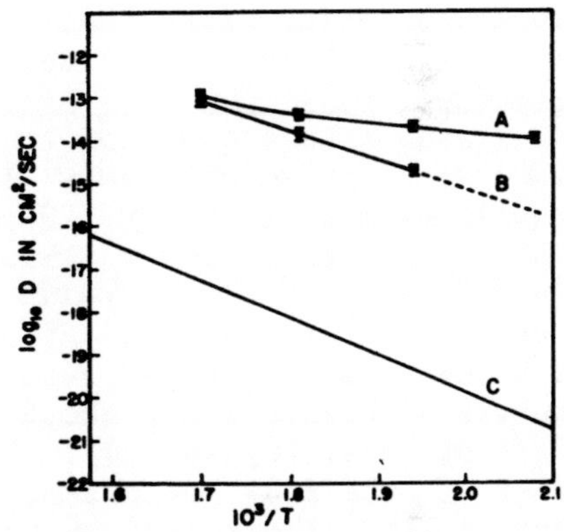

Fig. 8. Diffusion coefficients vs 1/T for 10% α-brass subjected to Ne^+ bombardment. A - irradiated samples; B - unirradiated samples; C - Extrapolation of high temperature diffusion data. (Reference 22).

and Sparks (24) have extended the X-ray study of short-range order in this alloy in considerable detail. They also observed an increase in short-range with neutron irradiation.

The selection of the neutron exposure for the X-ray investigation was based on prior work on this alloy by Wechsler and Kernohan (25). In an initial examination of the effects of neutron irradiation in this alloy they surveyed the effect of irradiation on single and polycrystalline specimens of different composition. Some of the results are illustrated in Figure 9. The resistance minima arise from the competition of two or more effects of the irradiation as discussed in the Introduction. The decrease is due

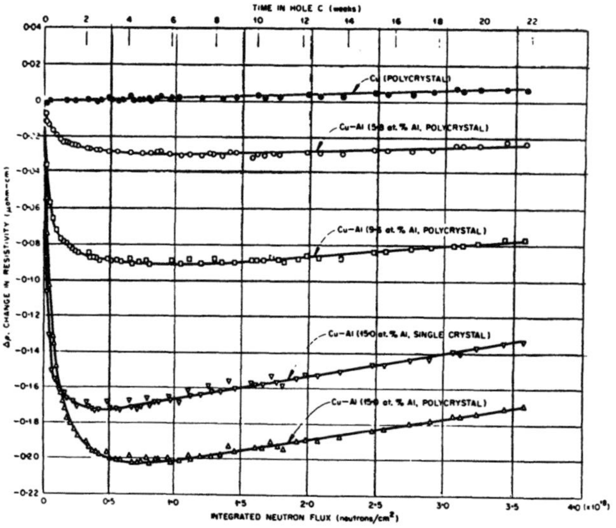

Fig. 9. Change in resistivity vs. neutron exposure at 350°C for Cu-Al alloys. (Reference 25).

to the ordering effect and the increase is caused by the introduction of another type of damage which is cumulative. These investigators found that the equilibrium resistivity of Cu - Al also decreased with decreasing temperature (26), and, as expected, irradiation at successively lower temperatures produces a lower equilibrium resistivity (27), as illustrated in Figure 10. When the curves of the lower three temperatures of Figure 10 are normalized they do not superimpose and the diffusion enhancement is not temperature independent as in the case of α-brass (see Fig.11). The activation energy resulting from the displacement of the curves of Figure 11 is about 0.4 eV, which the authors assign to $E_m^v/2$ of Eqns. (6) and (7). These equations also predict a square root of the flux dependence and this was investigated in some detail (10). The irradiations were performed in a variety of sites in a reactor and great care was exercised in monitoring the thermal and the fast flux. The expected displacements from both the fast flux and the recoil events were calculated and summed. The data are best represented by a slope of ½, in agreement with the simple linear plus recombination theory. It is not known why recombination is important in Cu - Al alloys at reactor neutron fluxes while much higher fluxes must be used to produce the same condition in α-brass. The implication is that there is a much greater concentration of sinks in α-brass than in Cu - Al, which may be the case because the dezincification in α-brass renders annealing difficult. A corollary of the simple theory is that any change in the sink concentration, α, will be reflected in a change in the radiation enhanced diffusion. α may be changed by cold work and, indeed, the radiation enhancement is decreased substantially upon cold-working (28,29) (since dislocations are introduced which act as sinks).

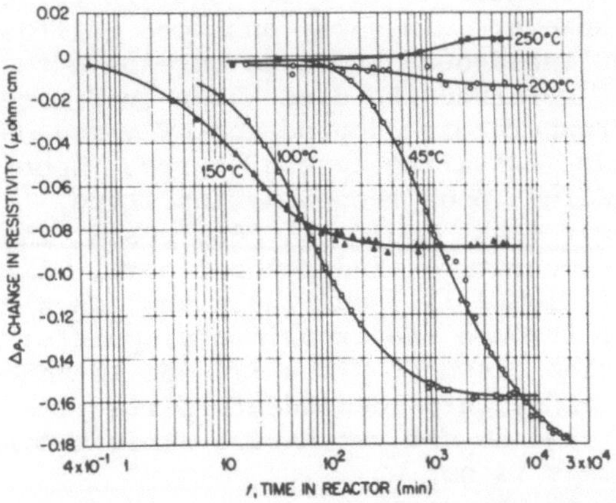

Fig. 10. Change in resistivity of Cu-Al (15 at .% Al) vs. irradiation times in the ORNL Graphite Reactor at the indicated temperatures. (Reference 27).

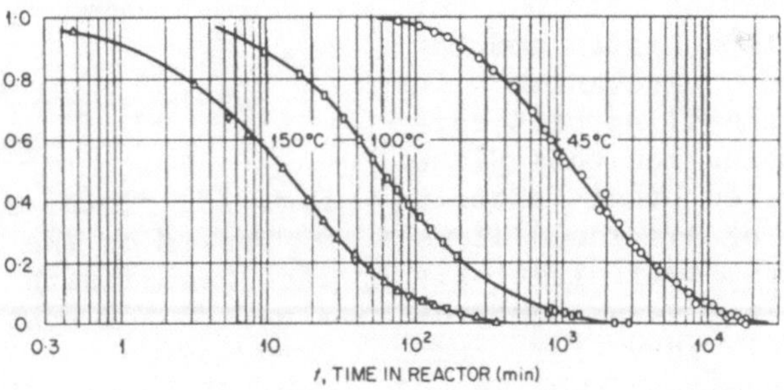

Fig. 11. Normalized decay curves from Fig. 10. (Reference 27)

So far the simple theory of radiation enhanced diffusion, as expressed in Equ. (4) and in Eqns. (6) and (7), has been able to describe the experimental data quite satisfactorily. Some deviations show up, however, in γ-ray enhanced diffusion in α-brass. From experiments analogous to those carried out in the reactor, Larsen and Damask obtained the data shown in Figure 12 (28). It is clear that the temperature dependence of γ-ray enhancement fits neither the linear annealing type of curve, labelled Case 1, nor the linear annealing plus recombination type, labelled Case 3. However, the γ-ray data could be decomposed into a temperature dependent and a temperature independent part, illustrated by the dashed curves of Figure 12, and their sum is given by the solid curves which pass through the experimental points. The temperature independent portion is most probably due to annealing to fixed sinks, but the temperature dependent parts yield an activation energy of 0.4 eV and do not yet have a satisfactory explanation. The flux dependence is not unequivocal because of the scatter of data, and arguments can be made either for linear dependence, as expected from pure annealing to fixed sinks, or for some admixture of recombination, which would result from Case 3 with a preponderance of linear annealing. The diffusion enhancement is very sensitive to cold work in that it can be decreased by a factor of 5 by a reduction in cross-sectional area of 0.5%. Since linear annealing plus random recombination could not match the data, Larsen and Damask proposed a mechanism in which linear annealing plus correlated recombination took place. It was suggested that since a large number of displaced atoms did not receive sufficient kinetic energy in a γ-ray bombardment to travel very far from their own vacancies, each would return and recombine with its own vacancy. This would be a first order process. The data could be described on this basis with 0.4 eV assigned to the interstitial migration energy.

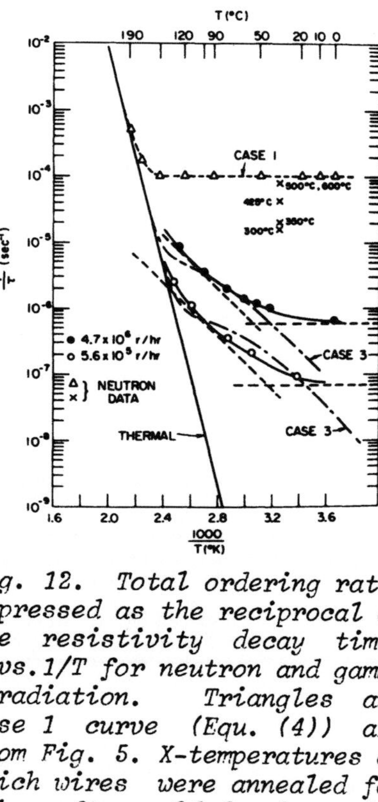

Fig. 12. Total ordering rate, expressed as the reciprocal of the resistivity decay time, τ vs. 1/T for neutron and gamma irradiation. Triangles and Case 1 curve (Equ. (4)) are from Fig. 5. X-temperatures at which wires were annealed for 1 hr. after cold drawing prior to irradiation. Case 3 (Eqns. (6) and (7)) curves are theoretical, thermal rate curve is experimental. (Reference 28).

This mechanism, proposed to explain the gamma results, has been criticized by both Barrett (30) and Heitkamp (31). Their arguments are effectively the same although their approaches differ. Heitkamp points out that with interstitial and vacancy migration energies of 0.4 eV and 0.8 eV respectively, the recombination part of the diffusion in this temperature range is too small to contribute noticeably to the diffusion. Barrett assumes that the lifetime of the vacancy escaping to sinks is much longer than the lifetime of the vacancy recombining, and on this basis also demonstrates that the recombining vacancy cannot contribute observably to diffusion. Approached another way, for the two types of vacancies to contribute observably would require that the recombining interstitial makes of the order of 10^9-10^{10} jumps before it recombines. This number of jumps could carry it by random movement 10^4-10^5 atom distances from the vacancy, and the likelihood of its return from so far away would be small. While one can argue around this on the basis of primary sublattice interstitialcy motion (32,33), it is difficult to conceive of a way to permit the interstitial to have 10^9 jumps and still return to its vacancy The gamma data must at this time remain without a satisfactory interpretation, and further experimental work on alloys is clearly required.

Considerable work has been done on magnetic alloys, a topic covered elsewhere in these proceedings (34). The mechanism again is radiation induced ordering.

Enhanced diffusion coefficients may of course lead to many different metallurgical reactions. In essence, radiation enhanced diffusion assists in the attainment of low temperature states in alloys which are thermally inaccessible because of low diffusion rates. Another interesting example is that of segregation in the Cu-Ni system promoted by radiation enhanced diffusion. Magnetic susceptibility measurements (35) indicate strongly that upon irradiation of the alloy at 80°C, the Ni atoms segregate and behave superparamagnetically. Resistivity experiments have shown that a resistivity decrease occurs upon irradiation and that the temperature coefficient of resistivity is increased (35,36). Both of these effects indicate segregation. Recent work at Oak Ridge, both with neutron and electron bombardment, points in the same direction (37).

Another effect of diffusion enhancement is the phenomenon of surface smoothing. Riehl et al. (38) observed that when a rough copper surface was irradiated with α-particles from a polonium source (5.2 MeV) to a dosage of 5×10^{15} α/cm^2, the roughness disappeared. From the degree of the smoothing it was estimated that about 9 copper atoms per incident α-particle were transported from the hills to the valleys on the surface. Further studies were made

(39) to try to determine the mechanism of smoothing. Sputtering craters were looked for but not found, and the investigators calculated that insufficient energy would be given to the surface atoms by the incident α-particle to cause surface redistribution. It was further observed that no visible smoothing occurred at 0°C with doses up to 10^{16} α/cm^2, although the effect is quite evident at 30°C and 70°C. To achieve the smoothing by thermal means a temperature of 250°C is required. Their conclusion is that radiation enhanced diffusion is occurring. The fact that this effect does not occur at 0°C suggests that interstitials are not the diffusing entity, since they are believed to be mobile in copper at even lower temperatures. The tentative conclusion reached by these investigators is that this is another manifestation of vacancy enhanced diffusion. This type of phenomenon was suggested by Spilners and Smoluchowski (40) to explain some catalytic data. They studied the catalytic rate of the decomposition of formic acid over copper. Radiation at ambient temperature with 350 MeV protons to a dose of about 4 x 10^{15} protons/cm^2 caused the activation energy of the catalytic rate to decrease by a factor of two and the overall rate to decrease by a similar amount. These investigators assigned this effect to the elimination of the reactive sites by radiation enhanced diffusion on the surface of the copper.

A summary of the radiation enhanced diffusion experiments discussed in this section is given in Table II.

Experiments on Semiconductors

The high sensitivity of semiconductor measurements has been used to study radiation enhanced diffusion on a scale involving thousands of interatomic distances. These experiments were done on silicon mostly with proton bombardment (41-43) (electron irradiation also produces the effect (43b)). As in other radiation enhanced processes, the proton irradiation creates a large density of point defects which are mobile at the appropriate irradiation temperatures. Because of the increased defect concentration the probability of impurity jumping is increased leading to radiation enhanced diffusion of an impurity atom.

The technique is to observe the motion of a p-n junction as illustrated in Figure 13. The thin silicon sample has a layer enriched with an electrically active substitutional impurity (P, Ga, B). This layer was prepared by diffusion prior to irradiation in order to produce a p-n junction. The position of the junction is determined by a staining technique and it represents an isoconcentration contour. The position of the junction is measured before and after irradiation, the irradiation temperature in a typical experiment being 600°C to 1200°C. In a typical experiment a movement of the junction of as much as 3μ can be observed.

TABLE II

Radiation Enhanced Diffusion Experiments in Metallic Systems

Material	Irradiation	Method of Measurement	Reference
Cu-Zn (30%)	Neutron	Resistivity	13, 18
"	Electron	"	17
"	Gamma-ray	"	28
Cu-Zn (10%)	Neon ions	Optical reflectivity	22
Fe-Si	Neutrons	Magnetic after-effect	19
Pb	α-particle	Self-diffusion	20
Ni^{63} in Cu	α-particle	Diffusion	21
Cu-Al	Neutrons	X-ray (ordering)	23, 24
"	"	Resistivity	25, 26,27
"	"	Lattice parameter	10
"	Electrons	Resistivity	10
"	Gamma-rays	"	10
Fe-Ni	Neutrons	Magnetic anisotropy	34
"	"	Electrical resistivity	34
Cu-Ni	Neutrons	Magnetic susceptibility	35
"	"	Resistivity	35, 36,37
"	Electrons	"	37
Cu	α-particle	Surface smoothing	38, 39
"	Protons	Catalytic effect	40

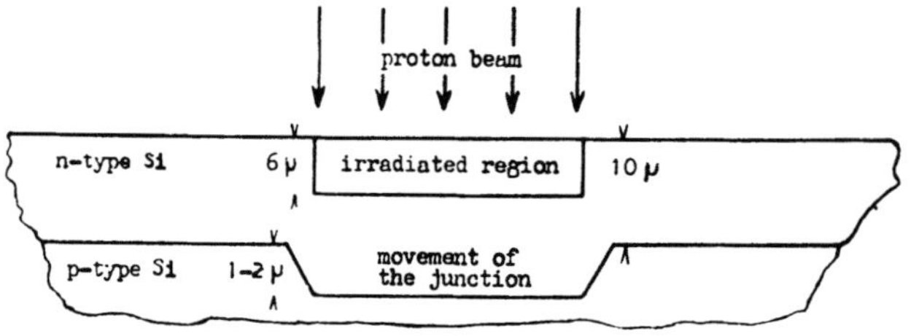

Fig. 13. Proton irradiation enhanced diffusion in silicon. (Reference 43).

Using the diffusion equation the measurement can be converted into a local enhanced diffusion coefficient. In the 900°C to 1100°C region the excess diffusion rate was found to be essentially independent of the temperature (43), indicating that a steady state concentration of defects was established and that the kinetically controlling step is the migration of defects to dislocations. These results appear quite analogous to the α-brass experiments discussed in the previous section.

Schmidt (44) reported some recent neutron irradiation experiments in which strongly enhanced diffusivity was observed at moderate temperatures. He observed a 20 - 30μ deep radiation enhanced diffusion tail of P and Cl in silicon slices exposed for one week at 100°C to 250°C to a flux of 1.2×10^{13} thermal and 1.5×10^{11} fast neutrons $cm^{-2}\ sec^{-1}$. This enhancement at moderate temperatures has not been observed during proton bombardment.

There is no question that radiation enhancement of diffusion occurs in semiconductors. There are not, however, sufficient quantitative data on hand to lead to any conclusion as to the identity of the diffusing defect and the mechanism of annealing. Further, the process is apparently complicated by impurity trapping effects (43b).

Experiments on Non-Metals

There have been no systematic studies of radiation enhanced diffusion in ionic or covalent solids. The situation in such crystals is, of course, considerably more complex than in the simple metals or semiconductors discussed so far. A variety of defects can be produced and even the stoichiometry of the crystal may be altered. Wehner (45) has demonstrated, for example, that in sputtering experiments with heavy ions Fe_2O_3 is reduced to Fe_3O_4 or FeO and finally to Fe.

Observations in the course of other experiments, made at Brookhaven National Laboratory showed that Al_2O_3 powder (46) and diamond dust (47) sinter under heavy neutron irradiation, presumably by some form of radiation enhanced diffusion. These observations prompted Damask (48) to carry out some radiation induced sintering experiments on silica sand. Silica sand was chosen because it is a likely constituent of the lunar surface and the main purpose of this experiment was to see whether solar radiation could compact the lunar surface by promoting sintering via diffusion.

The sand selected was crystalline silica and had a mean diameter of 0.015 cm. It was packed in a 1 cm. diameter quartz capsule with quartz wool, baked in a vacuum at 400°C for 24 hours and sealed with about 1 cm. of purified helium. A control specimen showed that this procedure did not cause sintering. The capsule was packed with a small piece of nickel wire, for monitoring the flux above 1 MeV, in a sealed aluminum can filled with fine aluminum powder for strength and weight. It was irradiated for one week in a reactor at Savannah River to a neutron dose of 6×10^{17} nvt (fast). In this reactor the samples are submerged in flowing D_2O, and prior studies have indicated that gamma heating in such an arrangement causes only a slight temperature rise in the sample. The sample assembly became highly radioactive and was left to cool for several weeks at room temperature before any attempt was made to open it. Upon opening, it was found that the aluminum powder had sintered into a solid cylinder, a portion of which is shown in the insert in Figure 14. This was removed by both mechanical and etching techniques and the nickel wire was counted to determine the fast flux. The capsule was found to have burst at the bottom, probably from expansion of the silica, and as a result the sand has loosened. However, at the center near the quartz wool the sand has sintered to itself, to the walls of the capsule and to the quartz wool. Figure 14 shows a piece of the sintered sand. It was not possible to make any strength measurements because the specimen crumbled when an attempt was made to make flat surfaces. The shearing strength of the sintered aluminum powder was measured and found to be 11% of an equivalent plate of aluminum. The block of powder shown was between the bottom of the aluminum can and the flat bottom of the quartz tube. The density was 1.6 gm/cm^3.

The lunar surface temperature varies between 100°K and 400°K, so it is likely that there is sufficient thermal energy to permit migration of interstitial defects in hard oxides (49), although probably not the vacancies. Since the migration energies of the

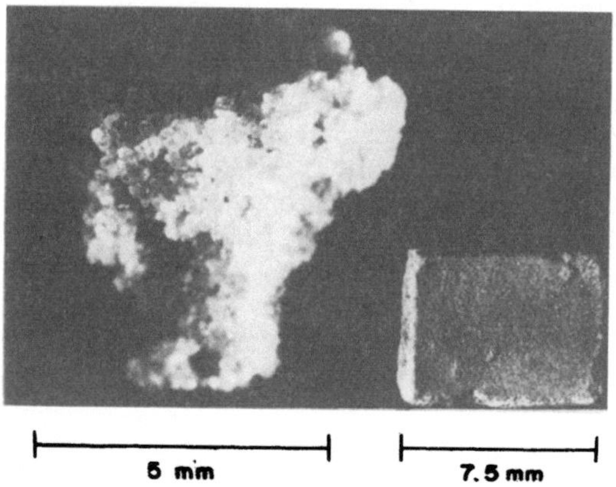

Fig. 14. Silica sand irradiated at ambient temperature to a neutron dose of 6×10^{17} nvt (fast). Insert-aluminum powder used to pack the sample tube in the reactor. (Reference 48).

different interstitials are not expected to be the same and the displacement probabilities of the silicon and oxygen atoms are known to be different, there is reason to believe that the surface bonding material is not stoichiometric. Also impurity atoms such as sodium may be preferentially displaced and migrate to the surface. Thus the phenomenon reported here is not necessarily comparable to thermal sintering.

Assuming that the radiation dose reported here corresponds to the threshold for sintering of silica grains of this size, a rough estimate can be made of the exposure time on the lunar surface required to achieve this threshold stage of sintering in .01 cm sand. Of the many charged particles which strike the lunar surface, the most important contributors to defect production are the solar wind protons with a mean energy of about 2 keV and a flux of 10^8 particles per square centimeter per sec. The equivalent proton exposure turns out to be 6×10^{14} protons. This is equivalent to 60 days exposure to the solar wind. For a reasonable strength to be obtained, the exposure time must, of course, be considerably longer than the threshold time. Moreover, the exposure time should be an undisturbed period, for premature motion of the particles would destroy the effect. On the basis of the flux and velocity of micrometeorites, the shortest estimate for a complete turnover of the upper 1 cm layer of dust is about 10^4 years. Thus it seems that there is sufficient time for the solar wind to cause bonds to be formed that are considerably stronger than the threshold effect described here.

Defect Trapping

As mentioned in the Introduction, small solute atoms, particularly interstitial atoms, may become trapped by point defects, particularly by vacancies. Such trapping inhibits the normal diffusion of the solute atom. The migrating point defect is also trapped, of course. An early observation of such trapping is that of Morin and Reiss (50) on the trapping of lithium by electron irradiation produced vacancies in germanium. The lithium-vacancy complexes apparently serve as precipitation nuclei. Quantitative experimentation in this field started with studies of the iron system. A great deal is known about the interstitial diffusion of carbon in iron. Irradiation is not expected to enhance interstitial solute diffusion and, therefore, the iron-carbon system appeared very attractive for investigating trapping and precipitation effects.

A series of experiments was carried out at Brookhaven on the iron-carbon system. α-iron containing carbon, usually 0.011 wt.%, was solution quenched and then subjected to a variety of irradiations, annealings, and measurements. The quenched iron-carbon was irradiated for 12 days in the Brookhaven reactor at 175°K in a fast flux of 10^{11} nv. The Snoek internal friction peak, which measures the concentration of isolated interstitial carbon atoms in the iron lattice, was found to decrease about 1000 times faster than in unirradiated samples (51). This is illustrated by the left and right hand curves of Figure 15 (the center curve will be discussed later). The kinetics of the decay in the irradiated material is bimolecular. Irradiation of less than 12 days also exhibits bimolecular kinetics, but not all of the carbon left the solution, i.e., the Snoek peak did not decay completely. A normalized set of decay curves is shown in Figure 16. Irradiations for periods longer than 12 days also obeyed bimolecular kinetics. It seemed apparent that carbon atoms were being trapped by a point defect, presumably the vacancy in a one-to-one correspondence. The activation energy of the process is 0.86 eV, the migration energy of the carbon atom in α-iron, so clearly carbon is the migrating species in this 50°C temperature region.

Some results from electrical resistivity measurements (52) are shown in Fig. 17. Unirradiated iron-carbon exhibits two decay steps, one at 170°C, corresponding to the formation of metastable ε-carbide, and one at 250°C, corresponding to the formation of the Fe_3C, cementite, precipitate. In contrast, an irradiated specimen shows 5 decay steps, -30°C, 50°C, 150°C, 260°C and 300°C. Calculations have shown that the iron interstitial can be bound to a carbon atom (53), and it is believed that the -30°C step may correspond to the dissociation of these complexes formed by the mobile iron interstitial during irradiation. This step has not

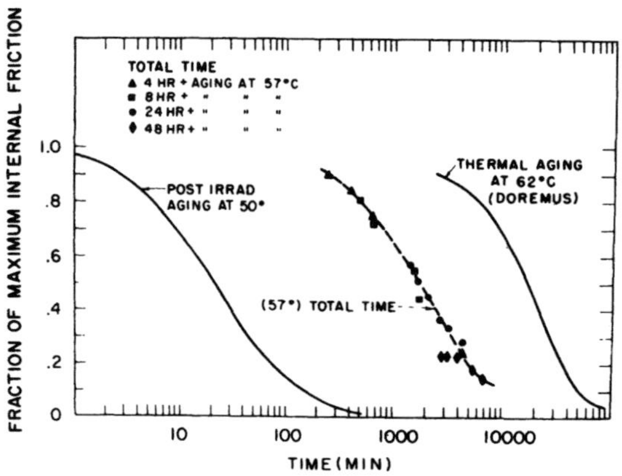

Fig. 15. Normalized plot of the height of the internal friction peak for C in Fe as a function of time for various irradiation and annealing conditions. The left-hand curve was obtained following low temperature irradiation, the center curve for ambient temperature irradiation, and the right-hand curve for thermal aging alone (after Doremus). (Reference 51).

Fig. 16. Normalized plot of internal friction peak height as a function of time for various annealing temperatures. All irradiations were done at -100°C. (Reference 51).

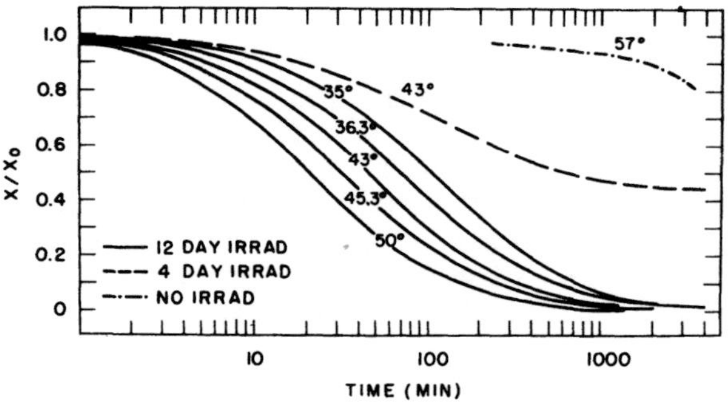

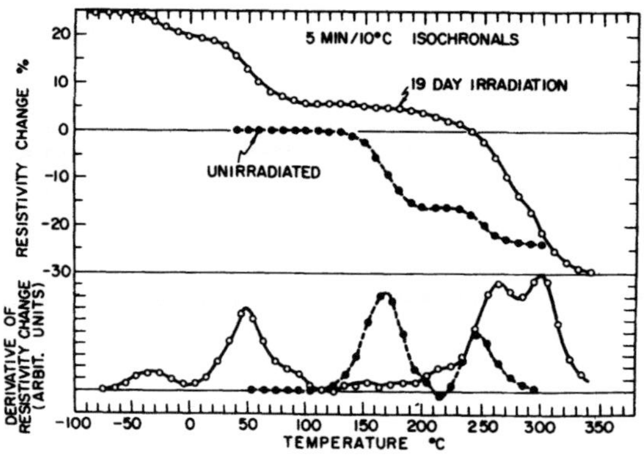

Fig. 17. Resistivity change for Fe-C alloy irradiated at -100°C for 19 days. Shown for comparison is the resistivity change for an unirradiated specimen. The lower curves are derivatives taken from the upper curves. (Reference 52).

been studied in further detail. The 50°C step has the same order of reaction and activation energy as that of the internal friction measurements, and is believed to correspond to migration of carbon and its being trapped by vacancies. The small 150°C step would then arise from a residual amount of untrapped carbon precipitating normally into the ε-carbide. The trapped carbon would leave the trap when sufficient thermal energy became available and precipitate at 260°C into Fe_3C. The 300°C step may arise from the migration of vacancies or their clusters.

The energy release associated with the 50°C step was found by calorimetric measurement (54) to be 0.41 eV per carbon atom. The energy per carbon associated with ε-carbon precipitation is 0.27 eV and that for Fe_3C somewhere between 0.52 and 0.89 eV. Thus a carbon atom trapped in a vacancy with an energy of 0.41 eV would not be subsequently trapped with 0.27 eV when it has sufficient thermal energy to leave the deeper trap. It would be expected, then, that in irradiated iron-carbon the ε-carbide phase should be suppressed. Electron micrographs (55) were taken of irradiated and unirradiated specimens at temperatures corresponding to those of Figure 17, and it was shown that, indeed, no ε-carbide ever appears in the irradiated alloy. In fact, no precipitates are apparent in the irradiated samples at any annealings prior to 230°C, at which temperature some Fe_3C precipitates begin to appear.

Since no precipitation is observable in the temperature region in which the indirect measurements indicate that the carbon disappears from solution, it was concluded that the carbon has been trapped by a point defect, presumably the vacancy. The possibility that the carbon is in the form of precipitates too small to be visible was rejected because of the one-to-one correspondence between carbon atoms and point defects required by the kinetics of the electrical resistivity and internal friction changes, and the incompletion of the 50°C internal friction decay after a 4 day irradiation (Fig. 16). The trapping of the carbon and the relative energy of the trap compared to the metastable carbide is also confirmed, since the metastable carbide does not appear in the alloy irradiated to a dosage sufficient to trap all of the carbon atoms.

Using the 0.41 eV binding energy of a carbon atom to a vacancy, the carbon migration energy of 0.86 eV, and the activation volume for carbon diffusion in α-iron, Johnson et al. (56) constructed a carbon-iron interaction potential. Using this potential the position of the carbon atom in the iron vacancy was calculated (53). This is shown in Figure 18 where the carbon atom is seen to be attached to one of the six equivalent sites of the vacancy. Although the carbon in such a position appears to be strain free, Johnson (57) found in a further calculation that a

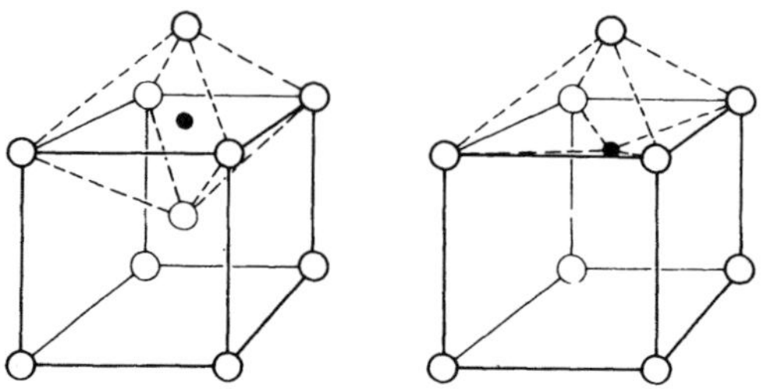

Fig. 18. Schematics of the most stable position of a carbon atom in an iron lattice. Left-hand - normal position. Right-hand - carbon in an iron vacancy. (Reference 53).

slight strain was associated with the configuration, which arises from the slight distortion of the position of the iron atoms nearest to the carbon atom. Wagenblast and Swartz (58), using an extremely sensitive internal friction apparatus, were able to detect such a relaxation of the predicted strength in an irradiated iron-carbon specimen annealed so as to produce the maximum number of carbon-vacancy complexes.

The trapping of carbon atoms by vacancies occurs after a low temperature irradiation, low enough to prevent carbon or the vacancies from moving during the irradiation. In this way, apparently, a sufficient number of vacancies are produced so that there is a high probability of a carbon atom encountering one before it reaches a precipitation site. When irradiations are performed at temperatures at which the carbon atoms are mobile, the trapping is not observed. Rather, the acceleration of carbon depletion during irradiation at ambient temperature, as shown by the center curve of Figure 15, is an enhanced nucleation phenomenon. This effect was reported by Hull and Mogford (59) based on electron microscope measurements of the number and size of the precipitate particles. These workers found that the number of precipitate particles formed by irradiation is about one order of magnitude greater than the number formed in about 5 hrs. of thermal aging. Further irradiation does not cause an increase in this number, and the growth in size may arise from thermal conditions alone.

This enhancement of nucleation is clearly demonstrated by the internal friction measurements of Figure 15. Torsion pendulum internal friction measurements were made, with the frequency adjusted so that the Snoek peak, which measures the interstitial carbon concentration, occurred at 57°C, this being the reactor temperature. In this way internal friction as a function of aging time could be measured outside the reactor at 57°C, following irradiations at this temperature. Thus the only difference experienced by the specimens during their history following quench was that they received different neutron exposures. An acceleration of about a factor of 10 was found for precipitation for all irradiated samples. These results are shown by the center curve of Figure 15, where total time at 57°C, i.e., time inside plus time outside the reactor, is plotted. It is seen that a saturation of the nucleation enhancement occurred at or before the 4 hr. irradiation. The reason for the saturation of the effect is not known at this time.

One would expect nitrogen to behave quite similarly to carbon, but Stanley's (60) work indicates that the situation is somewhat more complex in this case because the nucleation and trapping processes are not as easy to separate. Enhanced nucleation upon neutron irradiation is certainly observed, but the nature of these nucleation sites is not known. Recent work with electron irradiation (61) has shown quite clearly that electron irradiation does not enhance the rate of metastable nitride precipitation, but may, in fact, slow it down due to trapping. When iron-nitrogen specimens were irradiated with 2-MeV electrons at low temperatures to doses sufficient to produce defect concentrations equal to that of nitrogen atoms in solution, a certain fraction of nitrogen atoms disappeared from solution below 35°C. These nitrogen atoms were presumably trapped by the radiation produced defects. Internal friction measurements upon solution annealing of the electron-irradiated samples indicate that the nitrogen trap is rather deep, since detrapping occurs in the 370°C to 400°C temperature range.

ACKNOWLEDGMENTS

This work was performed under the auspices of the U.S. Atomic Energy Commission.

REFERENCES

1. *G.J. Dienes and G.H. Vineyard, Radiation Effects in Solids, Interscience, New York, 1957.*

2. *D.S. Billington and J.H. Crawford, Radiation Damage in Solids, Princeton University Press, Princeton, 1961.*

3. *For channeling effects, see C. Erginsoy, Proceedings of Cairo Solid State Conference, September 3, 1966.*

4. *J.B. Gibson, A.N. Goland, M. Milgram and G.H. Vineyard, Phys. Rev. 120, 1229 (1960).*

5. *J.A. Brinkman, Am. J. Phys. 24, 146 (1956); in "Int. Inst. of*

Physics Enrico Fermi", Course 18, "Radiation Damage in Solids" Academic Press, New York, 1962, pp. 830-860.

6. *For a recent review, see A.N. Goland in "Studies in Radiation Effects", Vol. I (Ed. G.J. Dienes), Gordon and Breach, 1966.*

7. *A.C. Damask in "Studies in Radiation Effects", Vol. II (Ed. G. J. Dienes), Gordon and Breach, 1966.*

8. *A.C. Damask and G.J. Dienes, "Point Defects in Metals", Gordon and Breach, 1963.*

9. *A.C. Damask in "Radiation Damage in Solids" II. IAEA, Vienna, 1962, pp. 3-18.*

10. *M.S. Wechsler and R.H.Kernohan in "Radiation Damage in Solids" II. IAEA, Vienna, 1962, pp. 81-103.*

11. *"Radiation Damage in Reactor Materials", IAEA, Vienna, 1963; particularly pp. 3-163.*

12. *W.M. Lomer, AERE T/R 1540 Harwell Report (1954) (Unpublished).*

13. *G.J. Dienes and A.C. Damask, J. Appl. Phys. 29, 1713 (1958).*

14. *D.S. Billington, R.H. Kernohan and A.B. Lewis, ORNL-380, p. 15 (1949); ORNL-1852, p. 51 (1955).*

15. *D.B. Rosenblatt, R. Smoluchowski and G. J. Dienes, J. Appl. Phys. 26, 1044 (1955).*

16. *A.C. Damask, J. Appl. Phys. 27, 610 (1956).*

17. *A.C. Damask, J. Phys. Chem. Solids, 4, 177 (1958).*

18. *R.E. Larsen, G.J. Dienes and A.C. Damask, Acta Met. 12, 1141 (1964).*

19. *H.D. Dietze and E. Balthesen, Nukleonik, 3, 8 (1961).*

20. *W. Bierman and D. Heitkamp, Zeits. Phys. Chem. 34, 265 (1962).*

21. *H. Bonzel, Acta Met. 13, 1084 (1965).*

22. *R.A. Arndt and R.L. Hines, J. Appl. Phys. 32, 1913 (1961).*

23. *C.R. Houska and B.L. Averbach, J. Appl. Phys. 30, 1525 (1959).*

24. *B. Borie and C.J. Sparks, Acta Cryst. 17, 827 (1964).*

25. *M.S. Wechsler and R.H. Kernohan, J. Phys. Chem. Solids, 7, 307 (1958).*

26. *M.S. Wechsler and R.H. Kernohan, Acta Met., 7, 599 (1959).*

27. *R.H. Kernohan and M.S. Wechsler, J. Phys. Chem. Solids, 18, 175 (1961).*

28. *R.E. Larsen and A.C. Damask, Acta Met. 12, 1131 (1964).*

29. *ORNL-3676 (1964).*

30. *J.H. Barrett, Acta Met., 14, 64 (1966).*

31. *D. Heitkamp, Acta Met., 14, 65 (1966).*

32. *J.A. Brinkman, C.E. Dixon and C.J. Meechan, Acta Met., 2, 38 (1954).*

33. *A.C. Damask, Acta Met., 13, 1104 (1965).*

34. *D. Dautreppe, Proceedings of Cairo Solid State Conference, September 3, 1966.*

35. *A. Ascoli, in "Radiation Damage in Solids" II. IAEA, Vienna, 1962.*

36. *A. Ascoli, M.G. Belloni and G.T. Queirolo, Phys. Letters, 9, 305 (1964).*

37. *ORNL-3878 (1965), p. 12.*

38. *N. Riehl, W. Marth and R. Sizmann, in "Reactivity of Solids" (Ed. J.H. de Boer), Elsevier, Amsterdam, 1961, p. 446.*

39. *R. Sizmann and U. Däunert, in "Radiation Damage in Solids", I IAEA, Vienna, 1962, p. 351.*

40. *A. Spilners and R.Smoluchowski, in "Reactivity of Solids" (Ed. J.H. de Boer), Elsevier, Amsterdam, 1961, p. 475.*

41. *P. Baruch, C. Constantin, J.C. Pfister and R. Saintesprit, Disc. Faraday Soc. No. 31 (1962), p. 76.*

42. *P. Baruch and J.C. Pfister, in "Radiation Damage in Solids" III, IAEA, Vienna, 1963, pp. 43-48.*

43a. J.C. Pfister and R. Baruch, J. Phys. Soc. Jap., 18. Suppl., III 251 (1963)

43b. J.C. Pfister, in "Radiation Damage in Semiconductors", Paris-Rogaumont, Academic Press, 1964, pp. 281-285.

44. P.F. Schmidt, Appl. Phys. Letters, 8, 264 (1966).

45. G.K.Wehner, in "The Lunar Surface Layer" (Eds. J.W. Salisbury and P.E. Glaser), Academic Press, New York, 1964, p. 313.

46. P.W. Levy, Private Communication.

47. D.T. Keating, Private Communication.

48. A.C. Damask, to be published.

49. R. Smoluchowski, J. Geophys. Res., 71, 1569, (1966).

50. F.J. Morin and H. Reiss, J. Phys. Chem. Solids, 3, 196 (1957).

51. H. Wagenblast and A.C. Damask, J. Phys. Chem. Solids, 23, 221 (1962).

52. F.A. Fujita and A.C. Damask, Acta Met., 12, 331 (1964).

53. R.A. Johnson and A.C. Damask, Acta Met., 12, 443 (1964).

54. R.A. Arndt and A.C. Damask, Acta Met., 12, 341 (1964).

55. H. Wagenblast, F.E. Fujita and A.C. Damask, Acta Met., 12, 347 (1964).

56. R.A. Johnson, G.J. Dienes and A.C.Damask, Acta Met., 12, 1215 (1964).

57. R.A. Johnson, Private Communication.

58. H. Wagenblast and J.C. Swartz, Acta Met., 13, 42 (1965).

59. D. Hull and I.L. Mogford, Phil. Mag., 6, 535 (1961).

60. J.T. Stanley, in "Diffusion in Body Centered Cubic Metals", The American Society for Metals, Metals Park, Ohio, 1965, pp. 349-356.

61. J.T. Stanley and W.E. Brundage, ORNL-3878, (1965), pp. 2-11.

DISCUSSIONS

A. FERRO (Istituto Elettrotecnico Nazionale, Torino, Italy): *Is the value of the vacancy concentration which can be evaluated from the Damask et al experiments of enhanced rate of diffusion of carbon in neutron irradiated iron in agreement with other possible evaluations of this value?*

G.I. DIENES: *Yes, in order of magnitude.*

F.H. HAMMAD (Atomic Energy Establishment, U.A.R.): *What is the effect of varying dislocation density on enhancing diffusion in α-brass and in Cu-Al alloys (particularly on the temperature dependence)?*

G.I. DIENES: *Increasing the dislocation density decreases very markedly the radiation enhancement. This is well established for both Cu-Zn and Cu-Al by cold working, to a rather small degree, the wire samples. No detailed studies are available of the temperature dependence of enhanced diffusion in cold worked samples.*

F.H. HAMMAD: *Do you think that the difference in stacking fault energy between Cu-Al and α-brass might account for the difference between their behavior?*

G.I. DIENES: *There is no information on this point.*

A. GUINIER (Faculte de Sciences d'Orsay, Paris, France): *Are there some long range diffusion experiments on irradiated metals?*

G.I. DIENES: *Yes, very precise experiments are needed because the effect is small. Bierman and Keitkamp have succeeded recently in measuring radiation enhanced self-diffusion in lead, and for the diffusion of Ni^{63} into copper, both under α-particle bombardment.*

A. AMMAR (Atomic Energy Establishment, Inchas, U.A.R.): *I would like to know the water content in the silica and alumina powders sintered under the influence of reactor irradiation. I would also like to know the sintering temperature.*

G.I. DIENES: *The silica sand was sintered by neutron irradiation at about 50°C (and certainly no higher than 75°C). The temperature of exposure for Al_2O_3 and diamond was about 100°C. There is no information on the water content. The samples were sealed under vacuum.*

M. RAFIK SOLIMAN (Central Laboratories of Mineral Researchs, Cairo, U.A.R.): *Was there any change in the physical properties of the sintered silica or alumina after irradiation?*

G.I. DIENES: *The sintered, or compacted specimen was too small and fragile for physical property measurements.*

MAGNETIC VISCOSITY INDUCED BY FAST NEUTRON IRRADIATION IN HIGH PERMEABILITY PURE IRON

A. Ferro, G. Montalenti and G.P. Soardo

Istituto Elettrotecnico Nazionale Galileo Ferraris

Torino, ITALY

This paper deals with a study of the nature of defects introduced by fast neutron irradiation responsible for magnetic permeability variations in high permeability pure iron. By comparing the magnetization curves of pure iron specimens irradiated with and without an applied magnetic field, it is shown that permeability changes are due to an induced anisotropy energy associated with anisotropic defects. The mobility and recovery kinetics are studied by taking magnetic viscosity measurements from 20°C to 400°C. An activation energy of 0.85 ± 0.05 eV and a relaxation time τ' of about 10^{-11}-10^{-12} sec are found for the diffusion of the point defects which are identified as impurity-vacancy couples or possibly as divacancies. These defects disappear at temperatures between 100°C and 250°C. The average distance between precipitation centers, evaluated from the time constant of viscosity field recovery, is of the order of 10^{-4} cm, which is in agreement with the data on the distance between defect clusters in irradiated iron obtained by other authors. Experiments of hardness recovery on the same specimens show that the changes in mechanical properties produced by irradiation do not depend on the defects responsible for the magnetic property changes, but on other defects whose recovery occurs between 250°C and 400°C.

INTRODUCTION

The first experiments on the effect of fast neutron irradiation on the magnetic properties of iron were performed on demagnetized specimens with doses of the order of 10^{18} n/cm^2 (E ⩾1 MeV) at about 50°C (1,2). A decrease of about 30% of the maximum permeability was observed and complete recovery was obtained by annealing for about 1 hour at 250°C. In another paper (3) it was shown that these effects were not related to interstitial carbon, whose precipitation might have been activated by the irradiation.

The decrease of maximum permeability could be related either to the presence of large irradiation defects (spikes) which can hinder the domain wall motion, or to an anisotropy energy induced in each domain due to the interaction of some anisotropic point defects produced by irradiation with the magnetization.

In the latter case, the effect of irradiation on the magnetic permeability would essentially correspond to the magnetic after-effect observed around room temperature by Moser et Al. (4,5,6) on neutron irradiated pure iron samples. Four after-effect zones were actually found by these authors between 77°K and 400°K after low temperature irradiation, each zone being characterized by several viscosity peaks. Similar experiments were recently performed by Balthesen et Al. (7). Minor discrepancies between the results of the two groups should be accounted for by the difference in the purity of the iron samples used (8).

Both groups of authors suggest that the viscosity effects are due to the orientation of anisotropic point defects introduced by the irradiation. In particular, the room temperature viscosity peaks observed by Moser et Al. are suggested to be related to the diffusion of vacancy-impurity defects. It has, however, been also suggested (9) that single vacancies might be responsible for this viscosity effect according to the Dietze model (10), which assumes that the vacancy magnetic energy depends on the interaction with the magnetic crystal anisotropy energy. Magnetic after-effect should then result from the migration of vacancies to within the Bloch walls to reduce their energy.

The aims of this paper are the following:

a) To show that the decrease of permeability observed in the previous works is due to an induced anisotropy energy, and that the defects responsible for this change are anisotropic point defects, which also yield the viscosity effect around room temperature.

b) To evaluate through viscosity field measurements the activation energy diffusion constant τ_0 of these defects.

c) To evaluate the order of magnitude of the number of defects introduced by the irradiation.

d) To study the recovery kinetics of the introduced defects.

e) To show that the irradiation defects which influence the magnetic properties at room temperature do not contribute to the variation of the mechanical properties.

EXPERIMENTAL

Specimens and Irradiation

Permeability variations induced by neutron irradiation can be observed in pure iron with sufficient precision only in very high permeability specimens. In order to obtain μ_{max} of at least 100,000, rather bulky samples must be used. Massive toroids (o.d. 35, i.d. 18, thickness 7 mm) were prepared from electrolytic iron melted in vacuo. After annealing at 1480°C for 24 hr in pure hydrogen and at 880°C for 18 hr, μ_{max} is between 150,000 and 220,000. Total impurity content is less than 200 ppm. Internal friction shows C and N content to be less than 2 ppm: some other 8 ppm can be estimated to be anchored at grain boundaries, thus not appearing through internal friction analysis. The specimens were irradiated in the Avogadro swimming pool reactor of SORIN, with a dose of about 10^{18} n/cm^2, at about 50°C. Four out of six samples were irradiated under a magnetic field of the order of 50 A/m. The remaining two were irradiated in the demagnetized condition.

Magnetic Measurements After Irradiation

The magnetization curves before and after irradiation are compared in Figure 1a and 1b for different samples. All magnetization curves were traced at room temperature and immediately after demagnetization. As shown in Figure 1, specimens irradiated under magnetic field show after irradiation an increase of about 20% of the maximum permeability. The opposite occurs for specimens irradiated in the demagnetized condition.

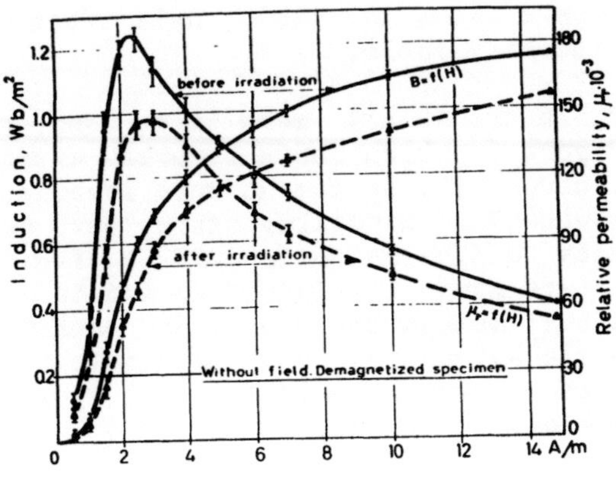

Fig. 1. Magnetization and permeability curves before and after irradiation. Fig. 1a: specimen irradiated in demagnetized condition.

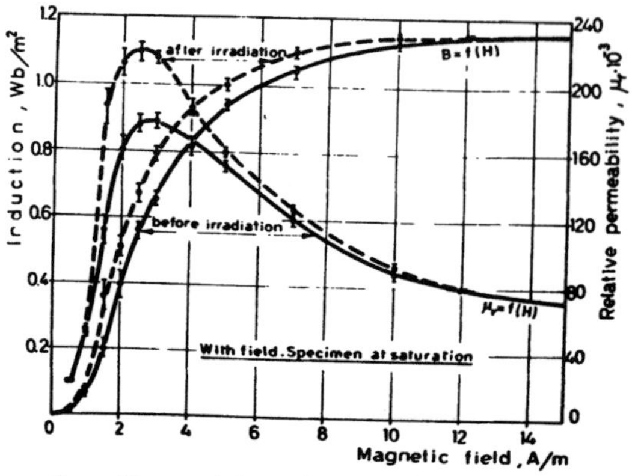

Fig. 1. Magnetization and permeability curves before and after irradiation. Fig. 1b: specimen irradiated under saturation magnetic field.

Magnetic viscosity was measured by comparing magnetization curves traced immediately after demagnetization with curves traced after annealing the specimen in the demagnetized condition. The shifts along the field axis of the magnetization curves after the annealing treatments are taken as the magnetic viscosity field values. Magnetization curves were automatically traced by means of an electronic hysteresigraph (11) which provided a direct record of the viscosity field, with a precision better than 2%, and a somewhat poorer reproducibility (about 10%), due to material instability, over repeated measurements. A typical set of viscosity field measurements is shown in Fig. 2 as directly recorded by the hysteresigraph at 70°C. Each constricted loop is obtained after annealing the specimens in the demagnetized condition for the time shown to the left of the loop in the figure. The straight, low permeability part of the constricted loop corresponds to the shift of the magnetization curve along the field axis after annealing, and thus gives the viscosity field.

Isochronal stages (2 hr, 16 hr, 48 hr) were first used, and the results are reported in Figure 3 which shows recovery effects above 90°C. After complete recovery the specimens were irradiated again. After irradiation, isothermal annealing stages were used to carefully study the defects diffusion below 90°C. Typical viscosity field measurements are shown in Figures 4a and 4b as a function of annealing time after demagnetization. No definite saturation is observed: to obtain saturation in a reasonable time, comparatively high annealing temperature should be used, at which recovery effects would be discernible.

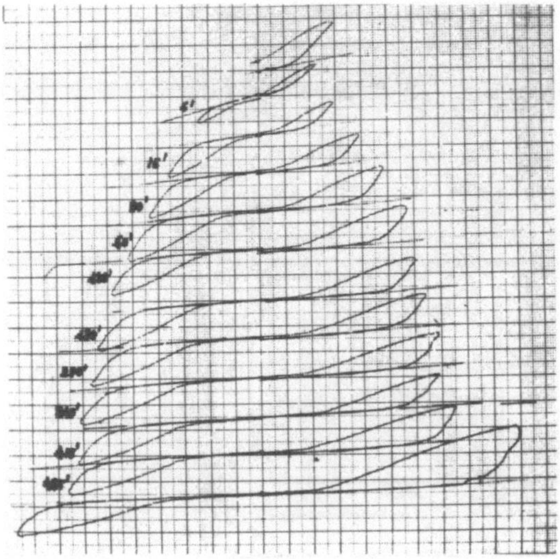

Fig. 2. A typical set of viscosity field measurements as directly traced by means of electronic hysteresigraph.

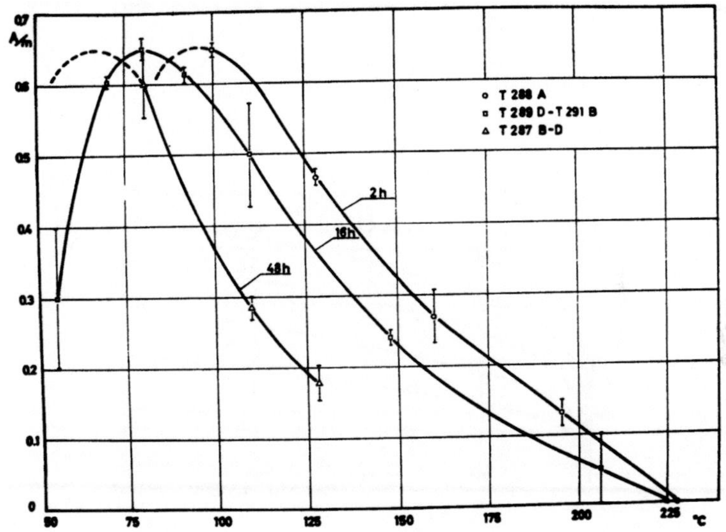

Fig. 3. Isochronal annealing experiments: viscosity field in A/m versus annealing temperature in °C.

(a)

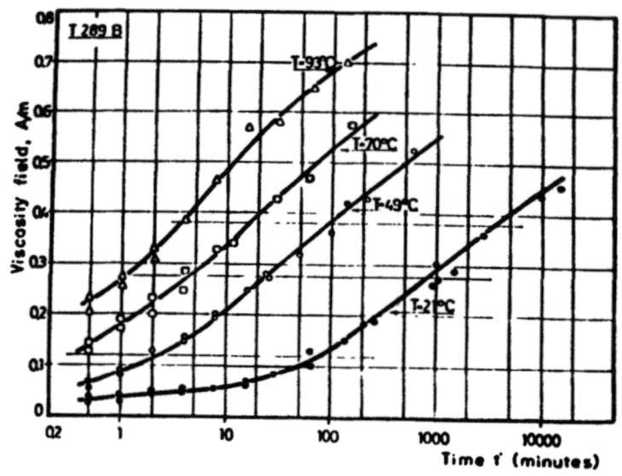

Fig. 4. Isothermal annealing experiments: viscosity field in A/m versus annealing time in minutes. Logarithmic time scale is used in Fig. 4a.

(b)

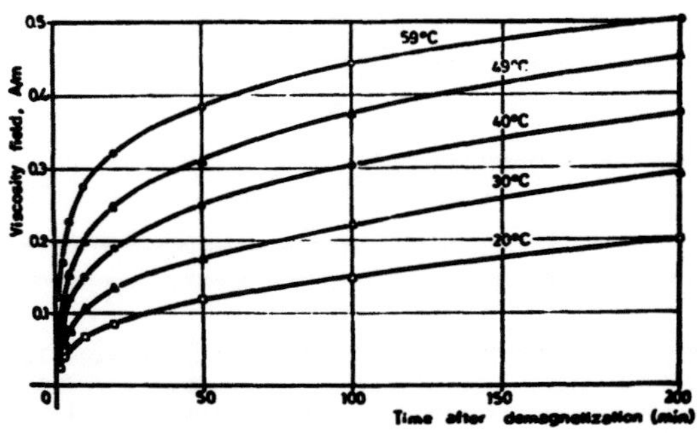

Fig. 4. Isothermal annealing experiments: viscosity field in A/m versus annealing time in minutes. Linear time scale is used in Fig. 4b to show earlier stages of relaxation.

The recovery of the magnetic damage was also followed by means of viscosity field measurements. The main part of recovery occurs below 160°C, and is almost complete at 250°C, some minor viscosity effects being, however, observed up to 400°C.

Hardness Measurements

After the second irradiation, hardness recovery was followed by means of isochronal treatments on three specimens. Hardness recovery occurs between 250°C and 400°C. This range of temperature is in agreement with the data of Kunz and Holden (12) on the recovery of yield strength of iron after neutron irradiation.

ANALYSIS OF EXPERIMENTAL RESULTS

Defect Anisotropy

The results in Figure 1 show that the permeability changes induced by the irradiation are due to an induced anisotropy energy (interaction of the irradiation defects with the magnetization). Furthermore, these results show that a model of isotropic defects like simple vacancies cannot hold: in fact, under this hypothesis specimens irradiated under magnetic field (in the absence of Bloch walls) should not show any variation of the magnetization curve before and after irradiation.

Permeability Change and Magnetic Viscosity

The common origin of these effects is shown by the fair agreement on the values, obtained in different ways, of the induced anisotropy energy W; W can be evaluated either as the area between the magnetization curves before and after irradiation (within a factor close to 1) (13), or as the product J_s, where H_∞ is the saturation viscosity field (from Figure 3, $H_\infty \simeq 1$ A/m) and J_s = 2.16 Wb/m^2 is the saturation magnetization. W is of the order of 2 Joule/m^3 from both evaluations.

Activation Energy and Relaxation Time

From the temperature shift of the isochronal curves one obtains as a rough estimate: Q = 0.9 - 1.0 eV. A more precise evaluation is obtained from the isothermal data. Each curve is given by the superposition of several relaxation components, which can be tentatively assumed to be characterized by the same activation energy. The viscosity field dependence from the annealing time t' after demagnetization, in the absence of recovery, can then be written

$$H(t',T) = H_\infty \left[1 - g(t' e^{-Q/kT})\right] \tag{1}$$

where g is a function which decreases from 1 to 0 in the course of time t'. To obtain Q one observes that a given amount of relaxation corresponds to a constant ($t' e^{-Q/kT}$) (14). Therefore:

$$\ln t' = \text{const.} + Q/kT \quad (2)$$

In evaluating Q the isothermal curves were renormalized with respect to a 1/T factor to take into account the dependence of H_∞ on T, which is of the form (15): H_∞ = const./T.

From Fig. 5 one finds Q = 0.85 ± 0.05 eV. The good linearity and parallelism of the lines corresponding to different amounts of relaxation confirm the validity of the assumption of a single activation energy for the earlier stages of relaxation. However, as is clear from Fig. 4a, later stages seem to be characterized by a somewhat higher Q (∿ 1 eV). The interception of lines of Fig. 5 with the 1/T = 0 axis provides an estimate for the average τ_0'. The value $\tau_0' = 10^{-11}$-10^{-12} sec confirms that magnetic viscosity is due to the orientation of point defects.

Nature of the Defects

By plotting the activation energy Q of the stage III of resistivity recovery of b.c.c. and f.c.c. metals after neutron irradiation, cold work and quenching, versus the melting temperature T_m (Figure 6) one obtains a closely linear relationship between Q and T_m. A similar relationship holds for the recovery temperatures. This suggests that the recovery stage III is in all cases associated with the migration of the same type of defects, which are generally identified as vacancies. In the present case the values obtained for Q and the recovery temperature fit well the linear relationship of Figure 6. Since the defects are anisotropic it seems reasonable to identify them, in agreement with Moser, as vacancies associated with impurities, or possibly divacancies.

Defects Concentration

The defect concentration is related to H_∞ by (15):

$$H_\infty = (c\, w^2)/(3\ kT\ J_s) \quad (3)$$

Assuming for w, interaction energy of the defect with J_s, a value of 1.6 - 3.0x10^{-16} erg/defect, since this is always the order of magnitude for different solute atoms in iron (15), and for H_∞ a value of 1 A/m, one obtains: c = 0.5 - 2.0x10-4 defects/atom. From the neutron flux data one obtains: c = 0.9x10-3 Frenkel-pairs/atom. Taking into account the room temperature partial recombination, the two estimates of c seem to be in fair agreement.

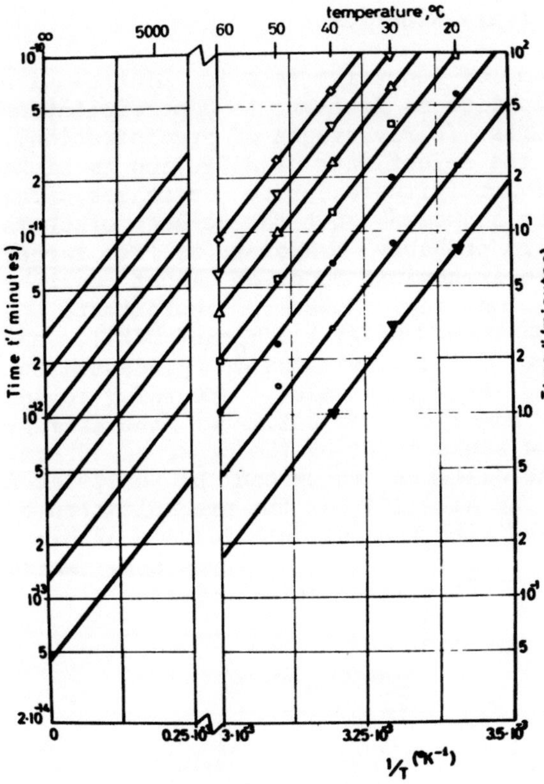

Fig. 5. Determination of Q and τ'_0 from isothermal data.

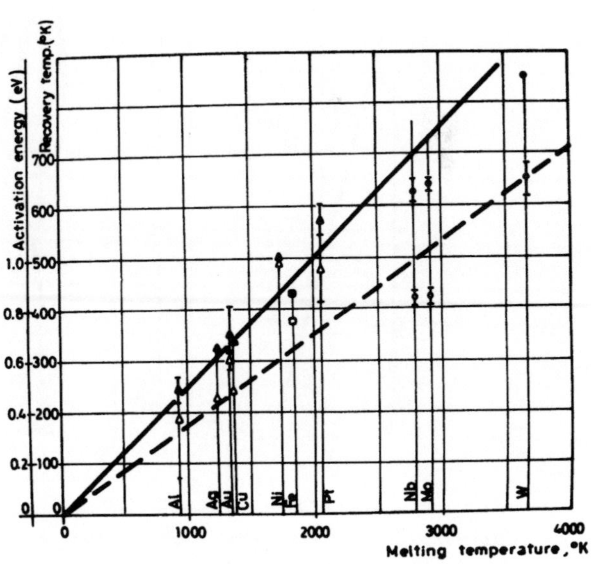

Fig. 6. Resistivity recovery Stage III for b.c.c.(dots) and f.c.c. (triangles) metals after irradiation, cold work, or quenching. Activation energy (black dots and triangles, full line) and recovery temperature (open dots and triangles, dashed line) versus melting temperature in °K. Squares show data obtained for irradiated iron from magnetic measurements.

Recovery Kinetics

In Figure 7 the logarithm of recovery rate is plotted as a function of 1/T. Recovery kinetics is similar to a precipitation kinetics. At lower temperatures (early stages of precipitation), the process is controlled by the defect rate of diffusion to sinks. At higher temperatures, the precipitation rate decreases after reaching a maximum. It can be supposed that the defect clusters acting as precipitation centers probably evaporate to form larger defects. Therefore, a simple Arrhenius equation cannot describe the whole recovery process. One may, however, approximate the recovery at early stages by an equation $1/\tau = 1/\tau_0 \exp(-Q/kT)$, and assume $Q = 0.85$ eV as for the defect orientation process. One finds (Fig. 7): $\tau_0 = 10^{-5}$ sec. From this value, according to the analytical solution given by Ham (16) one estimates the average distance between precipitation sinks to be of the order of 10^{-4}cm. This is in agreement with the distance which can be calculated from the data of Wagenblast et Al. (17) on the precipitation of carbon in irradiated iron, and with the average distance of black spots observed by Eyre (18) using the electron microscope, which were interpreted as clusters of irradiation point defects.

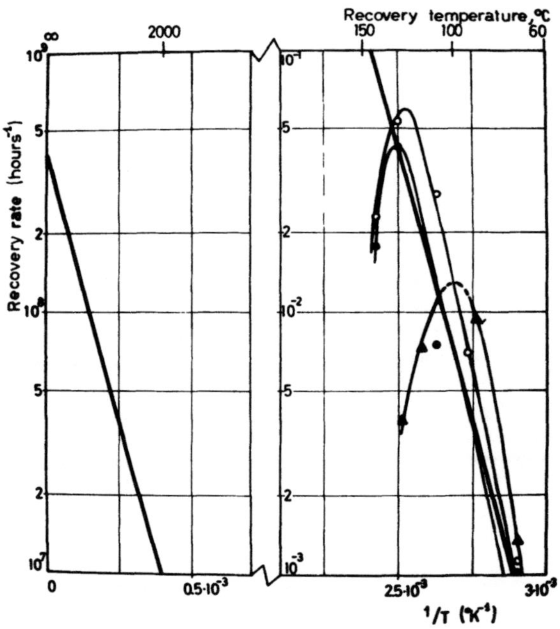

Fig. 7. Logarithm of rate of recovery versus reciprocal of recovery temperature for different samples. Straight line corresponds to Q=0.85 eV.

Comparison of Magnetic and Hardness Recovery

This comparison shows that no relation exists between irradiation defects responsible for magnetic and mechanical properties changes. Hardness recovery after irradiation was also compared to hardness recovery after cold work: the former occurs between 250 and 400°C; the latter between 400 and 500°C. It is, therefore, clear that no direct relation exists between defects responsible for hardness change in the two cases.

CONCLUSIONS

In the present work, clear evidence was obtained to show that the permeability change observed earlier in irradiated iron is due to an anisotropy energy. This energy originates from the interaction with the magnetization of irradiation point defects, whose anisotropic nature was clearly demonstrated. It was shown that these defects were observed somewhat above room temperature. By means of viscosity field measurements, an activation energy of 0.85 eV was evaluated for the earlier stages of the defects orientation diffusion. Later stages seem to be characterized by somewhat higher values of Q($\sim$1 eV). The evaluation of Q allowed us to identify the defects in agreement with Moser, as vacancy-impurity couples or possibly as divacancies. From the maximum viscosity field value, which is of the order of 1 A/m, a concentration of about 10^{-4} defect/atom was estimated, which seems in fair agreement with the concentration evaluated from the neutron flux data, taking into account room temperature recombination. The study of the magnetic damage recovery showed that this is similar to a precipitation kinetics. Earlier recovery stages are controlled by point defect diffusion to sinks, while the sharp decrease recovery rate in later stages may be associated with the evaporation of annihilation centers. An average distance between precipitation sinks of the order of 10^{-4} cm was estimated from recovery data, in agreement with the distance which can be calculated from other authors' data. Finally it was shown that the irradiation defects which influence the magnetic properties are not responsible for the irradiation induced change of hardness, whose recovery occurs in a higher temperature range.

ACKNOWLEDGMENT

This research was sponsored by European Energy Community (EURATOM), in the framework of the joint USAEC - EURATOM.

REFERENCES

1. *G. Biorci, A. Ferro and C.Montalenti, J. Appl. Phys., Vol. 31, No. 11, pp. 2046-7, 1960.*

2. *G. Biorci, A. Ferro and G. Montalenti, Nuovo Cimento, Vol. 20, No. 5, pp. 617-37, 1961.*

3. *A. Ferro and G. Montalenti, Nuovo Cimento, Vol. 28, No. 6, pp. 1512-6, 1963.*

4. *P. Moser, D. Dautreppe and P. Brissonneau, C. R. Acad. Sci., (France), Vol. 250, No. 24, pp. 3963-5, 1960 (In French).*

5. *P. Moser and D. Dautreppe, J. Phys. Radium, Vol. 24, No. 7, pp. 516-21, 1963 (In French).*

6. *P. Moser, P. Peretto, D. Dautreppe and P. Vigier, J. Appl., Phys., Vol. 36, No. 3 (part 2), pp. 1227-8, 1965.*

7. *E. Balthesen, K. Isebeck and H. Wenzl, Phys. Stat. Sol., Vol. 8, No. 2, pp. 603-11, 1965.*

8. *P. Vigier and P. Moser, C. R. Acad. Sci. (France), Vol. 260, No. 9, pp. 2443-6, 1965 (In French).*

9. *E. Balthesen, K.Isebeck and H. Wenzl, Phys. Stat. Sol., Vol. 8 No. 2, pp. 593-602, 1965.*

10. *H. D. Dietze, Techn. Mitt. Krupp, Vol. 17, No. 2, pp. 67-81, 1959 (In German).*

11. *P. Mazzetti and P. Soardo, Rev. Sci. Instr., Vol. 37, No. 5, pp. 548-52, 1966.*

12. *F.W. Kunz and A.N. Holden, Acta Metallurgica, Vol. 2, No. 11, pp. 816-22, 1954.*

13. *G. Biorci, A. Ferro and C. Montalenti, J. Appl. Phys., Vol.31, No. 12, pp. 2121-5, 1960; J. Appl. Phys., Vol. 32, No. 4, pp. 630-5, 1961.*

14. *A.C. Damask and C.J.Dienes, pp. 250-1, Gordon and Breach, New York (1963).*

15. *A. Ferro, P. Mazzetti and G. Montalenti, Nuovo Cimento, Vol. 23, No. 2, pp. 280-91, 1962.*

16. *F.S. Ham, J. Phys. Chem. Solids, Vol. 6, No. 4, pp. 335-51, 1958.*

17. *H. Wagenblast and A.C. Damask, J. Phys. Chem. Solids, Vol. 23, No. 3, pp. 221-7, 1962.*

18. *B.L. Eyre, Phil. Mag. Vol. 7, No. 12, pp. 2107-13, 1962.*

DISCUSSIONS

D. DAUTREPPE (Centre d'Etudes Nucleaires, Grenoble, FRANCE): *What is the impurity content of your pure iron?*

A. FERRO: *About ≤ 200 ppm of substitutional impurities. The impurities are mainly silicon and copper.*

LATTICE PARAMETER CHANGE BY POINT DEFECTS IN IRRADIATED ALUMINUM

H. Peisl and W. Waidelich

I. Physikalisches Institut der Technischen Hochschule

Darmstadt, GERMANY

The distortion of a crystal lattice by point defects was investigated by measuring the lattice parameter change of irradiated crystals. Pure aluminum foils were irradiated at 4.6°K with neutrons in the low temperature irradiation facility of the Munich Research Reactor at Garching. Without warming up, the irradiated samples were transferred to a special helium cryostat for the X-ray measurements. The relative lattice parameter change after irradiation with 2×10^{18} *nvt was* $\Delta a/a = (4.47 \pm 0.4) \times 10^{-4}$. *This irradiation induced lattice parameter change anneals completely during an isochronal annealing run in three distinct stages: (1) 10°K - 50°K; (2) 50°K - 120°K; and (3) 150°K - 250°K. Comparing these results with recovery measurements of the residual electric resistivity, stored energy, and, length change, one can determine the volume change of one Frenkel defect to be 1.4 atomic volumes. Since* $\Delta a/a$ *is equal to* $\Delta l/l$, *new lattice sites cannot be produced during the thermal annealing of the defects. This means that each annealing stage must be due to the recombination of interstitials and vacancies in different configurations.*

INTRODUCTION

The production of point lattice defects in a crystal changes many of its physical properties. In principle each of these properties can be investigated to study the number and the nature of the defects. In the investigation of lattice defects in metals after high energetic particle bombardment most of the information has been obtained from residual electrical resistivity measurements. This is mainly because it is experimentally not very difficult and furthermore it is one of the most sensitive properties.

From this and some other techniques very good information about the number of defects can be obtained, but no direct determination of the defect type is possible. Perhaps one of the most direct methods of investigating lattice defects is to determine the distortion of the crystal lattice by the defects. This distortion causes a uniform lattice parameter change $\Delta a/a$, which is directly proportional to the defect concentration c_F.

If one succeeds in measuring the lattice parameter change very accurately, i.e. with an accuracy of 10^{-5} or better, one has a very straightforward measure for the defect concentration. Furthermore, the type of the defects can be determined from the comparison of the lattice parameter change and length change.

EXPERIMENTAL

For the first time, lattice defects produced after low temperature neutron irradiation were investigated by lattice parameter measurements. Pure aluminum foils were irradiated at 4.6°K in the low temperature irradiation facility of the Munich Research Reactor FRM by a fast flux of 10^{13} neutrons / cm^2 sec. (E > 0.1 MeV) (1). After irradiation at 4.6°K for about 60 hours and decay of the short-lived isotopes, the sample can be lifted without warming up into a transport cryostat which is shown in Figure 1. The inner container can be filled with liquid helium and then the sample can be brought inside it and immersed in the liquid helium. The liquid helium filling lasts for half an hour, which is enough to dismount the transport cryostat from the low temperature irradiation facility and take it to another place. There, it is mounted in a liquid helium container and the sample can be emersed into liquid helium. The liquid helium container with the sample inside it can be transported to any place for decay of longer-lived radioactive isotopes and for performing the X-ray measurements.

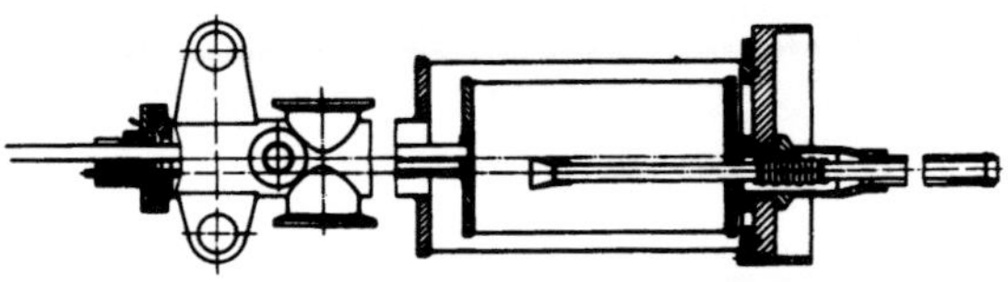

Fig. 1. Transport cryostat.

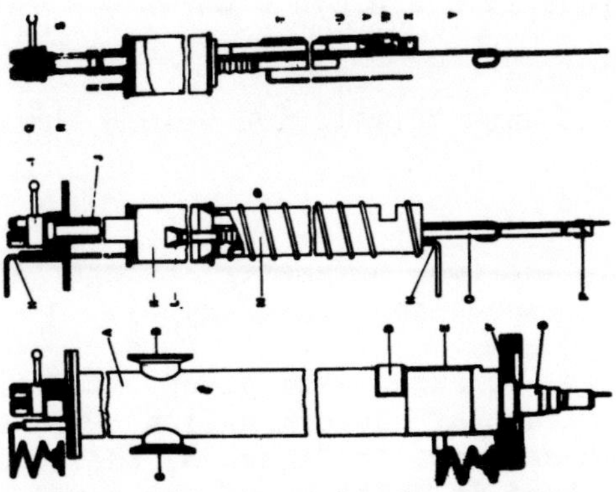

Fig. 2. Measuring cryostat.

The same principle as that used with the transport cryostat is used again to bring the sample into the measuring cryostat (Fig. 2) for the X-ray measurements. The container K is filled with liquid helium and then the sample is brought into the measuring position V where it is kept at liquid helium temperature in a He-exchange gas atmosphere which is cooled by a heat exchanger. The amount of liquid helium pumped through the heat exchanger is controlled by an arrangement consisting of a thermocouple, potential measuring bridge, Galvanometer, photo resistance and electromagnetic valves. Thus every temperature between 4^oK and room temperature can be obtained, and with an additional heating arrangement X temperatures above room temperature can also be obtained. Two windows of beryllium permit entrance and exit of the X-rays. The lattice parameter change is determined from the shift of the X-ray diffraction peak in the back reflection region.

To start with, the thermal expansion of aluminum was measured in order to examine the apparatus and furthermore to test the thermal and mechanical stability of the cryostat. Comparison with literature data was very satisfactory. The whole arrangement permits an accuracy of $\Delta a/a = 5 \times 10^{-6}$.

RESULTS AND DISCUSSION

After irradiation at 4.6^oK with 2×10^{18} neutrons/cm^2 the relative lattice parameter change of aluminum is

$$\Delta a_o/a = (4.47 \pm 0.4) \times 10^{-4}$$

The defect concentration was found from the change of the residual electrical resistivity to be $c_F = 9.4 \times 10^{-4}$. From this we can deduce the volume change per defect:

$(\Delta V/V)/c_F = (3\ \Delta a/a)/\ c_F = f_F = 1.4$

The distortion coefficient is $C = 0.075\ v$ where $v = (a/2)^3$, the atomic volume.

The ratio of resistivity change to relative volume change is:

$\Delta\rho/(\Delta V/V) = 2.4 \times 10^{-4}\ \Omega\text{cm}$

The irradiation induced lattice parameter change anneals completely during an isochronal annealing run (Fig. 3).

Annealing pulses of 10 min duration were given at different temperatures. The lattice parameter determination was always performed at liquid helium temperature. In Figure 3, the remaining lattice parameter change is given in percent of the total change $\Delta a_0/a = (4.47 \pm 0.4) \times 10^{-4}$. This total change is obtained as the difference of the lattice parameters after irradiation and after annealing. The annealing takes place in three distinct recovery stages:

Stage I	:	10 - 50°K
Stage II	:	50 - 120°K
Stage III	:	150 - 250°K

Annealing of the lattice parameter change means the disappearance of defects which cause a remarkable lattice distortion. Change of the number of lattice sites would not be observable as a lattice parameter change. More information can be obtained from comparison with other measurements.

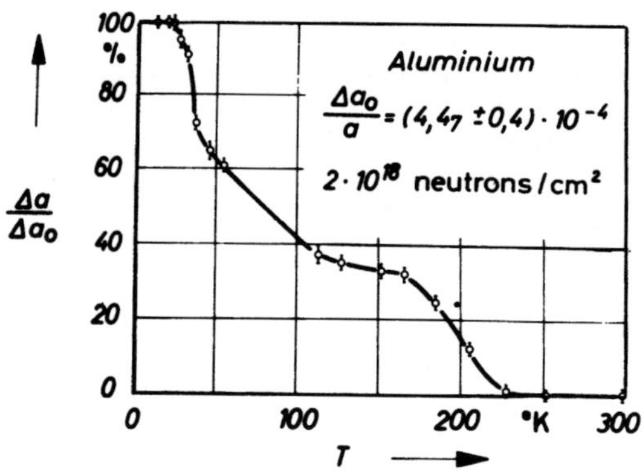

Fig. 3. Isochronal annealing of the relative lattice parameter change of pure aluminum after low-temperature neutron irradiation.

Figure 4 shows, for comparison, the annealing behavior of the length change, the residual electrical resistivity and the stored energy measured by WENZL et al. (2). For low-temperature neutron irradiated aluminum all quantities are essentially equal in all recovery stages.

If no new lattice sites are created during irradiation and thermal annealing of the defects, the relative length change $\Delta l/l$ should be equal to the lattice parameter change $\Delta a/a$ (3). Figure 4 shows that within experimental error both quantities are equal during the whole annealing procedure, and therefore recovery in each stage must be due to the recombination of interstitials and vacancies, whatever the original configuration of these defects. Nothing can be said about which defect is mobile in the different stages.

Figure 4 also shows that the relative lattice parameter and residual electrical resistivity change have the same annealing behavior. The annealing volume change $\Delta V/V = 3\ \Delta a/a$ divided by the annealing resistivity change $\frac{\Delta V/V}{\Delta\rho} = 0.4$ is the same for all three stages. The volume changes by an interstitial and a vacancy are different, the volume change by the interstitial being large and positive, that of a vacancy being small and likely negative. The resistivity changes are of the same order of magnitude for vacancies and interstitials. To obtain the same ratio in each stage it is necessary that the same annealing mechanism takes place, that is, recombination of interstitial vacancy pairs. Hence it is not possible that a single defect anneals in one stage and another defect in the next stage, since it must recombine with its complementary defect.

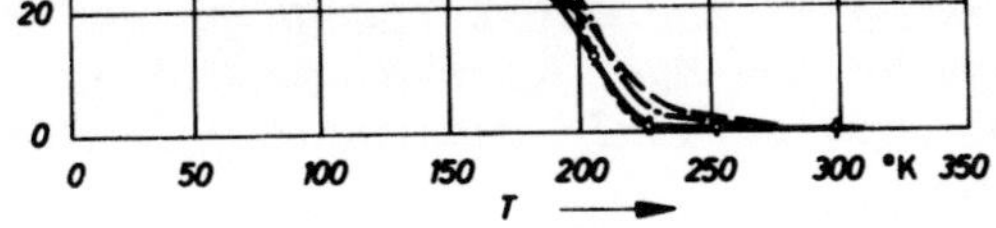

Fig. 4. Annealing of lattice defects in aluminum after low-temperature neutron irradiation (+ Results from WENZL et al. (2)).

ACKNOWLEDGMENTS

This experiment was performed together with A. Sepp and U. Himmler, I. Physikalisches Institut der TH Darmstadt. We thank them and H. Wenzl, Physics Department of the TH München, for the excellent cooperation. We gratefully acknowledge the support of Prof. Dr. H. Maier-Leibnitz and Prof. Dr. N.Riehl, Physics Department TH München, and the help always given by the reactor staff at FRM.

Funds were given by the Bundesministerium fur Wissenschaftliche Forschung and the Deutsche Forschungsgemeinschaft.

REFERENCES

1. *R. Doll, H. Meissner, N. Riehl, W. Schilling and F. Schmeisner, Z. Angew. Phys. 17, 321-329 (1964).*
H. Meissner, W. Schilling and H. Wenzl, Euronuclear 2, 277-280 (1965).

2. *H. Wenzl, W. Schilling and K.Isebeck, International Conference on Electron Diffraction and Crystal Defects, Melbourne 1965.*

3. *R.O. Simmons, Radiation Damage in Solids; pp. 568-593, New York, Academic Press Inc., 1962.*

DISCUSSIONS

A. D. ROSSIN (Argonne National Laboratory, U.S.A.): *Various annealing peaks have been noted in several metals at a particular fraction of the absolute melting temperature. Each of these is attributed to a different mechanism. Would you comment?*

H. PEISL: *The mechanism could be different but the last step always has to be the recombination of a vacancy - interstitial pair. In all three stages they disappear simultaneously.*

P. LUCASSON (Faculte des Sciences, Paris, FRANCE): *If one considers various irradiated metals, it is not quite right that all the recovery stages occur at temperatures proportional to that of the melting point. This is true for stages III, IV, V, but Stage I is the exception. It is possible to correlate the temperature of Stage I with the γ Gruneisen constant, which is related to the atomic elastic properties of the metal (Lucasson and Welkes-1962 - Phys. Rev.). This might be due to the existence of an elastic interaction between interstitial and vacancy in Stage I. However, this is no way a proof.*

H. PEISL: *No further answer to this problem can be given from our measurements besides the remarks given in the answer to the question of Dr. Rossin.*

F.H. HAMMAD (Atomic Energy Establishment, U.A.R.): *What is the purity of the Al you have used?*

H. PEISL: > 99,9999%.

A METHOD FOR STUDYING POINT DEFECTS - SPECIALLY INTERSTITIAL ATOMS IN METALS

Claude Budin and Pierre Lucasson

Faculte des Sciences de Paris

Orsay, FRANCE

A difficulty in the identification and in the study of interstitial atoms in metals is that there is no way to create them without producing other species of crystalline defects at the same time. Damage formation curves versus incident particles flux, as well as isochronal or isothermal recovery data, makes it possible to identify the domains of mobility of interstitial atoms compared to that of vacancies in the same metal, by combining electron irradiation and quenching experiments. It was also possible to obtain values of intrinsic properties such as formation energy, resistivity and formation volume.

Studies of the electrical resistivity behavior in quenched and quenched-irradiated aluminum and copper are reported. Recently obtained results concerning identification, movement energy and intrinsic resistivity of interstitial atoms are given.

INTRODUCTION

During the past decade, the properties of point defects - namely, vacancies and interstitial atoms - have very often been studied in F.C.C. metals by means of three fundamental production techniques: a) high temperature, b) irradiation, c) plastic deformation.

In spite of the large amount of work done in this field, one can but acknowledge that, as far as vacancies are concerned, the number of reliable conclusions which have been reached so far is small, and is almost nil as regards interstitials. One possibility of overcoming the limitations inherent in each of these production techniques is to use more sophisticated methods. In what follows, the features and limitations of each fundamental technique will be briefly reviewed. It will also be shown that dual experiments, in which quenches from high temperatures are associated with electron irradiation, can be used to obtain information on vacancies and interstitial atoms in metals, provided that experimental conditions permitting quantitative analysis of the raw data are fulfilled. Finally, some results obtained in such a series of dual experiments on aluminum will be discussed.

EXPERIMENTAL

Limitations of the Fundamental Production Techniques

High temperature: It is an important feature of F.C.C. metals that due to the rather high value of the interstitial formation energy compared with that of a vacancy, only vacancy type defects are produced in appreciable concentrations when the temperature is raised. This is the reason why vacancies are the best known of point defects.

The most widely used technique for studying these thermally generated vacancies is that of quenching, in which the high temperature equilibrium state is tentatively frozen by rapid cooling to a sufficiently low temperature, where precision measurements can be conveniently made. The major drawback in this process comes from the finite quenching rate compared to the high vacancy mobility at elevated temperatures, resulting in the formation of di, tri... multivacancies which are frozen together with single vacancies. This does not matter too much in formation energy determination, but when anneals are subsequently performed, the reaction kinetics is affected, due to the mobilities of di and tri vacancies which are higher than that of singles. For example, in aluminum (2,3), isochronal anneals show that the quench induced resistivity disappears at temperatures which are higher when the quenches have been conducted at lower temperature, due to the growing scarcity of di and tri vacancies in the population of singles. Moreover, isothermal anneals show that quenching temperatures higher than 400°C result in rather intricate annealing processes whose analysis is involved (3). On the other hand, in the case of quenches from 300°C or under, divacancy formation is almost entirely avoided, thus resulting in recovery processes whose order is close to one with an apparent activation energy not too different from that for vacancy motion E^v_m (4). Quite generally, however, the precise determination of E^v_m in quenched metals is a difficult task.

Irradiation techniques: Among these, irradiation with electrons in the MeV range probably affords the best way of studying point defects. This is mainly because in this case only pairs of single vacancies and interstitials are created at random in the specimen's lattice. It is well known that the irradiation induced resistivity recovers in five stages, the first four being due to point defect migration. The drawback, in this method, is that identification of the migrating defect is difficult, and still a subject of controversies in the best known cases (5,6). This is essentially the reason why the properties of interstitial atoms are not well known, although irradiated metals have been extensively studied.

Plastic deformation has been used to produce point defects of both kinds in metals. However, the study of the defects produced by this method is even more difficult than in irradiated metals, because the respective concentrations of interstitials and vacancies are not known, and, moreover, numerous dislocations are created at the same time.

Use of Dual Technique

Dual techniques, associating two production processes, have sometimes been used to study point defects in metals (see for example (7)). Rather surprisingly, they have not been used for defect identification purposes, and rarely have they allowed quantitative treatments of experimental data. Both of these aims are attainable, however, particularly by associating quenches and electron irradiations.

For example, let us suppose that after a quench, an irradiation is performed at a temperature where interstitial atoms are likely to be mobile. Their identification should be easy, because reactions such as the following should occur:

$$\underset{\text{mobile}}{I} + \underset{\text{from quench}}{V_n} \longrightarrow V_{n-1} \qquad (1)$$

I : interstitial atom
V_n: vacancy agglomerate containing n singles

where n = 1, and also 2, 3,... if the quench is from high temperature. This reaction may thus result in both a vacancy concentration dependence of the Frenkel pair production rate, and an increased purity of the remnant single vacancy population.

In a general way such a dual technique will allow the study of both the influence of the preliminary quench on the reaction kinetics of an irradiation recovery stage, and the effect of irradiation on the recovery of quenched defects. This technique will accordingly provide two sets of experimental data which will be readily available for identification purposes and qualitative discussions. For a quantitative discussion to be possible, however, it is necessary that the *experimental* procedure be such that each process is simple enough to allow a mathematical analysis. Thus the quenches should be performed from the lowest possible temperature, and one of the irradiated metal recovery stages should also be studied. A most promising choice is that of Stage III, which has several interesting features: its kinetics is well known and simple, with an order of reaction 2 and a well defined activation energy. Moreover, it is likely to correspond to an annihilation process, thus allowing the use of the experimental data for making quantitative estimates of a given intrinsic property of the interstitialcy involved in the reaction, in terms of the corresponding property of a vacancy. Such an estimate is made for the resistivity of an aluminum interstitial atom in the next section. It is worth mentioning, however, that the formulas established would apply as well for any other specific property such as formation energy or volume.

Quench + Irradiation Study of Defects in Aluminum

The samples used in these experiments were 0.1 mm diameter wires, furnished by Johnson-Mattheys. They had been annealed to 350°C for several hours in a helium atmosphere prior to quenching or irradiation. The samples were quenched down to liquid nitrogen temperature from temperatures between 300°C and 380°C; the rate of cooling being about 7000°/sec. Irradiations were performed at -170 ± 1°C, using 1.45 MeV electrons. The point defects produced by either of these techniques were studied by electrical resistivity measurements at liquid nitrogen temperature, using a potentiometric method. The precision attained was $\pm 5 \times 10^{-12}$ Ωcm. The apparatus has been described previously (9,8). It is worth noting that the *same* samples were used for both quenches and irradiations; all of the experiments were performed with the same sample holder.

Heating was produced by passing a direct current through the samples: this results in a heterogeneous temperature distribution, due to the thermal conductivity of the potential leads. As a consequence, the concentration of vacancy type defects quenched is not constant along the sample, while the irradiation process gives a constant Frenckel pair concentration distribution. This difference has to be taken into account in the quantitative analysis of the raw data. As the procedure used is described in detail elsewhere (1), unless specified only corrected results are given in this article.

RESULTS

Quenching Experiments

A number of isochronal annealing runs were performed after quenching. The curves (a) and (b) of Figure 1 show typical results obtained after quenching from 300°C and 380°C respectively (increases in resistivity were 1.21 and 5.6 x $10^{-9}\Omega$cm). Note how the main recovery stage (Stage IV) shifts towards low temperatures when the quenching temperature is raised. These results are in general agreement with previously published data (2,3). In this work, we will take care of Stage IV only, since Federighi and Panseri (2) have shown that it is the only stage corresponding to the migration of quenched defects.

Isothermal studies of the kinetics were performed by the authors in the Stage IV temperature range, together with activation energy determinations by Palmer's method. They show that after quenches from 300°C, the process is first order with an apparent activation energy $E^*_{iv} = 0.67 \pm 0.03$ eV. For a quench from 380°C, the order is between 1 and 2, and the apparent activation energy increases from 0.42 eV to 0.54 eV while the defect concentration decreases.

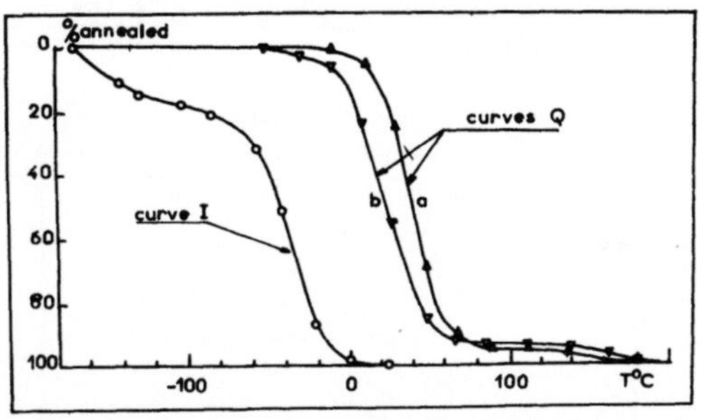

Fig. 1. Isochronal recovery (5 minutes anneals) of the electrical resistivity induced by quench (curves (Q)) and by 1.45 MeV electron irradiation (curve (I)).

Δ quench from 300°C ($\Delta\rho^o_Q = 1.21 \times 10^{-9}\,\Omega$ cm)

∇ quench from 380°C ($\Delta\rho^o_Q = 5.6 \times 10^{-9}\,\Omega$ cm)

o irradiation at -170°C ($\Delta\rho^o_I = 2.7 \times 10^{-9}\,\Omega$ cm)

Irradiation Experiments

Defect production. The curve (I) in Figure 2 is a plot of resistivity increase versus electron flux in several irradiation runs (the flux values given are uncorrected raw experimental data).

Stage III recovery process. Numerous isochronal and isothermal annealing runs were performed after different values of resistivity increases $\Delta\rho_I^o$ were produced by electron irradiation. The curve I in Figure 1 shows a typical isochronal curve corresponding to a value of $\Delta\rho_I^o = 2.7 \times 10^{-9}\Omega cm$ just intermediate between the values of $\Delta\rho_Q^o$ of the curves a and b drawn in the same diagram. One sees that the temperature domain of Stage III is widely different (by $\sim$ 60°C) from that of the quenched vacancy migration.

Prior to performing isothermal anneals or activation energy determinations, a - 80°C anneal was performed in order to get rid of the defects migrating in Stage II. Curve (I) in Figure 3 is an isothermal recovery curve $\Delta\rho = f(t)$ at a temperature T = 41°C. From this, plotting log $d\Delta\rho/dt$ versus log $\Delta\rho$, one can see that the reaction order is 2 except for 10% of the total stage at the beginning of the stage, where the order is higher. This may be explained either by the existence of a specific substage III_a, or by "Waite's effect", stemming from an unrandom distribution of Frenckel pairs. We will not discuss this point here, and wish only to point out that the quantitative analysis is to be applied to the main part of the stage only.

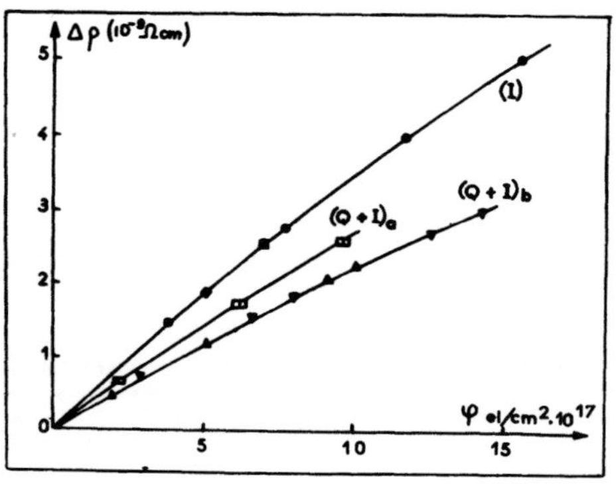

Fig. 2. Production curve for sample irradiated by 1.45 MeV electron at -170°C. Similar dots correspond to the same irradiation run.
Curve (I) : Irradiation only
Curve $(Q+I)_a$: Irradiation with previous quench $\Delta\rho_Q^o = 1.29 \times 10^{-9}\Omega cm$
Curve $(Q+I)_b$: Irradiation with previous quench $\Delta\rho_Q^o = 2.24 \times 10^{-9}\Omega cm$

Several activation energy determinations have been performed both by the slope change method and by Palmer's method. The first one, using isothermal anneals at - 36°C and - 49.5°C, gave $E_{III} = 0.57 \pm 0.02$ eV. The second one, using in turn anneals at temperatures of - 29°C and - 41.5°C, showed $E_{III} = 0.59 \pm 0.02$ eV. Thus E_{III} will be taken as 0.58 ± 0.03 eV.

Quench-Irradiation Dual Fxperiments

Defect production. The defect production curves $(Q + I)_a$ and $(Q + I)_b$ shown in Figure 2 have been obtained by irradiating the same specimen used to draw the curve (I), but only after resistivity increases of 1.29×10^{-9}Ωcm and 2.24×10^{-3}Ωcm were respectively produced by quenching. One can see that the stronger the quench, the lower $d\Delta\rho/d\psi$ is for a given $\Delta\rho$ or ψ value. Let us notice that, in this set of experiments, irradiations (I) and dual experiments (Q + I) were performed in turn, thus getting rid of any possible unnoticed change in the sample properties.

Recovery studies of stages III and IV. Four isochronal recovery curves were drawn after dual quench + irradiation treatment (Q + I). Two of them are shown in Figures 4 and 5, corresponding to experiments where the resistivities added by irradiation were the same: in Figure 4 the quench was from a low temperature (T_Q = 300°C), and in Figure 5, the quench was from an intermediate temperature (T_Q = 380°C). These results show that a preliminary quench always lowers the temperature domain of Stage III. As regards Stage IV, one sees that if the quench is from a low temperature, there is no visible change; but the recovery curve is shifted towards *high* temperatures if the quench is from a higher temperature.

In the Stage III temperature range, two isothermal recovery curves after (Q + I) dual experiments are shown in Figure 3. The curve $(Q + I)_a$ corresponds to a preliminary quench from 300°C, and the curve $(Q + I)_b$ corresponds to a quench from 325°C. The values of resistivity increases due to irradiation were approximately equal in both cases. The results show that the Stage III recovery process is accelerated by a preliminary quench, and that the effect is stronger the stronger the quench. A log $d\Delta\rho/dt$ versus log $\Delta\rho$ plot shows that the order of reaction is no longer 2: it is always smaller and tends towards 1 when the remaining resistivity increase tends towards zero.

With respect to Stage IV, the apparent activation energy was determined by Palmer's method after a quench from 325°C followed by electron irradiation. This quenching temperature was chosen in order to obtain a concentration of divacancies small but detectable in the isochronal runs. E^*_{IV} values increasing steadily

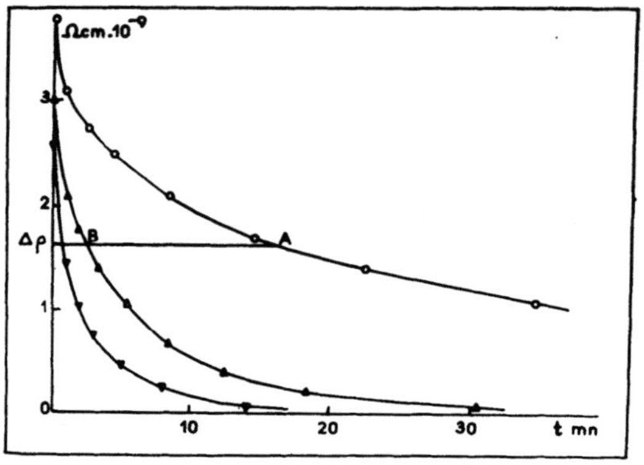

Fig. 3. Isothermal recovery of the electrical resistivity induced by irradiation only (curve (I)) and by irradiation after a quench (curve (Q+I)). The annealing temperature was-41°C.

o Curve (I)

Δ Curve (Q+I): $\Delta\rho^o_Q$ = 1.18 x $10^{-9}\Omega$ cm

∇ Curve (Q+I): $\Delta\rho^o_Q$ = 2.32 x $10^{-9}\Omega$ cm

Fig. 4. Isochronal recovery (5 minutes anneals) of the electrical resistivity induced by electron irradiation after a preliminary quench ($\Delta\rho^o_Q$ = 1.29x$10^{-9}\Omega$cm). (Raw experimental data).
—o— experimental curve (Q+I)
— — — theoretical curve ((Q)+(I)).

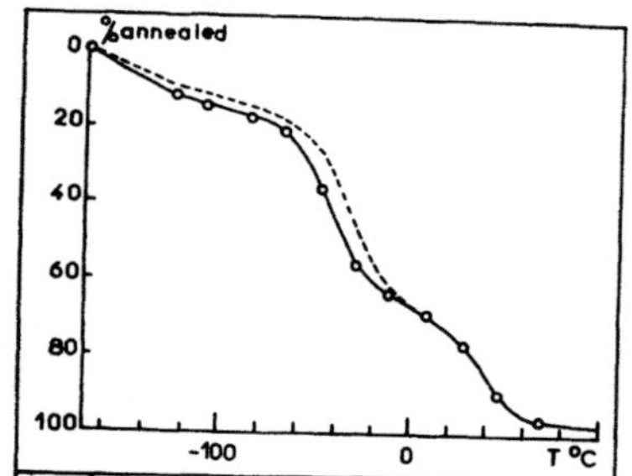

Fig. 5. Isochronal recovery (5 minutes anneals) of the electrical resistivity induced by irradiation after a quench ($\Delta\rho^o_Q$ = 5.6x$10^{-9}\Omega$cm).
—o— experimental curve (Q+I)
— — — theoretical curve ((Q)+(I))

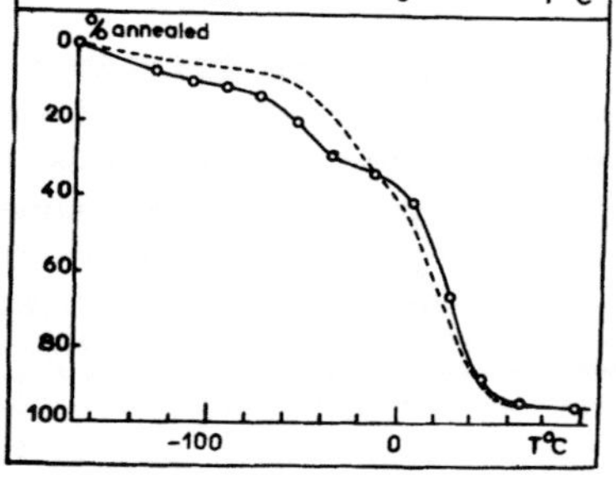

from 0.625 to 0.72 eV were obtained. As after the same quench, *not* followed by irradiation, values ranging from 0.61 eV to 0.69 eV had been obtained in the same sample, by the same method, it can safely be concluded that the effect of irradiation on Stage IV is to increase the apparent activation energy.

DISCUSSION AND ANALYSIS OF EXPERIMENTAL RESULTS

The results obtained in our quenching (Q) experiments agree with previously published results (3,4,10). However, with respect to the Stage III recovery kinetics after irradiation, the results of this work disagree with the electron irradiation results previously published (14). On the other hand, they are in good agreement with neutron irradiation data (11,12,13,16).

Qualitatively, all the dual experiment results can be explained by the assumption that:

1) quenches from 300°C and lower produce very small numbers of divacancies, but their concentration increases rapidly with the quenching temperature.

2) at - 170°C, during the electron irradiations, interstitial atoms are mobile and can combine with vacancy type defects.

3) as for Stage III, all the results can be explained qualitatively and quantitatively (see next paragraph) by assuming that it corresponds to the free migration of interstitial atoms, so that when the preliminary quench is from 300°C or lower, the process is almost purely:

$$\underset{\text{mobile}}{I} + \underset{\text{frozen}}{V} \longrightarrow 0 \qquad (2) \qquad \begin{cases} I : \text{interstitial atom} \\ V : \text{vacancy} \end{cases}$$

On the other hand, the effects of irradiation of Stage IV, in (Q + I) experiments are not compatible with the hypothesis of vacancies migrating in Stage III (12.b). It might be argued that the differences in activation energies and in migration temperature domains of Stages III and IV exist because quenched vacancies are bound to sinks, while irradiation produces unbound vacancies, which can move at a lower temperature. If that is the case, irradiation should *lower* the annealing temperature of a preliminary quench. This is not the observed effect.

The hypothesis of mobile divacancies or of mobile di-interstitials is discussed at length elsewhere (1). It should be mentioned, however, that the presence of mobile divacancies would not explain the experimental facts, while the mobile di-interstitials would be quite compatible both with the experimental data and with the quantitative analysis given below.

Quantitative Analysis of Stage III Results

We assume that Stage III corresponds to the reaction 2 (annihilation of Frenckel pairs). The resistivities per vacancy and per interstitial are unknown parameters. The ratio ρ_i/ρ_v of the resistivity of an interstitial atom (ρ_i), to that of a vacancy (ρ_v) can be determined from the experimental data obtained either a) in the Stage III isothermal recovery runs, or b) in the isochronal runs, or c) from the study of the production curve, provided only single vacancies are produced by quenching.

Use of isothermal data. Let C_i be the interstitial concentration, C_v that of vacancies produced by irradiation, and C_v^o that of vacancies produced by the preliminary quench. Reaction (2) gives:

$$P = \frac{dC_i}{dt} = \frac{dC_v}{dt} = - K\, C_i\, (C_v + C_v^o)\, e^{-E/kT} \tag{4}$$

Let us call $P_{(I)}$ and $P_{(Q+I)}$ two values of P obtained after a simple irradiation (I) and a dual experiment (Q + I) respectively. For equal values of remaining stages (points A and B, in Fig. 3), we have:

$$\alpha = \frac{P_{(I)}}{P_{(Q+I)}} = \frac{K\cdot C_i\cdot C_v\; e^{-E/kT}}{K\cdot C_i\cdot (C_v+C_v^o)\; e^{-E/kT}} = \frac{C_v}{C_v + C_v^o} \tag{5}$$

If $\Delta\rho$ is the resistivity increase corresponding to A and B, as $C_i = C_v$

$$\Delta\rho = \rho_i\, C_i + \rho_v\, C_v = (\rho_i + \rho_v)\, C_v$$

and as $\Delta\rho_v^o = C_v^o\, \rho_v$, one has:

$$\frac{\rho_i}{\rho_v} = \frac{\Delta\rho\,(1-\alpha)}{\Delta\rho_v^o\cdot\alpha} -1 \tag{6}$$

Applying this relation to the curves of Figure 3, one readily obtains:

$$\frac{\rho_i}{\rho_v} = 2.0 \pm 0.3$$

Interpretation of isochronal results. Numerical calculations (1) show that an isochronal recovery curve is almost identical to a tempering curve obtained by continuous increase of the temperature, provided a convenient rate of heating is chosen. In the case of our experiments, where 5 minute anneals were performed at temperature intervals of 20°C, the proper coefficient was found to be: A = 1.47°C/min. With this assumption, it is easy to show that at the center point of the isochronal curve (where inflexion occurs) for a process such as (2), if

$$\frac{dC_i}{dt} = \frac{dC_V}{dt} = -KC_iC_Ve^{-E/kT} \quad \text{one has:} \; C_{i(c)} + C_{v(c)} = \frac{EA}{kK}\,\frac{e^{E/kT_{(c)}}}{T^2_{(c)}} \tag{7}$$

where (c) has been added to all the values of the variables C and T at the *center* point, and where $C_V = C_v + C_v^o$, and K is the Bolzmann constant.

Now, let us consider two experiments, the first one, of dual type (Q+I), will give:

$$C_{i(c)1} + C_{v(c)1} + C_v^o = \frac{EA}{kK}\,\frac{e^{E/kT(c)1}}{T^2_{(c)1}} \tag{8}$$

and the second experiment, a single irradiation (I), will give:

$$C_{i(c)2} + C_{v(c)2} = \frac{EA}{kK}\,\frac{e^{E/kT(c)2}}{T^2_{(c)2}} \tag{9}$$

Combining equations 4, 8 and 9 gives:

$$\frac{\rho_i}{\rho_v} = 2\,\frac{\Delta\rho_{(c)}}{\Delta\rho_v}\left[\frac{T^2_{(c)2}\; e^{E/kT_{(c)1}}}{T^2_{(c)1}\; e^{E/kT_{(c)2}}} - 1\right]^{-1} \tag{10}$$

where $\Delta\rho_{(c)}$ is the resistivity left in Stage III at the central point. Applying iteratively this relation to the four (Q+I) Stage III isochronal recovery results together with the corresponding (I) curves gives:

$$\frac{\rho_i}{\rho_v} = 1.9 \pm 0.1$$

(It is necessary to proceed iteratively because of the existence of the Substage III_a, and due to the difficulty of locating precisely the center point of Stage III_b).

Analysis of the production curves. The combination of the production curves (Fig. 2) and of an array of isochronal annealing curves allows the study of the increases of Stage II (notation: $\Delta\rho_{II}(\psi)$) and of Stage III (notation: $\Delta\rho_{III}(\psi)$) versus the electron flux ψ. One finds that $\Delta\rho_{II}$ is just proportional to ψ in our irradiation range, and $\Delta\rho_{II}(\psi)$ is independent of any previous quench. As for $\Delta\rho_{III}(\psi)$, the curvature is always negative. Experimental results are represented satisfactorily by the relations.

$$\Delta\rho = \Delta\rho_{II} + \Delta\rho_{III} \tag{11}$$

$$\frac{d\Delta\rho_{III}}{d\psi} = (\rho_i + \rho_v)\,\sigma_{III}(1 - KC_v) \tag{12}$$

$$\frac{d\Delta\rho_{II}}{d\psi} = (\rho_i + \rho_v)\,\sigma_{II} \tag{13}$$

For a production curve (I), the solutions give:

$$\Delta\rho = \frac{\rho_i + \rho_v}{K}\left(1 + \frac{\sigma_{III}}{\sigma_{II}}\right)\left(1 - e^{-K\sigma_{III}\psi}\right) \tag{14}$$

and for a production curve (Q + I):

$$\Delta\rho = \frac{\rho_i + \rho_v}{K}\left(1 + \frac{\sigma_{III}}{\sigma_{II}} - KC_v^o\right)\left(1 - e^{-K\sigma_{III}\psi}\right) \tag{15}$$

These formulas fit the experimental results very well.

One can derive from (14) and (15):

$$\frac{\rho_i}{\rho_v} = \frac{\alpha\beta}{\Delta\rho_v^o} - 1 \tag{16}$$

where: $\alpha = \dfrac{\rho_i + \rho_v}{K}$

$\beta = KC_v^o$

are easily deduced from the curvatures and the slopes of experimental curves.

With $\Delta\rho_v^o = 1.2 \times 10^{-9}\Omega$cm one finds: $\dfrac{\rho_i}{\rho_v} = 2.05$

and with $\Delta\rho_v^o = 2.24 \times 10^{-9}\Omega$cm , $\dfrac{\rho_i}{\rho_v} = 1.61$

These values are in fair agreement with those found by the two previous methods, although the phenomenon used for their determination is quite different. It is also worthwhile to note that the lowest value of ρ_i/ρ_v is obtained after a stronger quench, which is quite likely to produce some divacancies.

It should also be mentioned that equation (7) gives the positions of the isochronal center points in the assumption of a second order process in Stage III. The experimental results are plotted in Figure 6, assuming that $\rho_i/\rho_v = 2$ and $\rho_v = 2.2\mu\Omega cm$. The agreement with the calculated curve is quite good.

CONCLUSION

This work shows that dual experiments associating quenches and electron irradiation are a fruitful way of studying point defects in metals. The study of single vacancies is especially made easier by this method than by plain quenching. Furthermore, the use of combined experimental conditions such that Frenckel pair mutual annihilation occurs, allows the determination of any physical property of interstitial atoms versus that of a vacancy, provided the changes of this property in the metal remain proportional to the concentration of defects. In the present work, the analysis of different processes, by three different experimental techniques, consistently showed that in aluminum, the resistivity of an interstitial atom is twice that of a vacancy.

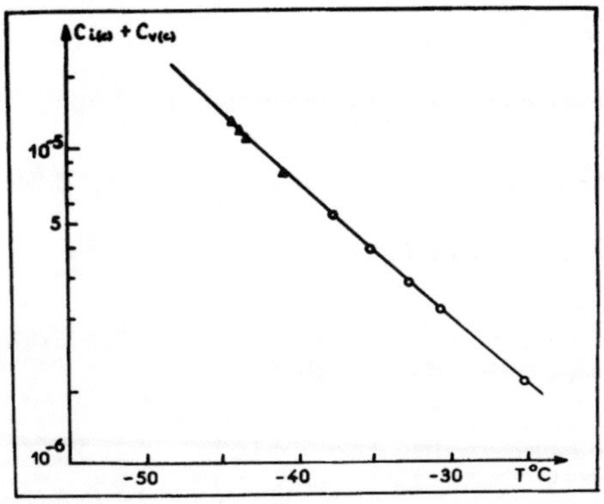

Fig. 6. Total concentration of defects $C_{i(c)} + C_{V(c)}$ at the "center temperature" T_c of Stage III versus T_c. (T_c is the temperature in isochronal annealing curve at which the inflexion occurs).

_____ theoretical curve $C_{i(c)} + C_{V(c)} = \frac{EA}{kK} \frac{e^{E/kT(c)}}{T^2(c)}$

o *experimental, deduced from Curves (I)*

Δ *experimental, deduced from Curves (Q+I)*

ACKNOWLEDGMENT

The authors are grateful to Professor Y. Cauchois, Director of the Laboratoire de Chimie-Physique, who encouraged this work and made the electron accelerator available. Thanks are also due to the accelerator team, headed by M. Boivin, for the electron bombardment.

This work is part of Claude Budin's Doctorat d' Etat thesis, to be submitted to the University of Paris in March 1967. (C.N.R.S. registration number: A.O. 1058)

REFERENCES

1. *C. Budin, Thesis, University of Paris (1966).*
2. *C. Panseri, T.Federighi, Phil. Mag. 3, (35), 1223-1240 (1958).*
3. *M. Doyama, J.S. Koehler, Phys. Rev. 134, 2A, A522-A529 (1964).*
4. *W. De Sorbo, D. Turnbull, Phys. Rev. 115, 3, 560-563 (1959).*
5. *A.Seeger, Radiation Damage in Solids, 101-127, I.A.E.A. Vienna (1962).*
6. *H.B. Huntington, Am. J. Phys. 32, 10, 775-786 (1964).*
7. *R. M. Walker, Radiation Damage in Solids, 609-618, Academic Press, New York, (1962).*
8. *C. Budin, P. Lucasson, A. Lucasson, J. Phys. 26, 9-14 (1965).*
9. *C. Budin, P. Lucasson, Phys. Letters, 16, 229 (1965).*
10. *S. Kabemoto, Jap. J. Appl. Phys. 4, No. 11, 896-905 (1965).*
11. *A. W. Mc Reynolds, W. Augustiniak, M. Mc Keown and D.B. Rosen Blatt, Phys. Rev., 98, 418 (1965).*
12. *S. Ceresara, T. Federighi and Gelli, a) 11 nuovo Cimento, 29, 1244-1261, (1963); b) Phil. Mag. 12, 120, 1093, 1103 (1965).*
13. *C. Frois and M. Dimitrov, C.R. Acad. Sc. Paris, 258, 5647-5650 (1964).*
14. *A. Sosin and L.H. Rachal, Phys. Rev. 130, No. 6, 2238 (1963).*
15. *T.S. Hutchison, Phys. Letters, 6, 22 (1965).*
16. *S. Ceresara, H. Elkholy, T. Federighi, F. Pieragostini, Phys. Letters, 16, 8 (1965).*

17. *R.D. Simmons, J.S. Koehler and R.W. Balluffi, Symposium on Radiation Damage in Solids and Reactor Materials, Venice (1962), Radiation Damage in Solids I (International Atomic Energy Agency, Vienna) 155.*

18. *P. Lucasson, R.M. Walker, Bull. Am. Ph. Soc. II, 7, 171 (1962).*

DISCUSSIONS

G.J. DIENES (Brookhaven National Laboratory, Upton, New York, U.S.A.): *Does the annihilation reaction in the quenched samples follow a bimolecular recombination kinetics?*

P. LUCASSON: *Yes, the annihilation reaction of the quenched and irradiated samples follow a bimolecular kinetics. However, if you write the reaction rate without taking into account the roles of interstitials and vacancies, i.e.*

$$\frac{d\Delta\rho}{dt} = -K_o \, \Delta\rho^{\gamma} \, exp \, \frac{-E}{kT}$$

you find that γ *is lower than 2 and tends towards 1 at the end of the stage. This is what you would expect, actually, in a bimolecular process where the concentrations of each reactant are not equal.*

O. HAUSER (Nuclear Research Center, Dresden, Germany): *Could you calculate from your experiments the activation energies of the different annealing stages so that it could be possible to get some informations about the migration and influence of single and divacancies?*

P. LUCASSON: *Yes, we have performed activation energy determination for both Stages III and IV. (Values are given in the text). As concerns Stage III, we find* $E_{iv} = 0.58 \pm 0.03$ *eV. Our results in Stage IV agree quite well with those of other workers. An interesting feature of our method is that it achieves a kind of "purification" of the vacancy population by increasing the proportion of singles. For example, if you quench from 325°C, the Palmer's method will give a spectrum of apparent activation energy values from say 0.62 to 0.68 eV. Now, if quenching is followed by irradiation, the same method will give values from 0.64 to 0.72 eV. This is, actually, what you would find after a quench from 300°C or less. However, in our method, you have more microhms to anneal, and this is very convenient, experimentally.*

DEFECT STUDIES IN IRRADIATED FERROMAGNETIC METALS OR ALLOYS

D. Dautreppe, A. Chamberod, W. Chambron, V. Hivert,
C. Minier-Cassayre, P. Moser, J. Pauleve, P. Peretto,
J. Verdone, P. Vigier

Centre d'Etudes Nucleaires de Grenoble, FRANCE

The different models proposed for stage I recovery are discussed on the basis of recent results of coupled experiments after electron irradiation at 20°K. The coupled measurements are magnetic after effect and electrical resistivity which are measured simultaneously during an isochronal annealing. The discussion is also based on different measurements after neutron irradiation at 28°K, such as magnetic after effect, electrical resistivity, stored energy, internal friction and length variation.

Different results on pure iron-nickel alloys neutron or electron irradiated at 78°K or 20°K are presented. The induced magnetic anisotropy produced by the enhanced diffusion during an isochronal annealing is measured. This physical property is very sensitive since only 2 ppm defects are necessary to give rise to very large effects. Different peaks of enhanced diffusion are observed and discussed.

INTRODUCTION

The experimental approach to the complex state of radiation damage in solids implies measurement of a large number of physical properties on equivalent samples. Magnetic materials (compared to non-magnetic ones) present a certain number of additional properties which can be powerful tools in defect investigation. Two of them have been successfully used in the authors' laboratory.

The first tool, applied to pure ferromagnetic metals, is the reorientation magnetic after-effect which takes place when a defect is coupled to the magnetization. A typical example is given by carbon in α Fe. The magnetic coupling tends to transform the isotropic distribution of the carbon atoms over the three equivalent sites, into an anisotropic distribution. This change occurs with a time constant

$$\tau = \tau_0 \exp\frac{W}{kT}$$

where W is the activation energy of migration of the carbon atoms. The anisotropic distribution tends to impede the motion of the Bloch walls causing a time decrease of the initial permeability with a time constant

$$\theta = \theta_0 \exp\frac{W}{kT}$$

From the study of this time decrease at different temperatures, W can be deduced and indications about its spectrum are given. Different defects created by irradiation give rise to important magnetic after-effects, particularly in the vicinity of stage I, and from these measurements strong arguments are given for free interstitial migration in this stage.

The second property is the induced magnetic anisotropy or directional order, which can be applied only to alloys. Let us take as an example the case of Ni-Fe (50%-50%). We use a polycrystalline sample with a great number of grains; in this way our sample is isotropic as far as magnetic properties are concerned, since the easy axes of magnetization are randomly distributed. If heat treatment is performed with an applied saturating magnetic field (of course at a temperature below the Curie point) and then the sample is quenched, an easy axis of magnetization is obtained in the direction of the treatment field. This phenomenon can be explained assuming magnetic coupling between the Ni-Ni, Fe-Fe and Fe-Ni bonds and the magnetization (1,2). This coupling tends to produce, as diffusion proceeds, a bond distribution with cylindrical symmetry around the magnetization which is frozen in by quenching. This distribution stabilizes the magnetization in the treatment field direction, explaining the observed easy axes. After irradiation at low temperature, an isochronal annealing in an applied magnetic field can produce the same kind of directional order depending on those defects which can enhance diffusion. From such experiments strong evidence was obtained for enhanced diffusion by free interstitial migration.

FERROMAGNETIC PURE METALS

Zone refined Ni and Fe were studied after electron and neutron irradiation at low temperature below Stage I.

The following measurements have been performed on Nickel : Electrical resistivity (3,4) Magnetic after-effect (3,5) and Electron microscopy (6).

The following studies have been made on Iron: Electrical resistivity (7,8,9,10), Magnetic after-effect (11,12,13,14,15), internal friction at low frequency (15,16), Stored Energy (15,17) and Length Change (18).

The effect of small amounts of known impurities on both Nickel and Iron has also been studied.

Finally, coupled experiments have been conducted to relate electrical resistivity and magnetic after-effect results. In this special type of experiment two identical samples are irradiated together. Then, during an isochronal anneal, where both samples are at the same temperature at any time, one sample is used for magnetic after-effect measurements and the other for electrical resistivity measurements.

Nickel

Figure 1 shows the isochronal annealing curve of the electrical resistivity after a 3 MeV electron irradiation at 20°K. Above 95°K we have a plateau and then very small Stages III and IV. During Stage I 85% of the damage is recovered. After a neutron irradiation at 28°K Substages 1B and 1C are not resolved and Stage I represents only 40% of the total resistivity increase.

Figure 2 shows the magnetic after-effect zones which appear after the same 3 MeV electron irradiation. After a neutron irradiation Zones 1B and 1C are much smaller.

The coupled experiments have shown that Zone IE anneals out during Substage IE, though this substage is not well resolved in electrical resistivity measurement. The activation energy for reorientation is equal to the activation energy for annihilation. The spectrum is very narrow, and the mean number of jumps for annihilation is 300 after electron irradiation. The kinetics is second order. It is concluded that the defect responsible for this zone must be the free interstitial.

The presence of a magnetic after-effect zone implies the following requirements:

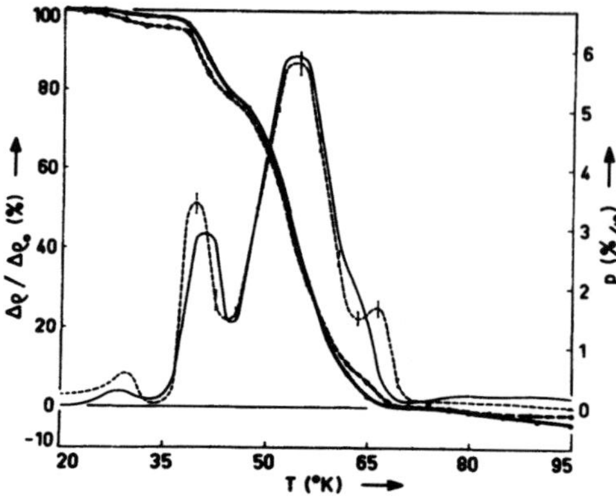

Fig. 1. Isochronal annealing of the electrical resistivity of electron irradiated nickel at 20°K
full curve : 3 MeV - 1.4×10^{18} elec/cm^2-isochrone 3° - 10 min.
dotted curve : 2.5 MeV - 10^{18} elec/cm^2-isochrone 3° - 20 min.
The dot above each peak indicates from left to right successively 1B, 1C, 1D, and 1E.

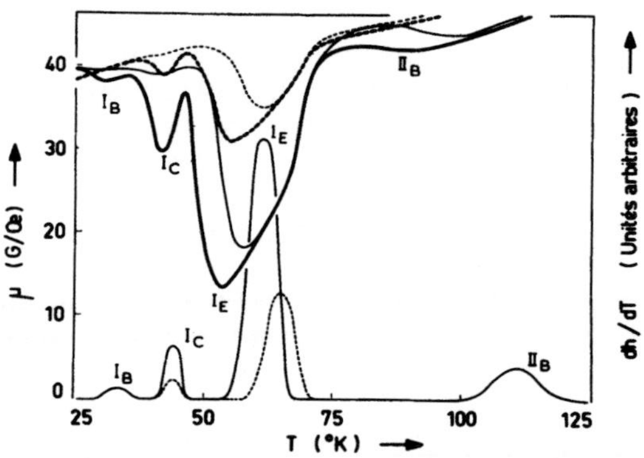

Fig. 2. Magnetic after effect zones in electron irradiated nickel.
full curve : 2.5 MeV
dotted curve : 1.22 MeV
The lower curves indicate the annealing temperature range of the magnetic after effect zones.

1. Some defect is able to be coupled with the magnetization.

2. There are equivalent sites for this defect.

The possible equivalent sites for an interstitial near a vacancy are shown in Figure 3. Two conditions may be added for the observation of magnetic after-effect.

3. Among the equivalent sites, some must be first neighbors (we shall say that this particular assembly of sites forms a subgroup of the group of equivalent sites).

4. The potential barrier between the sites of a subgroup must be slightly lower than the potential barrier towards the vacancy.

It was concluded from these results (3) that 1B and 1C can be attributed to close pair annihilation, the interstitial lying respectively on Subgroup 3 and 5 for instance. As no magnetic after-effect is associated with Substage ID, it cannot be attributed to the correlated annihilation of interstitials into vacancies, as proposed by Corbett et al (19) for Copper. This substage can be due to the annihilation of close pairs whose interstitials are lying on sites possessing no equivalent first neighbors such as group 6, or on subgroups which do not fulfil Condition 4.

Other coupled experiments on Iron doped Nickel show additional effects (15) which are perfectly compatible with the preceding model.

A model for Stage II, III and IV was proposed based on electrical resistivity, magnetic after-effect measurements and electron microscopy observations (6,20). Interstitial clusters are involved in Stages II and III, and vacancies in Stage IV.

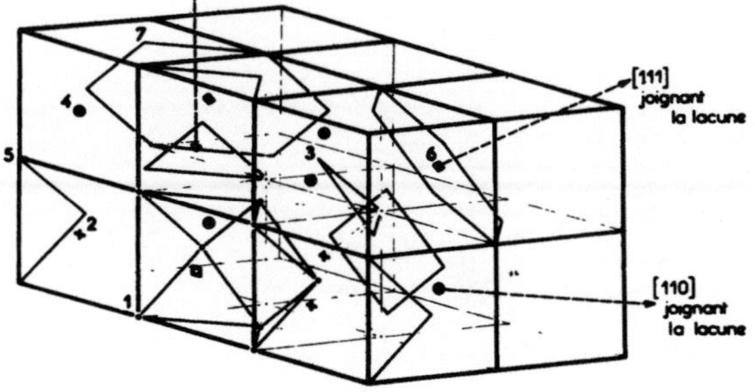

Fig. 3. Groups of equivalent sites around a vacancy in f.c.c. metals
- • : *1st neighbors -*
- + : *2nd neighbors -*
- Δ : *subgroups of 3rd neighbors -*
- ● : *4th neighbors -*
- □ : *subgroups of 5th neighbors -*
- ⊞ : *6th neighbors -*
- ⬭ : *subgroups of 7th neighbors.*

Iron

Similar experiments have been performed on Iron, but with the additional measurements indicated above. Figure 4 shows the isochronal annealing curves of the electrical resistivity after a 3 MeV electron irradiation at 20°K and after a neutron irradiation at 28°K. As in the case of Nickel, though Substage IE appears as a shoulder on the first peak, it corresponds to the annealing of a large magnetic after-effect zone. This shoulder appears in the same way on the stored energy curve (17) and on the length change curve (18). A well defined internal friction peak can be associated with the magnetic after-effect (16).

The magnetic after-effects associated with Substages IB and IC, which are resolved only after electron irradiation, are extremely weak. This can be understood on the basis of the same model used in the nickel case. In the b.c.c. structure of iron, there is no subgroup containing more than one site, then magnetic after-effect associated with close-pair reorientation implies the possibility of jump for the interstitial on a second neighbor equivalent site, and the probability of this event may be small.

The effect of small amounts of Ni, Si and Be on the magnetic after-effect zone, on the stored energy peak, on the internal friction peak and on electrical resistivity stages (21) can be explained easily assuming close pair annihilation in Substage ID, and free interstitial migration in IE. New effects related to each type of impurity appear. These can be explained for some impurities, by assuming a trapping of the interstitial by the impurity atom and then a release of the impurity atom as a free migrating impurity interstitial.

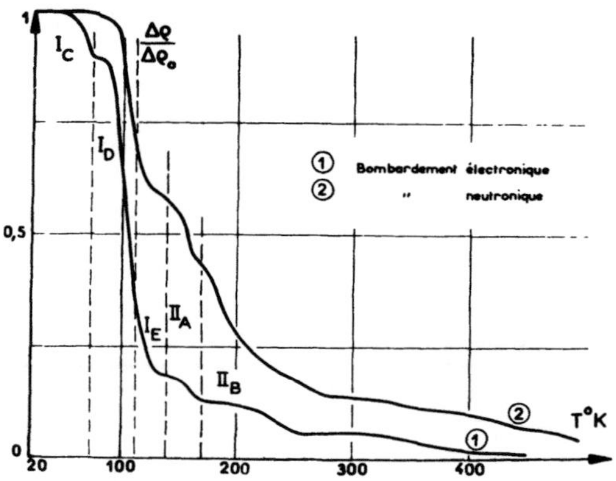

Fig. 4. Isochronal annealing curve of electron irradiated and neutron irradiated iron.

After electron irradiation, a new magnetic after-effect zone appears during the annealing of the first zone (Figure 5). This zone can be practically suppressed, while others appear when Be is added as an impurity. So this 80°K zone can be due to di-interstitials. After neutron irradiation a great number of zones appear which anneal out in the temperature ranges indicated in Figure 6. Interstitial clusters directly formed during the neutron irradiation can be responsible for these zones.

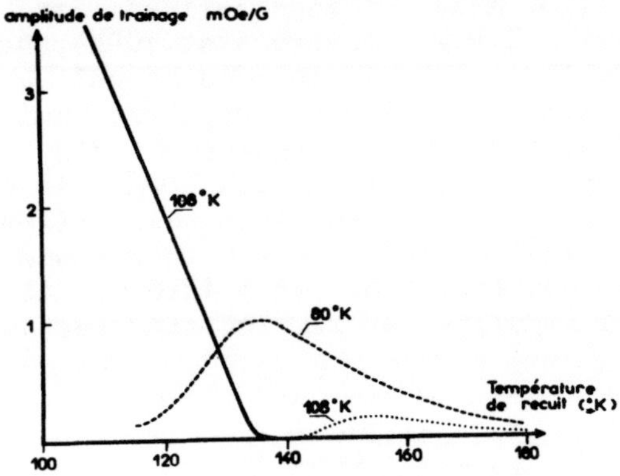

Fig. 5. Isochronal annealing of the magnetic after effect zones on electron irradiated iron at 20°K.

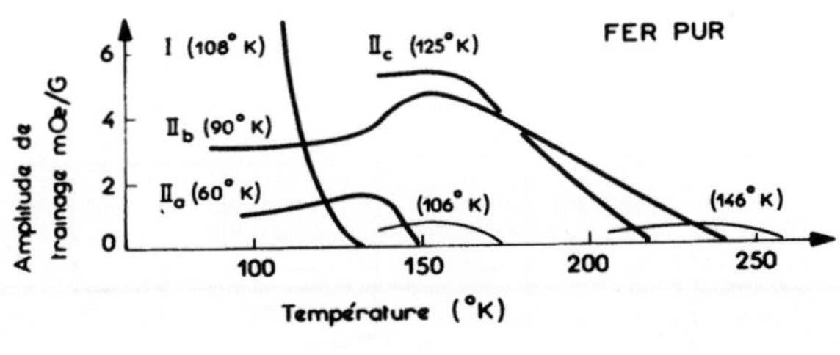

Fig. 6. Isochronal annealing of the magnetic after effect zones on neutron irradiation iron.

FERROMAGNETIC ALLOYS

Different Fe-Ni alloys and Ni-Co alloys (50%-50%) were studied. Directional order measurements have led to the discovery of a new order-disorder transformation in the Fe-Ni (50%-50%) alloy (22). This new transformation has been studied very carefully in single crystals (23,24,25). Attention will be given to low temperature irradiations which give information about the different types of defect which can enhance diffusion. Samples of Fe-Ni (50%-50%) alloy made of zone refined Iron and Nickel have been irradiated with 2 MeV electrons at 20°K and also with neutrons at 77°K. During a 10°C-10 mn isochronal annealing with an applied magnetic field, the induced magnetic anisotropy was measured at 77°K after each step. Figures 7 and 8 show the induced magnetic anisotropy increase after each step. As can be seen, the sensitivity is very high, since noticeable effects are obtained with a dose of 5.2×10^{15} elect/cm^2 which corresponds approximately to a total defect concentration of 0.4×10^{-6}. At such a low concentration, defect impurity interaction can play an important role.

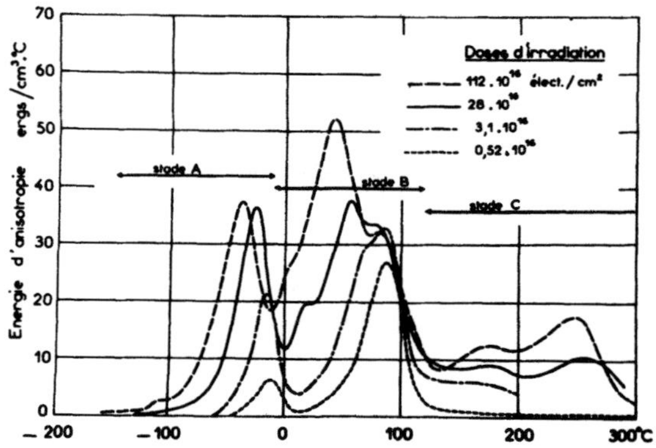

Fig. 7. Enhanced diffusion peaks on electron irradiated Fe-Ni (50%-50%).

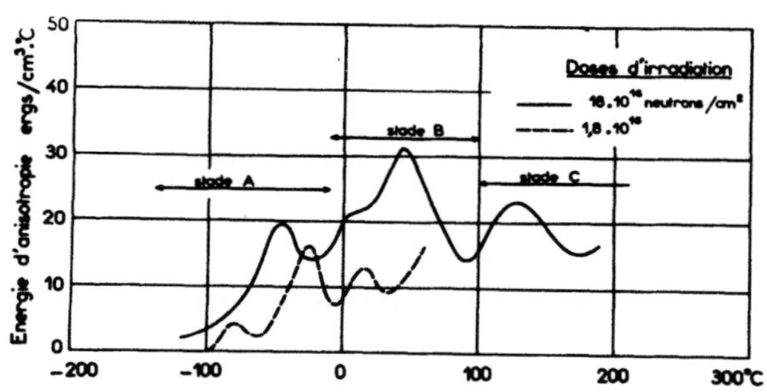

Fig. 8. Enhanced diffusion peaks on neutron irradiated Fe-Ni (50%-50%).

Three main stages of enhanced diffusion have been found.

Stage A is found after electron irradiation as well as after neutron irradiation. No other stages seem to occur at a lower temperature, since a 77°K neutron irradiation with an applied magnetic field give no directional ordering. Several stages of recovery of the electrical resistivity take place below Stage A. They can be attributed to close pair annihilation. No Stage A can be observed after quenching (26). Assuming that Stage A is due to interstitials which anneal out at fixed sinks and in vacancies, one can work out the behavior of the induced anisotropy during the isochronal annealing (27) for the different electron doses (26). The displacement of the peak maximum is explained and the experimental points fit the theoretical curve quite well except for the higher dose. An activation energy of 0.48 eV is found for the migration of the interstitials.

Stage B seems to be partly due to vacancies, since it is present in both types of irradiation and has been found in quenched samples exactly at the same place as it appears on the low electron dose curve (26). The fine structure which appears at higher doses may be due to a more important formation and dissocation of di-interstitials whose nature differs according to the nature of the components. Equally, interstitials trapped by some impurities may be released, leading to new substages of enhanced diffusion.

Stage C is practically absent at low electron doses, but is present after neutron irradiation. This stage can be due to the destruction of vacancy clusters formed during neutron irradiation. After electron irradiation, small clusters may be formed during the diffusion of single vacancies.

Stored energy measurements, and irradiation of doped samples have been started to try to work out the impurity effects, but in any case it seems impossible to explain the results only with vacancy-enhanced diffusion.

REFERENCES

1. *L. Neel, J. Phys. Rad. 15 (1954), p. 225.*

2. *S.Taniguchi, Sc. Rep. Res. Inst. Tohoku Univ. A7 (1955) p. 269.*

3. *P. Peretto, J.L. Oddou, C. Minier-Cassayre, D. Dautreppe and P. Moser, Phys. Stat. Sol. 16 (1966), p. 281.*

4. *J.L. Oddou, C. Minier-Cassayre and P. Moser to appear in Phys. Stat. Sol.*

5. *J. Verdone, P. Peretto, P. Moser, D. Dautreppe and J. Verdier, C.R. Acad. Sci. V. 260 (1965), p. 5209.*

6. *P. Peretto, A. Bourret, P. Moser, C. Minier-Cassayre and D. Dautreppe, Communication au X° Colloque de Metallurgie de Saclay (Juin 1966).*

7. *C. Cassayre, D. Dautreppe and J. Pauleve, C. R. Acad. Sci. V. 251, (1960) p. 370.*

8. *C. Minier-Cassayre, and D. Dautreppe, C. R. Acad. Sci. V. 256 (1963), p. 2368.*

9. *C. Minier-Cassayre, C.R. Acad. Sci. V. 258 (1964), p. 541.*

10. *C. Minier-Cassayre, C.R. Acad. Sci. V. 260 (1965), p. 2163.*

11. *P. Moser, D. Dautreppe and P. Brissonneau, C. R. Acad. Sci. V. 250 (1960), p. 3963.*

12. *P. Moser and D. Dautreppe, Symposium on Radiation Damage in Solids, I.A.E.A. Venise 1962.*

13. *P. Moser and D. Dautreppe, J. Phys. Rad. V. 24 (1965), p. 516.*

14. *P. Vigier and P. Moser, C.R. Acad. Sci. V. 260 (1965), p. 2443.*

15. *P. Vigier, V. Hivert, P. Moser and E. Bonjour, C.R. Acad. Sci. V. 260 (1965), p. 3359.*

16. *D. Dautreppe, V. Hivert, P. Moser and A. Salvi, C.R. Acad. Sci. V. 258 (1964), p. 4539.*

17. *E. Bonjour and P.Moser, C.R. Acad. Sci. V. 257 (1963) p. 1256.*

18. *E. Bonjour, P. Vigier and P. Moser, C. R. Acad. Sci. (sous presse).*

19. *J.W. Corbett, R.B. Smith and R.M. Walker, Phys. Rev. V. 114, (1959) p. 1452.*

20. *P. Peretto, D. Dautreppe and P. Moser, Phys. Stat. Sol. V. 13, (1966) p. 325.*

21. *P. Vigier, These Grenoble 1966 (en cours).*

22. *J. Pauleve, D. Dautreppe, J. Laugier and L. Neel, C.R. Acad. Sci. V. 255 (1962), p. 2037.*

23. *L. Neel, J. Pauleve, R. Pauthenet, J. Laugier and D. Dautreppe, J. Appl. Phys. 35, No. 3 (1964), p. 871.*

24. *J. Pauleve, K. Krebs, A. Chamberod and J. Laugier, C. R. Acad. Sci. V. 260 (1965), p. 2439.*

25. *J. Pauleve, A. Chamberod, K. Krebs and A. Marchand, Communication à Intermag, Stuttgart (Avril 1966).*

26. *A. Chamberod, W. Chambron and J. Pauleve, C. R. Acad. Sci. (sous presse).*

27. *W. Chambron, A. Chamberod, J. Pauleve and D. Dautreppe, Communication à la Réunion sur la Diffusion (Baden-Baden, Juin 1966).*

DISCUSSIONS

A. FERRO (Istituto Elettrotecnico Nazionale, Torino, ITALY): *What are the magnetic properties of the Nickel used?*

D. DAUTREPPE: *The Nickel used is zone refined Nickel with an initial permeability of about 300.*

ELECTRON MICROSCOPY OF HEAVY ION DAMAGE IN ALUMINUM AND NICKEL

D.I.R. Norris

Central Electricity Generating Board

Berkeley, Gloucestershire, England

Defect clusters formed by the bombardment of thin foils of aluminum and nickel with 150 keV xenon ions at 300°K have been studied by electron microscopy. Black spot images and black-white images yield information on a) whether a cluster is collapsed to a dislocation loop or is three-dimensional, b) its Burgers vector if it is a loop, c) whether it is of vacancy or of interstitial character, d) whether it is near the ion entrance or the ion exit surface of the foil, and, e) the number of point defects making up the cluster. In aluminum and nickel prismatic loops, Frank loops and three-dimensional clusters are present and they are of vacancy character. In aluminum in a direction close to [112], the density of visible clusters decreases in a distance 2200Å to a fraction 0.06 of its initial value. The largest loops are formed from about 500 vacancies in aluminum and about 250 in nickel. These results are consistent with the view that each cluster corresponds to the site of a displacement cascade.

INTRODUCTION

A common feature of irradiated metals is the presence of defect clusters that can be observed in the electron microscope as black spot images and black-white images. Since the first observation of these images in neutron-irradiated copper (1), they have been found in other cases also and are believed to play an important role in irradiation hardening. Any quantitative study of these defect clusters requires well defined irradiation conditions, monoenergetic heavy ion bombardment being very suitable. It is also desirable to work with doses sufficiently low that images of different clusters do not overlap.

To extract most of the information contained in an image in an electron micrograph, it is necessary to compare it with theoretical images which can be computed by means of the dynamical theory of electron diffraction for a selected model of a strained region in a crystal lattice. Such computations have been made by Ashby and Brown (2,3) for images close to the Bragg angle, their models being firstly a misfitting sphere and secondly a dislocation loop. They show that a black-white image is produced by a cluster near the surface of a specimen. A cluster near the center produces an image resembling a black spot divided by a narrow line of no contrast normal to the reflection vector g. The absence of such a line in a black spot image is attributed to the limited resolution of the microscope. From these two types of image it is possible to obtain information on a) whether a cluster is collapsed to a dislocation loop or is three-dimensional, b) its Burgers vector if it is a loop, c) whether it is of vacancy or of interstitial character, d) whether it is near the top or the botton surface of the specimen, and e) the number of point defects making up the cluster.

Several workers have applied image contrast theory to defect clusters in irradiated metals. Essmann and Wilkens (4) have. studied the black-white images in neutron irradiated copper and have reported them as being prismatic and Frank loops. Hesketh and Rickards (5) have studied exhaustively the clusters in ion bombarded copper at 300°K. By performing successive irradiations on the same foil they have shown that clusters are formed at their final size and that there is no general growth during subsequent irradiation. Prismatic loops, Frank loops and three-dimensional clusters are all present. The maximum loop size corresponds to 0.8 of the vacancies that could be created in a cascade. There are approximately 3 clusters for every ion stopped in the foil and the attenuation length of the damage is 3000A°, evidence of [011] channeling. Howe, McGurn and Gilbert (6) have performed ion bombardments at 20°K, finding similar electron microscope images in copper, gold and aluminum. This again disproves the possibility that a diffusion process dominates cluster formation. They identify the black-white images in copper and gold as prismatic and Frank loops. They find more Frank loops than prismatic loops, as do Hesketh and Rickards, but do not report three-dimensional clusters. Cluster structures remain controversial, but the view is now gaining acceptance that each cluster corresponds to the site of a displacement cascade, from which the interstitials are distributed relatively widely and at which there is therefore an accumulation of vacancies.

This paper reports results from an investigation in progress of damage in aluminum and in nickel by heavy ion bombardment. Beevers and Nelson (7) have reported black spot images in ion bombarded aluminum, but their images are dose rate dependent, are formed at very high dose rates and do involve diffusion. Greenfield and Wilsdorf (8) have reported black spot images in nickel neutron irradiated at 500°K. The present work applies image contrast theory to clusters formed in aluminum and nickel by low dose heavy ion bombardment at 300°K.

EXPERIMENTAL

Thin foils suitable for electron microscopy have been prepared from annealed high purity aluminum and nickel. These have been bombarded with various heavy ions, principally xenon, at energies up to a maximum of 150 keV and at dose rates of the order of 10^{12} ions $cm^{-2}sec^{-1}$. Particularly with aluminum the defects observed are smaller and fainter than those found in copper. Xenon has been used as the bombarding ion because the damage obtained with lighter elements was too small for quantitative work. The bombardment was nominally conducted at room temperature. However, temperatures may have risen to a maximum of a few tens of degrees in the thinnest parts of the foils during bombardment. The foils were examined in a Philips EM200 both before and after the bombardment, during which they remained in the electron microscope specimen holder. In the subsequent examinations it was possible to reproduce fairly well the diffraction conditions existing previously, and micrographs of identical areas were taken at the Bragg angle in each case. Except where otherwise stated, the results quoted in this paper refer to specimens bombarded with 150 keV xenon ions for doses between $3x10^{11}$ and $3x10^{12}$ ions cm^{-2}.

After bombardment, black spot images and black-white images were observed in aluminum (Fig. 1) and in nickel (Fig. 2). The black spot images were scanned with a microdensitometer, measurements being restricted to those close to the Bragg angle that also satisfied the criteria: a) of being sufficiently intense to avoid confusion with background intensity variations, b) of appearing to be symmetrical, and c) of being isolated from other images. Widths were measured at a level corresponding to a 20% decrease in electron intensity from the background level. The black-white images were classified according to their orientations and senses, i.e. whether black to white was closer to the direction +g or -g.

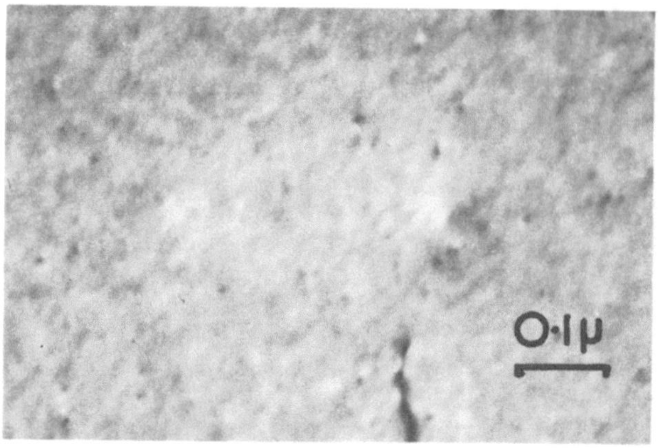

Fig. 1. Dark field micrograph showing black spot and black-white images in aluminium bombarded with 10^{12} xenon ions per cm^2.

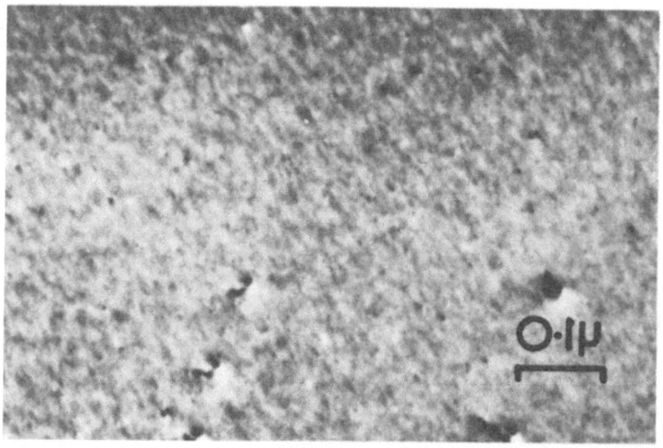

Fig. 2. Dark field micrograph showing black spot and black-white images in nickel bombarded with $3x10^{11}$ xenon ions per cm^2. Clusters are clearly distinguishable from larger images which are pre-existing dislocations.

RESULTS AND DISCUSSION

Image Analysis

The axis of the black-white image of a small loop is parallel to the Burgers vector (4), while that of a spherical cluster is parallel to the imaging vector (2). Thus by making counts of the orientations of these images in micrographs taken with differing reflections, it is possible to deduce the relative numbers present of prismatic loops, Frank loops and spherical clusters.

The sense of a small black-white image oscillates with distance of the cluster from a surface (9). In the first layer, 0-¼ξg (approximately) from a surface, a vacancy cluster has a dark field image on a positive print of black to white in the sense $+\underline{g}$ and an interstitial cluster the reverse; thus following the Ashby-Brown rule (ξg is the extinction distance). In the second layer, ¼ξg - 3/4 ξg, the opposite applies and in the third layer, 3/4 ξg - 5/4 ξg, the original rule applies. Thus to determine whether clusters are of vacancy or of interstitial type it is necessary to have information on their depth in the foil.

Black-white images of the same cluster in bright and dark field are similar in sense if the cluster is near the top surface of the foil and complementary if it is near the bottom (2). Thus one can determine the ratio of the number of clusters near the ion exit surface to those near the ion entrance surface, i.e. the attenuation of the damage.

The profile of the black spot image of a cluster depends on the number of point defects it contains, on its structure and on the metal. A cluster that is collapsed to a small dislocation loop gives an image that is known directly from the Ashby-Brown theory (3). The image width can then be related to the number of point defects that formed the loop for each possible Burgers vector. Figure 3 displays such relationships for the case of Frank loops with $\underline{g}$ = $[1\bar{1}1]$ the imaging reflection in four f.c.c. metals. Three-dimensional clusters could vary in structure from a void to a diffuse zone containing a large proportion of point defects, as proposed by Hesketh and Rickards (5). The image width in these cases can be deduced from the misfitting sphere computations of Ashby and Brown (2), being a function only of the total number of point defects in the cluster and not, on the simple model, of their density. The relationship of image width to number of point defects is less certain for three-dimensional clusters than for loops since it involves both the misfitting sphere approximation and an assumption about the strain of the cluster: the volume strain per vacancy is taken to be half an atomic volume. Figure 4 displays the relationship for the three cases of prismatic loops,

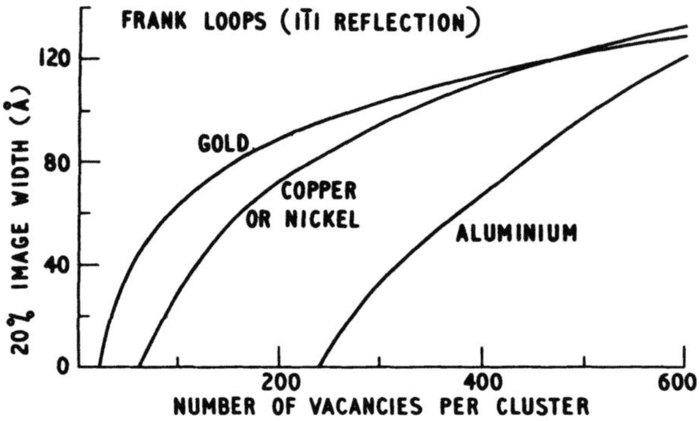

Fig. 3. The relationships between 20% electron image width and number of point defects per cluster, in gold, nickel, copper and aluminum for Frank loops and g = [1̄1̄1].

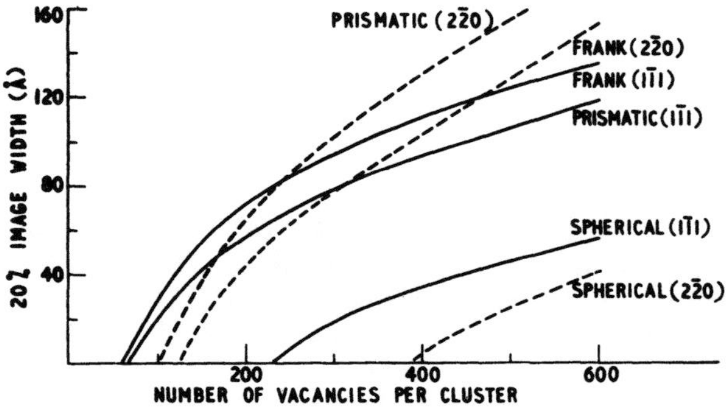

Fig. 4. The relationships between 20% electron image width and number of point defects per cluster in copper or nickel for prismatic loops, Frank loops and spherical clusters and g = [1̄1̄1] and [22̄0].

Frank loops and spherical clusters in nickel or copper for two different imaging reflections.

Several interesting points are illustrated in Figures 3 and 4. The minimum size of cluster visible in the electron microscope varies considerably with the metal concerned, depending on the ratio of extinction distance to lattice parameter. The visibility limit for Frank loops in gold corresponds to 20 point defects (if the theory is valid at such a size) compared with 240 in aluminum. The best reflection to image small clusters is the lowest order one. Spherical clusters have narrower images than corresponding loops.

Structure of the Clusters

Counts of black-white images for aluminum and for nickel give the results shown in Table I, where they are compared with those found by Hesketh and Rickards for copper bombarded with 150 keV zinc ions (5). All three types of cluster are found in each metal,

TABLE I

Comparative Numbers of Prismatic Loops, Frank Loops and Three-Dimensional Clusters Found in Aluminum, Nickel and Copper.

	Al	Ni	Cu (Ref. 5)
Prismatic loops	0.47	0.46	0.01
Frank loops	0.47	0.24	0.05
Three-dimensional	0.06	0.30	0.94
No. per stopped ion	0.03	0.30	3.0
No. measurable per stopped ion	0.003	0.03	

although the evidence for three-dimensional clusters in aluminum is less strong. Figure 5 shows a [110] nickel foil with images having axes along g = [002]; these can be identified unambiguously as three-dimensional clusters. In both aluminum and nickel, especially the former, there are likely to be very many clusters too small to be visible, so that the results for these two metals refer only to the largest clusters.

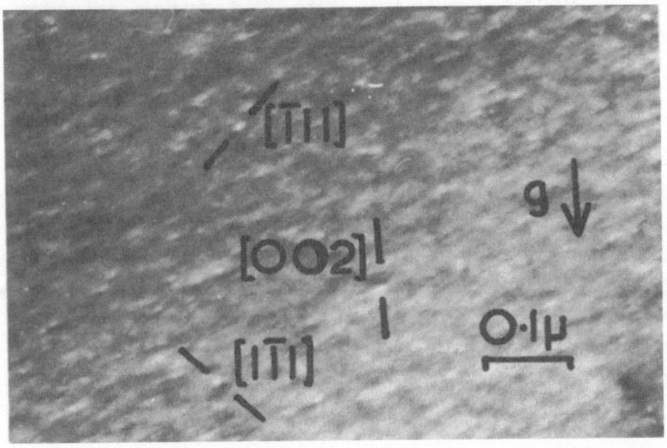

Fig. 5. Dark field micrograph with g = [002] of a [110] nickel foil irradiated with 8x10^{11}-xenon ions per cm^2. Prismatic loops are invisible, the black-white images of Frank loops lie along $[1\bar{1}1]$ and $[\bar{1}11]$ and those of three-dimensional clusters along [J02].

Sense of the Clusters

An unambiguous determination of whether the clusters are of vacancy or of interstitial type has been made for the case of aluminum bombarded with mercury ions at 15 keV in the [112] direction. Counts of 519 black-white images in dark field micrographs with **g** = [$1\bar{1}1$] of one such foil, Figure 6, gave a fraction 0.73 with black to white in the sense +g. The distribution of clusters with depth in the foil is deduced from experimental data on ranges of ions in aluminum (10), modified by theoretical means (11) to apply to the particular conditions of the present experiment. The ratio of mercury ions stopped in the first layer to those in the first and second combined is 0.54 (those in the third layer being negligible). The distribution of visible damage with depth in the foil is identical at worst with the final ion distribution and at best has a much more rapid decrease, giving a higher ratio. Therefore a fraction greater than 0.54 of the visible clusters obeys the Ashby-Brown rule. The experimental counts leave no doubt that the clusters are of vacancy type.

Attenuation of the Damage

Comparison of black-white images in bright and dark field for a [112] aluminum foil gives a ratio of clusters near the ion exit surface to those near the ion entrance surface of 0.06 for a total of 436 pairs of images. The thickness of the foil is 2800A° as determined by slip trace analysis, and the assumption is made that black-white images are within a distance ξg of a surface. Thus the density of visible clusters decreases in 2200A° to a fraction 0.06 of its initial value. Stopped ion distribution calculations in this case give a value of 0.05 for the ratio of ion fluxes at a distance ½ ξg of each surface.

Size of the Clusters

The distribution of image widths for similar conditions in aluminum and nickel is compared in Figure 7. Converting these image widths to numbers of vacancies per cluster gives the results shown in Figure 8 for the assumption that all clusters are Frank loops. These size distributions, as shown previously, vary according to the structure assumed for the clusters. The maximum sizes observed are of particular interest and are 250 for nickel and 500 for aluminum, assuming the largest images to be prismatic loops.

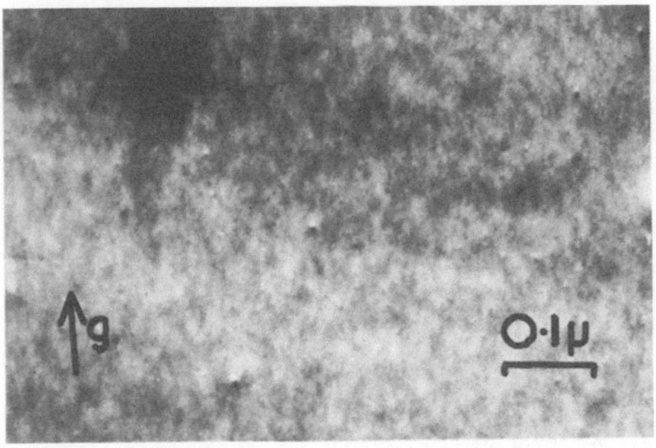

Fig. 6. Dark field micrograph of aluminium bombarded with 3x10^{12} mercury ions per cm^2 at 15 keV. The majority of the black-white images are black to white in the sense +g.

Fig. 7. Histogram showing the distribution of 20% electron image widths in aluminum and nickel for g = [1$\bar{1}\bar{1}$].

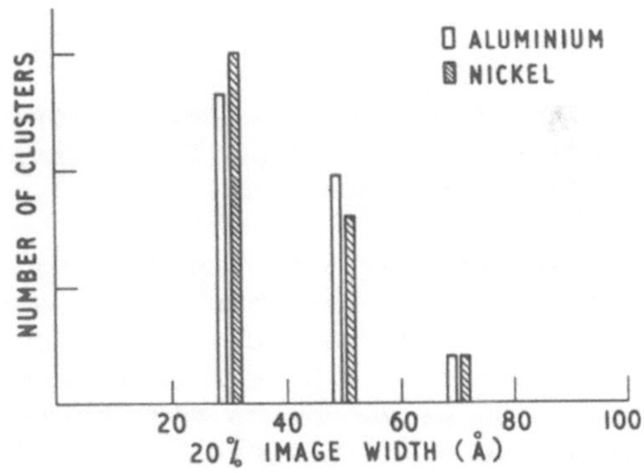

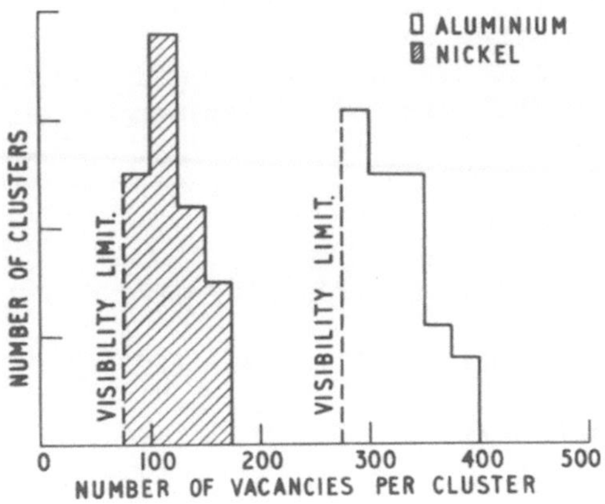

Fig. 8. Histogram showing the distribution of number of vacancies per cluster in aluminum and nickel for the assumption that all clusters are Frank loops.

CONCLUSION

Consideration of the structure of a displacement cascade suggests that on formation the vacancies are arranged in three dimensions. There will then be some re-arrangement, either to the most stable three-dimensional configuration or to a dislocation loop. From energy considerations the tendency will be for the largest clusters to be prismatic loops, the smallest to be three-dimensional clusters and those of intermediate size to be Frank loops. All three forms are observed in ion-bombarded copper, aluminum and nickel; the existence of three-dimensional clusters is proven.

The unambiguous result that the clusters in aluminum bombarded with 15 keV mercury ions are of vacancy character is of particular interest in view of the uncertainty of character determination in some work on irradiated metals. We suggest that, wherever the ratio of clusters to primary knock-ons is of the order of unity, they correspond to the sites of displacement cascades and will be found to be of vacancy character.

Detailed comparison of the distribution of clusters with depth in the foil with that of ion flux and ion energy would be most valuable. Work to date provides comparison at one depth only and with limited accuracy.

Rough estimates of the maximum number of atoms that could be displaced in a cascade are 600 in aluminum and 1100 in nickel for bombardment with 150 keV xenon ions. Comparison of these values with the largest clusters observed experimentally suggests that the fraction of vacancies and interstitials recombining is 0.2 in aluminum and 0.75 in nickel. That fewer clusters are observed in aluminum than in nickel for the same dose is accounted for by the much greater severity of the visibility condition in the former. That fewer clusters occur in nickel than in copper for the same dose is a real effect, to be explained by a higher recombination fraction. The observations are completely consistent with the view that each cluster corresponds to the site of a displacement cascade.

REFERENCES

1. *J. Silcox and P.B.Hirsch, Phil. Mag. 4 (48), 1356-1374 (1959).*

2. *M.F.Ashby and L.M. Brown, Phil. Mag. 8 (91), 1083-1103 (1963).*

3. *M.F.Ashby and L.M. Brown, Phil. Mag. 8 (94), 1649-1676 (1963).*

4. *U. Essmann and M. Wilkens, Phys. Stat. Sol. 4 (2), K53-55 (1964).*

5. *R.V. Hesketh and G.K. Rickards, Proc. Roy. Soc. A289 (1418), 353-365 (1966).*

6. *L.M. Howe, J.F. McGurn and R.W. Gilbert, Acta Met. 14 (7), 801-820 (1966).*

7. *C.J. Beevers and R. S. Nelson, Phil. Mag. 8 (91), 1189-1202 (1963).*

8. *I.G. Greenfield and H.G. F. Wilsdorf, J. Phys. Soc. Japan 18 (Supplement III), 20-26 (1963).*

9. *M. Rühle, M. Wilkens and U. Essmann, Phys. Stat. Sol. 11 (2), 819-829 (1965).*

10. *G.R. Piercy, M. McCargo, F. Brown and J.A. Davies, Can. J. Phys. 42 (6), 1116-1134 (1964).*

11. *J. Lindhard, M. Scharff and H.E. Schiøtt, Mat. Fys. Medd. Dan. Vid. Selsk. 33 (14), 1-42 (1963).*

DISCUSSIONS

F.H. HAMMAD (Atomic Energy Establishment, U.A.R.): *Since the temperature of irradiation and examination (e.g. room temp.) in terms of homologous temperature (T/T_m) is different for Al, Cu, Ni, does not this explain the difference in cluster size and distribution in the metals studied? (Because of the annealing behavior)*

D.I.R. NORRIS: *It would be interesting to perform the experiments at different temperatures. However, the difference in homologous temperature cannot explain the difference found between Nickel and Copper.*

F.H. HAMMAD: *Do you think that the stacking fault energy has an influence on the size of the 3 - dimensional cluster?*

D.I.R. NORRIS: *There is evidence that our three-dimensional clusters are zones with a very large fraction of vacant lattice sites, rather than stacking fault tetrahedra or voids. Which type of cluster is formed from a given number of vacancies depends on the relative energies of the possible configurations, that of a Frank loop involving the stacking fault energy, and on other factors in addition.*

NEUTRON RADIOGRAPHY AND MICRORADIOGRAPHY STUDIES

A.A. Fahmy

The American University in Cairo

Cairo, Egypt, U.A.R.

The technique of neutron radiography and microradiography depends on the attenuation of neutron beams by matter. The similarities and differences between this technique and x- and gamma-ray radiography are pointed out.

The physical principles on which the technique is based are discussed, and the different methods of obtaining the radiograph are described and compared. Resolution pieces have been made and used to ascertain the resolution of which the radiographic method is capable, independent of the nature of the object.

Specimens of boral, a composite material of B_4C *in an aluminum matrix, in the cast and rolled conditions, have been radiographed with x-rays and neutrons and the radiographs compared. By setting up conditions for high resolution and using an extremely fine grained film, it was possible to obtain neutron radiographs capable of 100x magnification, revealing microstructural features of the alloy.*

The technique is evaluated and found to be useful and practical in certain circumstances.

INTRODUCTION

Radiography is essentially a non-destructive test method in which penetrating radiations such as X-rays, γ-rays, or neutrons are passed through the object to be examined. The emerging beam forms an image, or more appropriately a shadow, on a suitable radiation detector such as a photographic film. On the resulting shadowgraph, or as it is known, radiograph, one may detect differences from point to point in the object, in thickness, density or composition, cracks, inclusions, blowholes and such like heterogeneities. The set up for such an examination is shown in Figure 1.

X-ray radiography started very soon after the discovery of x-rays in 1895. At first it was applied in the field of medicine, but later, as it became possible to produce harder or more energetic x-rays, it found wide application in industry. γ-ray radiography came into being with the availability of γ active radioisotopes. Now both X and γ-ray radiographic techniques are well established and widely used in industry, to the point where in some production methods such as welding, forging, and casting the radiographic test is almost routine. By comparison, however, neutron radiography is a technique still in its infancy and the present work is a small step in its development.

Microradiography is the adaption of radiographic techniques to the examination of a very small part of an object, in such a way that a microradiograph is obtained which discloses fine internal details such as, microcracks, seggregated impurities and phase distribution. In brief, microradiography is related to

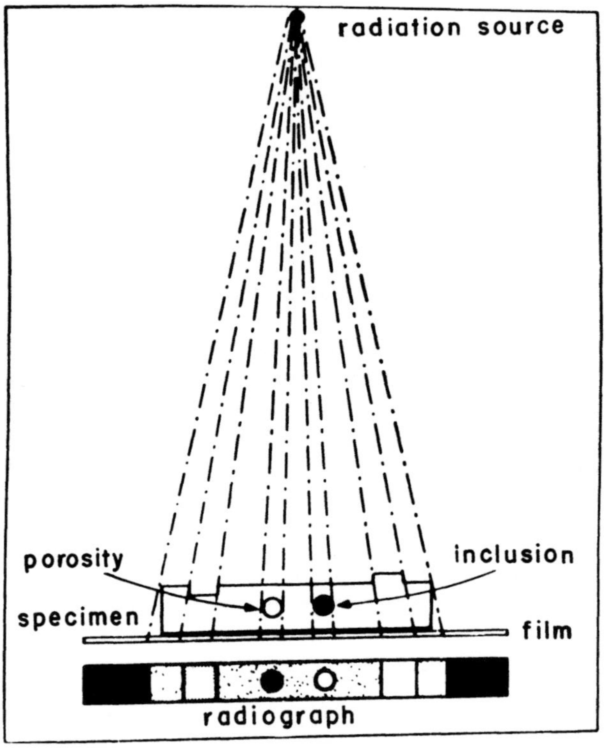

Fig. 1. A set up for radiographic examination.

radiography in the same way as micrography is to photography. The magnification effect required in microradiography may be obtained by one of two methods; either by geometrical spreading using a point source of radiation placed close to the object and a radiation detector placed some distance away, or by simply enlarging a high quality radiograph in which the required details are resolved. X-ray microradiography is a very special technique, the use of which has been severely limited by the exacting conditions necessary for high resolution. One purpose of the present work is to investigate the feasibility of neutron microradiography and the conditions favoring the high resolution required.

Physical Principles

Neutron microradiography like neutron radiography depends on the fact that when a neutron beam is passed through an object the neutrons are attenuated, both by true absorption and by scattering, to an extent that depends on the thickness or length of path, the density of the material, and more important, the kind of atoms that constitute such material. X and gamma rays also behave in much the same way, but the dependence of the absorption of X and gamma rays on the absorbing material is systematic and regular. In the case of neutrons, however, the absorption changes seemingly erratically and sometimes very drastically from one nuclide to another. Elements that are neighbors in the periodic table may cause very different absorptions of neutrons, and isotopes of the same element may interact quite differently with them. Figure 2 shows the systematic behavior of x-rays of 130 keV energy (shown by the continuous curve) compared with the behavior neutrons.

It can be seen that some heavy elements such as lead which are practically opaque to x-rays are practically transparent to neutrons and may easily be radiographed by them. Also the difference in neutron absorption between two elements close to each other in the periodic table, and so generally difficult to distinguish from one another by the use of X or gamma ray radiography, may have neutron absorption coefficients or neutron cross sections so different that the distinction between them by neutrons is very easy. Another situation where neutrons may be used with success while X and gamma rays usually fail, is the radiography of radioactive materials. Here the decay radiation of the radioactive material will confuse and may completely obliterate the radiographic image. With neutrons, however, a transfer method - described later - may be employed, that alleviates this difficulty.

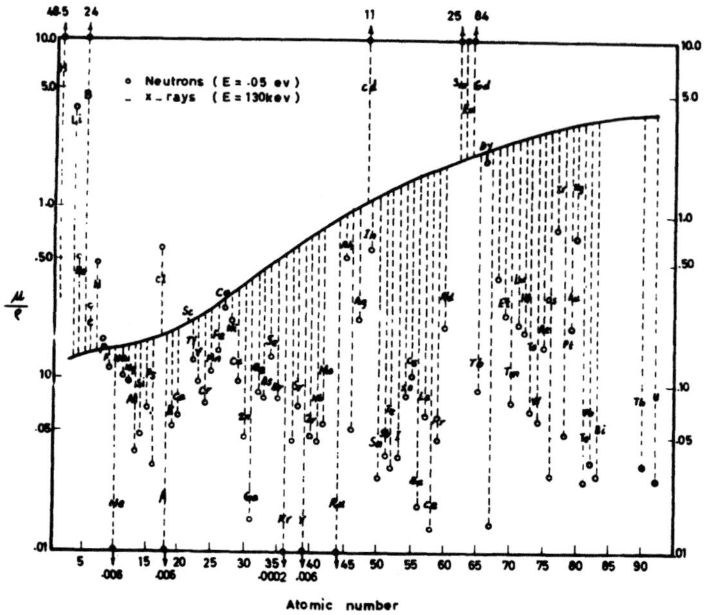

Fig. 2. Absorption of neutrons and X-rays by elements

Detection of Neutrons in Neutron Radiography

A neutron radiograph is made in much the same way as an X-ray radiograph. The object to be inspected is placed between the source of radiation and a radiation detector such as a photographic film. All nuclear radiation detectors including photographic films are based on the interaction of an energetic charged particle with matter. The direct photographic effect of neutrons is thus negligible. So, neutron "conversions" which result in charged particles or radiations are used. These fall in one of two categories: 1) Neutron induced transmutations in which the product particles or gamma ray photons make the detection of the neutron possible. Such transmutations are exemplified by (n,α), (n,p), (n, fission), and (n,γ) reactions. 2) Neutron induced transmutation resulting in radioactive product nuclides. Here the particles or radiations given off by the decay of such nuclides afford the detection. These are called activation conversions. Based on these two categories we have two categories of convertors. We can summarize the more promising of these as follows:

1) of the first category mentioned above ^{10}B and ^{6}Li have been used in neutron sensitive phosphors containing ^{10}B or ^{6}Li. One type is a fine mixture of ^{10}B and ZnS (Ag) in a plastic binder,

another is a mixture of LiF (96% enriched in ^6Li) and ZnS (Ag) in a similar screen. The thicker the powder the more is the luminescence given off, but beyond a certain thickness referred to as "infinite thickness", which in the case of ZnS (Ag) phosphor is about 80 mg/cm^2, there is practically no increase in luminescence.

2) of the second category, Dysprosium, Indium, and Gold are the best. They are all activation convertors. They may be used in direct exposure, like in the two cases mentioned above. They may also be left for sometime in contact with the film after the exposure to the neutrons has ended to get the additional benefit of the subsequent radioactive decay. Finally the metal screen may be exposed with the object without a film, whereby a radiative image of the object is left on it, then after the exposure, the screen is transferred to contact the film where the effect will be solely due to the radioactive decay. Three half life periods are usually allowed, and it is because the half life times of the three nuclides produced in the above materials are of suitable duration that they are preferred to others. The latter technique is very valuable, especially where the neutron beam contains considerable gamma radiation, and in the examination of the radioactive materials, for by this method the film is never exposed to the gamma rays which fog it.

EXPERIMENTAL WORK

To evaluate the overall resolving ability of a neutron radiographic technique and to test the applicability of neutron radiography to the examination of the fine structure of metals and alloys, a material known as Boral was selected for inspection both in the cast and hot rolled conditions, as well as a cast alloy containing 10% cadmium and 80% bismuth. The experiments were performed using the neutron inspection facility at the "Juggernaut" research reactor, at Argonne National Laboratory, USA. For this work a beam hole was used which had a high density concrete collimator, lined with stainless steel. The collimator limited the neutron beam size to 2.5 x 4 inches. The beam emerging from the hole was more or less parallel and was controlled by a shutter assembly within the reactor wall operated by remote control. The neutron beam intensity was estimated at 10^7 neutrons per cm^2 per second and the ratio of the thermal to the epithermal intensities was found to be 3.7:1.

The methods of neutron radiography used with each of the above mentioned materials were:

a) Direct exposure with 3 mil gadolinium foil convertor and an X-ray Kodak film type AA.

b) Direct exposure with 3 mil cadmium foil convertor and an X-ray Kodak film type AA.

c) Direct exposure with ^{10}B, ZnS (Ag) phosphor, "infinite", and an X-ray Kodak film type AA.

d) Transfer exposure with 10 mil dysprosium screen and an X-ray Kodak film type AA.

The same materials were also radiographed with X-rays at several X-ray tube voltages and examined metallographically.

Although some of the cadmium bismuth alloys showed shadows of cadmium dendrites, only aluminum boron alloys as cast and rolled, gave good enough contrast to make it possible to judge the resolution. It was also very clear from visual inspection that of all the techniques used, the direct exposure with gadolinium foil gave the best results, followed by the transfer exposure with dysprosium. Methods using the phosphor screen and the cadmium foil were deemed unsuitable. It was then decided to proceed with the work using direct exposure with gadolinium on the aluminum boron alloy.

The Effect of Thickness of Gadolinium Foil on Speed and Resolution

It was necessary for this investigation to obtain extremely thin foils of gadolinium. This was done by rolling foils 1 mil thick down to 0.1 mil. The foils produced were somewhat pyrophoric; in fact one foil flashed and burned out on touch due to static electricity spark. Care was then taken to ensure that everything the foil touched was grounded. A stack of 0.1 mil gadolinium foils was then arranged so that at one end it was one layer thick increasing to 2,3,4,5,10 and finally 15 layers at the other end. An exposure was taken of the stack and the optical density of the developed film was measured as a function of gadolinium foil thickness. Figure 3 shows this relationship. It can be seen that a foil thickness of about 0.4 mil gave the maximum speed. This is

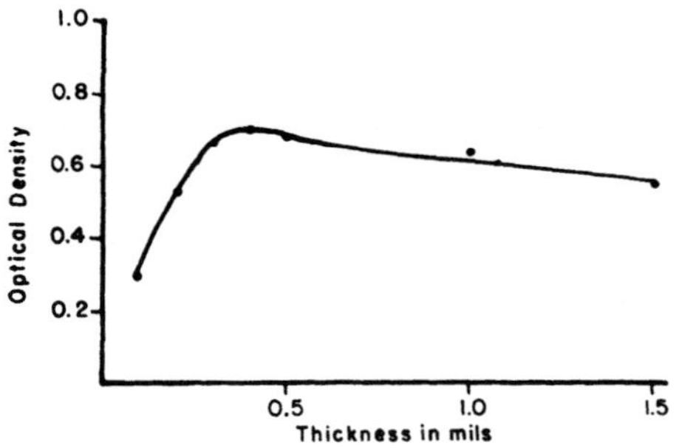

Fig. 3. Effect of gadolinium foil thickness on film density.

in fair agreement with calculation on the basis of known values of the neutron cross-section and the mass absorption coefficient of the gamma-ray produced in the (n,γ) reaction and the range of β particles produced in the decay process.

For resolution, radiographs of the aluminum boron cast alloy taken with gadolinium foils in the front position of 0.1, 0.5, 1 and 1.5 mil thickness were examined. Another type film, Kodak SA Type V-O was used because of its much finer grain. It was found that resolution deteriorated progressively as the thickness of the foil was increased. It seems thus that the use of a foil thicker than 0.4 or 0.5 mils is undesirable both from the point of view of speed and resolution. However, since the speed of the ultra-fine grain films used for high resolution is inevitably very low, a compromise has to be worked out between the speed and resolution, on basis of which the proper thickness of the foil may be chosen. Except where a very high degree of resolution is essential a 0.4 mil foil is adequate.

The Resolution Piece

It must be noted however, that it was impossible to separate the lack of resolution due to the technique used from that due to the three dimentional distribution of the boron containing phase in the alloy. A certain amount of fuzziness will have remained even if the resolution of the method was "perfect". Because of this, it was thought advisable to make our own sample for checking the resolution, "a resolution piece", where we can actually choose the shape, size, and distribution of the two phases. It was decided that the best configuration would be alternating parallel plates of one phase then the other, where the thickness of the plates is controlled. An overall resolution equivalent to the thickness of the plate will be achieved, if on the radiograph, the different plates appear distinct from one another. Such resolution piece should then be radiographed with the object to ensure whether a certain resolution has been achieved. A set of resolution pieces would be required to cover a range of resolutions. Three were made in this work with separations of 5,2, and 1 mil.

One resolution piece was made as follows: Two strips, one of cadmium, and the other of lead, each measuring 3 x 4 inch and approximately 1/8th inch thick, were thoroughly cleared, and rolled down to the desired thickness. The long rolled strips were cut up to suitable lengths and stacked on top of one another with foils of one metal alternating with foils of the other. The stack was then wrapped in aluminum foil and placed between two heavy plates of stainless steel. The whole assembly was heated to 250 deg. C., and then transferred quickly to a press where the foils were press welded together and their thickness slightly reduced.

The stack, now in one piece was then chopped up very carefully to pieces 1 x ½ inches on the sheet plane. One piece was sandwiched between two aluminum plates and the sandwich then clamped tight in a small vice and milled down with great care reducing the size to 1 x 3/16. The piece was then removed, and slightly twisted around an axis parallel to its long edge. It was then mounted in a lucite disc using a metallographic mounting press with the twisted sheets running more or less perpendicular to the plane of the disc. The lucite disc with the piece embedded in it, was then ground down manually with the minimum of pressure on a grinding wheel with fine emery paper. During grinding adhesive tape held the disc to the fingers, as the disc was too thin to hold directly. Grinding was performed on one surface of the disc and then the other alternately until the thickness of the disc reached 0.1 inch. Both sides of the disc were then polished on an ultrasonic vibratory polisher.

Operation of the Resolution Piece

Cadmium has a very high absorption cross section for thermal neutrons while lead is almost transparent to them. A stack of alternate sheets of cadmium and lead, provides easy passages for thermal neutrons travelling parallel to the sheet plane, separated by virtually impermeable walls. However, if the direction of the neutron beam is out by only a few degrees, the passage is completely blocked. A deviation of this amount can easily be expected to occur if the sheets in the resolution piece are plane but not quite perpendicular to the disc surface, or if the neutron beam is very slightly inclined to the surface of the disc and film. The twist in the resolution piece overcomes this difficulty, the neutrons always find a position on the twisted piece where the direction of the motion is along clear passages.

Choice of the Film

It was clear that the ordinary gelatino-silver bromide plates or cut films were not suitable for high resolution work because of their relatively course grains which limit the magnification the film can stand. Of the many Kodak films tried only two were found to possess suitable quality. These were Kodak spectroscopic cut films Type V-O and Type 649-O both of which were available only on a special order. Both are of extremely high contrast and extremely high resolution and the relative sensitivities of both films are very low. The relative sensitivity of the V-O is only 0.2 and that of the 649-O film is 0.0016 compared with that of the tri-X type B film of 25. The resolving power of the V-O film is better than 225 lines/mm while that of the 649-O film is better than 2000 lines/mm. The first was found to be quite suitable for macro work where the magnification did not exceed 10 x and the latter has been used successfully for micro work with magnifications of up to 200 X without any distracting effect due to the film grain.

DISCUSSION OF RADIOGRAPHS AND CONCLUSION

All the macroradiographs presented here were taken originally on spectroscopic cut film type V-0 and all microradiographs were taken on Kodak spectroscopic cut film type 649-0 and later photographed with a Leitz microscope at magnification of 100X.

Plate (1) shows a radiograph of the coarsest resolution piece (5 mil) on which can be seen the individual sheets resolved near the center of the piece where the plates happened to be parallel to the neutron beam direction.

Plate (2) shows a microradiograph of the 1 mil resolution piece taken at the position where lines are resolved.

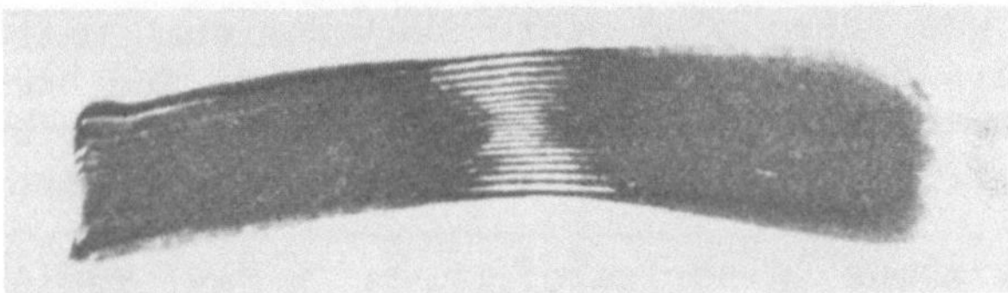

Plate (1)

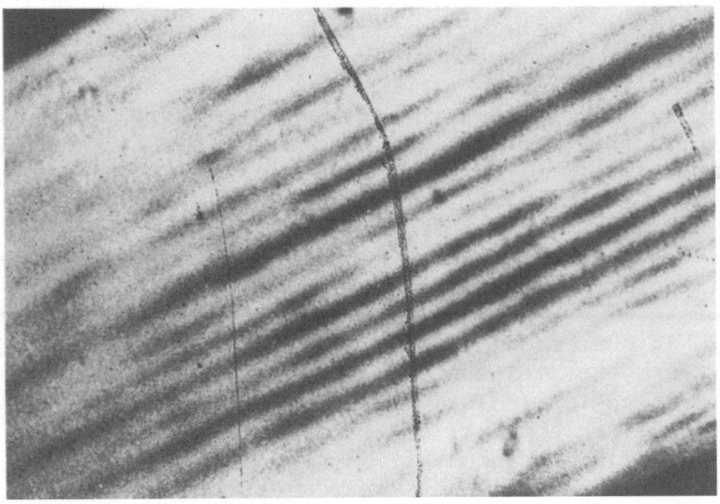

Plate (2)

As was mentioned above an alloy known as boral was selected to demonstrate the feasibility of the technique. Boral, which is a material often used in the control elements of nuclear reactors, is a composit material of boron carbide B_4C in an aluminum matrix. Boron carbide itself is a dense, hard, highly inert material containing 80% boron, which can be suspended in molten aluminum. The alloy is made by mixing powdered boron carbide with about half of the necessary aluminum, and then adding this powder mixture to the molten aluminum balance. The liquid mixture is cast into steel molds from a temperature of 670 C which is only ten degrees above the melting point of aluminum. The Boral used in this investigation contained only about 10% boron.

Plate (3) is a photograph (macrograph) of the cast boral alloy, in it can be seen big spherical holes that are open to the surface. These are not what is usually referred to as porosity in castings, they are not interconnected and are not due to gas absorption, more likely, they are due to entrained air or gas which could not escape while the alloy was molten. The boron carbide phase was introduced into the aluminum matrix by violent agitation, this being the source of the air, and because the temperature was so close to the melting point as has been mentioned before, much of the entrained air did not have chance to escape.

Plate (4) shows a series of three x-ray radiographs of the cast alloy, which were taken in an attempt to study the three dimensional distribution of the boron carbide and of the holes in the sample. What, in fact, they show clearly is the distribution of holes, which become more evident as the x-ray tube voltage is increased from 12 kilo volts in (a) to 14 kilo volts in (b) to 18 kilo volts in (c). These radiographs, however, fail to even reveal the presence of the boron carbide at all.

Plate (5) is a neutron macro-radiograph, where one can see very clearly the boron carbide phase, not only at the surface but throughout the sample. The holes here are faintly visible and they do not distract from boron carbide distribution.

Plate (6) represents the result of an attempt to utilise the (n, α) interaction in boron. Here no convertor foil was used and the print shown is due to the effect on the film of α-particles produced in the boron carbide phase by this interaction. This should give better resolution, but since the range of the α particles is very limited, the print represents the distribution on the surface which contacts the film. In this particular instance, it does not add new information regarding the structure of the alloy which is not obtainable by metallography, but where there is doubt as to the identity of the different phases, such technique may be quite useful.

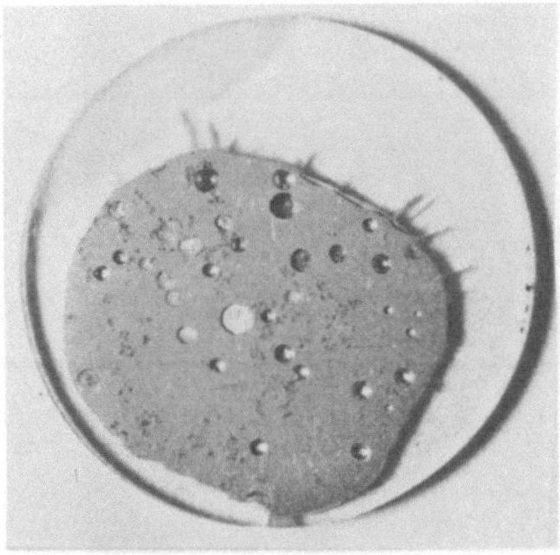

Plate (3)

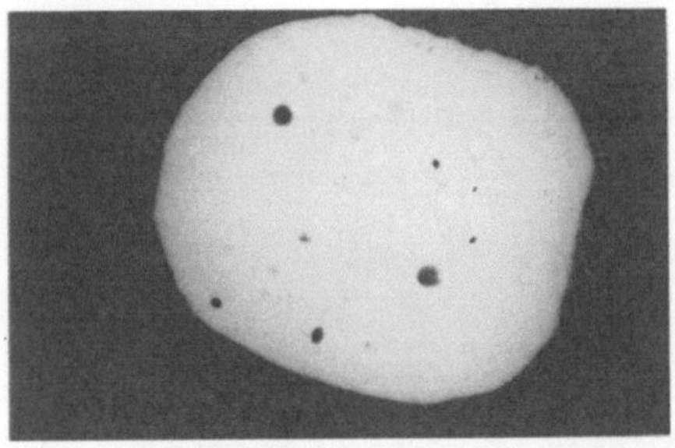

Plate (4a)

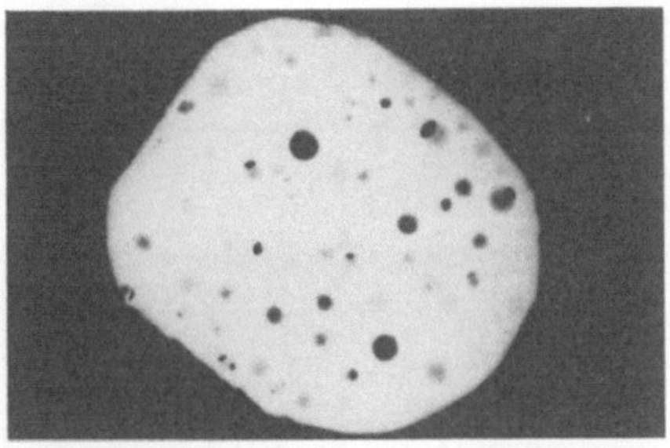

Plate (4b)

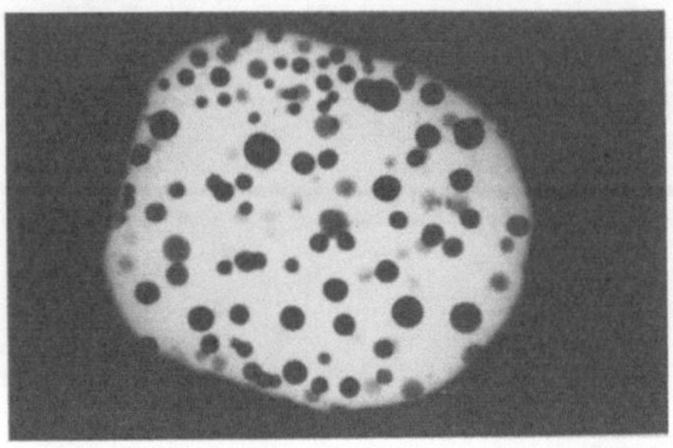

Plate (4c)

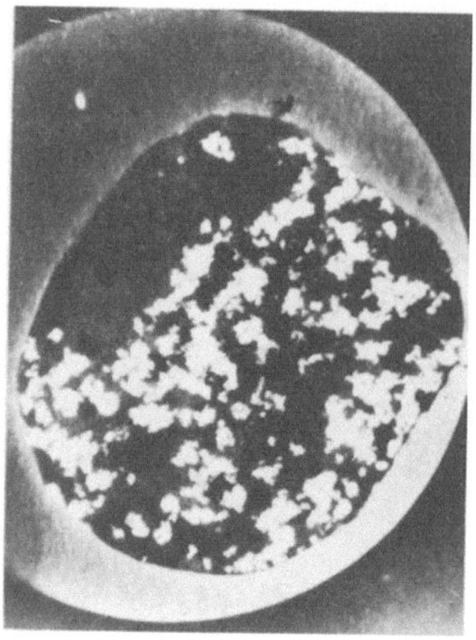

Plate (5)

Plate (6)

Plate (7) is a macrograph of rolled Boral where the streaks of the boron carbide phase can be seen.

Plate (8) is a radiograph of the rolled Boral showing again the streaky nature of the boron carbide distribution, except that here it shows it throughout the sample.

Plates (9) and (10) are neutron micro-radiographs of the cast Boral. We note that although the boron carbide phase appeared fairly massive in plates 5 and 6, here it is seen that it is actually made up of clusters of multitudes of very fine particles.

Plate (11) is a neutron microradiograph of rolled Boral where the fine particles of boron carbide are much more uniformly distributed than in the cast alloy.

In conclusion, we can say that it is feasible to use neutron beams for radiographing small objects and that the technique reveals structural details unobservable neither by micrography nor by X-rays.

ACKNOWLEDGMENT

The author wishes to acknowledge the help and encouragement of the director and staff of the Institute of Nuclear Science and Engineering of Argonne National Laboratory. The capable assistance of Mr. R. Friddle is gratefully acknowledged as is the cooperation of the Nondestructive Testing group of the metallurgy division. The author thanks are also due to Cairo University for providing the opportunity for this work to be done.

Plate (7)

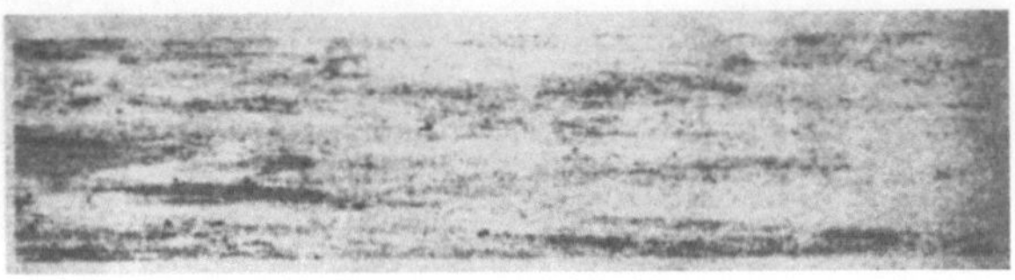

Plate (8)

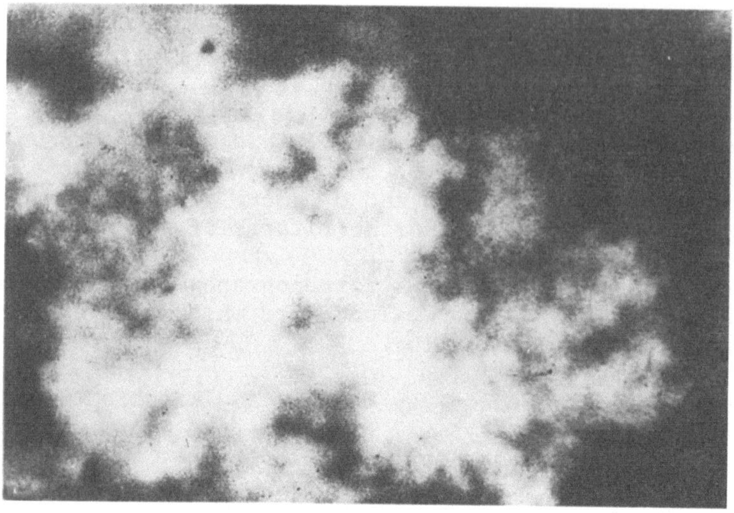

Plate (9)

Plate (10)

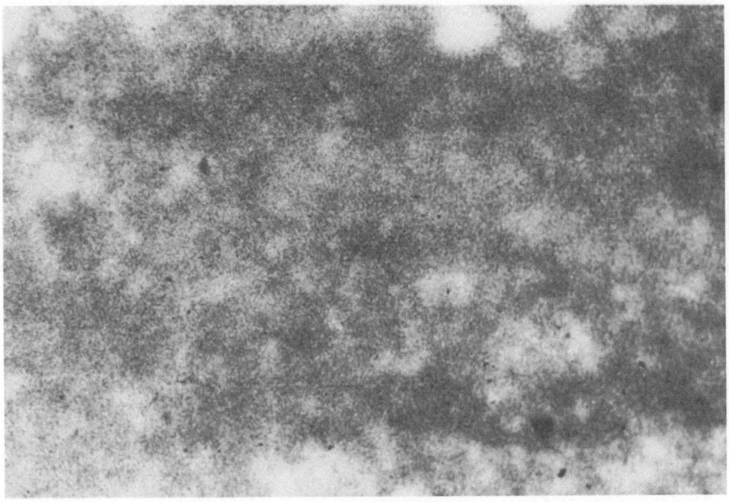

Plate (11)

REFERENCES

1. H. Kallmann, Research, Vol. 1, p. 254-60 (1947).

2. J. Thewlis, Brit. J. Applied Physics, Vol.7, p. 345-50 (1956).

3. J. Thewlis, in Progress in Nondestructive Testing, Published by Heywood, Ltd., London, 1958, p. 111-126.

4. H. Berger, J. Applied Physics, Vol. 33, p. 48-55 (1962).

5. H. Berger, Scientific American, No., 1962.

6. H. Berger and W.N. Beck, Nuclear Science and Engineering, Vol. 15, p. 411-414 (1963).

REPORTING NEUTRON EXPOSURE FOR RADIATION DAMAGE

A. David Rossin

Argonne National Laboratory

Argonne, Illinois, U.S.A.

Although reactor neutrons provide the physicist with a powerful tool for introducing defects into lattice structures, he would also like to know the number and distribution of these defects. The defect array formed by a neutron collision with a lattice atom is strongly dependent on the energy imparted by the neutron to the primary knock-on atom. Numerous studies deal with the resulting distribution as a function of the energy of the primary. The statistical distribution of primary knock-on energies can be calculated only if the energies of the neutrons are known.

This paper presents a method for determining and reporting the data needed to interpret neutron irradiation experiments. It makes it possible to calculate the number of atoms displaced by the bombardment, and most important, provides exposure data in units which eliminate the neutron spectrum as a variable. Thus, using these units, experimental results from entirely different reactors can be compared directly, even though the fast neutron spectra may be quite different.

Reactor spectra always differ from the pure fission spectrum, and these differences become more pronounced at energies below 2 MeV. It has been shown that comparing exposures by quoting $\Phi > 1$ MeV offers very limited accuracy. Usually these attempts were based on activation data from a threshold detecting foil assuming a pure fission spectrum for the fast neutrons, thus introducing another arbitrary error.

Recent experience shows that the reaction Fe^{54} (n,p) Mn^{54} is of great value for monitoring radiation damage. It is possible to *relate other activation rates and displacement production rates to measured iron activation rates, if the spectra are known. These ratios become a key in determining and reporting exposure.*

The use of realistic spectra and neutron-energy-dependent damage models was proposed in 1961 and their use is becoming more widespread. Several different units for reporting damage have been proposed: RDU by the writer, D_t by K. Shure of BAPL, the fluence notation, etc. The bases are similar, but the choice depends on what investigators understand and are willing to use. The paper will include sample calculations and exposure rate data for a number of reactor irradiation facilities.

INTRODUCTION

In the design and safety analysis of nuclear power plants it is necessary to predict what the behavior of an irradiated component will be, based on data obtained from test reactor irradiations. A substantial part of the uncertainty that exists in these applications could be eliminated by precise and standardized dosimetry. Because of arbitrary choices of parameters, safety factors of two and three have had to be allowed, whereas the methods described in this paper should be good to within ± 15%. Since a factor of two means the difference between a 20 year and a 40-year life for a nuclear power plant, this order of uncertainty is not acceptable.

Much work reported in this symposium deals with the concentrations of defects induced by radiation. Comparisons between neutrons and other irradiations will require precise dosimetry in order to develop adequate theoretical models to analyse the changes radiation causes in solids.

A suitable system for neutron dosimetry must take into account the following (1,2):

a) the kinetic energy of a struck atom is transferred by collision to other atoms, thus displacing them from their lattice sites. Changes may be a function of the total number of displaced atoms, or the number of regions of disorder (spikes), where a high concentration of displacements exists.

b) a fast neutron has enough energy to cause hundreds, or even thousands of displaced atoms. While the energy of the struck primary may be proportional to the incident neutron energy, the array of defects produced and its influence on properties might have an entirely different energy dependence.

c) since the energy of the struck primary is dependent on the neutron energy and scattering angle, there is no true "threshold energy" for damage production.

d) neutron spectra differ from reactor to reactor, and none has a true fission spectrum. Some do not even come near the fission spectrum shape, especially fast reactors and locations near pressure vessels in power reactors. Additional error is introduced if the fission spectrum is arbitrarily assumed to exist when it does not. Any measured flux or fluence, quoted as "neutrons > 1 MeV" makes an assumption of some spectral shape, since no foil detector has a threshold at 1 MeV.

The dosimetry techniques recommended in this paper lead to a single radiation damage unit (RDU) and provide enough data to characterize any irradiation facility. This permits the investigator to choose his conditions, select his irradiation period, and report his exposure directly. The Data Sheets in this paper are examples of the type of information that should soon be widely available.

DETERMINATION OF REALISTIC NEUTRON SPECTRA

Spectra can be calculated by multigroup neutron theory, a technique that has reached a high degree of sophistication by virtue of its use in the design of fast breeder reactors.

Spectrum determination by analysis of a number of neutron detecting foils runs into difficulties due to the lack of sufficient accurate data on activation cross-sections as a function of neutron energy in the ranges of interest (3). There are too few foils with suitable mechanical properties and half-lives, that have appropriate activation cross-sections to give statistically reliable data.

A practical method employed successfully by the author (1) employs the computer calculation to find the shape of the neutron spectrum, and uses measured foil activation rates to check the shape and fix the magnitude of the flux. Once the shape is established, activation rates and ratios can be calculated using the cross sections in Table I.

ENERGY-DEPENDENT DAMAGE MODELS

As knowledge of the behavior of energetic nuclei in solids grows, more sophisticated approaches are being made to the development of an energy-dependent damage model. When a fast neutron strikes a lattice atom it can give up a portion of its energy to the struck atom, depending heavily on the scattering angle. This process is described in detail by Dienes (4) and applied to the case of iron by Rossin (5), Beeler (6) and Shure (7). By simple

TABLE I

CROSS-SECTIONS FOR ACTIVATION AND RADIATION DAMAGE

Energy Group	$Fe^{54}(n,p)Mn^{54}$	$Ni^{58}(n,p)Co^{58}$	U^{238} fission	$Fe^{54}(n,\alpha)Cr^{51}$	RDU	Spike
1	0.665	0.6400	1.0400	0.031	1.2081	0.8781
2	0.590	0.6340	0.8400	0.017	1.1360	0.9417
3	0.440	0.5470	0.6000	0.010	1.0258	0.9290
4	0.250	0.3840	0.5670	0.005	0.9368	0.9163
5	0.210	0.2350	0.5720	0.000	0.9199	0.8908
6	0.1720	0.1340	0.5830		0.9962	0.8654
7	0.060	0.0540	0.5280		1.1657	0.8908
8	0.0	0.0140	0.2830		1.2293	0.9545
9		0.0040	0.0560		1.2802	1.0563
10		0.0005	0.0114		1.2081	1.0944
11		0.0	0.0027		1.0470	1.1962
12			0.0006		0.7842	1.2344
13			0.0		0.6486	1.1199
14					0.5087	1.0563
15					0.4239	1.1329
16					0.3264	1.2853
17					0.2628	1.3490
18					0.2120	1.3235
19					0.1865	1.2217
20					0.1568	1.0817

application of conservation of energy and momentum, it can be seen that the primary knock-on starts out with an energy between zero and 7% of the neutron energy. For a 1 MeV neutron, the average energy of the primary is 35 keV. Considering that it only takes about 25 eV to displace an iron atom, there is energy available to cause many hundreds of displacements.

Kinchin and Pease (8) and Snyder and Neufeld (9) gave models which predict the number of displacements resulting from a primary of a given energy, assuming that the kinetic energy not dissipated in ionization losses goes toward displacing atoms. These results were used to give the damage model labelled "RDU" in Table I.

The large digital computer has made possible a more realistic investigation of the behavior of energetic atoms in a metal lattice (10,11). Beeler uses mathematical models of the lattice in three dimensions, expresses the interatomic potentials using the most recent theoretical models, and calculates the array that results after neutron scattering. Beeler's work shows that vacancies and interstitials end up in clusters and complex groupings. Using Brinkman's terminology, large clusters are called displacent spikes. Studies of mechanical deformation indicate that single defects cannot effectively block dislocation movement, but spikes can. A function can be developed giving the relative probability of spike production due to a neutron of a given energy (12). This function is the basis for the damage model called "Spike" in Table I.

For any damage model, a unit may be chosen for reporting the total damaging neutron exposure. Each of the models considered by the author has been normalised, such that when weighted by a unit fission spectrum, the damage cross-section is unity. Thus if 10^{18} neutrons having a fission spectrum energy distribution bombard a 1 cm^3 iron specimen, the exposure is 10^{18} RDU, regardless of what mechanisms are used to depict the behavior of the atoms. (Of course, depending on the geometry of the specimen, a correction might have to be made for self-shielding).

All damage models assume that the cascade is complete when the struck atoms have less than that energy necessary to displace one atom from its site (about 25 eV). This is the type of behavior that would occur at absolute zero. In a more recent study by Beeler (13), the computer is asked to express the "annealing" of the damage, by allowing vacancies and interstitials to move in the lattice due to their thermal energy. If an interstitial and a vacancy meet, they eliminate one another. Ultimately one could hope to express the entire irradiation mathematically, and arrive at a figure for the net amount of damaging exposure as a result of an irradiation in a given facility for any period of time at a given temperature. The RDU concept can be adapted to any energy-dependent model.

The discussion above has referred entirely to the damage caused by lattice displacements only. Neutrons also cause embrittlement by transmutations. These can be due to thermal neutrons, i.e., [$B^{10}(n,\alpha)Li^{7}$], or fast neutron activations [$Fe^{54}(n,p)Mn^{54}$], leaving helium and other foreign atoms in the lattice. These changes contribute to embrittlement in their own way, and do not generally anneal out when the specimen is heated. The phenomena are discussed by Dr. J. Moteff (14) in this symposium. As detailed knowledge of cross-sections and spectra increases, it may become possible to predict the number of transmutations that will take place in a sample, using calculations of the same type as those used for lattice damage above.

REACTOR FACILITY DATA SHEETS

Data Sheets 1 and 2, and Table II are examples that illustrate the differences that exist between a pure fission spectrum, that of a very hard location in a thermal reactor (CP-5 Fuel Element) (15), and the core of a fast breeder reactor (EBR-II), (16).

The calculated spectrum is presented first. Its magnitude has been determined by analysis of the foil data; the fission spectrum (Table II) is normalized. The spectrum is easily pictured as a histogram, each bar representing the flux in one energy group. The values for flux above any arbitrary energy and the activation rates are calculated directly, using the cross-section data of Table I. From these, activation ratios are obtained, and are tabulated with reference (arbitrarily) to the $Fe^{54}(n,p)Mn^{54}$. Measured values of foil activation rates, if available, are recorded along with the ratio of calculated to measured rates.

It is convenient to report fluxes in terms of flux per unit power, that is, (n/cm^2-sec) per MW of reactor power. It is then easy to apply the reactor operating history to determine total exposure. Selected rates are summarized for full power, to aid in selection of irradiation facilities.

SPATIAL DISTRIBUTIONS

Each specimen of a set irradiated in a reactor sees a different flux intensity due to the spatial variation over the core. If the flux is carefully mapped, a few tests can indicate the dependence of the property being measured on specimen position. What appears to be scatter of data can be evaluated, and perhaps eliminated (18,19).

DATA SHEET

Reactor : CP-5 Argonne National Laboratory, Argonne, Illinois
Facility: VT-10 Fuel Element Central Thimble
Spectrum:

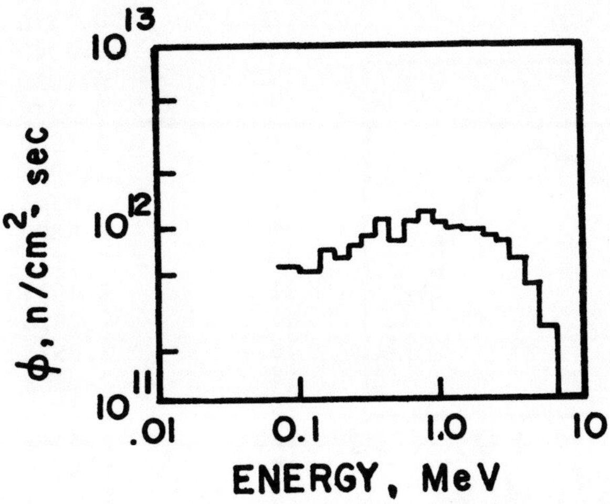

Group	Φ/MW n/cm^2-sec$\times10^{12}$
1	0.0416
2	0.1092
3	0.2983
4	0.4471
5	0.6543
6	0.9045
7	0.9516
8	1.0215
9	1.0312
10	1.0334
11	1.1002
12	1.1045
13	0.7723
14	1.0212
15	0.9307
16	0.8349
17	0.7230
18	0.8034
19	0.6649
20	0.7021

ACTIVATION RATES AND RATIOS

Reaction	Calculated		Measured (15)		Calc./ Meas.
	Act.Rate	Act./Fe^{54}	Act. Rate	Act./Fe^{54}	
Fe^{54}(n,p)	6.18×10^{11}	1.0	6.18×10^{11}	1.0	1.0
Ni^{58}(n,p)	7.00×10^{11}	1.13	7.89×10^{11}	1.28	0.89
U^{238} Fiss.	2.33×10^{12}	3.41	3.84×10^{12}	2.38	0.606
Other					
RDU	1.04×10^{13}	16.88			
Spike	1.49×10^{13}	24.06			

Summary: At Full Power of 4.6 MW:
Φ Thermal: 5×10^{13} n/cm^2-sec
$Fe^{54}(n,\alpha)$: 3.84×10^{10} $/cm^2$-sec
RDU : 4.78×10^{13} $/cm^2$-sec

DATA SHEET

Reactor : EBR-II, Argonne National Laboratory, Idaho Falls, Idaho
Facility: Row 4, Subassembly C1, at Midplane of Core
Spectrum:

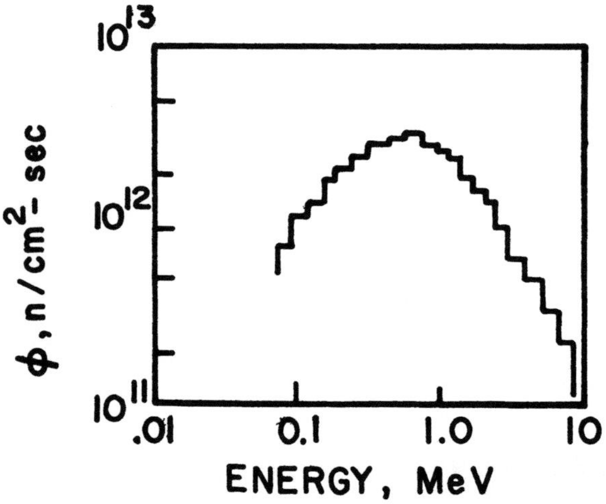

Group	Φ/MW n/cm^2-secx10^{12}
1	0.0440
2	0.1235
3	0.3042
4	0.4935
5	0.6975
6	1.1280
7	1.4025
8	1.6563
9	1.8800
10	2.1432
11	2.4064
12	2.7824
13	2.8576
14	2.6508
15	2.2748
16	1.9364
17	1.6544
18	1.3592
19	1.0904
20	0.8310

ACTIVATION RATES AND RATIOS

Reaction	Calculated		Measured		Calc./ Meas.
	Act.Rate	Act./Fe^{54}	Act.Rate	Act./Fe^{54}	
Fe^{54}(n,p)	$7.84x10^{11}$	1.0	$7.84x10^{11}$	1.0	1.0
Ni^{58}(n,p)	$8.86x10^{11}$	1.129	$9.81x10^{11}$	1.25	0.913
U^{238}Fiss.	$3.02x10^{12}$	3.846			
Other					
RDU	$2.195x10^{13}$	28.01			
Spike	$3.324x10^{13}$	42.46			

Summary: At Full Power of 4.5 MW:
Φ Thermal: 0
$Fe^{54}(n,\alpha)$: 4.03 x 10^{11}/cm^2-sec
RDU: 9.86 x 10^{14}/cm^2-sec

TABLE II

THE FISSION SPECTRUM AND PERTINENT RATIOS

Energy Group	E_L MeV	Spectrum	Ratios
1	7.788	0.0061	ψ > 3 MeV = 0.2082
2	6.065	0.0171	
3	4.724	0.0398	ψ > 1 MeV = 0.6901
4	3.679	0.0684	
5	2.865	0.0963	ψ > 0.5 MeV = 0.8672
6	2.231	0.1140	
7	1.738	0.1190	ψ > 0.10 MeV = 0.9861
8	1.353	0.1120	
9	1.054	0.0983	Ni^{58}/Fe^{54} = 1.140
10	0.821	0.0823	
11	0.639	0.0646	U^{238}/Fe^{54} = 3.198
12	0.498	0.0501	
13	0.388	0.0371	Fe^{54} : (n,α)/(n,p) = 0.0127
14	0.302	0.0276	
15	0.235	0.0197	RDU = 1.00
16	0.183	0.0143	
17	0.143	0.0100	RDU/Fe^{54} = 10.44
18	0.111	0.0071	
19	0.086	0.0049	$Spike/Fe^{54}$ = 10.40
20	0.067	0.0035	

REFERENCES

1. A.D. Rossin, ANL-6826 (March 1964).

2. D.K. Holmes, AIME Rad. Eff. Symp., Ashville, N.C., September 1965 (to be published).

3. R. Gold, Nucl. Sci. Eng., 20, 493 - 512 (1964).

4. G.J. Dienes and G.H. Vineyard, "Radiation Effects in Solids" Interscience (1957).

5. A.D. Rossin, Nuclear Science and Engineering: 9, 137 - 147 (1961).

6. J.R. Beeler, J. Appl. Phys., 37 No. 7, 2226 (July 1964).

7. K. Shure, Nucl. Applications, 2, 106 (April 1966).

8. G.H. Kinchin and R.S. Pease, Reports on Progress in Physics, 18, 1 (1965).

9. W.S. Snyder and J. Neufeld, Phys. Rev., 103, 862 (1956).

10. J.R. Beeler and J.L. McGurn, GEMP-101 (1961).

11. C. Erginsoy, Phys. Rev., 133, A595 (1964).

12. J.R. Beeler, Phys. Rev. (to be published).

13. J.R. Beeler, ASTM-STP, 380 (1965).

14. J. Moteff and M. Hoch, Cairo S.S. Conf. (1966), abstract.

15. A.D. Rossin and R.J. Armani, Neutron Dosimetry Vol. II, pp. 293 - 304.

16. F. Kirn, Argonne National Laboratory, Idaho Division (pers. comm.).

17. Reactor Physics Constants, ANL-5800 (1963).

18. A.D. Rossin, Nucl. Struct. Eng., 1, 1, 76 (January 1965).

19. A.D. Rossin, Thesis, Case Inst. of Tech. (1966).

THE SPECTROSCOPY OF GLASSES

R.W. Douglas

Department of Glass Technology

University of Sheffield, ENGLAND

The absorption of radiation from 2,000A° up to about 12μ by glasses will be discussed. In the far ultra-violet a very strong absorption is associated with alkali ions. Between this absorption and the vibrational absorption in the infrared there is a series of absorption bands associated with other ions, chiefly those of the transition metals and the rare earths.

Although the vibrational bands are impossible to interpret rigorously there is very strong experimental evidence which suggests that they may be associated with particular local groupings of atoms. The absorption due to the transition metal ions can be interpreted satisfactorily by means of ligand field theory.

INTRODUCTION

In this paper it is proposed to give an account of the absorption of radiation by glasses within the wavelengths 2,000 angstroms up to about 10 microns; where possible, interpretation in terms of energy transitions will be attempted. Other experimental techniques such as ESR, NMR and X-ray diffraction will only be referred to in indication of how far they agree or assist in the interpretation of the absorption observations.

The glasses to be discussed will be those based on silica, boric oxide and P_2O_5. The data are by no means complete and the interpretation has to be restricted to the behavior of various compositions rather than to attempt to elucidate the effect of composition on the absorptions. The absorptions to be discussed fall into two main groups, the electronic transitions and the absorptions arising from vibrational energy levels. There is an important subclass of the electronic transitions, those associated with the coloring oxides, chiefly the transition metal ions and the rare earth ions; the absorption spectra of these ions can be well elucidated in terms of ligand field theory, and these absorptions are probably better understood than any others.

THE ABSORPTION IN THE ULTRA-VIOLET

In pure silica the so-called ultra-violet cut-off edge which presumably corresponds to some valence-conduction band transition, or anion-cation electron transfer process, occurs at about 1500 Angstroms (Appendix I). The addition of other oxides causes this absorption edge to move to longer wavelengths. Borrelli (1) has shown that a linear increase in the cut off wavelength occurs on adding Na_2O to silica (Fig. 1).

For more than thirty years it has been customary to speak of the structure of glasses in terms of network forming oxides and network modifying cations. The oxygens in the glasses were divided into two classes, bridging and non-bridging oxygens. Formally this can be expressed symbolically as

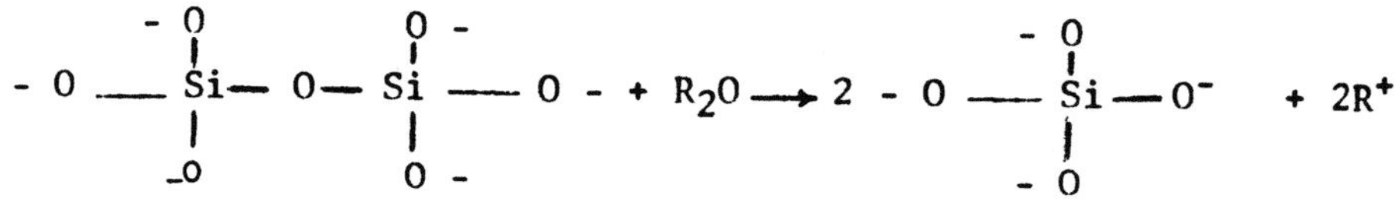

non-bridging oxygen

This nomenclature will be used throughout, without, at this stage, further comment on the accuracy of this structural model.

It is interesting to note that Stroud et al. (2), from consideration of absorption centers produced by irradiation suggested an electronic structure for soda silica glasses in which the introduction of alkali produced an electronic level in the forbidden energy gap of pure silica glass (Fig. 2).

In each section of this review the non-bridging oxygen will occur as a most important entity, as indeed it is in all the chemistry and physics of oxide glasses. Glasses are often supersaturated solutions of alkali in solvents which are the network forming oxides and it seems probable that in these solutions the concentration of non-bridging oxygens is as important as the hydrogen-ion concentration in aqueous solutions.

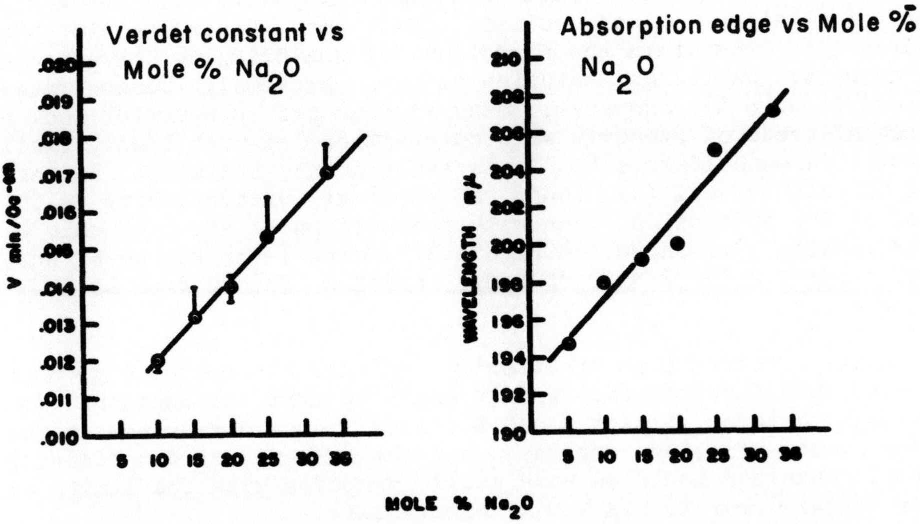

Fig. 1. a) Verdet constant versus mole % Na_2O in binary x $Na_2O.(1-x)$ SiO_2 glasses.
b) Absorption edge, defined as wavelength corresponding to $\alpha=163cm^{-1}$, versus mole % Na_2O in binary x $Na_2O.(1-x)$ SiO_2 glasses.

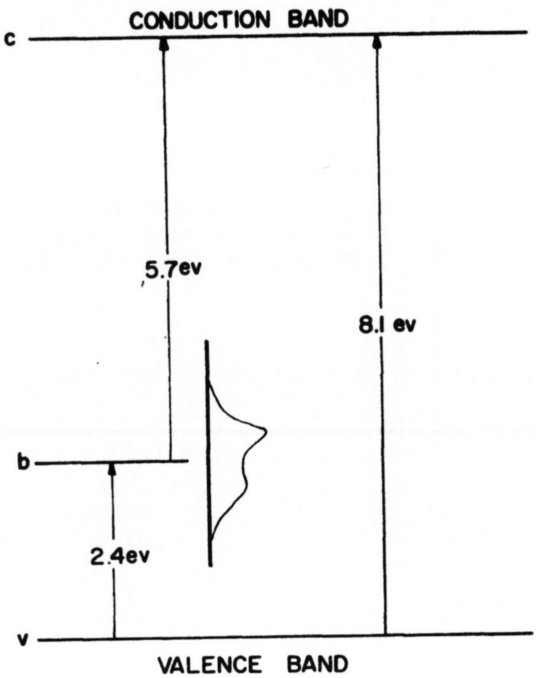

Fig. 2. Schematic electronic band structure of soda-silica glass.

Boric oxide glasses have been frequently studied. The addition of alkaline oxides to B_2O_3 does not, according to NMR studies (3), result in the formation of non-bridging oxygens in any quantity until the solution becomes extremely concentrated (around 35 mole %). However, many macroscopic properties show a change of trend of property with composition at much lower alkali contents (around 15 mole %). It is interesting, therefore, to note that McSwain et al. (4) found an apparent discontinuity in the trend of the absorption edge with composition at about 15 mole %. Their results are shown in Figure 3. Curve 1 is for pure B_2O_3 glass, curve 2 for 5 Na_2O 95 B_2O_3, curve 3 for 10 Na_2O 90 B_2O_3, and so on.

Merely observing an arbitrarily defined absorption edge does not give much information, and it could be that the apparent discontinuity between glasses 4 and 5 arises from a difference in the number of non-bridging oxygens, but the numbers of non-bridging oxygens concerned could be very small compared with the limits of experimental error in the N.M.R. experiments.

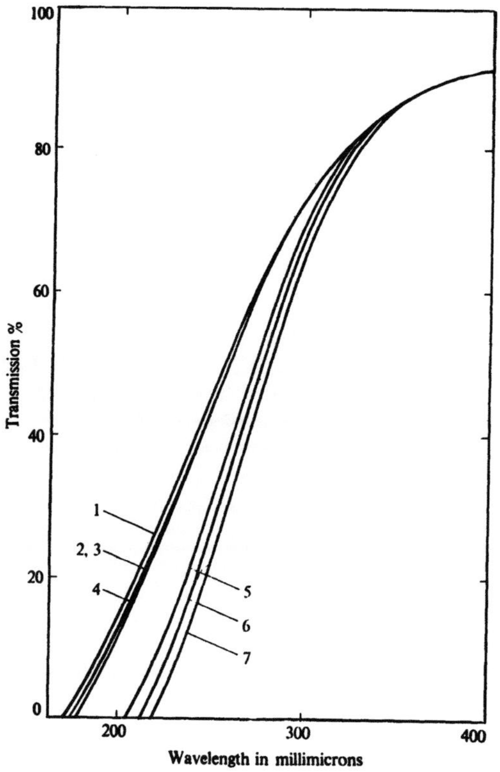

Fig. 3. Transmission of sodium borate glasses. Thickness of samples 4mm.

Charge Transfer Absorptions in the Ultra-Violet

Absorption due to Cr^{6+}. The charge transfer absorption bands of the Cr^{6+} ion are of great interest. In acidic glasses, i.e. glasses in which the number of non-bridging oxygens is small, their form is similar to that of the chromate ion in acidic solutions, and addition of alkaline oxide causes the absorption to change towards a similarity with the chromate ion in solutions of high pH. These studies are still in progress (5), but it is interesting to show now the behavior in boric oxide glasses. Typical absorption spectra are shown in Figure 4. It is to be specially noticed that in this optical phenomenon an apparent discontinuity occurs at about 25% Na_2O (Fig. 5).

The extinction coefficient found for the 370 mμ band for Cr^{6+} is 4200. It is customary to define the extinction coefficient ε by the equation

$$I = I_o \, 10^{-\varepsilon c d}$$

where d is the thickness in cms.
c is the concentration in moles/litre.

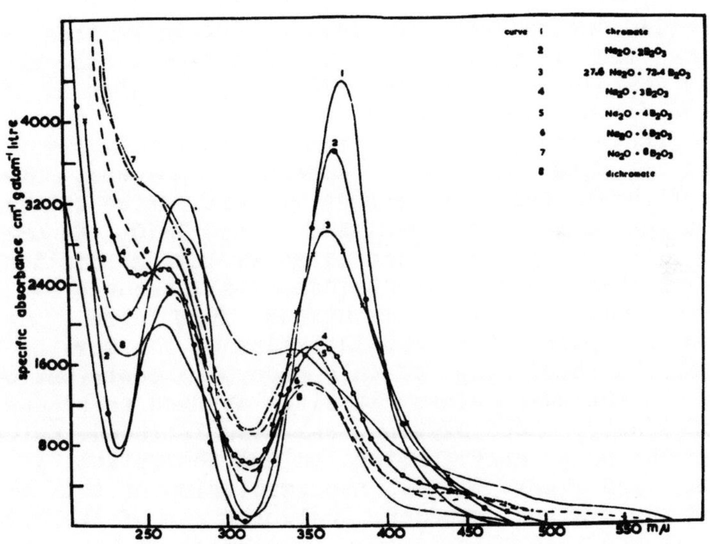

Fig. 4. Ultra-violet absorption of Cr^{6+} in Na_2O-B_2O_3 glasses.

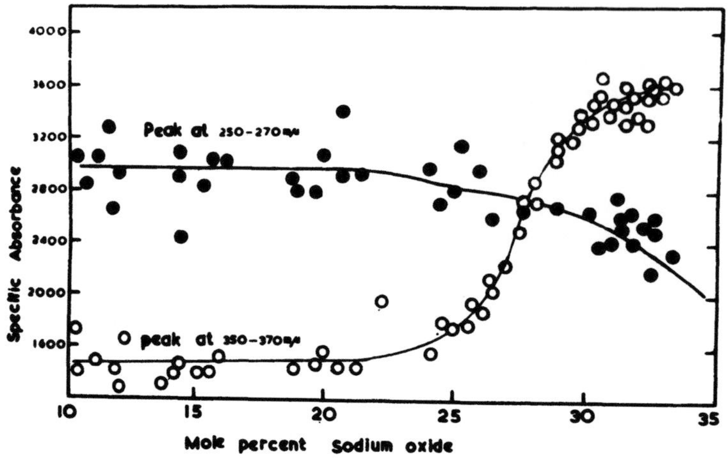

Fig. 5. Intensity of the Ultra-violet absorption of Cr^{6+} in Na_2O-B_2O_3 glasses.

Absorption due to Iron. In glasses containing iron the ultra-violet absorption is characterized by a cut-off edge which moves towards the visible region as the iron concentration or the thickness of the glass specimen is increased. This edge has been generally attributed to ferric iron but in the past has not been the subject of detailed study. Recently, however, Steele and Douglas (6,7) have been able to give a quantitative interpretation of iron absorption in three glasses, $3Na_2O \cdot 7SiO_2$, $Na_2O \cdot 9B_2O_3$ and $3Na_2O \cdot 7B_2O_3$. In their work, the ultra-violet absorption spectra of glasses in these systems containing small amounts of iron <0.5 mole % were analyzed for contributions from ferrous and ferric iron to the absorption. From consideration of the work of Borelli, mentioned at the beginning of this paper, a contribution to the absorption by the base glass itself, assumed to be due to the presence of singly-bonded oxygens, was also included. Calculations showed that the main contribution to the absorption is given by ferric iron, and Figure 6 shows the resolution of this absorption in each of the systems studied. Calculation of the base glass absorption showed that this is only important at very short wavelengths and may be presumed to have no influence above 220 mμ.

It was also found that ferrous iron has an absorption in this region in the $3Na_2O \cdot 7SiO_2$ and $Na_2O \cdot 9B_2O_3$ glasses, but this absorp-

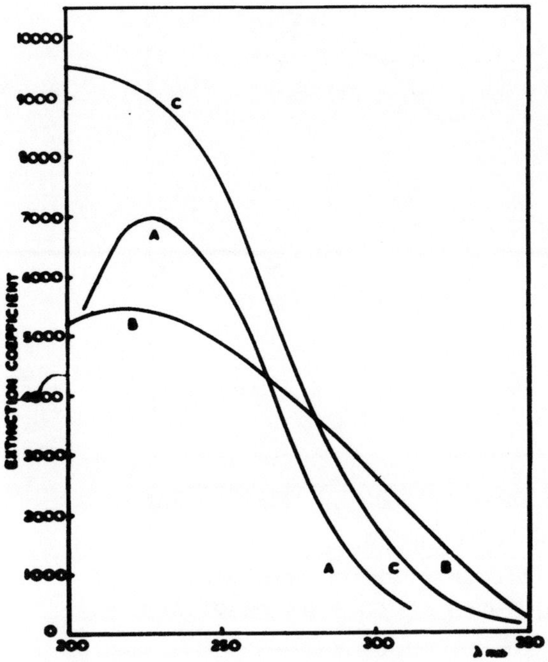

Fig. 6. Specific extinction curves for ferric iron in the ultra-violet.
A 3 $Na_2O \cdot 7\ SiO_2$ Glasses
B $Na_2O \cdot 9\ B_2O_3$ "
C 3 $Na_2O \cdot 7\ B_2O_3$ "

tion could not be resolved with any certainty. Figure 7 indicates the magnitude of the ferrous absorption in the $3Na_2O \cdot 7SiO_2$ system in comparison with the ferric absorption in the same region.

A further aspect of the work which is particularly interesting is that Urbach's Rule was shown to apply in part to the long wavelength tail of the ferric absorption in each system (Fig. 8). Urbach's Rule, which relates the absorption coefficient to frequency and which is temperature dependent, is obeyed by many ionic crystals and shows that the absorption must be due to some photon-phonon interaction, $\varepsilon = \varepsilon_0 \exp - \sigma (h\nu_0 - h\nu)/kT$. With glasses containing iron, however, σ varies with temperature and concentration. This variation appears to be associated with changes in coordination of the iron with concentration and the presence together of iron ions in both octahedral and tetrahedral coordination.

LIGAND FIELD BANDS

It has become customary to refer to the energy absorptions in which the only electrons involved are the d electrons of a transition metal ion in a glass as ligand field bands. Many excellent books in which ligand field theory is developed have been pub-

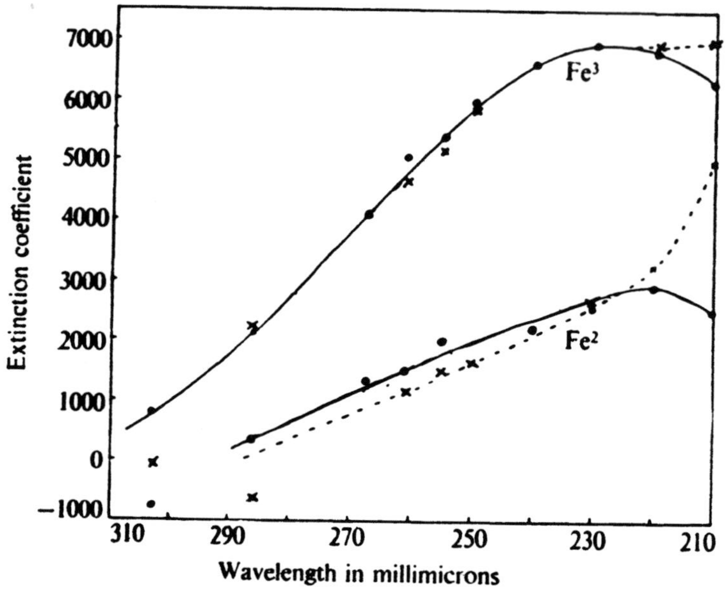

Fig. 7. Extinction coefficient versus wavelength for silicate glasses containing iron - ultra-violet.

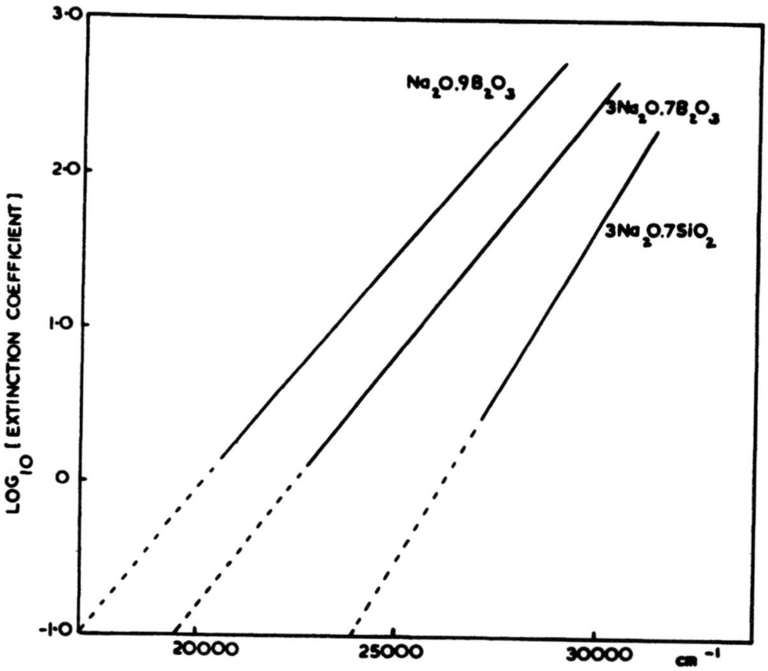

Fig. 8. Urbach's Rule in $3Na_2O{\cdot}7SiO_2$, $Na_2O{\cdot}9B_2O_3$ and $3Na_2O{\cdot}7B_2O_3$ glasses.

lished in the last decade; in order of increasing mathematical detail those of Orgel (8), Cotton (9), and Griffiths (10) may be mentioned.

The atomic levels of the free ion may be expressed in terms of the Racah parameters B and C. The splitting of these levels in a crystalline field of a particular symmetry can be shown on a diagram in which the energy, E, of a level is given as a function of the strength of the crystalline field Δ. If the unit of energy is taken as equal to the Racah parameter B, a diagram is obtained for a particular value of C/B applicable to any ion having that number of d electrons. In ligand field theory the values of B and C are allowed to change to take account of departure from purely ionic bonding, but in many uses of the theory in interpreting spectra it is hardly necessary to be concerned with the exact values of these constants. We start, therefore, by considering the diagram for the d^3 system in octahedral coordination calculated for B/C = 4.50 by Tanabe and Sugano (11) (Fig. 9).

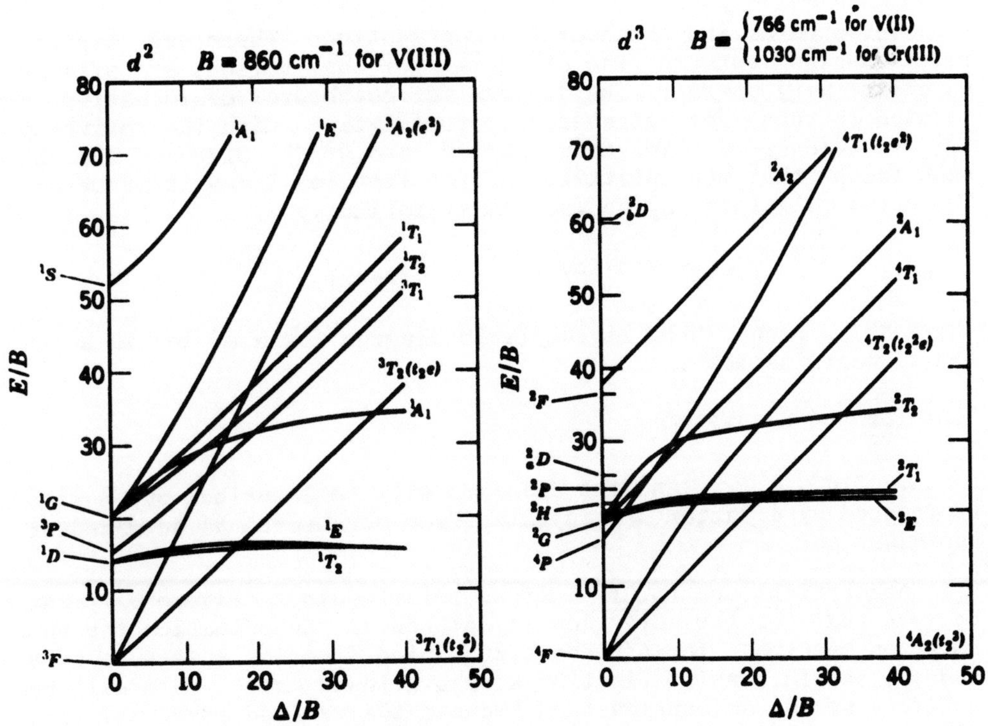

Fig. 9. Energy level diagrams, after Tanabe and Sugano, for d^2 d^3 systems in octahedral symmetry.

In this diagram the energy of the ^4F ground state is taken as the zero of energy. The energies of the absorption peaks for any particular value of Δ/B may be read off from the intercepts of the atomic energy levels with a vertical line erected at the value of Δ/B.

The intensities of the absorptions. Transitions $d^n - d^n$ are forbidden by the Laporte selection rules. However, if there is some small departure from symmetry the 3d and 4p orbitals may mix and there will be some relaxation of the selection rules. Tetrahedral complexes have no center of symmetry and the relaxation of the selection rules is greater. These considerations account for the observed variation of intensity in the ligand field bands. Typical order of magnitude of the extinction coefficients are

Tetrahedral coordination	Spin allowed	10^2
	Spin forbidden	1
Octahedral coordination	Spin allowed	10
	Spin forbidden	10^{-1}

Tetrahedral and octahedral coordination. There are several relationships between the diagrams for the various d^n systems which are very useful. The diagram for octahedral coordination is related to that for tetrahedral coordination with the splitting $\Delta_{tet} = -4/9\ \Delta_{oct}$. Similarly the d^n and d^{10-n} diagrams are the same except that the splitting of the free ion terms is inverted. These two relations may be written symbolically

$$d^n_{(oct)} \equiv d^{10-n}_{(tetr)} \text{ are inverse to } d^n_{(tetr)} \equiv d^{10-n}_{(oct)}$$

where d^n_{oct} means "the ligand field diagram for a d^n ion in octahedral coordination".

Thus for the d^5 system

$d^5_{(oct)} \equiv d^5_{(tetr)}$, and the diagrams will be identical but Δ will be reduced by a factor 4/9 in going from octahedral to tetrahedral coordination.

Absorptions due to Cr^{3+}. If we refer again to Figure 9 it can be seen that for the Cr^{3+} ion in octahedral coordination the absorption spectra, for Δ/B about 20, going from low energy to high energy, should consist first of a spin allowed band; this will be a fairly broad absorption band because the excited level diverges from the ground state in the ligand field diagram; near to this band should be two small spin forbidden bands which should be sharp because the excited levels are parallel to the ground state; next, there is another spin allowed excitation to the $^4T_1$ state.

These are in fact the bands observed in the visible spectrum as is shown in Figure 10 which gives the absorption curves for a series of sulphuric acid solutions all containing the same concentration of Cr^{3+}. (In this figure a different notation is used, in fact,

$^4\Gamma_4 = {}^4T_1$

$^4\Gamma_5 = {}^4T_2$

$^2\Gamma_3$, $^2\Gamma_4$ = 2E, $2T_1$

The spectrum shifts along the wavelength scale as the sulfuric acid concentration is increased corresponding to a lowering of Δ as sulphate ions replace water molecules as ligands.

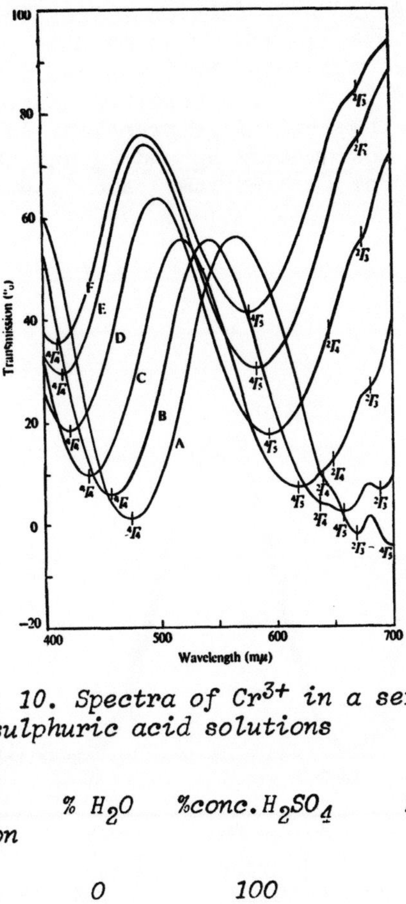

Fig. 10. Spectra of Cr^{3+} in a series of sulphuric acid solutions

Curve	Cr^{3+} concentration	% H_2O	%conc. H_2SO_4	Displacement of Curve
A	0.0033 M	0	100	-25%
B	"	20	80	-20%
C	"	40	60	-15%
D	"	60	40	-10%
E	"	80	20	- 5%
F	"	100	0	0

For the 80 and 100% H_2SO_4 it is seen that the $^4\Gamma_5$, $^2\Gamma_3$ and $^2\Gamma_4$ levels are very close together, indicating that Δ/B must be very near the value at which these levels cross in the diagram. In this region there must be mixing of these levels and this accounts for the change of the energy of $^2\Gamma$ transitions in these solutions.

The absorption curve for 80% H_2SO_4 was found to be almost identical with that for the same concentration of chromium in a soda-lime-silica glass as is shown in Figure 11.

Absorption due to Mn^{2+}. The most complete study of ligand field bands so far available is that of Bingham and Parke (12) for the manganous ion. The free Mn^{2+} ion has a ^6S ground state and quartet terms ^4G ^4P ^4D ^4F. The observed energies relative to the ground state, the theoretical expression for these energies in terms of the Racah parameters B_o and C_o and the calculated values of those energies for B_o = 950 cm^{-1} and C_o = 3280 cm^{-1} are given in Table I.

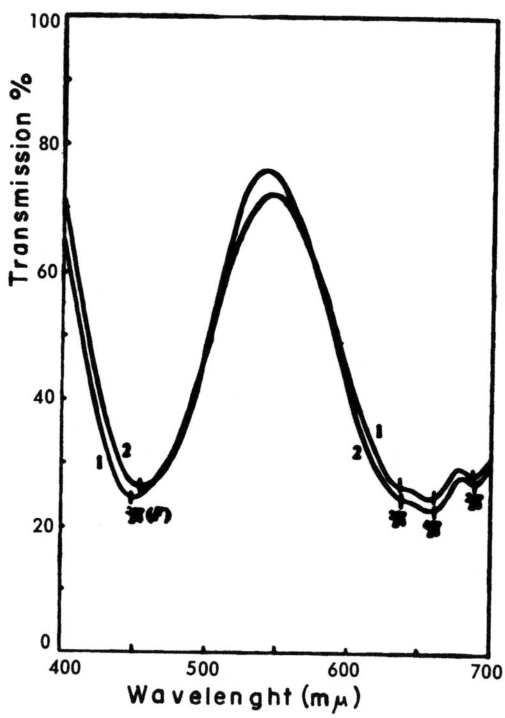

Fig. 11. Comparison of Cr^{3+} spectra in glass and in solution

Curve 1 Glass
Curve 2 Solution

TABLE I

Calculated Energies Relative to the Observed
Energies of the Free Mn^{2+} ion. B_o=950 cm^{-1}, C_o=3280 cm^{-1}

Term	Separation from the ground term ^6S	Observed (cm^{-1})	Calculated (cm^{-1})	Difference (cm^{-1})
^4G	10B + 5C	26,850	25,900	-950
^4P	7B + 7C	29,200	29,610	410
^4D	17B + 5C	32,340	32,550	210
^4F	22B + 7C	43,620	43,860	240

The energy level diagram for the Mn^{2+} (d^5) configuration is shown in Figure 12. The ^6S ground state is of different spin mul-

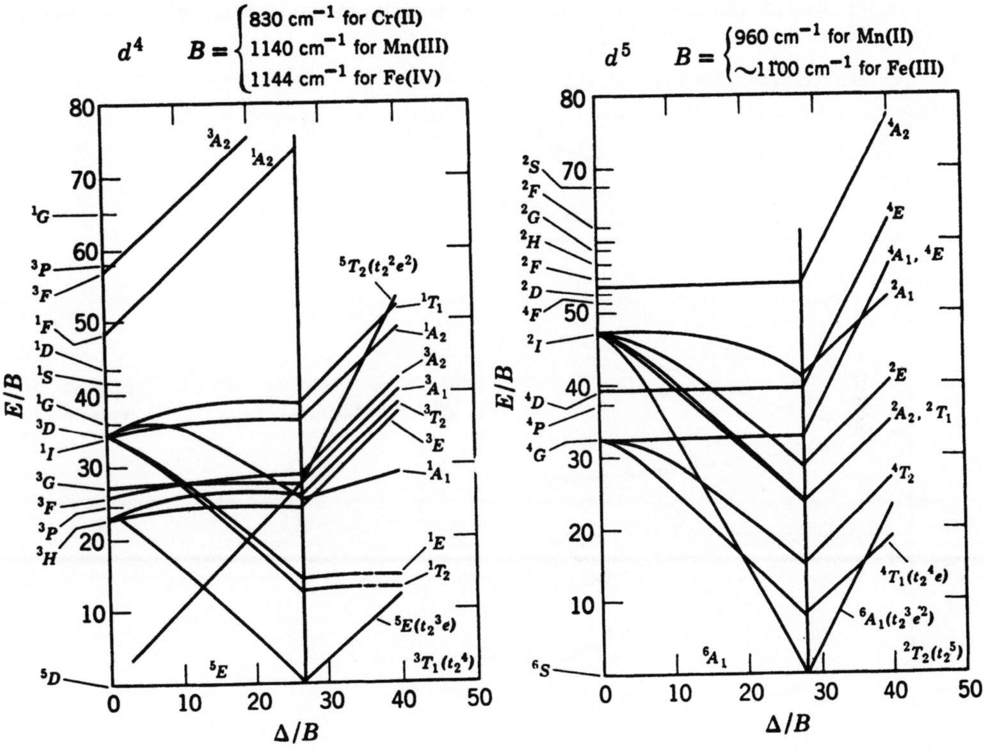

Fig. 12. Energy level diagrams, after Tanabe and Sugano, for d^4 d^5 systems in octahedral symmetry.

tiplicity from all the other levels and only absorption bands with $\varepsilon < 1$ can be expected. It will be noticed that the energies of the states ^4E, $^4A_1$ (G) and ^4E (D), $^4A_2$ (F) are independent of Δ; these states should therefore give rise to narrow bands in the absorption spectra. From Table I it can be seen that the separation of the ^4D term from the ^4G is 7B and that of the ^4G from the ground state is 10B + 5C. Thus if the energies of these terms can be obtained from the absorption curves, values of B and C can be estimated.

A simplified energy level diagram giving the atomic states of interest is shown in Figure 13. This has been calculated for C/B= 5.43, a value very near to that found experimentally for MnF_2, the absorption spectrum of which is given in Figure 14.

First it may be noted that the extinction coefficients are very small, $\varepsilon \approx 0.15$, corresponding to spin-forbidden transitions. Next there are two narrow bands at about 25,000 and 30,000 cm^{-1} which may be assigned to transitions to the $^4A_1$ (G) and ^4E_D levels respectively. The shoulder on the 25,000 band is thought to be due to the ^4E component of the ^4G term. The assignment of the broader bands presents no difficulty.

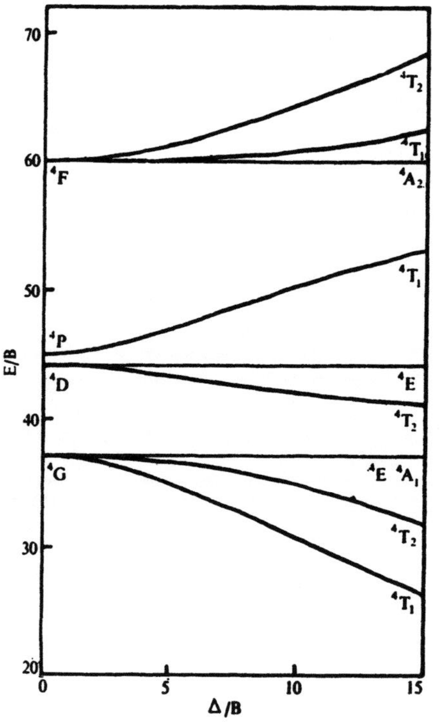

Fig. 13. Energy level diagram of a d^5 configuration (C/B = 5.43).

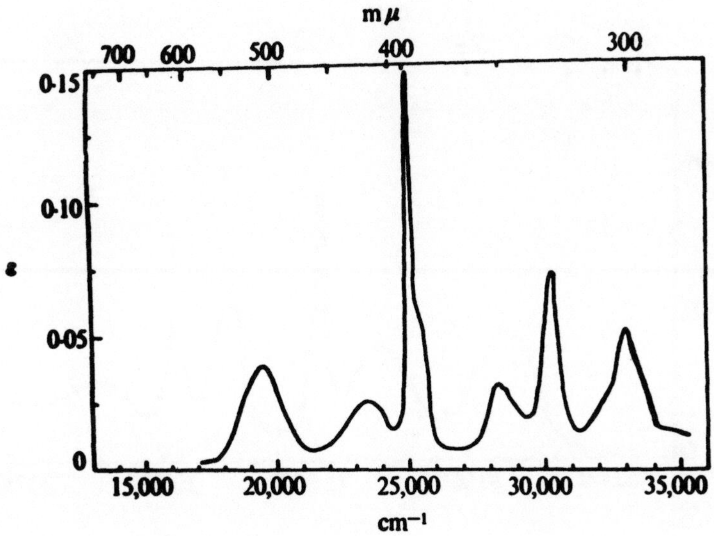

Fig. 14. Absorption spectrum of MnF_2 (after Stout). Unpolarised light propagated parallel to the crystallographic c-axis.

Comparison of the absorption spectrum of MnF_2 with that of a solution of $(Mn(H_2O)_6)^{2+}$ in water (Fig. 15) shows that the $^4A_1$ band, the most intense absorption, has shifted to a slightly lower energy.

The spectrum of Mn^{2+} in glasses shows less detail and, at higher energies, is obscured by the intrusion of the UV cut off. An absorption curve of Mn^{2+} in a glass of composition 3 K_2O.7 P_2O_5 is shown in Fig. 16, while Fig. 17 shows the absorption curve for the ion dissolved in 3 Na_2O.7 SiO_2. The general trends are the same but it may be noticed that the extinction coefficients are much higher for the soda glass $\varepsilon > 0.3$. This indicates that the Mn^{2+} is in tetrahedral coordination in the silicate glass and in octahedral coordination in potassium phosphate glass, an assignment made years ago by Weyl (13) by comparison of the colors of the glasses with those of crystals in which the coordination of the manganese was known from X-ray diffraction.

In Table II the results of the experiments and calculations are summarized. For each spectrum the energies of the absorptions associated with the transitions to $^4T_1$ (G) $^4T_2$ (G) $^4A_1$, ^4E (G) $^4T_2$ (D) ^4E (D) were tabulated. The difference between the two E levels is 7B; C could be calculated from the fact that the ^4E (G) level lies at 10B + 5C above the ground state. A value of Δ was

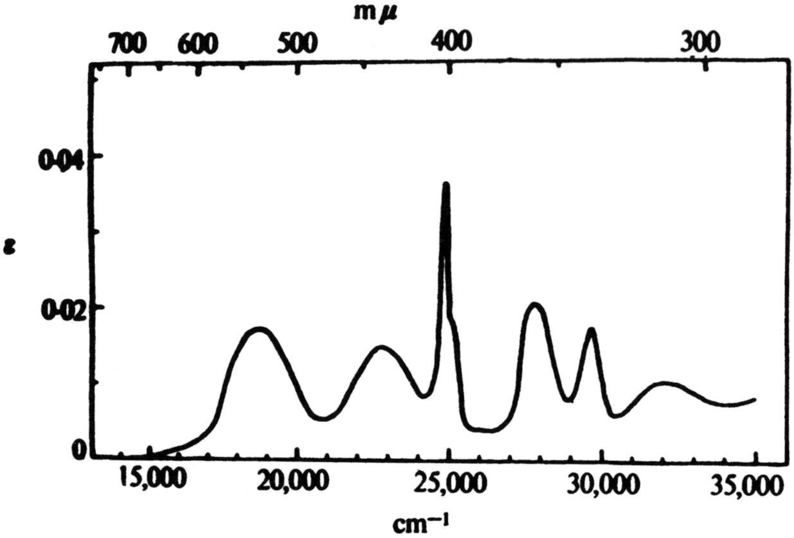

Fig. 15. Absorption spectrum of $(Mn(H_2O)_6)^{2+}$

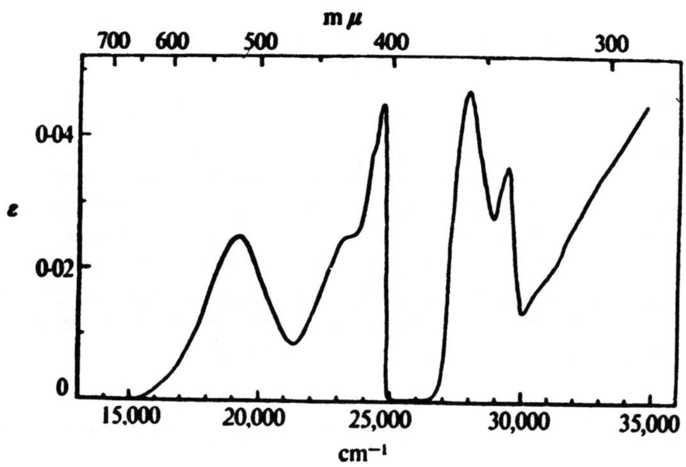

Fig. 16. Absorption spectrum of Mn^{2+} *in a* $3K_2O.7\ P_2O_5$ *glass.*

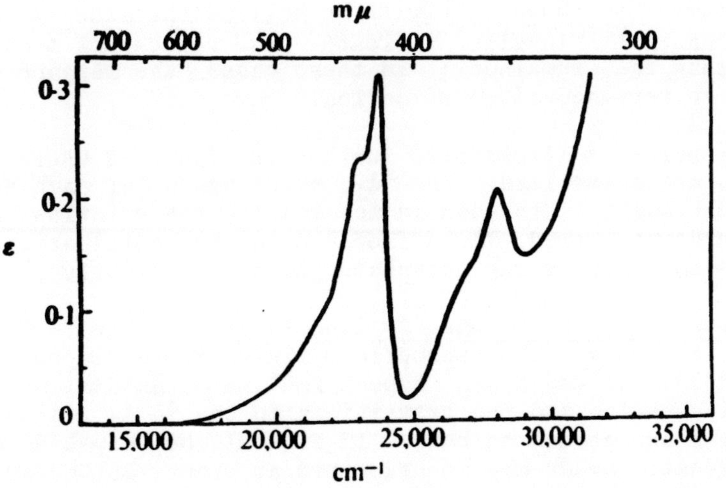

Fig. 17. Absorption spectrum of Mn^{2+} in a $3Na_2O.7\ SiO_2$ glass.

TABLE II

Values of the Field Strength, Δ, and Extinction Coefficients of the $^4A_1$, $^4E(G)$ band

System	$^4T_2(D)$	$^4T_2(G)$	$^4T_1(G)$	B	C	Δ	ε
Zn_2SiO_4:Mn	---	0	0	0.619	3.49	4.4	---
MnF_2	0.2	-0.4	0.4	0.676	3.75	8.4	0.150
$(Mn(H_2O)_6)^{2+}$	0.1	-0.4	0.4	0.683	3.62	8.3	0.037
$3K_2O.7P_2O_5$	0	-0.7	0.6	0.691	3.57	7.4	0.045
$3Na_2O.7P_2O_5$	-0.1	0.4	0.5	0.704	3.51	7.1	0.052
$CaO.P_2O_5$	-0.6	0.4	0.4	0.644	3.59	6.4	0.160
$Na_2O.4B_2O_3$	-0.5	---	0.3	0.630	3.61	6.8	0.230
$3K_2O.7SiO_2$	0.4	0.1	-0.1	0.619	3.49	4.5	0.320
$3Na_2O.7SiO_2$	0.6	0	0.1	0.619	3.49	4.1	0.310

Unit of energy: 10^3 cm-1

then chosen to give the best fit to the three remaining levels and in the table the difference between the calculated and the observed energy is shown. It will be noticed that the two silicate glasses and Mn^{2+} in zinc silicate show values of Δ distinctly smaller than the remainder; in these cases the manganese ion is probably in tetrahedral coordination.

This point is illustrated further in Figure 18 where the band positions are summarized. The $^4A_1$, $^4E(G)$ bands for each substance have been brought into coincidence keeping the relative position of the bands unaltered. The difference in the coordination of Mn^{2+} in potassium silicate and phosphate glasses is clearly shown.

Ligand field bands due to iron in glass. The Fe^{3+} ion like Mn^{2+} is a d^5 ion and the appropriate ligand field diagram is shown in Figure 13. We may again expect Fe^{3+} in octahedral or tetrahedral configuration and the transitions to the $^4E\ A_1$ (G) and ^4E (D) levels will be common to both. If as would be expected Δ is not too different, bands may be expected at about 25,000 and 28,000 cm^{-1}. For tetrahedral coordination the splitting of a level may be expressed as if $\Delta_{tet} = -4/9\ \Delta_{oct}$ and the $^4T_1$ $^4T_2$ absorptions would therefore be expected at higher energies in tetrahedral coordination than in octahedral. The predicted energies based on an observed absorption at 23,300 cm^{-1} are shown in Table III.

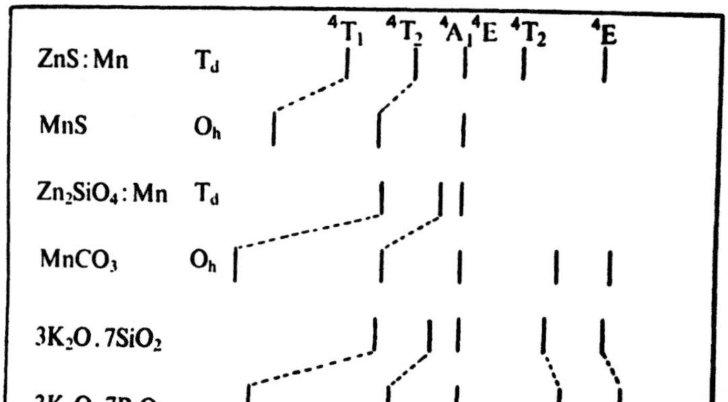

Fig. 18. Tetrahedral and Octahedral crystal spectra and the spectra of the glasses $3K_2O.7P_2O_5$ and $3K_2O.7SiO_2$.

TABLE III

Ligand Field Predictions for Ferric Iron

Calculated position mμ(cm^{-1})	Valence State	Transition	Co-ordination Symmetry
735(13,600)	Fe^{3+}	$^6\Gamma_1$-$^4\Gamma_4$(G)	Octahedral
570(17,500)	Fe^{3+}	$^6\Gamma_1$-$^4\Gamma_5$(G)	Octahedral
427(23,300)	Fe^{3+}	$^6\Gamma_1$-$^4\Gamma_1{}^4\Gamma_3$(G)	Octahedral
380(26,200)	Fe^{3+}	$^6\Gamma_1$-$^4\Gamma_5$(D)	Octahedral
500(20,000)	Fe^{3+}	$^6\Gamma_1$-$^4\Gamma_4$(G)	Tetrahedral
446(22,300)	Fe^{3+}	$^6\Gamma_1$-$^4\Gamma_5$(G)	Tetrahedral
427(23,300)	Fe^{3+}	$^6\Gamma_1$-$^4\Gamma_1$,$^4\Gamma_3$(G)	Tetrahedral
380(26,200)	Fe^{3+}	$^6\Gamma_1$-$^4\Gamma_5$(D)	Tetrahedral

Recent work on $3Na_2O.7SiO_2$, $Na_2O.9B_2O_3$, $3Na_2O.7B_2O_3$ glasses (7) not yet published, shows that the extinction coefficients arising from the tail of the ferric ion charge transfer bands are approximately

	0.001,	0.1,	0.02	at	600mμ
and	0.001,	0.43,	0.15	at	500mμ respectively.

That due to ferrous ion at 600mμ is 4.0 in $3Na_2O.7SiO_2$ glasses, and ≈ 2 in $3Na_2O.7B_2O_3$ glasses.

Thus if attempts are made to get all the iron in an oxidized state, it is very difficult to resolve the ferric ligand field bands for which extinction coefficients < 0.1 would be expected in the B_2O_3 glasses. Fortunately it appears that electron spin resonance may be called in aid.

In certain glasses some of these bands can be observed. For example Bamford (14) has given a transmission curve for a glass of base composition $1Na_2O.2SiO_2$ containing iron expressed as Fe_2O_3 at approximately 0.5%, 0.75% and 1%. Small absorption peaks superimposed on the tail of the UV charge transfer band can be seen at 380, 420 and 440 millimicrons (Fig. 19).

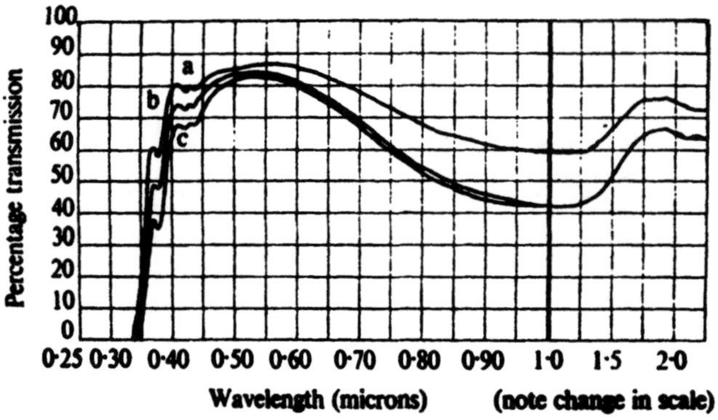

Fig. 19. Iron-containing sodium silicate glasses melted under oxidizing conditions.
Percentages by weight:
a Na_2O. 33%, SiO_2 67%, Fe_2O_3 0.496%
b Na_2O. 33%, SiO_2 67%, Fe_2O_3 0.769%
c Na_2O. 33%, SiO_2 67%, Fe_2O_3 1.007%

We may note that the glass was melted under oxidizing conditions. The ferrous peak can be seen at 1.0μ and we shall shortly see that this has an extinction coefficient of about 28; the extinction coefficients of these small peaks are therefore almost certainly < 0.1.

These observations appear to indicate that the Fe^{3+} is in tetrahedral coordination in this glass.

Ferrous iron. Ferrous iron in glasses usually gives a band in the near infrared at about 1 micron. If this is attributed to octahedral coordination, a band at about 4,000 cm^{-1} or 2.5μ would be predicted for tetrahedral coordination; a band has been observed by Bishay (15) in a phosphate glass at about this wavelength.

There are several points about the 1.0μ ferrous band which require further elucidation. For example the extinction coefficient and position of maximum absorption change drastically as the glass composition changes. Steele (6,7) has found that in a $1Na_2O.9B_2O_3$ glass the peak is at 980μ with $\varepsilon = 2.2$ while in $3Na_2O.7SiO_2$ glasses the peak is at 1.05μ and $\varepsilon = 29$. For $(Fe(H_2O)_6)^{2+}$ in crystals, McClure (16) found the maximum at 1.04μ and $\varepsilon = 1.8$. The band is broad and McClure attributes this to the Jahn-Teller effect which requires the surroundings to distort so that all the orbital degeneracy is removed from the ground state.

OXIDATION-REDUCTION PROCESSES IN GLASSES

The object of this brief review of the absorptions produced, mainly in the visible region of the spectrum, by the transition metal ions - the traditional coloring ions of the glass maker - has been to indicate the progress made in the last decade in developing quantitative understanding of the absorption processes. The chemistry of these ions in glasses has other interesting aspects, perhaps the most important of which is the information they can provide about the oxidation-reduction properties of glasses. One result will be quoted and this because it is of particular interest when taken together with other information given in this paper.

If a glass containing an ion of variable valency is heated long enough in a controlled atmosphere, it will reach equilibrium with the partial pressure of oxygen and the ratios of the different valence states will reach equilibrium values. A series of binary alkali silicate glasses containing small amounts of iron or cerium or chromium were brought to equilibrium with air at 1400°C (17). Figure 20 gives the resulting ratio of the oxidized to the reduced species. It will be seen that the proportion of the oxidized species increases as the alkali content increases and as the alkali changes in the order Li Na K. Here again we have a process controlled by the number and type of "non-bridging" oxygens.

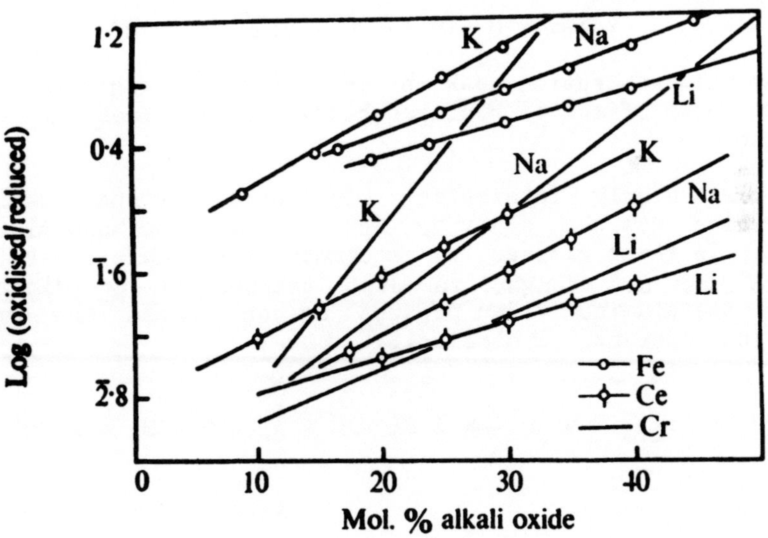

Fig. 20. Variation of Redox ratio with alkali concentration in binary alkali silicate glasses at 1400°C.

THE INFRA-RED ABSORPTIONS

The vibrational absorptions which have been studied can be divided into two classes, i.e. those due to minor constituents, the chief being H_2O, and those arising from vibrational levels associated with the network forming oxides. Normally the first class, is investigated with specimens 1 or 2 mm thick while for the second, reflection spectra or transmission through the equivalent of very thin films are necessary. Some Raman spectroscopy has been carried out; the results do not affect materially the generalizations which will be made in this short look at the present situation.

Water in glass. In pure silica, water is incorporated as hydroxyl groups in a manner which can be represented formally as

$$\geqslant Si - O - Si \leqslant \; + H_2O \longrightarrow 2 \geqslant Si - OH$$

A sharp peak due to the OH stretching vibration occurs at 2.73μ with an overtone at 1.38μ. This is illustrated in Figure 21 from Adams and Douglas (18) (1959). The band at 2.22 is probably a combination frequency.

In glasses containing alkali, hydrogen bonding occurs; again formally this may be written

$$2 \geqslant Si - OH + \geqslant Si - O - Si \leqslant + Na_2O \longrightarrow$$
$$2 \geqslant Si - O - H \text{ --- } {}^{-}O - Si \leqslant \; + 2\,Na^+$$

The main absorption peak shifts to 3.62μ as would be expected from the known effect of hydrogen bonding on the frequency of this vibration.

A particularly interesting application of spectroscopy in this region was the study by Scholze (19) of $Na_2O.Al_2O_3.SiO_2$ glasses. In these glasses, as is common in silicate crystals, the aluminum ions are thought to adopt tetrahedral coordination as long as the aluminum ions present do not exceed the number of sodium ions present. Symbolically

$$SiO_4 + Al_2O_3 + Na_2O \longrightarrow 2\; Si - O - \begin{array}{c} Si \\ | \\ O \\ | \\ Al \\ | \\ O \\ | \\ Si \end{array} - O - Si + 2\,Na^+$$

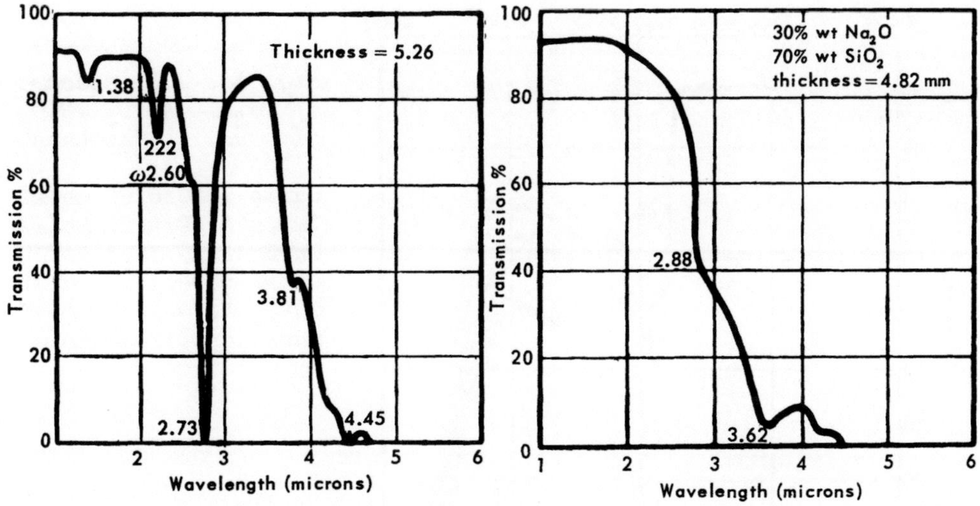

Fig. 21. a) The I.R. spectrum of (Corning) synthetic fused silica from 1-5μ.
b) The I.R. spectrum of sodium silicate glass.

Thus in the soda-alumina-silica glasses, no non-bridging oxygens can occur until Na_2O is present in excess of the Al_2O_3. Figure 22, after Scholze, shows how the hydrogen bonding of water revealed by infra-red spectroscopy confirms the generally held view of the structural role of aluminum (See Douglas (20)).

Network vibrational modes. The force constant approximation can be used with reasonable satisfaction in the interpretation of the infra-red spectra of isolated molecules in the gaseous phase. The interpretation in terms of atomic parameters of the infra-red absorption of condensed phases is much more difficult. However, there is experimental evidence which suggests that in glasses the absorption peaks occur at about the same frequencies as the vibrations in crystals of similar composition. In this sense infra-red absorption spectra can be used as a "finger-print" technique to infer the near neighbor coordination of the network-forming cations. An example of this technique at its best is given in the study of borate glasses by J. Krogh Moe (22). It will be illustrated here by reference to some studies of the $PbO-B_2O_3$ glasses (21) Figure 23. In pure B_2O_3 the strong absorption peaks are at 6.9 and 8μ and are ascribed to boron in triangular coordination. As PbO is added a broad band centered about 10μ develops which is thought to arise from BO_4 groups. It was concluded that the BO_4 groups began to decrease between 0.92 and 1.83 PbO and have nearly

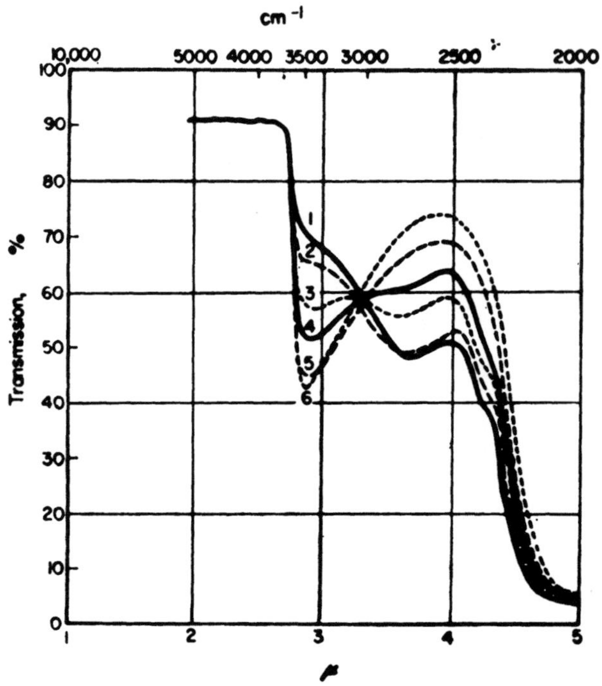

Fig. 22. I.R. spectra of Na_2O - Al_2O_3 -SiO_2 glasses. Thickness 0.103 cm.
Curve 1:20:0:80 mole%
Curve 2:20:5:75 mole%
Curve 3:20:10:70mole%
Curve 4:20:12:68mole%
Curve 5:20:15:65mole%
Curve 6:20:20:60mole%

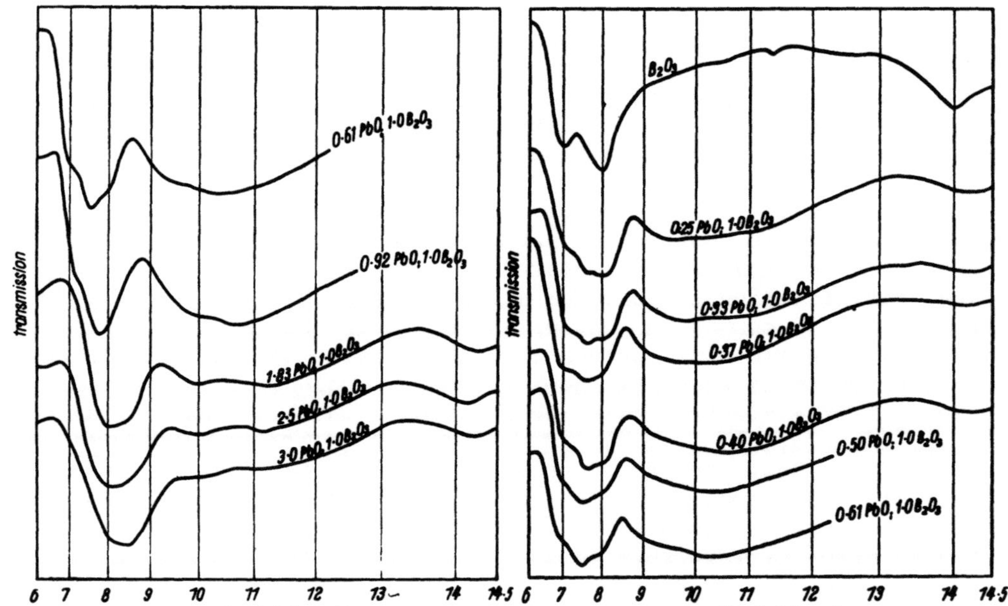

Fig. 23. I.R. spectra of thin films of B_2O_3 glass and some PbO-B_2O_3 glasses. Molecular PbO content 0.25-0.61 and 0.61-3.0. Curves are displaced vertically to avoid confusion.

disappeared completely at the composition 3.0 PbO 1.0 B_2O_3. This conclusion received subsequent confirmation by Bray (22) from NMR studies of the lead borate glasses.

Quantitative infra-red data. The absorptions associated with the atomic vibrations are very intense, and thin specimens, or some equivalent technique such as the KBr disc, must be used. Although all these methods can lead to the successful location of absorption peaks, it is very difficult to make them quantitative in investigating, for example, the effect of varying the quantity of alkaline oxide in a binary silicate glass forming system. The optical constants, n, the refractive index, and k, the absorption coefficient, can be extracted from data on thin films of varying thickness (23) and from the appropriate reflection experiments. It remains, of course, as difficult to interpret the optical constants in terms of atomic parameters as it is for any other macroscopic material constant.

THE OPTICAL CONSTANTS OF SOME Na_2O-SiO_2 GLASSES

Crozier and Douglas (24) have recently published some values for the optical constants of some $Na_2O.SiO_2$ glasses derived from thin film measurements. The values of the absorption constant nk are shown in Figure 24. The compositions of the glasses are given in Table IV.

TABLE IV

Na_2O Content of Glasses Investigated by Crozier and Douglas

Glass Number	Mole % Na_2O
Fused Silica	0
1	16.7
2	21.7
3	25.2
4	30.0
5	34.5
6	43.1

Although the curves clearly are comprised of at least three components it can be seen immediately that the peak at about 1100 cm^{-1} decreases as Na_2O is added to SiO_2, while a peak at 900 cm^{-1} grows. Estimates of the areas under component peaks have been made

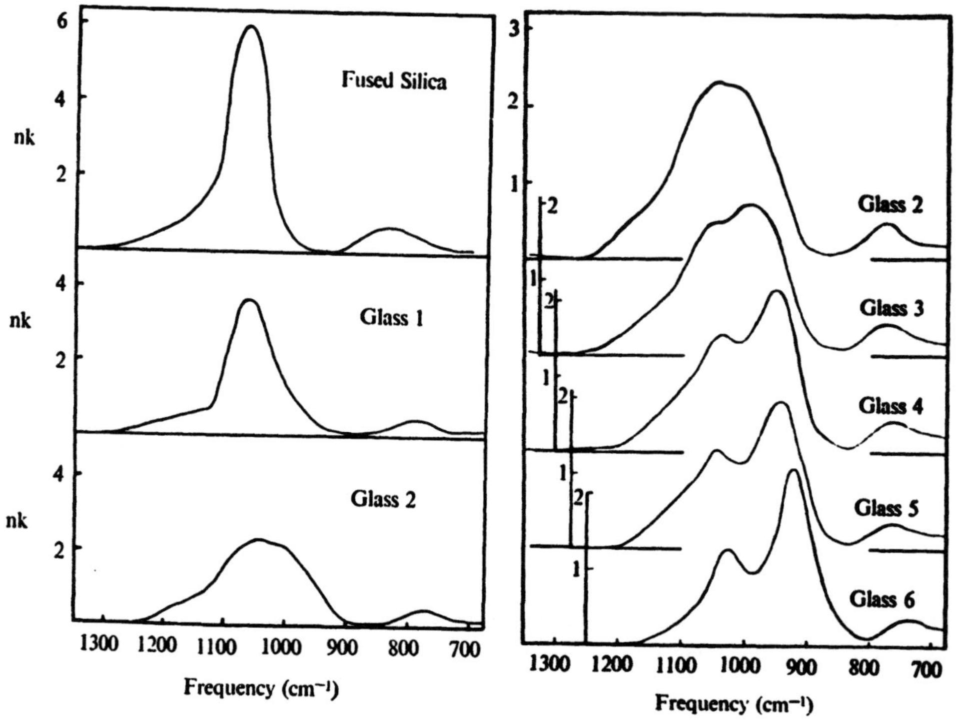

Fig. 24. Plots of nk against frequency for sodium silicate glasses.

and it appears that if the 900 cm^{-1} peak is to be attributed to the formation of non-bridging oxygens then either the oscillator strength varies or the ratio of the number of non-bridging oxygens to Na ions decreases as the Na_2O increases.

Reflection spectra published by Jellyman and Procter (25) (Fig. 25), although the optical constants were not extracted from the data, strongly suggest that the area under the peak at 900 cm^{-1} increases with alkali content and in the order Li Na K.

CONCLUSIONS

Ligand field theory has made it possible to determine the coordination of transition metal ions in glasses with some confidence while infra-red observations on glasses, coupled with similar observations on crystals of structure determined by X-ray diffraction, gives good indication of the coordination to be expected for network forming ions. In all spectroscopic observations, either directly or indirectly, there are contributions from the non-bridging oxygens and it is to be hoped that in the future these can be put on a more quantitative basis.

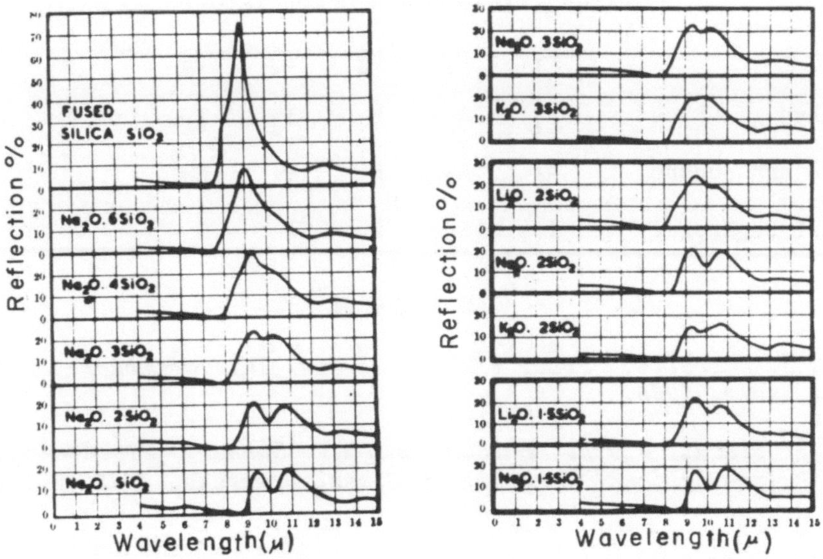

Fig. 25. Spectral reflection curves for glasses in various alkali silicate systems.

APPENDIX I

In this review paper illustrations have been taken from papers by several different authors who use various units to describe the energy of electromagnetic radiation; indeed some authors are not consistent in one paper. But there are good reasons for this, instruments are calibrated in different units and it is, in fact, not very difficult to carry approximate conversion factors in one's head. For those not working in this field the following conversion table may help.

Conversion Table

eV	cm^{-1}	λ microns μ	 millimicrons mμ	 Angestrom A°
8.0	64,600	-	155	1550
4.0	32,600	-	310	3100
2.0	16,300	-	620	6200
1.0	8,150	1.24	1240	12,400
0.1	815	12.4	-	-

APPENDIX II

Ligand Field State Nomenclature

Bethe	Γ_1	Γ_2	Γ_3	Γ_4	Γ_5
Mullchen	A_1	A_2	E	T_1	T_2
Tanabe and Sugano	A_1	A_2	E	F_1	F_2
Orbital Degeneracy	1	1	2	3	3

REFERENCES

1. *N. F. Borrelli, Proc. VIIe Congress International du Verre Brussels, 1965.*

2. *J.S. Stroud, J.W.H. Achreurs and R.F. Tucker, Proceedings of 7th International Congress on Glass, Brussels, 1965 (pp. 7-24).*

3. *P.J. Bray and J.G. O;Keefe, Phys. Chem. of Glasses, 4 (2), pp. 37-46, 1963.*

4. *B.D.McSwain, N.F.Borrelli and G.J.Su, Phys. Chem. of Glasses, 4 (1), pp. 1-10, 1963.*

5. *A. Paul, unpublished work, Department of Glass Technology, University of Sheffield.*

6. *F.N. Steele and R.W. Douglas, Phys. Chem. of Glasses, 6(6), pp. 246-52, 1965.*

7. *F.N. Steele, unpublished work, Department of Glass Technology, University of Sheffield.*

8. *L.E. Orgel, An Introduction to Transition Metal Chemistry, Methuen, 1960.*

9. *F.A. Cotton, Chemical Applications of Group Theory, Interscience Publishers, John Wiley 1963.*

10. *J.S. Griffith, Theory of Transition Metal Ions, Cambridge University Press, 1961.*

11. *Y. Tanabe and S.Sugano, J. Phys. Soc. Japan, 9 (5) pp. 753-79, 1954.*

12. *K. Bingham and S. Parke, Phys. Chem. of Glasses, 6 (6) pp. 224-32, 1965.*

13. *W.A. Weyl, Colored Glasses, Society of Glass Technology, Sheffield 1959.*

14. *C.R. Bamford, Phys. Chem. of Glasses, 2 (5) pp. 163-8, 1961.*

15. *A.M. Bishay and A. Kinawi, Physics of Non-Crystalline Solids, Proceedings of International Conference, Delft. 1964 pp. 589-605.*

16. *D.S. McClure, Solid State Physics, 9, pp. 399-525, 1959.*

17. *R.W. Douglas, P. Nath and A. Paul, Phys. Chem. of Glasses, 6 (6), pp. 216-223, 1965.*

18. *R.V. Adams and R.W. Douglas, Glastech. Ber. 32 K (7), pp. 12-24, 1959.*

19. *H. Scholze, Glastech. Ber., 32, (4) pp. 142-152, 1959.*

20. *R.W. Douglas, Progress in Cermanics, 1, pp. 200-223, 1961.*

21. *J. Krogh-Moe, Phys. Chem. of Glasses, 6 (2), pp. 46-54, 1965.*

22. *P.J.Bray, M.Leventhal and H.O.Hooper, Phys. Chem. of Glasses, 4 (2), pp. 47-66, 1963.*

23. *B.J. Blain and R.W. Douglas, Phys. Chem. of Glasses, 6 (6), pp. 233-39, 1965.*

24. *D.S. Crozier and R.W. Douglas, Phys. Chem. of Glasses, 6 (6), pp. 240-5, 1965.*

25. *P.E. Jellyman and J.P. Procter, J. Soc. Glass Tech., 39, pp. 173T-192T, 1955.*

THE YTTERBIUM ION AS A MICROPROBE IN OXIDE GLASSES

S.W. Barber

Owens-illinois Technical Center

Toledo, Ohio, U.S.A.

The relatively simple electronic spectrum of Yb^{3+} in oxide glasses has been found to respond sensitively to changes in both thermal history and composition. It can also detect the more subtle effects of temperature changes in both quenched and thermally stabilized glasses. The very interesting case of vitreous silica will be discussed.

Ligand field calculations assuming point charges have been made. They agree remarkably well with observed spectra when D_3 symmetry is assumed. When the host glass has been air quenched from 1700°C, the distances to the two oxygens on the z-axis must be assumed greater than the six others, while these distances must be assumed to be shorter for the same glass after stabilization for 24 hours at 1100°C. The probable meaning of this and some related experiments will be discussed.

The temperature dependences of these spectra are no less interesting. They faithfully parallel the increase in elastic constants with temperature and illuminate this and other anomalous behaviors of vitreous silica, such as its very low thermal expansion in spite of its anomalously high heat capacity at low temperatures.

INTRODUCTION

Casual inspection of rare earth spectra in oxide glasses excites imagination because of the extraordinary variety of details. The bands occur in groups or "manifolds" some of which contain only broad overlapping bands. Others contain one conspicuous sharp band and one to many broad bands. Some are skewed enough to suggest breakers racing toward a beach; others are beautifully symmetrical, suggesting mathematical Gaussian and Lorentzian forms. Some are obligingly isolated, while many are so overlapping as to defy resolution. Intensities (i.e., f-values) range from zero to $\sim 7x10^{-5}$. Some details, especially the needle-sharp bands, seem almost immune from effects of solid hosts; most are quite strongly affected. Is it too fanciful to imagine that such a rich array of detail is trying to tell us something about glass?

With respect to usefulness as probes, the richness of detail in rare earth spectra is a fault in all but the simplest cases. When more than one unpaired electron is involved, the number of bands in most manifolds is so large and they are so close and so overlapping, that complete resolution is out of the question in glasses. For this reason, we are forced to use Yb^{3+}, the only other single electron case, Ce^{3+}, being unavailable because all its 4f transitions are at frequencies to which the oxide glasses are opaque. Yb^{3+} has too few levels -- no more than three in the $^2F_{5/2}$ manifold and no more than four in the $^2F_{7/2}$ (ground) manifold, but we make up for this shortcoming by exploiting details, specifically:

1. Temperature dependence of intensity. To what extent does thermal motion lift the La Porte forbiddance? How are thermally excited shifts in electron population manifested?

2. Separation between levels in each manifold. What are the ratios between their magnitudes in each manifold? How are they affected by variations in the host due to temperature, composition, stress, etc.?

3. Band shapes. Why are some so sharp and some so broad in a host that might be expected to broaden all of them?

So far, we have studied pure vitreous silica in both quenched and stabilized forms, and alkali-silica glasses (1). Only the simplest case -- quenched silica -- will be discussed to illustrate the procedures and the efficacy of Yb^{3+} as a probe.

OBSERVATIONS

The pure silica samples could not be prepared by conventional techniques. Procedure used is described elsewhere (2). Two concentrations were studied; 700 $Yb/10^6SiO_2$ and 360 $Yb/10^6SiO_2$.

The Spectra

Absorbance spectra were taken on a Cary Model 14 recording

spectrophotometer fitted with a small heater and with a helium dewar. Nearly all information was read directly from the absorbance spectra between 0.8 and 1.1 microns including three of the four levels in the ground manifold. The fourth ground level was found in an emission spectrum which also confirmed the other three.

Figure 1 shows the absorbance of Yb^{3+} in quenched (air cooled from 1750°C) vitreous silica at various temperatures between liquid He and 873°K. The He - N_2 spectrum is simplest, and it is significant that nothing important changes between these two temperatures. To be noted are:

1. Positions of the three main bands. These belong to the $^2F_{5/2}$ manifold and are the only bands remaining at liquid N_2 temperature and lower. The ratio of intervals between them is 2.0 with the wider interval at the lower energies.

2. The lowest of the three main bands is very narrow and well isolated from the other two, both of which are much broader and overlapping. Both are noticeably skewed toward the sharp band.

3. As temperature increases the three main bands decrease in height, with their half-widths remaining constant, while other (hot) bands develop and become very pronounced at, say, 500°K and above. Both effects are due to thermal population of excited levels in the ground manifold as temperature rises. Spacings between the main bands and the hot bands are the spacings between the ground and excited levels of the $^2F_{7/2}$ manifold.

Positions of all the levels are shown in a conventional energy level diagram on the right in Figure 1.

4. Invariant points such as A (also B and C if they could be precisely located) have a special significance. Such "isosbestic" points (3) occur when intensities of neighboring bands change reciprocally due to the simple shifting of absorbers from one state to another without other effects on intensity or change in total number of absorbers. Here electrons are being shifted from the ground to the first excited level of the 7/2 manifold by thermal excitation. This simple explanation is confirmed in Figure 2 which shows that the height of the narrow band varies linearly with population of the ground level and extrapolates to 0.25 at zero intensity as it should because at that limit electrons are distributed equally among the four 7/2-levels.

Such perfect examples of isosbestic points are rarely seen except in titrations.

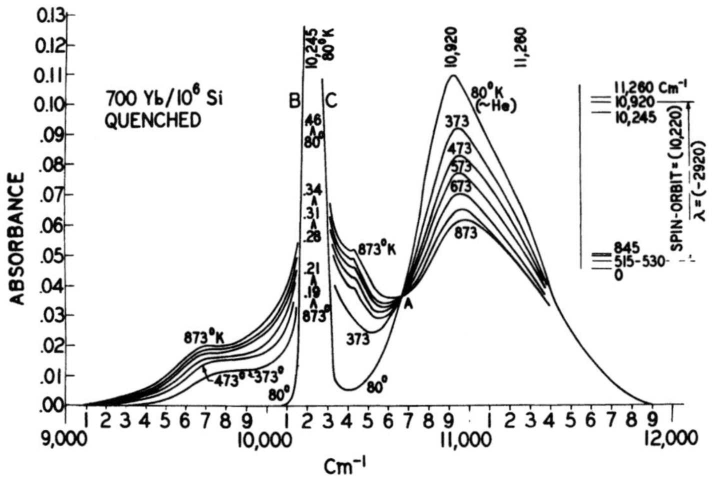

Fig. 1. Absorbance and energy level diagram of Yb^{3+} in quenched silica glass. "A" is an isosbestic point and there are probably others at "B" and "C". Curves for liquid N_2 and liquid He temperatures were indistinguishable.

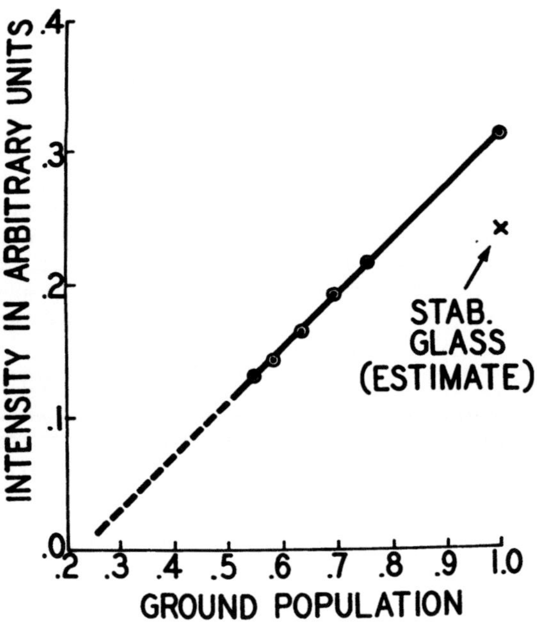

Fig. 2. Height of the 10,245-cm^{-1} band as function of population in the ground level. Ideally, extrapolation to zero intensity would give 0.25 corresponding to equal distribution among the four levels of the 7/2 manifold.

Figures 3 and 4 are included only to show how different are the spectra in stabilized silica and alkali-silica glasses.

Theory appropriate for the calculation of rare earth spectra from internal structures of the isolated ions and the superimposition of effects due to ligand fields has been made conveniently available by Wybourne (4). Watson and Freeman (5) have provided heretofore unavailable parameters of the free ions. Hutchings (6) has given us a manual for calculations wherein the ligands are treated as point charges. Since the laser directed attention rare-earthward there has been a rapid accumulation of experience in such calculations. Among the recent papers on ytterbium, references 7(a) through (f) are of special interest here. Many such studies start with field potentials of known geometry and seek the "lattice sums" empirically from observed spectra. Such works have done much to verify and refine the procedures to be used and to develop judgements regarding results, but ours is a different problem.

The symmetries of rare earth complexes in various oxide glasses are not yet known and may not be knowable from other sources. This study must start with assumed models, calculate the corresponding spectra and compare with observed spectra. Relatively large uncertainties in magnitude must be tolerated, even as large as twofold. This amounts to only an annoyance, however. As Wybourne (4) mentions, those aspects of the calculation which are determined by symmetry (such as the angular parts of our potentials and wave functions) are "capable of exact" evaluation in contrast to the radial part which can be evaluated only approximately..... As will be shown, relatively uncomplicated and very plausible assumptions reproduce observed spectra with remarkable fidelity to details of shape and relative position and with very acceptable magnitudes for intervals between levels and implications about relative intensity.

Choice of Models

Certain features of the ligand cage are known from independent sources and are expected to apply in all oxide hosts:

1. The eight oxygen ligands (7(a),8) around Yb^{3+}.

2. The average Yb-O distance in ytterbium gallium garnet(7a) is 4.35 atomic units (i.e., 2.30A^{o}). The average in other oxides (8) is similar within uncertainties in various radial parameters to be used. This value is therefore chosen.

3. As found by Perkins and Crosby (7c), the assumption of ionic interactions between Yb^{3+} and its oxygen ligands is much

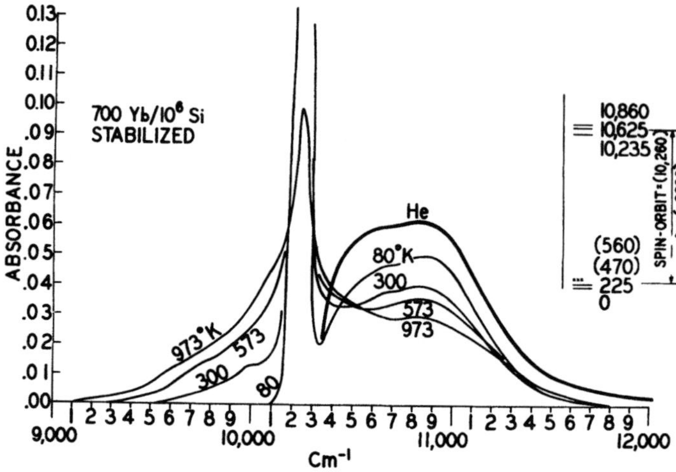

Fig. 3. Absorbance and energy level diagram of Yb^{3+} in stabilized silica, i.e., after 24 hr at 1100°C.

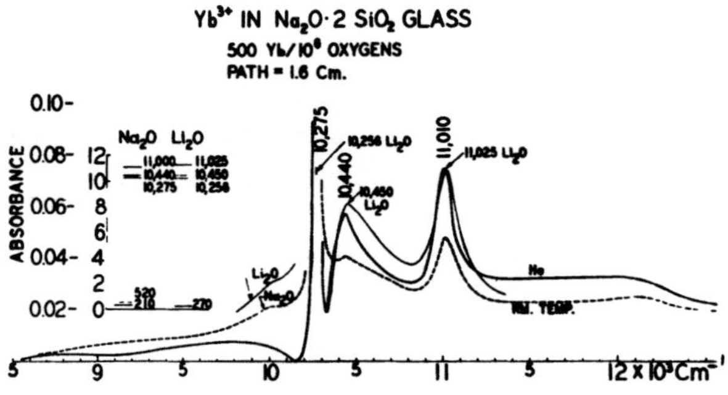

Fig. 4. Absorbance and energy level diagram of Yb^{3+} in annealed Na_2O-SiO_2 glass at liquid helium temperature (heavy solid line). The spectrum at room temperature (dashed line) and selected detail for Li_2O-SiO_2 glass (narrow solid line) are shown for comparison.

closer to the experimental situation than the assumption of covalent binding although neither could be omitted from a comprehensive discussion . With these features in mind, we consider probable geometries for the cage. The simplest arrangement of eight oxygens would be at the corners of a cube, but the appearance of three levels in the 5/2 manifold requires a lower symmetry. The wisdom of our trade leads us to seek the simplest and most symmetrical deviations from cubic symmetry that will satisfy the data.

There are two simple, highly symmetrical deformations of a cube:

Rhombohedral: Change the length of one body diagonal. This lowers the symmetry to D_3.

Tetragonal: Change four parallel edges equally. This lowers the symmetry to D_{4h}.

So far no cases have been found which require more complicated geometries except statistical deviations from these to be discussed next.

Both D_3 and D_{4h} geometries have inversion symmetry. If perfect, either one would completely forbid all f → f electric dipole transitions and these spectra would not be observed (La Porte Rule). Consequently, if one of these is chosen, statistical deviations from inversion symmetry must be assumed. Since intensities are not temperature dependent (Fig. 2), these deviations are not thermally excited (9) and must, therefore, be assumed to be static; a likely situation in glasses. The average of many cages can still be compatible with these high symmetries, however, and our calculations are not complicated by this seemingly incompatible assumption.

For both of these symmetries, the ligand field splitting for both the $^2F_{5/2}$ and the $^2F_{7/2}$ manifolds have been calculated as functions of a parameter, ρ, which expresses the degree of deviation from cubic symmetry in each case. For D_3, ρ is the ratio of average "polar" Yb-O distances along one body diagonal (z-axis) to the average of the other six. For D_{4h} it is the ratio of average O-O distances on the two square ends to the average of the other four O-O distances. Details of these calculations are much too long to be reproduced here. They are to be published elsewhere (10).

Figure 5 shows the D_3 splittings as functions of ρ and locates the values for which the observed splittings in quenched vitreous silica (ρ = 1.11) and in stabilized vitreous silica (ρ = 0.96) are reproduced. These values define the observed ratio of intervals between levels in both manifolds regardless of magnitudes of the intervals. Somewhat fortuitously they also define observed magnitudes in the 7/2 manifold, but indicate magnitudes 10-15% too small for the 5/2 manifold. If the 4f wave function were identical in both states, identical values of ρ should fit both manifolds equally well. But we cannot expect the wave function to be unchanged by the absorption of 10,000 cm^{-1} of energy. As Watson and Freeman (5a) have shown for the analogous case of Ce^{3+}, the 4f wave function expands and the shielding $5s^2$ and $5p^6$ shells contract, both effects contributing to increased interaction of 4f with the wave functions of oxygen. The increased splitting in the excited states, as indicated by dashed lines in Fig. 5, are about the magnitude to be expected in the case of Yb^{3+} based on analogy with Ce^{3+}.

Space does not permit discussion of the case for stabilized silica further or the alkali silica case at all. We pause only to note that ρ is definitely less than 1.00 for the stabilized silica and greater than 1.00 for quenched silica. This decision has implications about band shapes and relative intensities and can be verified by construction of the spectra from the models to be presented next.

CONSTRUCTION OF SPECTRA FROM MODELS

The band shapes and relative intensities as well as the spacings observed at He and N_2 temperatures can be reproduced from the calculated curves of Figure 5 by a simple set of geometrical projections, and the exercise is instructive. Figure 6 shows such a construction for both J-manifolds in quenched silica.

Clearly, if all ligand cages conformed to a single value of ρ, all bands would be needle sharp. Broadness of some bands requires assumption of a distribution in ρ. This is, in fact, already implied in the average deviations from inversion symmetry required above to account for observed intensities. To construct Figure 6 only the maximum and a small portion of the high energy side of the 11,260 cm^{-1} band need be projected from the observed spectrum to a distribution function, f(ρ). Only enough to determine one side of f(ρ) is needed. Then a symmetrical Gaussian-like curve can be constructed, and projections from it reproduce the entire spectrum in both manifolds, provided one more assumption regarding relative intensities is made. In this case the simplest possible is appropriate: all f-values are the same in a given manifold. This is a plausible assumption since all transitions

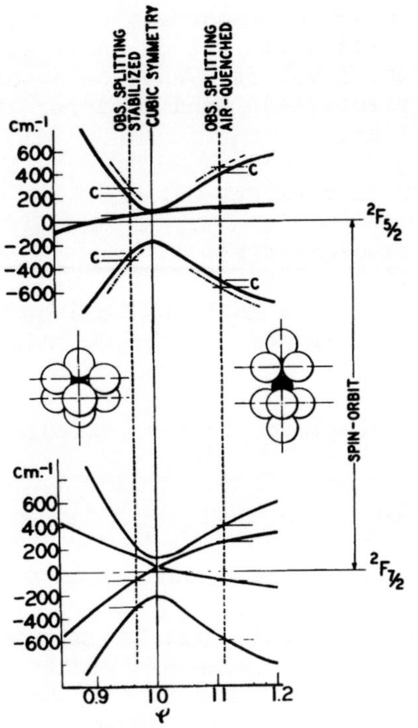

Fig. 5. Ligand field splittings in D_3 symmetry as function of deviation from cubic symmetry. ρ=(Av. polar Yb-O)/(Av. of other Yb-O). C = correction due to excitation of 4f wave function.

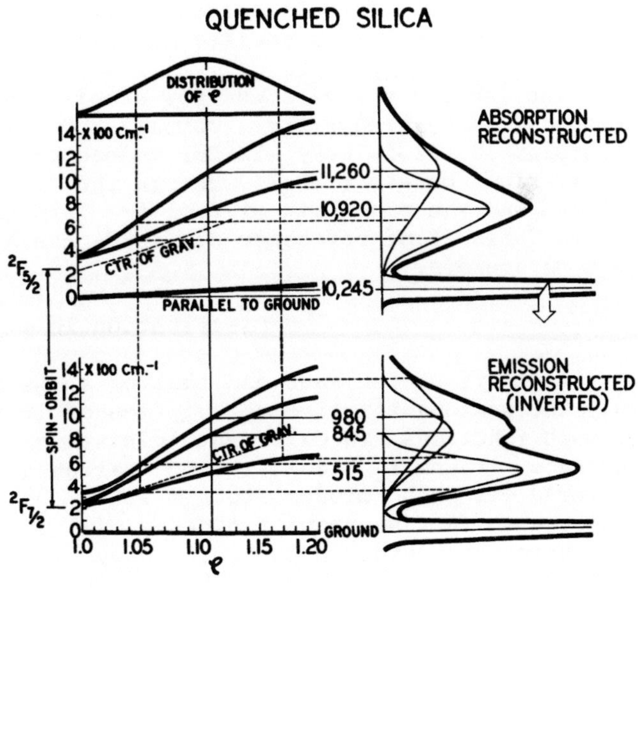

Fig. 6. Construction of Yb^{3+} spectra in quenched silica glass from calculated splittings in D_3 symmetry and a distribution function in ρ. The observed emission was essentially as constructed but imperfectly resolved.

are from common levels and terminate in states that would be indistinguishable without the small effects of the ligand field. Furthermore, Figure 2 shows that there are no important effects on relative intensities other than population in one of the transitions. It has been found that constant f-values cannot be assumed when $f(\rho)$ spans $\rho = 1.00$ at which intensities tend to zero, but this is not the case for quenched silica.

Instruction gained by this exercise consists of:

1. Semiquantitative width and shape of $f(\rho)$.

2. Reasons for sharpness in the 10,245 cm^{-1} band and broadness in the other two; i.e., different slopes in the calculated function of ρ.

3. Origin of skewness, i.e., curvature in the calculated function of ρ.

DISCUSSION

The Yb-O Complex in Silica

The diagonally stretched "cubic" ligand cage in quenched silica glass can be thought of as a $(YbO_6)^{9-}$ complex between two polar oxygens belonging to other complexes in the glass. Such a structure is vaguely similar to that found for lanthanum borate ($LaBO_3$) crystals by Newnham, Redman, and Santore (8). There two oxygens are at 2.74 A° and the other seven average at 2.51 A°, making $\rho = 1.09$. This is surprisingly close to our $\rho = 1.11$ considering that Yb^{3+} is a smaller ion with only 8 oxygen ligands.

The lower $\rho = 0.96$ found for stabilized vitreous silica corresponds to a uranyl-like complex, O-Yb-O. Should we call it ytterbianyl? It is very similar indeed to the structure (8) found for yttrium borate (YBO_3) wherein the two polar oxygens are at 2.21 A° and the other six at 2.41 A°. This gives $\rho = 0.92$. According to Table II of reference 8, $YbBO_3$ has almost identical structure parameters.

Distribution in ρ

A distribution such as that of ρ is no surprise in glass. Only the semiquantitative description of it is novel. Comparisons of such functions in different glasses can be interesting. For instance, thermal stabilization narrowed $f(\rho)$ considerably and centered it much closer to $\rho = 1.00$. On the other hand, mild exposure to fast neutrons broadened $f(\rho)$ for quenched silica without introducing other detectable changes in the spectrum.

In the case of vitreous silica, no vibronic effects were detected. Hence f(ρ) describes a distribution of static geometries. If only thermally excited deviations from D_3 symmetry had existed, they would have been equally well represented by f(ρ) and its temperature dependence would have been interesting. If both static and vibronic effects contribute, their effects might be separable by observing the temperature dependence of f(ρ).

A central problem in fluorescence of rare earths in glasses is the mechanism of radiationless energy transfer. A comprehensive treatment of this would seem to require distribution functions such as f(ρ) since the problem is basically one of phonon-photon interactions. Such considerations lead back to my earlier interest in infrared (11) and Raman spectra (12), and the low temperature heat capacities (i.e., phonon spectra) of glasses (13), The last is now being further pursued by Professor Leadbetter at the University of Bristol, and Dr. P. G. Klemens (14) is still interested in the related area of low temperature thermal conductivities (i.e., phonon-phonon interactions). Thus, we see the present study in a larger pattern of interest related to glasses.

In conclusion, reference is made to our first paragraph. It seems that rare earth spectra in glasses can indeed tell us something provided we find an appropriate language, and such a language is here proposed for consideration.

ACKNOWLEDGMENTS

Helpful discussions with J. I. Krugler and T. Bates, of this laboratory, are gratefully acknowledged. Much credit is also due to R.T. Obermyer for experimental assistance, and to him and R.J. Bieringer for techniques to handle liquid helium.

REFERENCES

1. *R.T.Obermyer; "Absorption Spectra Study of Trivalent Ytterbium Doped into Sodium Silicate Glasses", Master's Thesis, University of Toledo; 1966.*

2. *S. D. Brown; "Devitrification of High Temperature Glasses", Doctoral Thesis, University of Utah; August, 1957.*

3. *W. R. Brode; Chemical Spectroscopy; J. Wiley and Sons, Inc., New York, 1943.*

4. *B.C. Wybourne; Spectroscopic Properties of Rare Earths; Interscience Publishers, New York, 1965.*

5. *R.E. Watson and A.J. Freeman; (a) Phys. Rev. 127, No. 6, 2058-2075 (1962); (b) Phys. Rev. 139, No. 5A, A1606-A1615 (1965).*

6. *M. T. Hutchings; pp. 227-273, Solid State Physics, Vol. 16, Edited by F. Seitz and D. Turnbull, Academic Press, Inc., New York, 1964.*

7. *(a) R. Pappalardo and D. L. Wood; J. Chem. Phys. 33, No. 6, 1734-1742 (1960).*

(b) J.C.Eisenstein, J. Chem. Phys. 35, No. 6, 2097-2100 (1961)

(c) W.G. Perkins and G.A. Crosby; J. Chem. Phys. 42, No. 1, 407-414 (1965).

(d) D. L. Wood; J. Chem. Phys. 39, No. 7, 1671-1673 (1963). (NOTE: The technique used here to locate levels in the ground manifold is that used by us).

(e) W. Low; J. Chem. Phys. 37, No. 1, 30-33 (1962). (NOTE: Line shapes are Lorentzian in CaF_2 *at* 85^oK*).*

(f) M. T. Hutchings and W.P. Wolf; J. Chem. Phys. 41, No. 3, 617-626 (1964).

8. *R.E. Newnham, M.J. Redman and R.P. Santoro; J. Am. Ceram. Soc. 46, No. 6, 253-256 (1963).*

9. *C.J. Ballhausen; Ligand Field Theory; pp. 180-210; McGraw-Hill Book Company, Inc., New York, 1962.*

10. *R.T. Obermyer and S.W. Barber; "Trivalent Ytterbium as a Microprobe in Silica and Alkali-Silicate Glasses", Submitted to J. Chem. Phys.*

11. *(a) I. Simon; J. Am. Ceram. Soc. 40, No. 5, 150-153 (1957).*

(b) I. Simon and H.O. McMahon; J. Chem. Phys. 21, No. 1, 23-30 (1953).

(c) J. Reitzel; J. Chem. Phys. 23, No. 12, 2407-2409 (1955).

12. *(a) P.Flubacher, A. J. Leadbetter, J. A. Morrison and B. P. Stoicheff; J. Chem. Phys. Solids 12, No. 1, 53-65 (1959).*

(b) G.B. Wilmot; "Raman Spectra and Structure of Silica in the Soda-Silica Glass", Doctoral Thesis, Massachusetts Institute of Technology; 1954-1955.

13. (a) *S. W. Barber and B. Martin; J. Chem. Phys. Solids 9, Nos. 3/4, 198-213 (1959).*

(b) A.A. Antoniou and J.A. Morrison; J. Appl. Phys. 36, No. 6, 1873-1877 (1965).

14. *P. G. Klemens; pp. 67-8, Solid State Physics, Vol. 7 (1-98) Edited by F. Seitz and D. Turnbull, Academic Press, Inc., New York, 1958.*

DISCUSSIONS

VAN VLECK (Harvard University, Cambridge, Massachusetts, USA): *The spectacular agreement which Dr. Barber obtains with simple theory using a point charge model is in striking contrast to the rather dismal situation which I will present in my paper for rare earth salts which are perfect crystals and so naively should be expected to be more amenable to theory than glasses.*

Is there any chance of doping rare earth ions into the same sites as the iron ion discussed by Prof. Douglas, and so see how the ligand fields are modified by the presence of the paramagnetic ion itself?

S.W. BARBER: *I know of no way to substitute a rare earth for iron in an oxide structure. It probably cannot be done without changing the site quite drastically. The coordination numbers of rare earths in oxides are most likely to be 9 in the first half of the series and 8 in the second. I know of no iron compounds with similar coordination.*

VAN VLECK: *Prof. Douglas said enough iron could be doped into glasses to permit measurement of the magnetic susceptibility. Is this also possible for rare earths?*

S.W. BARBER: *Rare earth concentration up to about 1500 rare earth ions per million silicons are possible to obtain in pure silica glass. When alkali is added, concentrations up to 5% or above can be obtained. Susceptibilities at any of these concentrations can undoubtedly be measured with commercially available apparatus but we have not pursued this.*

R.A. WEEKS (Oak Ridge National Laboratory, U.S.A.): *At what doping level of Na are effects on the optical spectra of Yb detected?*

S.W. BARBER: *We have studied only two alkali - silica compositions,* $Na_2O \cdot 2\ SiO_2$ *and* $Na_2O \cdot 3\ SiO_2$*. There are no interesting differences between these. This suggests that both cases are characterized by the rare-earth-alkali-oxygen complex in the compound* $Na_2O \cdot 2\ SiO_2$*. It would be interesting to see at what concentration this characteristic spectrum begins to be replaced by that characteristic of silica glass but we have not done this.*

THE MECHANISM OF SPECTRAL ABSORPTION IN GLASSES USED FOR PASSIVE Q-SWITCHES

Emil Deeg

The American University in Cairo

Cairo, UAR

Bret and Gires observed a remarkable change in the transmission of the RG10 Schott glass with the intensity of a light beam emitted by a ruby laser. This property can be utilized to produce giant pulses of laser light in a similar way as is done with solutions of certain organic compounds.

The reason for this change in spectral transmission of RG 10 glass can be explained on the basis of Mie's theory. The transmission curve is determined by the size of colloidal particles of cadmium-sulfo-selenides. Under the assumption that these particles are of spherical shape, their efficiency factor for extinction is calculated as a function of the ratio particle size to wavelength. If a complex value of the refractive index of the colloidal particles is taken into consideration, it can be shown that for particles having approximately the size of the wavelength of the light, the efficiency factor for extinction decreases with increasing value of the refractive index. Such an increase can be caused by photoconductivity of the cadmium-sulfo-selenide particles.

INTRODUCTION

Colored glasses may be divided into two main groups because of the different origin of their specific spectral transmission. One of these groups includes color centers associated with different ions as mentioned in the paper given by N. J. Kreidl during this conference. The second group includes segregated submicroscopic crystalline or amorphous particles. The optical effect of these particles depends on their composition as well as on their size. Well known examples are the different colors of gold and copper ruby glasses although they have approximately the same particle size distribution, and the different colors of copper ruby glasses of the same composition but different thermal history.

In order to explain the absorption edge of the second group, the so called temperature colored glasses, Mie's scattering theory has been applied in a simplified form and proved to be very helpful. However, this theory of temperature colored glasses did not consider the effect of low but finite conductivity of the particles embedded in the glass. This paper will consider the effect of these conducting particles and some conclusions will be drawn from the results.

EXTINCTION OF GLASSES CONTAINING CONDUCTIVE, SMALL PARTICLES

For the mathematical treatment of the problem the following assumptions are made:

(a) The particles embedded in the glass are of spherical shape.
(b) The diameter of the spheres is in the order of the wavelength of the light.
(c) The particles are far apart from each other.

Since the shape of the particle is of minor importance if only the spectral transmission of a temperature colored glass for unpolarized light is considered, assumption a) is allowed for our discussions. Assumptions b) and c) are also allowed, because the glasses become opaque if the diameter of the particles reaches values higher than the wavelength of the light and the particles are located close to each other. This has been proved by electron microscopy. Electron micrographs also give evidence of the almost spherical shape of most particles in temperature colored glasses.

Let a sphere, having a refractive index n, be embedded in a transparent medium of a refractive index n_o. For a description of the interaction of electromagnetic waves with the sphere, the relative refractive index

$$m = n/n_o$$

must be considered. For conducting or semiconducting spheres the refractive index n is a complex number. n_o can be considered a real number within the visible spectrum. Therefore it is possible to put: $m = m' - im'' = m' \cdot (1 - ik)$,

where
$$m' = n'/n_o ;$$
$$m'' = n''/n_o ;$$
$$k = n''/n' ;$$

n' = real part of the refractive index n
n'' = imaginary part of the refractive index n.

n' as well as n'' depend on the wavelength of light.

There will be an effective cross section of extinction C_e for electromagnetic waves falling on the sphere. C_e is composed of an effective cross section for scattering (= C_s) and absorption (= C_a) according to

$$C_e = C_s + C_a$$

Using the geometrical cross section of the sphere (= $r^2\pi$), a dimensionless efficiency factor for extinction Q_e can be defined:

$$Q_e = \frac{C_e}{r^2\pi}$$

r = radius of the sphere.

Since there is a great number of spheres present in a glass sample, the total coefficient of extinction γ has to be found:

$$\gamma = \frac{1}{t} \cdot \ln \left\{\frac{I_o}{I}\right\}$$

where t = thickness of a plane parallel sample
I_o = intensity of incident beam of light
I = intensity of beam leaving the sample; I is correct-ted for reflection losses.

If N (r) is the specific number of particles having the radius r, the total coefficient of extinction γ can be expressed as

$$\gamma = \int N(r) \cdot Q_e \cdot r^2\pi \cdot dr$$
$$= \int N(r) \cdot C_e \cdot dr.$$

The problem is now to determine the distribution function N (r) and the efficiency factor Q_e. There is evidence, based on elect-ron micrographs that N (r) varies only in a small range of r if sufficient care is taken during the production of a temperature colored glass.

Q_e can be determined on the basis of the classical scattering theory (1, 2, 3). For spherical particles, the efficiency factor of extinction is given by

$$Q_e = \frac{4}{\rho^2} \cdot \mathrm{Re}\{S_o\}$$

where

$$\rho = \frac{2\pi}{\lambda} \cdot r \cdot n_o$$

S_o = scattering function for foreward scattered light,
$\mathrm{Re}\{S_o\}$ = real part of the scattering function.

According to Mie's theory (1),

$$S_o = \tfrac{1}{2} \cdot \sum_{\nu=1}^{\infty} (2\nu + 1)\cdot(a_\nu + b_\nu).$$

a_ν and b_ν are the Mie coefficients, given by

$$a_\nu = \left[\frac{\Psi_\nu'(\tau)\cdot\Psi_\nu(\rho) - m\cdot\Psi_\nu(\tau)\cdot\Psi_\nu'(\rho)}{\Psi_\nu'(\tau)\cdot\xi_\nu(\rho) - m\cdot\Psi_\nu(\tau)\cdot\xi_\nu'(\rho)}\right]$$

$$-b_\nu = \left[\frac{\Psi_\nu(\tau)\cdot\Psi_\nu'(\rho) - m\cdot\Psi_\nu'(\tau)\cdot\Psi_\nu(\rho)}{\Psi_\nu(\tau)\cdot\xi_\nu'(\rho) - m\cdot\Psi_\nu'(\tau)\cdot\xi_\nu(\rho)}\right]$$

where

$$\tau = m\cdot\rho \quad (1a) \qquad \text{and } \rho = \frac{2\pi}{\lambda}\cdot r\cdot n_o \quad (1b)$$

$$\Psi_\nu(\tau) = \sqrt{\frac{\pi\cdot\tau}{2}} \cdot J_{\nu+\frac{1}{2}}(\tau); \quad \chi_\nu(\tau) = -\sqrt{\frac{\pi\,\tau}{2}} \cdot N_{\nu+\frac{1}{2}}(\tau);$$

$$\xi_\nu(\tau) = \Psi_\nu(\tau) + i\chi_\nu(\tau);$$

$J_\nu(\tau)$ = Bessel function;

$N_\nu(\tau)$ = Neumann function;

If the refractive indices of the matrix glass and the spheres are close to each other, the efficiency factor can be calculated from an approximation formula

$$Q_e = 4,\left\{\tfrac{1}{2}-e^{-y.\tan\xi}.\ \frac{\cos\xi}{y}\cdot\left[\sin(y-\xi)+\frac{\cos\xi}{y}\ .\ \cdot\left(e^{y.\tan\xi}.\cos 2\xi-\cos(y-2\xi)\right)\right]\right\} \quad (2)$$

where $\tan\xi = x = \frac{m''}{m'-1} = \frac{n''}{n'-n_o}$; $\xi = \arc\tan x \approx x$;

$$y = 2\rho\cdot(m'-1) = 2\pi.\frac{2r}{\lambda}\ .\ (n'-n_o)$$

This formula can only be applied if the condition $|m-1| \ll 1$ is met. The physical meaning of such a condition is that the real part of the refractive index of the sphere is approximately the same as the refractive index of the glass matrix. Furthermore the imaginary part of the refractive index of the spheres must be very small.

A discussion of eq. (2) shows, that usually $\frac{dQ_e}{dx} > 0$ or, since x is proportional to n'' the expression $\frac{dQ_e}{dn''} > 0$. However, for certain values of y also a negative slope of $Q_e(n'')$ is possible. This means that by increasing absorption of the spheres a decrease in the efficiency factor of extinction can be produced. Before this result is applied to the problem of the spectral transmission of certain temperature colored glasses, it has to be checked if a negative slope of the function Q_e (n'') is also to be expected without the limiting condition $| m - 1 | \ll 1$.

It can be shown that, within certain ranges of ρ, determined by the value of n', such a behavior of the function Q_e (n'') is given (see Fig. 1).

TRANSMISSION OF GLASSES CONTAINING SELENIDES AS SCATTERING PARTICLES

One group of temperature colored glasses consists of a matrix of an optical glass whose refractive index is approximately 1.7 and contains particles of semiconducting material. The latter are mainly produced by heat treatment of a glass whose batch contained a mixture of sulfides, selenides and sometimes also tellurides of Cd and/or Zn. All these materials can exhibit photoconductivity.

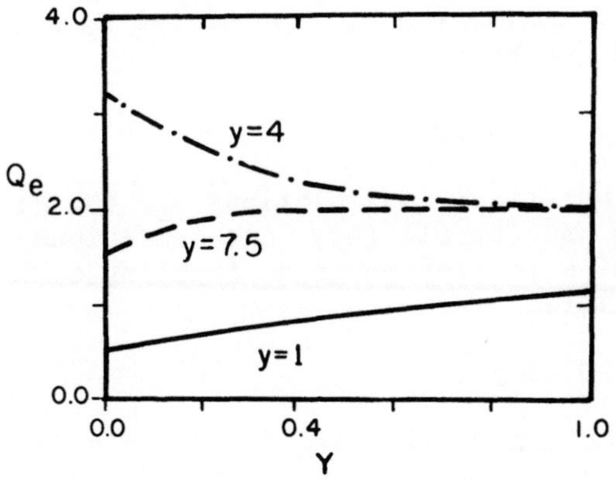

Fig. 1. Efficiency factor of extinction Q_e as a function of $x = n''/(n'-n_o)$, for three different values of the parameter $y = \{4\pi r.(n' - n_o)\}/\lambda$

Since the imaginary part of the refractive index is proportional to the conductivity, one has to expect a change in the n"- or k- values under illumination. Such a change may not affect the total coefficient of extinction γ as can be seen from Fig. 1 in the case of

$$y = 7.5 \quad \text{and} \quad x > 0.4$$

However, there are also possibilities that Q_e increases under illumination (e.g. $y = 7.5$ and $0<x<0.2$ or $y = 1.0$ and $x > 0$), or that Q_e decreases (e.g. $y = 4$ and $0 < x < 0.8$). In order to explain such a decrease certain assumptions about the nature of the particles have to be made.

Unfortunately, there is no available information about direct crystallographic or chemical analysis of the particles present in these glasses. It seems to be reasonable to assume that these particles are mixtures of Cd, S, Se and Te. If the glass would contain particles of amorphous Se only, the photo-threshold had to be expected at a wavelength of approximately 500 mμ. Addition of S produces a very slight shift towards shorter wavelength. Addition of Te, however, results in a remarkable shift of the photo-threshold towards longer wavelengths (6). At a Se/Te ratio of approximately 7/3, the threshold is to be expected at about 700mμ. At the wavelength of the ruby laser radiation a value of $k \simeq 0.05$ was found for a mixture of Cd, Se and Te having such a composition. Since the refractive index of the sample is approximately 2.2, an approximate value for m' results:

$$m' = \frac{n'}{n_o} \simeq \frac{2.2}{1.7} \approx 1.3$$

In addition to the values for $m = m'.\ (1 - i\ k)$
it is necessary to also know the parameter ρ. According to eq. (1b) and using the value $n_o = 1.7$, we have

$$\rho = 5.34\ \frac{d}{\lambda}\ ; \qquad d = 2r$$

Making use of the results of numerical calculations of the Mie coefficients by Johnson, Eldridge and Terrell (4), the efficiency factor Q_e for different values of ρ is plotted as a function of k in Fig. 2. As a result, the condition

$$2r = (\ 1.25 \pm 0.25\ )\ \lambda$$

for a negative slope of the function Q_e (k) was found (Fig.2b). This is in agreement with condition (b) mentioned above. However, it is also shown that the existence of a negative slope of Q_e (k) is very sensitive to variations in the size of the particles. On the other hand, the dependence of Q_e on ρ and m' is less sensitive.

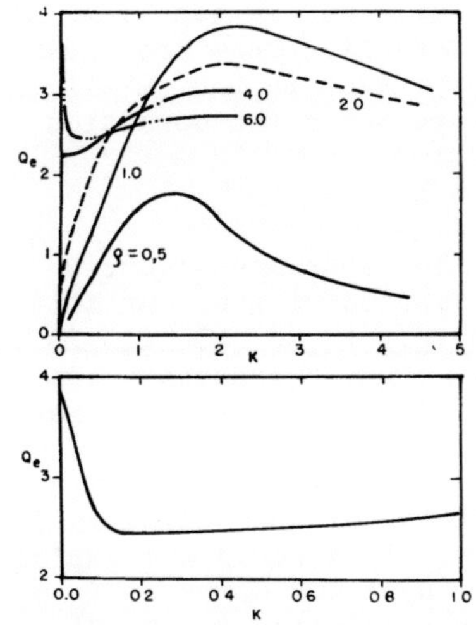

Fig. 2. Efficiency factor of extinction Q_e as a function of $k = n''/n'$ for a relative refractive index $m \approx 1.3 \cdot (1 - ik)$.
a) different values of ρ_i
b) for $6.0 < \rho < 8.0$
$\rho = (2\pi r.n_0) / \lambda$.

The latter determines the spectral transmission, or, in other words, the shape and position of the absorption band of the colored glass.

CONCLUSION

About two years ago, G. Bret and F. Gires (5) observed that a temperature colored glass of a certain type can change its transmission at 694.3 with the intensity of light. The measurements were done with light emitted by a ruby laser. The authors found that for a glass sample of 3mm thickness, the transmission at 694.30mμ can vary between 0.2% (low intensity) and 50% (high intensity); 10 $^{W}/cm^{2}$, equivalent to an electric field of 60 k V/cm). This change in transmission corresponds to a change in the coefficient of extinction γ of $\frac{6.2}{0.7} \approx 9$. In order to explain such an effect, one has to assume that γ depends on the intensity of the incident radiation. Since there was no such effect observed in temperature colored glasses of the same composition as the base glass, it must be due to the nature of the particles embedded in the glass. This conclusion is also supported by the fact that two samples of the same commercial glass showed the change to a different extent. During the melting process of such glasses (on a technical scale), slight variations in the composition of the glass cannot be avoided. In particular the S-, Se- and Te- contents may vary due to volatilization and changes in the furnace atmosphere. These changes have only a very slight effect on the refractive index of the particles produced in the glass during heat treatment. However, they should have a rather high influence on the imaginary part of the refractive index of the particles, if these are semiconductive.

In order to explain the findings by Bret and Gires, it was assumed that the particles present in the glass in question are photoconductors. The high intensity of the laser beam produces a change in conductivity, resulting in an increase in the m''- or k-values, respectively. If the particle size meets the condition

$$2r = (1.25 \pm 0.25) \cdot \lambda$$

one can expect an increase in Q_e during the experiment because of the negative slope of Q_e (k). The composition of the particles, as calculated from the batch is approximately 40 Cd, 40 Se, 20 Te (in wt.%). Therefore, the threshold of the photo-response can be expected to be close to 700mμ which is slightly above the wavelength of the ruby laser radiation.

From Fig. 2b, a change in Q_e of a factor of approximately 1.6 can be expected if our assumption is true. This would not explain, however, the large variation observed by Bret and Gires. If one calculates the total coefficient γ, for a relation for Q_e ($^r/_\lambda$) as given in Fig. 3 and a particle size distribution N (r) according to Fig. 4 for $r_1 \leq r \leq r_2$, one finds the approximate relation

$$\gamma_\ell = 8.5 \cdot \gamma_h$$

where γ_h = γ for high intensity,
γ_ℓ = γ for low intensity.

This result is in good agreement with the measurements. It is now possible to understand the fact that other temperature colored glass , whose absorption band is of the same shape and almost the same position and of a very similar composition, cannot be used as passive Q-switches. The shape and position of the absorption band in temperature colored glasses are determined mainly

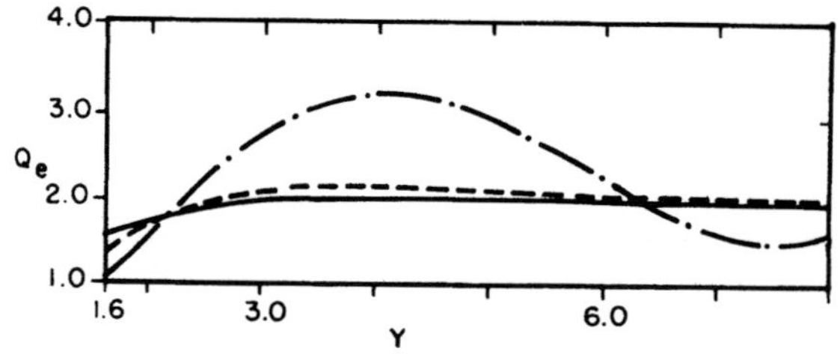

Fig. 3. Efficiency factor of extinction Q_e as a function of $y = 4\pi \cdot r \cdot (n' - n_o)/\lambda$. A first order approximation of $Q_e(y)$ between y = 2.5 and y = 4.7 is given by $Q_e(y) = 2 + 0.6 \cdot y$.

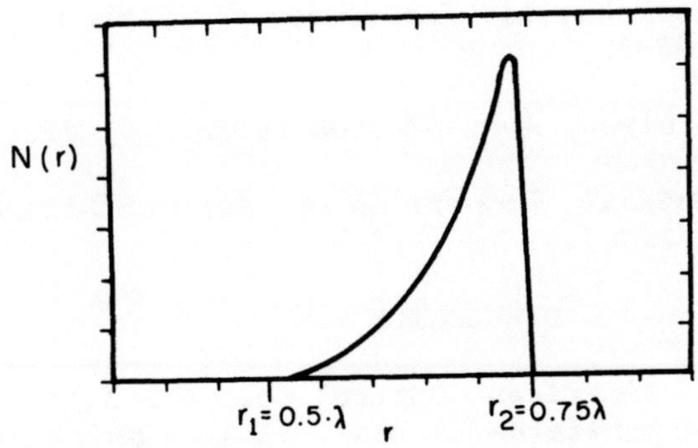

Fig. 4. Assumed particle size distribution N(r). The curve is represented by the approximation

$$N(r)\begin{cases} 0 \text{ for } r<r_1 \\ N_o\,(r - r_1)^2 \text{ for } r_1 \leq r \leq r_2 \\ 0 \text{ for } r > r_2. \end{cases}$$

by the real part of the refractive index, the size and the size distribution of the particles. The influence of the imaginary part of m has not been studied until now if semiconducting particles are present in temperature colored glasses.

In addition to the explanation of the results of Bret and Gires one can use this theory to find new temperature colored glasses which are phototropic or photochromic. Furthermore, these discussions may make it possible to explain the yellow color of many glasses of high index of refraction; e.g. high lead containing glasses. It is known that these glasses exhibit the well known microphase separation where the optical nature of the two phases depends strongly on the melting conditions. The effect of heat treatment on the color of such glasses can also possibly be explained on the basis of the theory given above. One has to consider the fact that during heat treatment a change takes place in the size and the distribution of the micro phase components.

REFERENCES

1. G. Mie, Ann. Physik 25, 377 (1908).

2. P. Debye, Ann. Physik 30, 59 (1909).

3. H. C. Van De Hulst, "Light Scattering by Small Particles", N.Y. (1957), Pub. by John Wiley & Sons, Inc.

4. *J. C. Johnson, R. G. Eldridge and J. R. Terrell, Scientific report No.4, Massachusetts Institute of Technology, Dept. of Meteorology (1954).*

5. *G. Bret and F. Gires, App. Physics Letters 4, 175 (1964).*

6. *T. S. Moss, "Optical Properties of Semiconductors", Butterworth, London (1960).*

DISCUSSION

R.F. WOODCOCK (American Optical Co., U.S.A.): *What is the composition of the particles in lead glasses which produced the yellow color with heat treatment?*

E. DEEG: *In addition to Cd, S and Se in the glass in question some Te is present. The exact composition of the particles is not known, and as we know today from studies of copper ruby glass one has to be extremely careful accepting compositions which are indirectly derived.*

R.F. WOODCOCK: *What are the changes needed in your glass composition so that it can be used for passive Q-switching at 1.06μ?*

E. DEEG: *It can be expected that an increase in Te content shifts the critical energy of the photoresponse of the particles towards lower values. This would result in a shift of the critical wavelength towards higher values. In addition it may become necessary to change the size of the particles.*

CHARLES GREENE (College of Ceramics at Alfred University, U.S.A.): *The cadmium sulfoselenide glasses show a pronounced reversible change in color as the temperature changes, becoming darker when heated and lighter when cooled. How does your theory explain this observation?*

E. DEEG: *To explain this effect one has to take the solution given for low intensities. From this solution the spectral transmission $\gamma(\lambda)$ has to be calculated and the temperature dependence of the conductivity of the scattering particles as well as the temperature dependence of the real parts of the two refractive indices n and n_0 must be considered. Such calculations can only be done with the use of computers. Therefore, it is not possible to give a quantitative result right away. However, such calculations are presently carried out by our group.*

LUMINESCENCE PHENOMENA IN MAGNESIUM OXIDE

J.E. Wertz, L.C. Hall, J. Helgeson, C.C. Chao, and W.S. Dykoski

University of Minnesota, Minneapolis

Minnesota, U.S.A.

Both intrinsic and impurity centers contribute to X-ray fluorescence and thermoluminescence phenomena in magnesium oxide. We have shown earlier that a thermoluminescent formation of Cr^{2+} from Cr^{3+} is accompanied by a conversion of Fe^{1+} to Fe^{2+} or Fe^{3+}. A red X-ray fluorescence has been shown to correspond to the main-line R-emission of Cr^{3+} observed on optical stimulation. A red thermoluminescence peak at 110°C shows the principal R-lines, and in fact closely resembles the X-ray induced R-line fluorescence at the same temperature. A blue fluorescence or phosphorescence is seen at low temperature, and numerous blue thermoluminescence peaks are observed. Their intensity and wavelength distribution are dependent upon sample treatment. Parallel studies of bleaching of trapped hole (V_1) centers and low temperature blue luminescence will be reported. We have shown that a characteristic absorption band at 2.3 eV is due to the V_1 centers and is responsible for the purple coloration of x- or UV-irradiated magnesium oxide samples.

INTRODUCTION

Both intrinsic and impurity centers contribute to x-ray fluorescence and thermoluminescence phenomena in magnesium oxide. Luminescence phenomena of particular relevance to this paper include the following:

1. Fluorescence of the d^3 ions Cr^{3+} and V^{2+} under ultraviolet or x-ray excitation. In MgO each ion in octahedral symmetry gives a single sharp line (R-line) due to transitions from the ^2E to the $^4A_{2g}$ state, irrespective of the level to which absorption occurs. The line widths vary greatly with temperature. For Cr^{3+}, the R-

line occurs at 14325 cm-1 (1), while in V^{2+} it occurs at 11498 cm-1 (2). In Al_2O_3 there are two closely-spaced R-lines for each ion due to the splitting of the ^2E state by the trigonal component of electric field. In MgO one also observes lines at 14302 and 14208 cm-1, due to Cr^{3+} ions with a tetragonal component of electric field symmetry arising from an associated vacancy along a [001]-type direction (3). From ESR measurements, the splitting of the $^4A_{2g}$ state is found to be 0.16 cm-1 (4), and thus one must assign practically all the tetragonal splitting of the emission line pair to that of the ^2E level. Another pair of lines at 14308 and 14217 cm-1 are attributed to chromium pairs (3). Peaks due to phonon-assisted transitions are seen on the low energy side of the several line systems. Under some conditions one observes them on the high energy side of the V^{2+} R-line also (5). The large amplitude of the R lines of Cr^{3+} and of V^{2+} observed upon stimulation in one of the absorption bands at higher energy (pumping) makes it feasible to locate absorption bands by observing the excitation spectrum. It has been possible to locate an additional absorption band each for Cr^{3+} (at 34700 cm-1) and for V^{2+} (32300 cm-1)(6). Additionally, doubt is cast upon a band at 29400cm-1 reported (7) for Cr^{3+}, since no corresponding excitation peak is observed. A pair of lines occurring at 15010 and 15079 cm-1 have been stated to be R-lines of Mn^{4+} (8). This system has been more extensively studied by Henderson (9).

2. Broad-band fluorescence of MgO in the vicinity of 24000 to 29000 cm-1 upon x- or ultraviolet-excitation. Eisenstein (10) reports a band at 3.6 eV (∿29000cm-1) which shows prolonged phosphorescence. The intensity is very strongly dependent upon thermal treatment, increasing greatly upon heating in vacuum or in hydrogen at temperatures of 1475°K or more. There may be a single band or several, the number and location depending upon the sample and its past thermal treatment (11).

3. Thermoluminescence peaks in MgO at several temperatures and at various energies (12-14). Hansler and Segelken reported peaks at 363 and 433°K in one crystal (12). Their glow curves for other crystals show peaks at about 350-360, 415 and 465°K. Their observations on the emitted light are largely qualitative, but to fit an energy level scheme they found it convenient to assume values of 2.0 eV for the red light and variously used 2.7, 2.9 and 3.6 eV for "blue" luminescence. However, they did not demonstrate the actual occurrence of luminescence peaks with maxima at these energy values. Thomas and Houston (13), who make no mention of the earlier work of Hansler and Segelken, report peaks at 410 and 520°K. They did not state the wavelength distribution of the emitted light.

A blue thermoluminescence (∿25300 cm-1) which occurs below

390°K in x-irradiated MgO and which is intense enough to be observed in a semi-darkened room has been ascribed to the process $Cr^{3+} + e \rightarrow Cr^{2+}$ (14). The rate-determining step which governs the appearance of the thermoluminescence peak is ascribed to the dissociation of Fe^{1+} produced upon x-irradiation: $Fe^{1+} \rightarrow Fe^{2+} + e$. It has been shown by ESR measurements that the initially large Fe^{1+} concentration ($\sim 4 \times 10^{16}$ cm^{-3}) before thermoluminescence occurs, decays dramatically with the thermoluminescence. Simultaneously, the ESR intensity of Cr^{3+} lines decreases. The presence of Cr^{2+} is inferred after the x-irradiation; its ESR spectrum in MgO has not yet been conclusively established. However, it has been shown by microwave ultrasonic resonance experiments that Cr^{2+} ions are present both in Cr-doped MgO before irradiation and in undoped crystals after x-irradiation (15).

EXPERIMENTAL ARRANGEMENTS

A variety of single magnesium oxide crystals from the Norton Company, Semi-Elements Inc., and the Muscle Shoals Electrochemical Company have been used for luminescence studies. X-irradiation was usually carried out with a Westinghouse tube operated at 80 kV, for periods ranging from 10 min to several hours. Samples for low temperature study were irradiated in cryostats equipped with thin aluminum foil windows ($\sim$0.02 mm). X-fluorescence was excited with a portable Machlett or General Electric tube normally used for diffraction studies. By using a crystal mount equipped with a heater, it was possible to study x-fluorescence at elevated temperatures. Excitation voltage was about 40 kV, with currents up to 18 ma. Detailed analysis of the spectrum is made with a 75 cm. Spex scanning spectrometer, using an EMI 9558Q or 6256B photomultiplier cooled to 190°K to minimize the dark current. These (selected) tubes had very low ($\sim$1 nA) dark current at 20°C. The 6256B tube was especially advantageous for blue luminescence, since its response peaks in the region of interest. When studying thermoluminescence in the red or blue regions, it was convenient to use the appropriate Optics Technology type 60 variable bandpass filters to isolate this region. In some instances MgO crystals were irradiated at liquid hydrogen temperatures and their fluorescence was observed at these temperatures. For this purpose a Cryotip refrigerator (Air Products and Chemicals Co., Allentown, Pa.), which employs liquid nitrogen to precool tank hydrogen gas below its inversion temperature, was used.

Many of the thermoluminescence measurements have been carried out with a hyperbolic rather than linear heating rate. If T_0 represents the absolute temperature at time t=0, then $T^{-1} = T_0^{-1} - \beta_0 t$. The programmed current supply is adapted from the design of Halperin, Leibovitz and Schlesinger (16). The advantage of this procedure is the great simplification in determination of energy of activation from a single thermoluminescence curve. Two precautions are in order for accurate determination of peak positions. First, one must insure an adequate rate of heat transfer between the fine (0.12 mm diam.) platinum wire used for heating and the crystal mount. The early design used by the authors had the wire coiled into a helix which was then wound upon an aluminum rod recessed to hold the crystal. The heat transfer is considerably improved by winding the wire (in fiberglass sleeving) directly around a 5 cm aluminum tube terminating in a recessed 2 x 2 cm cross section crystal mount. Use of a heat transfer agent (Dow Corning 340) improves thermal contact of wire with mount. Secondly, one must take care to insure the best possible thermal contact between crystal and crystal mount. Even with cleaved surfaces in contact with a "flat" metal surface, the transfer is surprisingly poor. It was found that use of sheet indium at temperatures below 410°K gives luminescence peaks lower by 10-30° than in its absence.

It is of interest to note the simple relation between the thermoluminescence maximum observed for a linear and for a hyperbolic heating rate. If I represents the intensity of the emission at time t, n the number of trapped electrons or holes which are to be released, S is the pre-exponential factor for a first-order decay process and E the activation energy, we write: $I \propto -dn/dt = nS \exp(-E/kT)$ (1). Upon substituting $T^{-1} = T_o^{-1} - \beta_o t$, and solving for the temperature T_{mh} of the maximum for a hyperbolic heating rate, one obtains:

$$T_{mh}^{-1} = -\frac{k}{E} \ln \frac{E\beta_o}{kS} . \qquad (2)$$

This relation will be used to solve for S, which will later be substituted into an expression for the temperature of maximum emission, using a linear heating rate. If we now write for this latter case:

$$I \propto -dn/dt = nS \exp\left[-E/k(\gamma t + T_o)\right], \qquad (3)$$

one may derive an expression for the temperature T_{ml} of the maximum in the linear heating rate thermoluminescence curve. One finds:

$$T_{ml}^{2} = \frac{\gamma E}{Sk} \exp(E/kT).$$

Eliminating S between the two expressions, one obtains:

$$T_{mh}^{-1} = \frac{k}{E} \left(\frac{E}{kT_{ml}} - 2 \ln T_{ml} - \ln \frac{\beta_o}{\gamma} \right).$$

The condition that one should obtain the same temperature for the peak in the thermoluminescence curves by linear or by hyperbolic heating is then:

$$\beta T_{m1}^{2} = \gamma$$

RESULTS

Red Thermoluminescence

The high sensitivity of the EMI 9558Q tube and the low dark current of the selected tube made it possible to observe line structure in the spectrum of the red emission. This spectrum is displayed in the uppermost curve of Figure 1, taken at the peak of the thermoluminescence curve at 358°K. This value is obtained upon heating hyperbolically with $\beta = 5 \times 10^{-7}$ sec^{-1} deg^{-1}. The activation energy for this luminescence was found to be 0.9 eV. Additional red emission peaks were observed at 500, 630 and 670°K in some crystals.

The observation of structure in the vicinity of the R-group led the authors to investigate the appearance of the x-fluorescence as the temperature of the magnesium oxide is increased. Starting at 77°K, the R-group spectrum under x-excitation is displayed as the lowest curve in Figure 1. Additional x-fluorescence spectra are given for sample temperatures of 295 and 358°K as the two intermediate curves of Figure 1. One notes the striking similarity between the peaks and shoulders of the x-fluorescence spectrum at 358°K and the spectrum of the thermoluminescence in the same temperature region. A marked reduction in intensity of red fluorescence is observed upon heating in O_2 at ~1400°K.

Upon cessation of x-excitation, the red emission decays over a period of minutes at room temperature. One may, in fact scan the spectrum of the phosphorescence and with a tenfold increase in gain observe all of the major peaks ordinarily seen in the fluorescence. This observation contrasts with the report of Eisenstein (10) that persistent emission occurred only at 3.6 eV (29000 cm^{-1}). The difference is readily understood as one of sensitivity of the photomultipliers used, i.e., that of the EMI 9558Q against that of the 1P28.

Blue Fluorescence and Thermoluminescence

The number, location, and intensity of emission bands between 24000 and 36000 cm^{-1} is very strongly dependent upon the crystal of MgO chosen, upon its thermal treatments, and past irradiations. For most crystals the principal emission band in this region is a

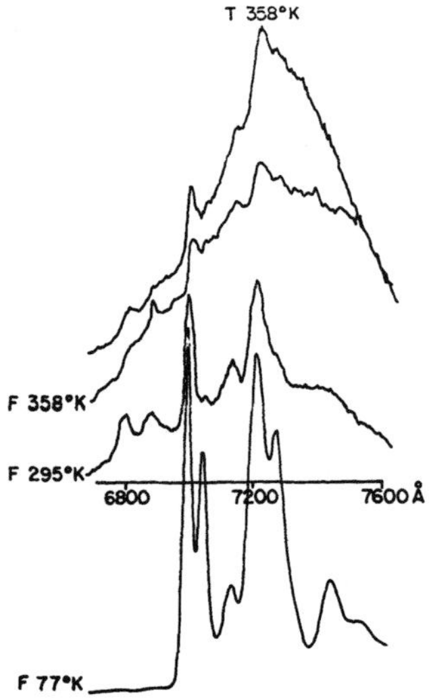

Fig. 1. Comparison of thermoluminescence and x-fluorescence spectra of MgO in the R-line region. The line labelled T refers to the spectrum of the red luminescence at 358°K. Lines labelled with F refer to x-fluorescence at the temperature indicated. Correspondence of sets of peak positions at 358°K for thermoluminescence and R-group fluorescence gives evidence that the thermoluminescence is R-emission from Cr^{3+}.

Gaussian centered near 26300 cm^{-1}, with a half-width of about 8000 cm^{-1}. The apparently common property of all fluorescence bands in this region is that their intensity diminishes during continuing excitation with x-rays. The initial exponential decay rate during irradiation for the fluorescence band at 26300 cm^{-1} is of the order of 0.04 min^{-1}. The decay of emission may be followed after cessation of irradiation at room temperature; the intensity drops to a very low value if then one cools to 77°K. Besides this peak, the position of which is somewhat variable, one sometimes observes either fluorescence shoulders or peaks at 23300 and 20800 cm^{-1}. For simplicity, these are referred to as blue peaks. Just after a crystal has been heated to 1475°K, one observes an emission peak at about 36300 cm^{-1} which disappears during a short period of irradiation. The intensity of the blue emission may be increased by 40-fold by heating in vacuum at 1475°K. The thermoluminescence in the blue region variously peaks between 24400 and 27800 cm^{-1}, with a usual value being 25600 cm^{-1}. The number of thermoluminescence peaks observed depends upon the sample, and upon its thermal and radiation history. If irradiation is carried out at room temperature, one may see the principal blue peak at 358°K; from hyperbolic heating studies, one finds the activation energy for this peak to be 0.85 eV. We have observed additional peaks at 490 and at 670°K in some crystals. If x-irradiation is carried out at 77°K, one may observe additional blue peaks near 150, 180, 240, 290 and 305°K in some crystals.

An unusually sharp set of four emission lines under x-ray excitation is seen for some samples on the low energy side of the usual blue fluorescence peak. These are shown in Figure 2. Their peak values correspond to 20325, 20505, 20725 and 20905 cm^{-1}. The intensity of these lines diminishes with continued x-excitation with a half-period of about 15 min; this decay is faster than that of the blue band upon which it is superimposed. They are not normally observed upon excitation at room temperature, but show up readily at 77°K. At 20°K their amplitude is much enhanced, presumably due to the narrowing which occurs. However, the maximum intensity was observed after the sample had first been heated at 675°K for several hours, x-irradiated at room temperature for five minutes at 80 kV, then cooled to 77°K and excited with x-rays. After decay of these lines during irradiation, one may restore their intensity temporarily again by repetition of the cycle. In order to check the origin of these lines after Ni^{2+} was suspected, x-fluorescence was observed at 77°K of an MgO: Ni crystal doped so lightly that only a faint greenish tinge was detected. The four lines of Fig. 2 were observed with much enhanced intensity. Additional weaker lines were observed at 21295, and 20150 cm^{-1}. Here the lines do not obviously ride upon a broad blue emission peak; instead they appear to rise from the base line. If the blue emission is present, it is very weak in this sample. An undoped crystal which gave a peak in the 20800 cm^{-1} region was observed in x-fluorescence at 20°K. At this temperature one sees partially resolved peaks at the above positions.

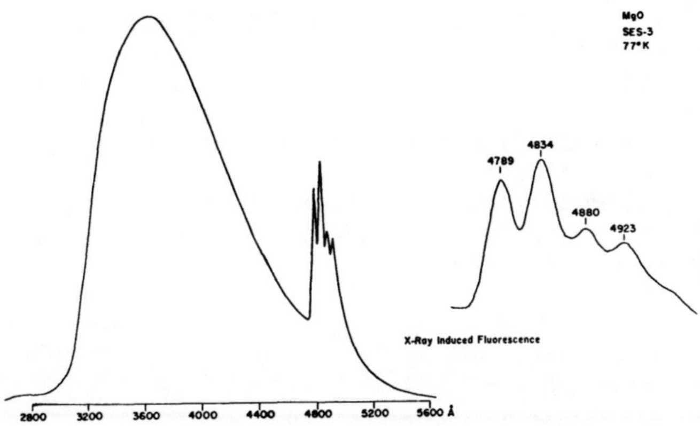

Fig. 2. X-fluorescence spectrum of one crystal of MgO in the blue and near-UV region at 77°K, showing a set of four sharp lines. These are repeated at a lower scanning rate, and peak wavelength values are shown.

DISCUSSION OF RESULTS

It had previously been suggested (12) that the thermoluminescence in the vicinity of 14000 cm^{-1} might be due to the process:

$$Fe^{2+} + \text{free hole} \rightarrow Fe^{3+}.$$

The basis for the suggestion was the increase from 2.9 to 4.4×10^{17} Fe^{3+} ions per cm^3 in the interval between ultraviolet irradiation and heating through the first thermoluminescence peak, presumably around 100°C. It would appear that the close correspondence between the structure of the red thermoluminescence peak shown in Figure 1 and that for x-fluorescence over a range of temperatures conclusively indicates that the luminescence arises from Cr^{3+} formed in the excited state. We shall consider the possible precursors shortly. An additional type of center which is induced by ionizing radiation and which was unknown in making the assignment to Fe^{2+} is the V_1 center (hole trapped adjacent to a positive ion vacancy). These centers can readily be produced at concentrations of the order of 10^{17} cm^{-3}. They decay slowly in the dark at room temperature and more rapidly in the presence of visible light. Such thermally- or optically-released holes interacting with Fe^{2+} could account for the twofold increase in Fe^{3+} concentration after heating through the first thermoluminescence peak. We have ourselves erred in too hastily surmising a correlation between qualitative trends in ESR spectra and emission or absorption bands.

R-line phosphorescence and thermoluminescence has been observed after x-irradiation of ruby (17). Direct measurements have been made of R-line decay following pulses of 30 MeV electrons lasting from 10 to 4500 n sec (18). The emission is attributed by Compton et al. to "recombination of radiation-produced electrons and holes at the Cr^{3+} ions". Owing to the long life involved in thermoluminescence, it is preferable to consider the interaction with Cr^{3+} as occurring in stages. If the first stage is addition of an electron, one gets

$$Cr^{3+} + e \rightarrow Cr^{2+} \; ; \; Cr^{2+} + h \rightarrow Cr^{3+*} \rightarrow R \text{ emission.}$$

The alternative is

$$Cr^{3+} + h \rightarrow Cr^{4+} \; ; \; Cr^{4+} + e \rightarrow Cr^{3+*} \rightarrow R \text{ emission.}$$

The orange color of irradiated ruby, considered to be characteristic of Cr^{4+}, disappears upon heating as thermoluminescence occurs, giving rise to R-line emission. Observation of the Cr^{4+} ESR spectrum confirms its presence in Al_2O_3 (19). This is regarded by Compton as favoring the second process (20). We have observed an orange color to appear after green MgO:Cr is x-irradiated. Quadrivalent ions are not unknown in MgO; the Mn^{4+} ion occurs during x-irradiation (8,9), and it may be stabilized in powders by incorporation of monovalent ions (21). Ti^{4+} is almost surely the stable form occurring substitutionally in MgO, with the trivalent ion being produced by x-irradiation (22). Hence it is possible that in MgO Cr^{4+} could be involved in the R emission process. However, Cr^{2+} is known from microwave ultrasonic measurements to occur in MgO after x-irradiation; both ions must therefore be seriously considered as possible precursors of excited Cr^{3+} responsible for the red fluorescence.

The blue emission represents a rather more complex process than the red; the identity of the centers whose decay gives rise to the various blue peaks in the x-fluorescence or thermoluminescence spectra is not known. It is remarkable that both red and blue emission peaks occur at the temperature of 358°K. This revised value is near one of the values of Hansler and Segelken, but well below the value of 393°K used earlier (14) for blue luminescence or the 410°K value quoted by Thomas and Houston. Both high values are probably due to inadequate heat transfer to the crystal; this problem would be accentuated with the high heating rate of 1.5° sec^{-1} (13). The activation energies for the 358°K peaks are also very similar, being respectively 0.9 and 0.85 eV. The earlier hypothesis (13) of the process.

$$Cr^{3+} + e \rightarrow Cr^{2+}$$

for the blue thermoluminescence as well as x-fluorescence emission at 358°K still appears to be consistent with known data. During x-ray excitation, it is known that Cr^{2+} is being formed and except for reversal of the process, the blue fluorescence intensity should decrease as the Cr^{3+} concentration decreases. The Cr^{2+} is continuously reacting with free holes to form Cr^{3+} in the excited state, with subsequent R-group emission. It is proposed that the effect of heating in oxygen at 1400°K may be due to the formation of large numbers of Fe^{3+} ions from the ever-abundant Fe^{2+}. The usual quenching effect of Fe^{3+} is well known.

The origin of the set of sharp lines shown in Fig. 2 appears clearly to be due to Ni^{2+}. This ion gives a blue-green emission in MgF_2 at very low concentrations (23), and for the increased crystal field in MgO, one should expect a displacement to higher energies in the latter. More direct evidence comes from cathodoluminescence studies (24) on MgO:Ni; these show the four lines of same relative intensities. The most compelling evidence comes from the great intensity of these lines in a single MgO:Ni crystal. Upon x-irradiation, Ni^{2+} is converted to Ni^{3+} (25), and hence one should observe an emission decay during excitation, as observed. The restoration of emission upon standing or upon heating is also consistent with the assignment to Ni^{2+}. In some cases where these lines are not actually resolved, but there is a fairly sharp peak in x-fluorescence near 20500 cm^{-1} at 77°K one may suspect that the peak is again due to traces of Ni^{2+}. This was shown in one instance by cooling to 20°K and observing partially resolved peaks at the expected Ni^{2+} positions.

ACKNOWLEDGMENT

It is a pleasure to acknowledge helpful discussions with Drs. D.M.J. Compton and L. F. Johnson. We thank Dr. J.E. Ralph for kindly sending us a report on cathodoluminescence in MgO in advance of publication.

This work was supported by the U. S. Air Force Office of Scientific Research under Grants 200-63 and 200-66.

REFERENCES

1. *O. Deutschbein, Ann. Physik 20, 828-842 (1934).*

2. *M.D. Sturge, Phys. Rev. 130, 639-46 (1963).*

3. *G.F. Imbusch, A.L. Schawlow, A.D. May and S.Sugano, Phys. Rev. 140, A830-838 (1965).*

4. *J.E. Wertz and P. Auzins, Phys. Rev. 106, 484-88 (1957).*

5. *B. DiBartolo and R. Peccei, Phys. Rev. 137, A1770-76 (1965).*

6. *J.E. Wertz, R.L. Weber and L.C. Hall, to be published.*

7. *W. Low, Phys. Rev. 105, 801-6 (1957).*

8. *M. Nakada, K. Awazu, S. Ibuki, Y. Miyako and M. Date, J. Phys. Soc. Japan, 19, 781 (1964).*

9. *B. Henderson, to be published.*

10. *A.S. Eisenstein, Phys. Rev. 93, 1017 (1953).*

11. *L.C. Hall and J.E. Wertz, Bull. Am. Phys. Soc. 9, No. 7, 707 (1964). Several statements in this abstract require modification. With adequate sensitivity, one can see both red and blue fluorescence for any crystal, including doped samples; heating in vacuum enhances blue luminescence and extended heating in oxygen suppresses it.*

12. *R.L. Hansler and W. Segelken, J. Phys. Chem. Solids, 13, 124 (1960).*

13. B. Thomas and E. Houston, Brit. J. Appl. Phys. 15, 953-58 (1964).

14. J. Helgeson and J.E. Wertz, Bull. Am. Phys. Soc. 9, No. 7,707 (1964); J.E. Wertz and R.L. Coffman, J. Appl. Phys. 36, 2959-66 (1965).

15. J.R. Fletcher, F.G. Marshall, V.W. Rampton, P.M. Powell and K.W.H. Stevens, Proc. Phys. Soc. (London) 88, 127-30 (1966).

16. A. Halperin, M. Leibovitz and M. Schlesinger, Rev. Sci. Inst. 33, 1168-70 (1962).

17. W. Low, Appl. Phys. Letters, 5, 35-37 (1964).

18. D.M.J. Compton, J.F. Bryant and R. A. Cesena in Physics of Quantum Electronics, McGraw-Hill, N.Y. (1966) p. 305-14.

19. R.H. Hoskins and B.H. Soffer, Phys. Rev. 133, A490-93 (1964).

20. D.M.J. Compton, private communication.

21. J. S. Prener, J. Chem. Phys. 21, 160-61 (1953); W.D. Hershberger and H.M. Leifer, Phys. Rev. 88, 714-20 (1952).

22. J.E. Wertz, G.S. Saville, L. Hall and P. Auzins, Proc. Brit. Ceram. Soc. 1, 59-70 (1964).

23. L.F. Johnson, private communication.

24. J.E. Ralph, private communication.

25. U. Hochli, K. A. Muller and P. Wysling, Phys. Letters 15, 5 (1965).

26. R. Pappelardo, D.L. Wood and R.C. Linares, Jr., J. Chem. Phys. 35, 1460-78 (1961).

PHASE CHANGE IN TRANSMISSION THROUGH SILVER LAYERS AND ITS VARIATION WITH THICKNESS FOR VARIABLE RADIATION

N. Barakat and F.G. Abou Zakhm

Ain Shams University

Cairo, U.A.R.

Phase change of visible radiation in transmission through thin evaporated Ag layers over the range from 50 A° to 400 A° was determined using Rayleigh refractometer with improved accuracy in measuring both fringe shifts and thickness of the thin metallic films. The direction of the phase shift was also determined experimentally and the phase change in transmission was utilized in redistributing the intensity in multiple beam interference fringes at reflection. The phase curve showed an abrupt change at a thickness t of about 80 A° for λ5461 A° then went through a minimum at t = 106 A°. For t > 200 it increased steadily approaching a constant value of 0.2 λ.

INTRODUCTION

The measurement of the changes of phase resulting from transmission through thin films furnished useful data in understanding their optical properties.

Ishiguro, 1950 (1), 1951 (2) and 1953 (3) measured the phase changes resulting from transmission through thin metallic films of Ag of thickness 180, 270 and 300 A° only using a Rayleigh-interferometer. The thin films were prepared by evaporation and their thickness determined by weighing the tungsten filament before and after evaporation assuming that the film was uniform and had the same density as the bulk material. It is obvious that the accuracy reached in such measurements and based on such assumptions is not high.

The aim of the present work is (a) to determine the phase

change in transmission through thin evaporated Ag layers over the whole range from 50A° to 400A° with improved accuracy in measuring both fringe shifts and thickness of the thin metallic films. (b) to utilize the phase change in transmission to redistribute the intensity in the multiple beam reflected system. (c) to determine the direction of the phase change experimentally.

Figure 1 shows the passage of light in Rayleigh refractometer. Light from a slit source S is made parallel by a lens L_1 and the two wide slits S_1, S_2 are illuminated at normal incidence. After passing respectively through the cells C_1, C_2 and through the glass plates G_1, G_2, the lens L_2 brings the beams to a focus at F where interference fringes are viewed with a cylindrical lens.

The cells C_1 and C_2 only cover half the slit length and produce fringes, F_1 above the half way line. Light passes below this line and produces a fixed fringe system F_2, which acts as a reference to which F_1 is compared. The inclined glass plate H is a simple device for bringing up the lower reference system to touch the upper system and avoid the obstruction due to the wall thickness of the cells.

The two cells supplied with the apparatus were thoroughly cleaned. They were then introduced in their appropriate position and white light pattern of fringes was observed. The head governing the inclination of the compensator was then adjusted so that no shift was observed for the zero order dark fringe. The fringes were then recorded photographically and examined using an Abbé comparator to ensure that no shift took place using the green spectral line of Hgλ = 5461 A°. The two cells were introduced in the evaporation plant and by masking one of them with a freshly cleaved mica sheet, a thin evaporated metallic film coated the other cell. This was the film whose phase change in transmission was determined. The fringe shift due to the phase change resulting from transmission through the metallic coating was observed as the two cells were introduced. The resulting monochromatic system was recorded photographically.

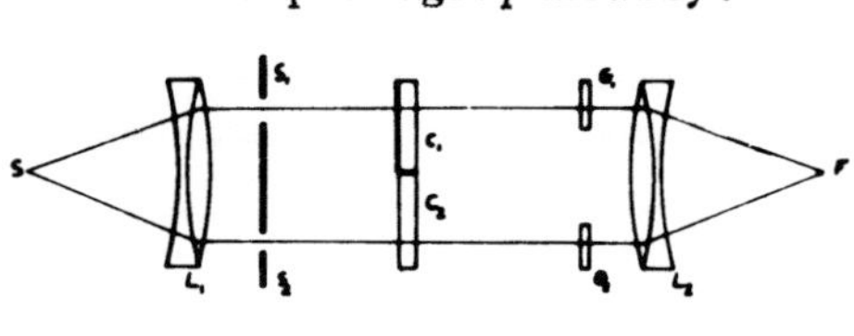

Fig. 1. The optical system in Rayleigh refractometer.

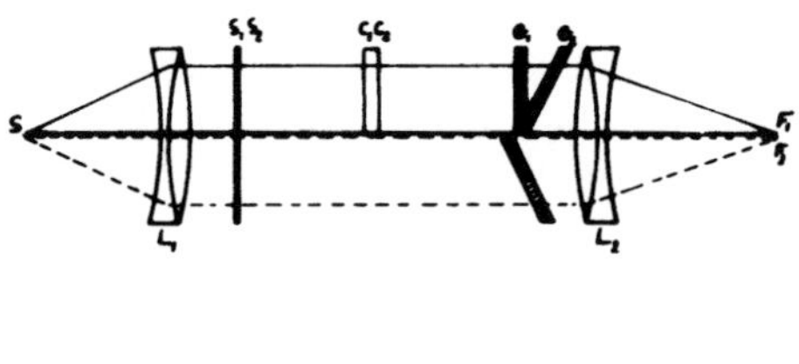

Plate (1) shows the appearance of the two systems of interference resulting as the two cells were introduced in the path of the beams, a thin evaporated silver film coated one of the two cells. The lower system was the reference, the two beams responsible for producing it, passed through air while the upper system was shifted to the right due to the path difference introduced as one of the beams passed through the thin silver films. The fringe shift was measured by an Abbé comparator.

Measurement of thickness of the evaporated thin film was performed by applying multiple beam interference fringes at reflection, as developed by Tolansky (4). In this work an optical flat was introduced side by side with the glass cell to be coated, during evaporation, with a silver layer of the same thickness. The film on the optical flat had a sharp edge resulting from masking part of the flat. An opaque layer of silver was then deposited by evaporation on both the coated and uncoated parts of the optical flat. This served as one of the components of a silvered air wedge interferometer, the upper component facing the light was another optical flat coated with a thin Ag film of about 80% reflectivity.

The edge of the wedge was adjusted perpendicular to the step. Parallel monochromatic light of λ5461 A° illuminated the wedge and the fringe shift x revealed the step height given by $\Delta x/x \cdot \lambda/2$, x being the separation between any two successive fringes. Plate (2) shows the fringes crossing a step of height 360 A°.

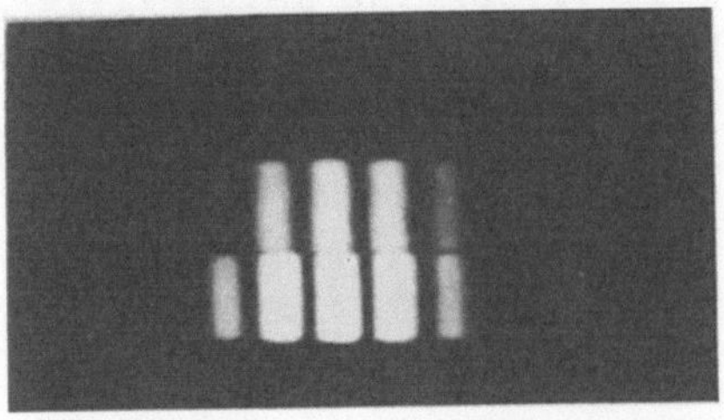

late 1. The appearance of the wo systems of interference ith a thin evaporated Ag film oating one of the two cells.

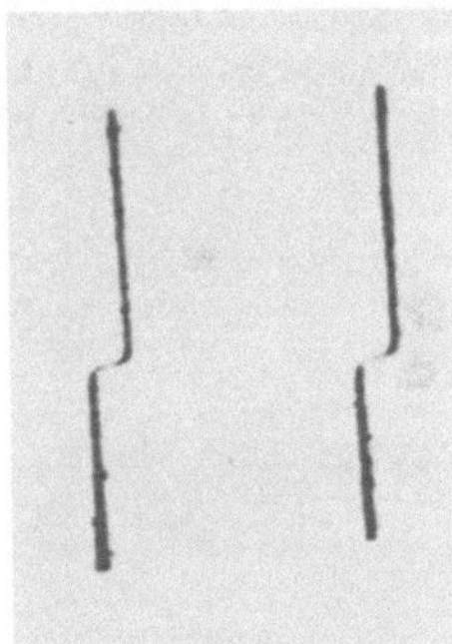

Plate 2. Multiple beam Fizeau fringes at reflection crossing a step of height 360 A°.

RESULTS AND DISCUSSION

Figure 2 shows the result of determining the phase change in transmission through thin evaporated Ag layers using Rayleigh refractometer. The thickness of the Ag layers ranged from 50 to 400 A° for λ5461 A°. The curve showed an abrupt change at a thickness of about 80 A° and went through a minimum at t = 106 A°. As the thickness increased further, the phase change increased rapidly at first till t = 200 A° then it increased approaching a constant value of 0.2λ. It was not possible to measure γ for Ag layers of thickness more than 400 A° because of the poor transmissivity of the films.

For normal incidence Ishiguro's three values of the phase change γ for Ag films of thickness t = 180, 210 and 300 A° for λ5890 are 0.25π, 0.29π and 0.34π respectively. The corresponding values of γ as determined from the present work for the three values of thickness for λ5461A° are 0.288π, 0.32π and 0.368π which show close agreement in the order of magnitude.

Determination of Direction of Phase Introduced as the Result of Transmission Through Silver Layers

In a recent publication by the authors (5) (1965) the phase change in transmission through thin metallic films was utilized to redistribute the intensity in multiple beam reflected system. The result of introducing three Ag layers of thickness 160, 180 and 260 A° in the passage of the first beam changed the intensity distribution in the interference pattern at reflection characterised by sharp dark lines on a bright background to bright lines on a dark background. In the present work we are only concerned with the determination of the direction of the phase. The experimental

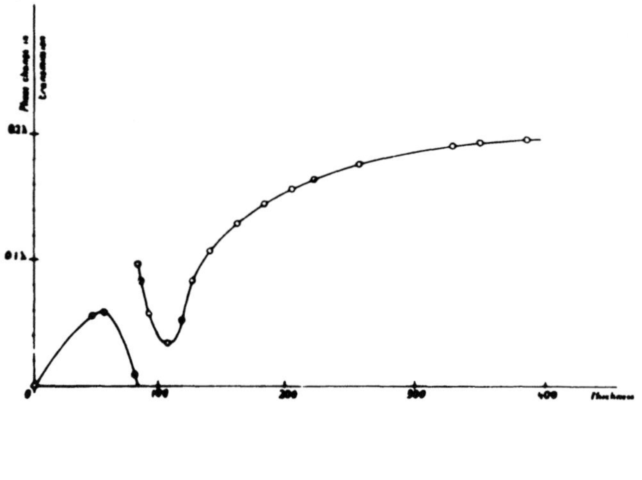

Fig. 2. The curve showing the result of determining the phase change in transmission through the Ag layers of thickness from 50 to 400 A°.

arrangement is as shown in Figure 3. A silver layer on a glass substrate of thickness 260 A° was introduced in the passage of the first of the series of images formed at the focal plane of the objective. A change in the intensity distribution of the reflected system was observed and recorded as shown in Plate (3). It is to be noted that the plate introduced was adjusted such that the rest of the images passed through the glass substrate which was an optical parallel. The additional phase was due to the passage of the first beam through the metallic layer. The corresponding value of the phase change is 0.356π. Figure 4 shows the diagramatic summation of the two vectors $\vec{r_1}$ and $\vec{A_T}$ when additional phase angle is ± 0.356π and r_1 = 0.456, where $\vec{r_1}$ represents the first beam and $\vec{A_T}$ the resultant of the rest of the beams. Figure 5 shows the variation of the amplitude of the resultant for Θ = + 0.356π and - 0.356π as Δ changes from - π to + π, the continuous curve represents the former case. The direction of the phase introduced as the result of transmission through the silver layer was investigated by observing whether the direction identified by a maximum followed by the adjacent minimum was away from the apex of the wedge or towards it. The former is the case resulting from adding a phase angle of + 0.356π to the first beam. This means that the change of phase in transmission is a retardation, a result reached by Ishiguro using a different method.

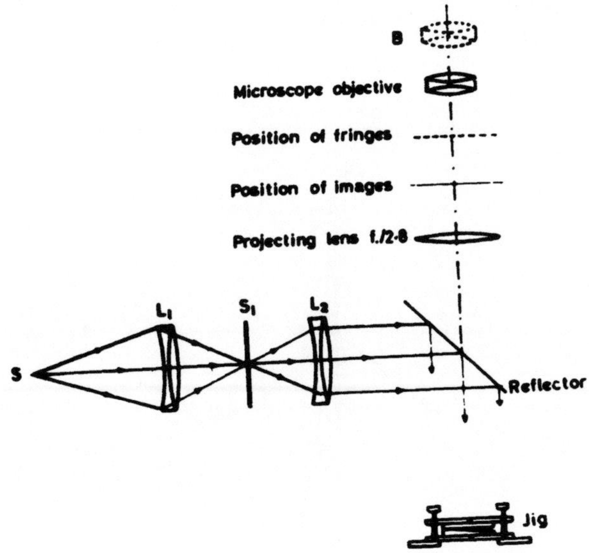

Fig.3. Experimental arrangement used to investigate the effect of introducing an additional phase to the first beam in multiple beam reflected system.

Plate 3. The change in intensity distribution of the reflected system resulting from introducing a silver layer of thickness 260 A°.

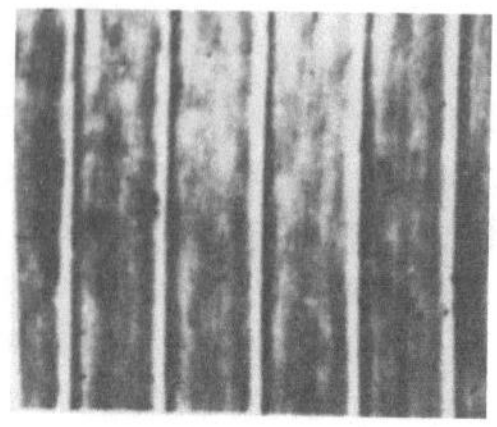

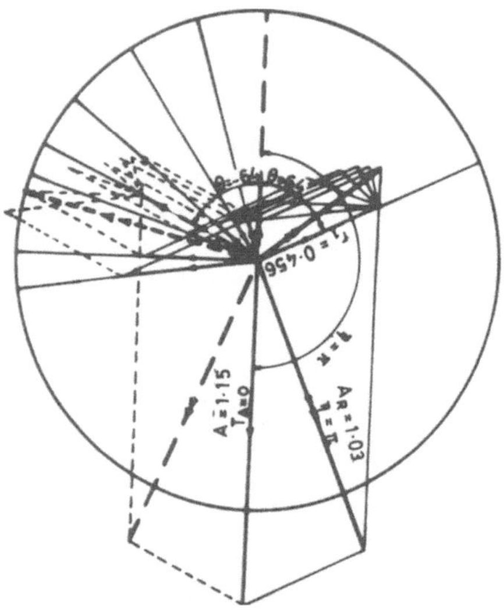

Fig.4. Diagramatic summation of the two vectors r_1 and A_T with $\Theta = 64°$ and $r_1 = 0.456$.

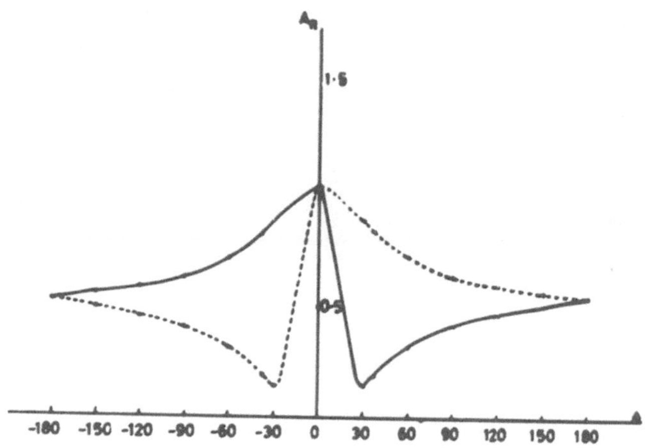

Fig. 5. The variation of the amplitude of the resultant for $\Theta = +64°$ (continuous curve) and $\Theta = -64°$, (dotted curve) as the phase difference Δ changed from $-\pi$ to $+\pi$.

REFERENCES

1. *K. Ishiguro, J. Phys. Soc. Japan, 5, 187 (1950).*
2. *K. Ishiguro, J. Phys. Soc. Japan, 6, 1 (1951).*
3. *K. Ishiguro, J.O.S.A., 43, 5, p. 365 (1953).*
4. *S. Tolansky, "Surface Microtopography" p. 62 - 65, Longmans (1960).*
5. *N. Barakat and F. Abou Zakhm, Optica Acta, 12, 4, p. 321-332 (1965).*

STUDY OF COTTON FIBER STRUCTURE BY INFRA-RED IRRADIATION

M.I. Nasser and M.A. Moharram

National Research Center

Cairo, U.A.R.

When raw cotton and some cotton derivatives were irradiated by infrared radiation, a number of absorption bands were observed.

The bands in the region from 2 - 4μ are characteristic of the hydroxyl groups. The study of these absorption bands has given considerable information about the hydrogen bonding in the structure.

It was also possible by means of infrared irradiation to study the degree of crystallinity of cotton fibers quantitatively. The results indicated that infrared spectroscopy provides an independent self-calibrating method of the percent crystallinity.

Changes in the infrared spectra of cotton fibers during progressive hydrolysis by hydrochloric acid have been studied. This study shows that acid hydrolysis increases the degree of crystallinity of this material. It was noticed that during the initial stages of this heterogeneous hydrolysis a sharp change in the degree of crystallinity took place. It has been further shown that continued acid action results in a slower rate of changes in the degree of crystallinity. Infrared irradiation of cotton fibers can thus follow the changes in their crystallinity during the mild acid hydrolysis.

The infrared irradiation was also used for studying the action of the oxidizing agents on cotton and on hydrocotton prepared by mild acid hydrolysis. Oxidation changes can only be followed when such a process results in the formation of c=o groups in the oxidized samples. It was found that for any period of treatment the intensity of the c=o groups for cotton is always higher than that for hydrocotton. The ratio of the absorbances of the c=o stretching groups in the two cases has been found to be approximately 1.26.

INTRODUCTION

The preparation of thin and uniform samples of a fibrous material for I.R. spectral investigations has always given rise to some difficulties. Cotton cannot be dissolved in a solvent suitable for absorption measurements, at least, without considerable modifications. Thus, common solvent techniques for that purpose are not available, unless some measurements over very limited portions of the spectra are required. Other techniques, such as mulling with a mineral oil, or the preparation of thin films cast on glass plates for measuring the spectra of insoluble materials are not entirely satisfactory. In 1965 M.R. Harvey (1) and others have applied the KBr disc technique to the infrared spectral measurements of relatively large solid particles. In 1957 O'Connor et al. (2) described the use of the KBr disc technique for obtaining the spectra of cellulose and modified celluloses. Forziati and co-workers in 1951 (3, 4) carried out some studies on the crystalline structure of cotton and celluloses. Brown et al. (1951) published some results (5) about one of the very few investigations of cotton in the fundamental O-H vibration region of the I.R. spectra (2.5 - 4μ). Marrinan and Mann (1954) studied the deuteration of various celluloses in some detail (6). Tsubio (1957) studied the I.R. spectra of cellulose using polarized radiation and also examined the deuteration effects on its spectrum (7). In 1961 Higgins and Harrington (8) made a comparison between the I.R. spectra of native cellulose and those of various closely related materials. In 1963 R. Jeffries (9) made an infrared study of the OH-OD exchange reaction in films of cellulose derivatives.

EXPERIMENTAL

A special die was designed by the authors and was manufactured in the workshop of the National Research Center. By means of this die discs were prepared for the infrared measurements.

The measurements were carried on a double beam I.R. spectrophotometer (UR 10) from 2 - 25μ.

RESULTS AND DISCUSSION

The results of the investigation of the hydrogen bonding in cotton by means of its I.R. spectra are shown in Figure 1.

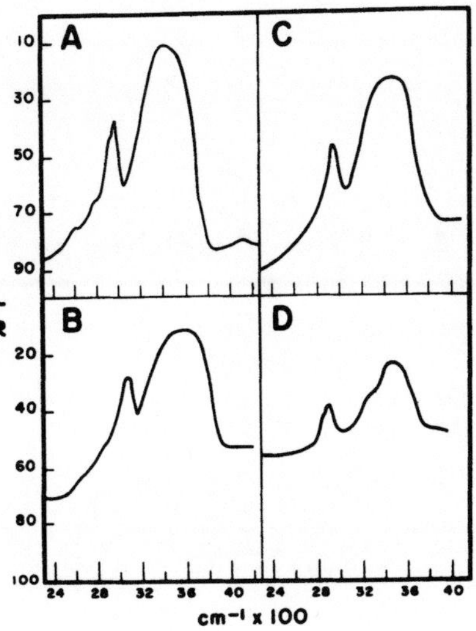

Fig. 1. Infrared spectra of Cotton fibers in the OH region
A untreated
B after acelytation
C after carboxymethylation
D after mercerization

It could be noticed that the original band of O-H stretching vibration in row cotton occurs at 3405 cm^{-1} (2.93µ). As a result of carboxymethylation, this band is shifted to the position 3455 cm^{-1} (2.89µ). Mercerization caused a shift of the same band to a new position at 3475 cm^{-1} (2.87µ). Acetylation again produces a further shift of this band to the position 3575 cm^{-1} (2.8µ). In other words, partial caboxymethylation, mercerization and acetylation of cotton result in a weakening of the degree of hydrogen bonding involved with the consequent shifts of their absorption bands. It seems probable that the reactions involve the amorphous regions of the cotton and remove a greater proportion of the more highly hydrogen bonded O-H groups in this region. The result, as observed, is a shift in the position of the unresolved band towards the less highly bonded crystalline O-H groups at shorter wavelengths.

In Figure 2 the absorption bands are shown after a series of grinding. The band at 900 cm^{-1} (11.11µ) appeared in the 1st. stages as a shoulder and then progressively increased in intensity with the increase of the time of grinding. The increase of the grinding time has caused a detectable effect on the bands at 1035 cm^{-1} (9.65µ), 1060 cm^{-1} (9.43µ) and 1115 cm^{-1} (8.89µ). These bands appeared at 1st. as weak sharp peaks and on decrystallization they gradually disappeared, giving a broad absorption band. It can be readily observed that the bands at 1245 cm^{-1} (8.03µ), 1285 cm^{-1} (7.78µ), 1345cm^{-1} (7.43µ) and 1435cm^{-1} (6.96µ) gradually decreased in intensity and finally disappeared.

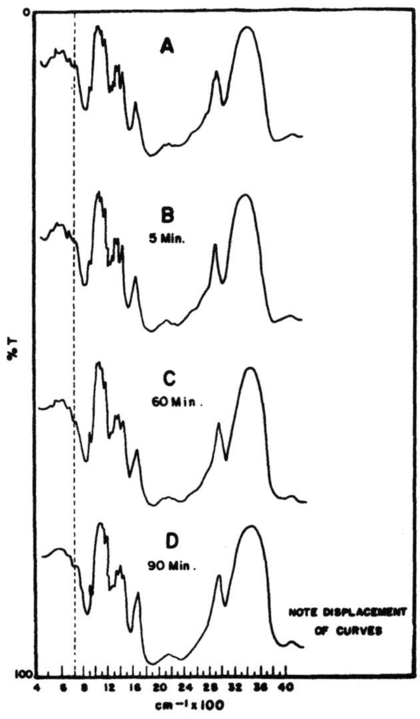

Fig. 2. Effect of grinding on the absorption bands.

A ≡ control sample
B ≡ 5 min. grinding
C ≡ 60 min. "
D ≡ 90 min. "

Definite absorption maxima of nearly equal intensity were found at 1325 cm^{-1} (7.54μ) and 1380 cm^{-1} (7.24μ) and with grinding these two bands became more distinct and the absorption at 1380 cm^{-1} (7.24μ) became considerably more intense than that at 1325 cm^{-1} (7.54μ).

The two absorption bands at 1435 cm^{-1} (6.96μ) and 900 cm^{-1} (11.11μ) have been chosen for calculations to represent crystalline and amorphous contents, respectively. The absorbances/gm of each band were plotted against the grinding time (Fig. 3).

From this figure it is shown that, during the initial stages of grinding a sharp increase in the intensity of the 11.11μ band (amorphous) takes place. Continued grinding results in a slower rate of change. On the other hand, a sharp decrease in the intensity of the 6.96μ band (crystalline) is shown in the initial stages of grinding, continued grinding results also in a slower rate of change.

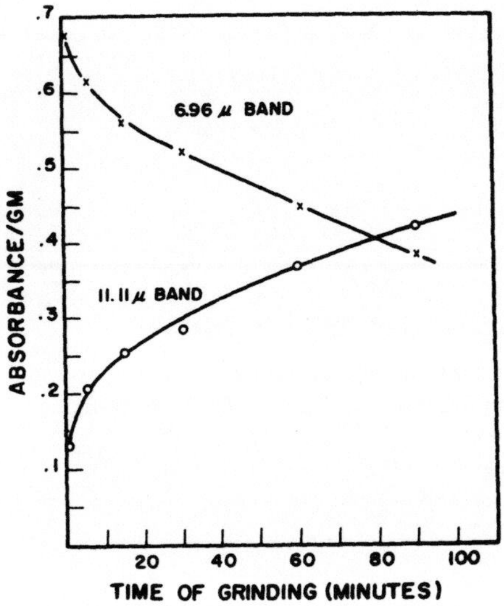

Fig. 3. Relation between time of grinding and absorbances/gm for 1435 and 900 cm^{-1} after grinding.

Moreover, the absorbances/gm of these two bands were plotted against each other as shown in Figure 4. A straight line was obtained. This phenomenon indicated that I.R. spectroscopy provides an independent self-calibrating measurement of percent crystallinity.

In Fig. 5 the effect of acid hydrolysis on the infrared bands of cotton fibers could be observed. In the spectrum of raw cotton the absorption band at 900 cm^{-1} (11.11μ) became less distinct as the acid concentration increased till it finally disappeared.

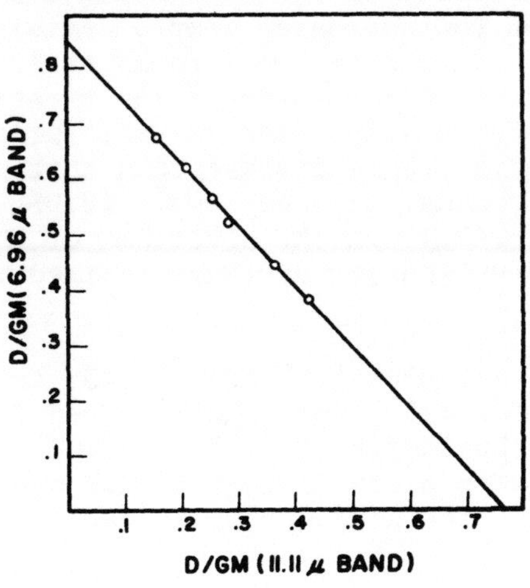

Fig. 4. Relation between absorbances/gm of the two bands 1435 and 900 cm^{-1}.

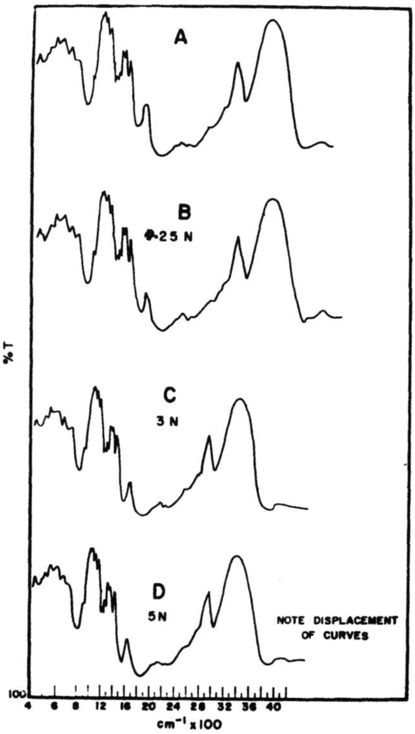

Fig. 5. Effect of acid hydrolysis on the absorption bands.

A = control sample
B = hydrolysis by 0.25N acid
C = " by 3N "
D = " by 5N "

The spectra of the treated samples showed sharper and more clearly defined absorption bands at 1035 cm^{-1} (9.65μ), 1060 cm^{-1} (9.43μ) and 1115 cm^{-1} (8.89μ) than those in the spectrum of untreated sample. In the spectrum of the untreated sample definite absorption maxima of nearly equal intensity were found at 1325 cm^{-1} (7.54μ) and 1380 cm^{-1} (7.24μ). In the spectra of the treated samples, these two bands became less distinct and the absorption at 1325 cm^{-1} (7.54μ) became considerably more intense than that at 1380 cm^{-1} (7.21μ). The absorption bands at 1245 cm^{-1} (8.03μ), 1285 cm^{-1} (7.78μ) and 1345 cm^{-1} (7.47μ) in the spectrum of the untreated sample became considerably sharper and more defined in the spectra of the treated samples.

The crystalline and amorphous absorption bands at 1435 cm^{-1} (6.96μ) and 900 cm^{-1} (11.11μ) respectively were again chosen for calculations. The determined absorbances/gm of each band and the corresponding acid concentrations were plotted against each other in Figure 6. In the initial stages, the acid action is rapid and results in:

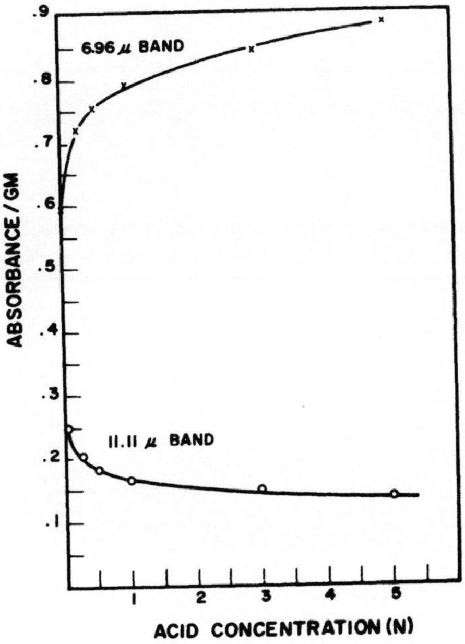

Fig. 6. Relation between acid concentration and absorbances/ gm of 1435 and 900 cm^{-1} bands.

a) Sharp increase in the intensity of the 6.96μ band (crystalline).

b) Sharp decrease in the intensity of the 11.11μ band (amorphous).

Continued increase of acid concentration results in a slower rate of changes.

The absorbances of the 6.96 and 11.11μ bands representing the crystalline and amorphous contents have been plotted against each other in Figure 7. The points fall on a straight line which means that, the crystallinity of cotton fiber can be measured with great sensitivity by means of infrared absorption spectra.

Oxidation of cotton cellulose and hydrocellulose prepared from it, was studied using a mixture of potassium dichromate and sulphuric acid. The spectra of these oxidized samples showed an absorption band at 5.73μ (1745 cm^{-1}) which was identified as C = O stretching band, Figures 8 and 9.

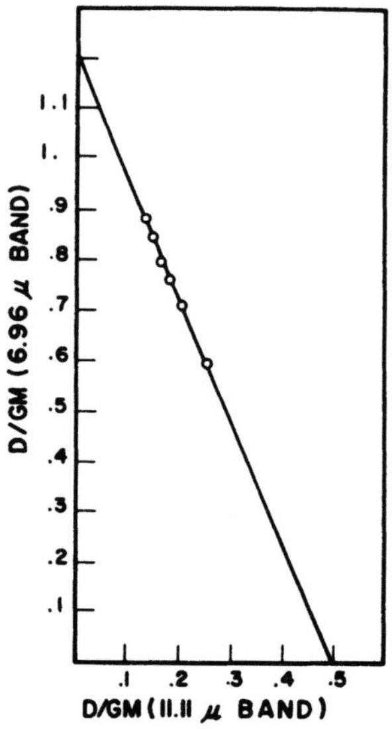

Fig. 7. Relation between absorbances/gm of the two bands 1435 and 900 cm-1 after hydrolysis.

Fig. 8. Effect of oxidation on the absorption bands of cotton.

A = 1 hr oxidation
B = 3 hr "
C = 19 hr "
D = 24 hr "

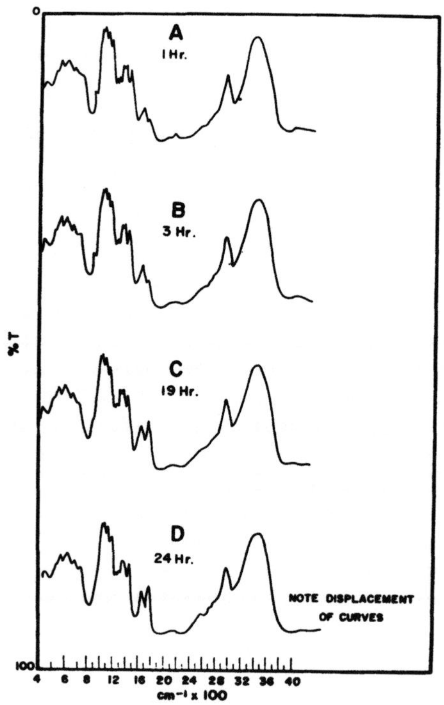

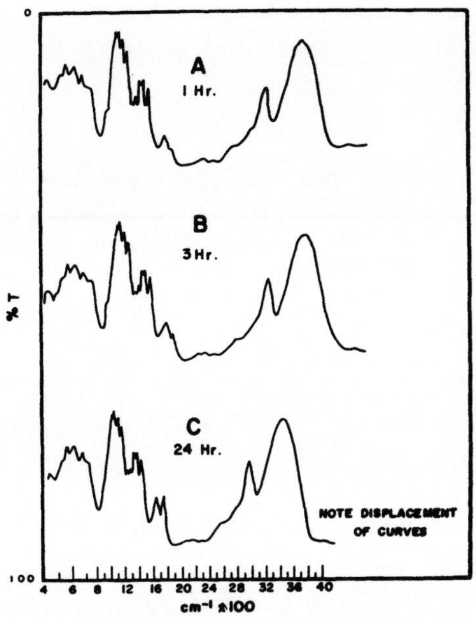

Fig. 9. Effect of oxidation on the absorption bands of hydrocotton.

A = 1 hr oxidation
B = 3 hr "
C = 24 hr "

The relation between the absorbance of this band and the period of treatment with the oxidizing agent for both cotton cellulose and hydrocellulose is shown in Figure 10. It can be seen that both cotton cellulose and hydrocellulose show the same behavior. However, for any given period of treatment, the intensity of the C=O stretching group for cotton cellulose is always higher than that for hydrocellulose.

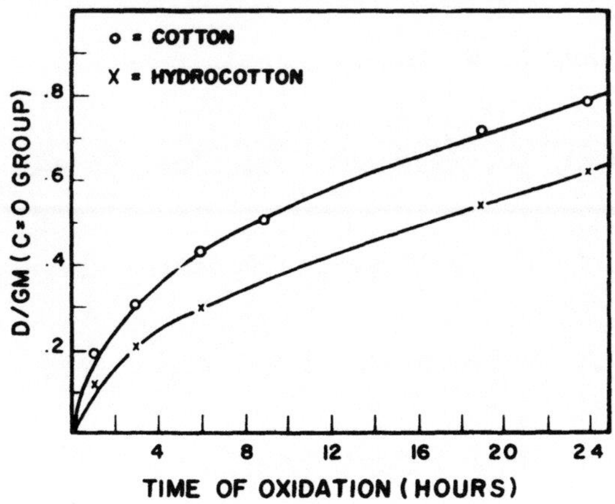

Fig. 10. Relation between time of oxidation and absorbances of the 1745 cm^{-1} (C=O stretching band) for cotton and hydrocotton.

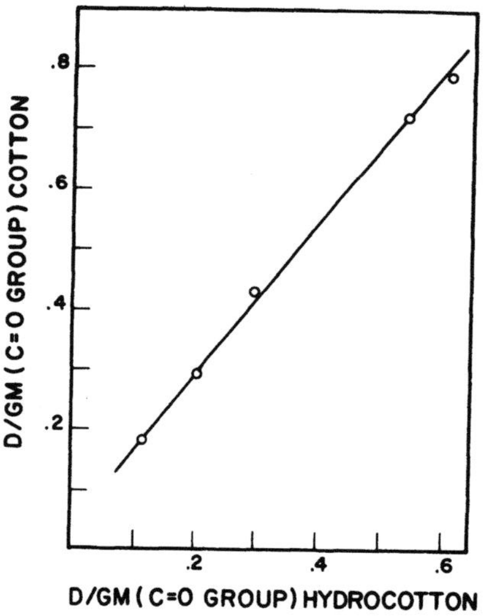

Fig. 11. Relation between absorbances/gm of c=o band for cotton and hydrocotton after oxidation.

The absorbances of both oxidized cotton cellulose and hydrocellulose for different periods of times are plotted against each other in Fig. 11. The linear relationship obtained indicates that a similar mechanism of oxidation reactions takes place with both fiber substances, but rates of reaction for the two fibers are different. The ratio of the absorbances of oxidized cotton cellulose and hydrocellulose (prepared from it) is approximately 1.26.

REFERENCES

1. *M.R. Harvey, J.E. Stewart and B.G. Achhammer, J. Research Nat. Bur. Standards, 56, 255 (1956).*

2. *R. T. O'Connor, E. F. Dupac, and E. R. Mecall, Anal. Chem. 29, 289 (1957).*

3. *F. H. Forziati, and J. W. Rawen, J. Research Nat. Bur. Standards 46, 38 (1951).*

4. *F. H. Forziati, J. W. Rawen and E. K. Player, J.Research Nat. Bur. Standards 46, 288 (1951).*

5. *L. Brown, P. Holliday, and I. F. Trotter,J. Chem. Soc. 1532 (1951).*

6. *H. J. Marrinan and J. Mann, J. Appl. Chem. (London) 4, 204 (1954).*

7. M. Tsubio, J. Polymer Sci. 25, 159 (1957).

8. H. G. Higgins, C. M. Stewart, and K. J. Harrington, J. Polymer Sci. 51, 59 (1961).

9. R. Jeffries, J. Polymer Sci., 4, 375 (1963).

THE EVOLUTION OF CRYSTAL FIELD PARAMETERS FOR RARE EARTH SALTS

J.H. Van Vleck

Harvard University

Cambridge, Massachusetts, U.S.A.

The status of crystal field parameters is reviewed for the rare earth ethyl sulphates, chlorides, and garnets, the three types of compounds for which the analysis is most complete. From the magnetic and spectroscopic data a reasonably satisfactory set of empirical theories have been evolved by various workers for the ethyl sulphates and chlorides. No set has yet been proposed which describes both the spectroscopic and magnetic properties of the garnets.

Another difficulty is that the observed second and sixth order terms of the crystalline potential differ by an order of magnitude from what one calculates by pure theory with a point charge model.

INTRODUCTION

Of all applications of the theory of magnetism to solids, probably the simplest is that to magnetically dilute rare earth salts. We assume throughout that this dilution is so great that all interaction between elementary magnets can be neglected. In the rare earths, in distinction from the iron group, only a single J manifold need be considered in the first approximation (except for Sm^{+++} and Eu^{+++} which have narrow multiplet intervals). Although we will focus our attention to a considerable extent on magnetic properties, we do so mainly to find out the nature of the energy levels in which the lowest J manifold is split by the crystalline Stark effect. Transitions between the different Stark levels are observable by microwave and infra-red techniques, and so a study of them falls within the conference topic of the interaction of radiation with solids. The big problem is to find a model which will explain both the magnetic and optical properties.

The study of glasses is a subject on which particular emphasis has been placed at the American University in Cairo. Some very interesting and practical questions regarding crystalline energy levels arise when small amounts of rare earths are introduced into glasses. For instance energy transfer between two kinds of rare earth impurities in glasses, can implement laser action as discussed in other papers of the conference. I will, however, confine my attention to materials more simple than glasses. Hopefully this kind of investigation may ultimately throw light on the more complicated behavior of rare earth energy levels in a glass matrix.

Since we are dealing with materials in which all cooperative actions between the elementary magnets can be disregarded, we have to deal only with a one-atom problem. Perhaps I should say one-ion problem, as all the rare earth atoms with which I shall be concerned are triply ionized. When I speak of a rare earth, a trivalent ion is to be understood throughout. The effect of other atoms in the crystals on the rare earth ion can, at least in the first approximation, be represented by a crystalline potential of the form

$$V = \Sigma_i V(x_i, y_i, z_i), \tag{1}$$

where the x_i, y_i, z_i are the coordinates of the 4f electrons whose incomplete shell is responsible for the magnetism. Since this potential can be developed in spherical harmonics, we have

$$V = \Sigma_i \Sigma_{n,m} V_n^m (r_i) Y_n^m (\theta_i, \psi_i) \tag{2}$$

where $V_n^m(r_i) \sim r_i^n$ as long as one assumes that V satisfied Laplace's equation. We assume that the normalization of the Y's, and correspondingly the units in which the V_n^m are expressed are the same as employed by Stevens and elaborated by Baker and Bleaney, as their operator equivalent usage has become quite standard (1). As long as we are dealing with a single configuration, we may replace $V_n^m(r)$ by its expectation value, and when we give numerical values of V_n^m, we mean really those of $(V_n^m(r))$.

In the rare earths, the azimuthal quantum number of the 4f electrons has the value 3, and it is readily shown that for this reason the summation in equation 2 need not be extended to values of n higher than 6 so that the problem of the crystalline potential is greatly simplified. (In the iron group, the magnetic electrons are 3d instead of 4f, and so it is not necessary there to go beyond n = 4, but the resulting simplification is more than offset by the fact that it is essential to consider more than one component of a spin-orbit multiplet).

The advent of quantum mechanics furnished the machinery for calculating energy levels and magnetic susceptibilities with the crystalline field potential (1). To include magnetic effects, one merely adds to the Hamiltonian function a term $g_J \beta J_m \cdot H_m$, so that the secular problem is simply to determine the eigenvalues of the Hamiltonian

$$\mathcal{H} = V_{crys} + g_J \beta J_m \cdot H_m \ . \tag{3}$$

As long as we restrict ourselves to a given J manifold, this is only the problem of finding the roots of an algebraic problem of finite degree. The work of doing this is greatly facilitated by the Kramers theorem on double degeneracy when there is an odd number of electrons, and by group theory, as first applied to crystalline energy levels by Bethe.

In the early 1930's there was not enough experimental information to warrant use of more than the most primitive forms of equation 3. The natural assumption to make was that there was reasonably good convergence in powers of the ratio of the 4f radius r to the lattice spacing R, or in other words, it should be a good approximation to keep only the non-vanishing terms of lowest order in r/R for the symmetry characteristic of the given crystal. Thus for some of the most common types of symmetry, the form of equation 2 was taken to be

axial (or hexagonal) $$V(x_i, y_i, z_i) = A(x_i^2 + y_i^2 - 2z_i^2) \tag{4}$$

rhombic or monoclinic $$V = Ax_i^2 + By_i^2 - (A+B)z_i^2 \tag{5}$$

cubic $$V = D(x_i^4 + y_i^4 + z_i^4 - \frac{3}{5} r_i^4 \tag{6}$$

It was on this basis that Penney and Schlapp (2), in their pioneer paper in 1932, gave a primitive explanation of the magnetic behavior of a number of rare earth salts at low temperatures, mainly Cabrera's measurements on hydrated sulphates. They found that an axial or rhombic potential (equation 4 or 5) would not describe the data satisfactorialy, but that on the other hand a fourth order cubic potential (equation 6) would work quite well. In some ways this seems rather surprising, for the crystals of the form $(R)_2(SO_4)_3 8H_2O$ have monoclinic rather than cubic symmetry. However, it has become increasingly clear over the years that often the crystalline fields are nearly cubic, corresponding to a nearly octahedral or tetrahedral coordination of the neighboring atoms, even though the actual symmetry is lower. Also powder susceptibilities, the main type of data available in 1932, are rather insensitive to deviations from cubic symmetry.

During the late 1930's and of course the period of the world war, comparatively little progress was made in the spectroscopy of rare earth salts. After the war the availability of microwave equipment, thanks to radar, made a new technique available. Also much more work has been performed in recent years on straight optical spectroscopy, even though this was a method that could have been pursued with vigor 35 years ago. The percentage of physicists interested may not be much greater today, but there are so many more physicists that in any case the momentum is greater. We will for simplicity confine our attention to three types of rare earth salts, viz., the ethyl sulphates, the chlorides and the garnets. These are the three for which the parameters of the crystalline field have been examined the most thoroughly, although there also exists a considerable amount of data for other salts such as the acetates, double nitrates, chelates, and perovskites. In principle, the simplest rare earth salts should be the nitrides RN, as they have a simple rock salt structure, but I am told that single crystals of them are hard to prepare--in any case the literature is not as extensive as for the three salts that I will discuss. A considerable amount of work has been done on rare earth ions doped into CaF_2, which has a cubic structure. However, the fact that divalent calcium is replaced by a trivalent rare earth introduces problems of charge compensation, in virtue of which the cubic character of the environment is usually spoiled. The sulphate octahydrates which furnished the bulk of the magnetic data in the prewar era are now rather in disrepute and neglected, for they have eight atoms in the unit cell, and only monoclinic symmetry (3). In 1932 the crystallographic, magnetic, and spectroscopic data were so lacking that Penney and Schlapp were able to base their calculations on one atom in a cubic unit cell, and with this simple model they were able to fit the existing magnetic data, which were so sketchy that the parameters of the crystalline field were underdetermined. Nowadays for the materials to be discussed, one is usually in a situation in which they are overdetermined, so that real consistency tests are possible.

THE RARE EARTH ETHYL SULPHATES

The ethyl sulphates have the structure $R(C_2H_5SO_4)_3 9H_2O$ and possess the great merit of having only one atom in the unit cell. It is for this reason that the magnetic resonance work at Oxford after the war concentrated heavily on this material. The local symmetry at the rare earth site is C_{3h}, so that the general formula 2 for the crystalline potential reduces to

$$V = V_2^0 Y_2^0 + V_4^0 Y_4^0 + V_6^0 Y_6^0 + V_6^6 (Y_6^6 + Y_6^{-6}) \tag{7}$$

and so involves only four constants V_2^0, V_4^0, V_6^0, V_6^6, proportional respectively to (r^2), (r^4), (r^6), (r^6), but it is not feasible to try and calculate them from pure theory. The following six kinds of experimental measurements can be used to evaluate them:

a) the positions of crystalline Stark energy levels as obtained directly from infra-red spectroscopy;

b) the positions of these levels as deduced indirectly from optical spectroscopy;

c) the g-factors associated with the Zeeman effect as obtained by microwave resonance;

d) determinations of the magnetic susceptibility;

e) specific heats;

f) activation energies associated with the Orbach process in paramagnetic relaxation.

All six methods have their limitations. In (a) or (e), the absorption or specific heat associated with the transition between Stark components must be resolved from a background arising from phonon effects, achieved only by a difficult comparison with a diamagnetic crystal in which the rare earth ion is replaced by a non-magnetic ingredient of similar size. In (b) the bulk of the energy change is associated with the excitation of an electron out of a 4f state to a much higher level, in which the Stark splitting arising from the crystalline potential merely gives a modulation or fine structure. It is not easy to determine or isolate the contribution of the lower or 4f state. Lines which arise from phonon coupling and which are spurious for our purposes have plagued the theorist interpreting the experimental results ever since the 1930's. Method (c) is exceedingly informative when it can be applied, but it is of only limited scope, since the magnetic resonance method can be used only for the lowest state. The upper states are excited only at high temperatures where the lines become excessively wide. Also in an even ion the lowest level may be non-degenerate and show no resonance. Method (d) is much more instructive if the susceptibility measurements are made on a single crystal rather than a powder. However, the theoretical expression for the susceptibility is a statistical expression involving the properties of a mixture of states. Consequently the individual levels cannot be studied separately, except, the lowest state, which can, of course, be isolated by going to very low temperatures. Method (f) is the most indirect and uncertain of all.

The first attempt to interpret the microwave and other data extensively for the various rare earth ethyl sulphates was made by Elliott and Stevens (4). They suggested that the crystalline field parameters might vary with the atomic number Z according to a simple power law, $V_n^m \sim (Z-55)^{-n/4}$. It has subsequently turned out that actually the dependence on Z is more complicated. The most comprehensive analysis which has been published to date on crystal parameters in the rare earths is that by Hüfner (5). His graph of how the parameters vary as one passes from ion to ion is shown in Figure 1. It is surprising that the curves for the different coefficients are so different. At least the curves seem to be fairly smooth and regular. Unfortunately some of the regularity is spoiled if one uses instead the parameters deduced subsequently by Hill (6) at Yale on the basis of a careful analysis of infrared measurements, which is, as already intimated, a more straightforward and less ambiguous procedure than using frequency differences from optical spectroscopy.

Some idea of the degree of accuracy with which the parameters of the crystalline potential have so far been deduced can be found by comparing the crystalline values obtained by Hüfner with those of Hill. This is done in Table I. Also included are some selected values reported by various investigators for other ions. The results given for Ce are earlier than for the other ions and are based mainly on magnetic rather than optical data, as but little work has been done on this ion recently. Despite having only one electron, it is difficult to treat because of the proximity of the 4f and 5d levels.

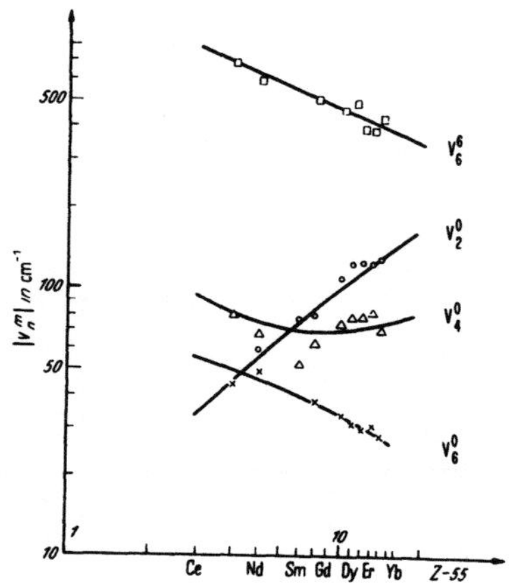

Fig. 1. Crystal field parameters for rare earth ethyl sulphates, according to Hüfner (5).

TABLE I

Parameters of the Crystalline Potential in cm^{-1} (equation 7)

Ion	V_2^0	V_4^0	V_6^0	V_6^6	Investigator
Ce	-15 ± 10	-40 ± 12	-92 ± 2	660	Elliott and Stevens (4)
Pr	23	-80	-44	695	Hufner (5)
Nd	60	-69	-50	600	Hufner (5)
Sm	78	-53	---	---	Hufner (5)
Sm	61	-64	-40	575	Hill (6)
Eu	126	-82	-31	387	Judd (7)
Tb	110	-75	-34	465	Hufner (5)
Dy	124	-79	-31	492	Hufner (5)
Dy	144	-85	-33	535	Hill (6)
Ho	125	-79	-30	376	Hufner (5)
Er	119	-74	-30	376	Hill (6)
Er	126	-81	-31	387	Erath (8)

Generally the various workers agree on the signs of the terms and their approximate size. What is comforting is that the crystalline field parameters, as deduced from optical or infra-red spectroscopy, yield curves for the susceptibility of single crystals as a function of temperature which are in reasonably good accord with experiment. The needed susceptibility measurements have been wanting until recently, but data on the ethyl sulphate of Pr have been obtained by Hellwege et al. (9), and of Nd and Er by Cooke, lazenby, and Leask (10). The curves calculated from the optically determined parameters agree gratifying well with observation. This is shown in Figure 2 for erbium ethyl sulphate. There is also reasonably good agreement with the magnetic resonance data for these three ions, as well as with specific heats, of which measurements are available only for Er. Unfortunately, similar comparisons with susceptibility curves are not possible for the other ions (except Ce), as the requisite measurements have not been made. However, in view of the consistency obtained for the other three ions it would appear that, apart from relatively minor

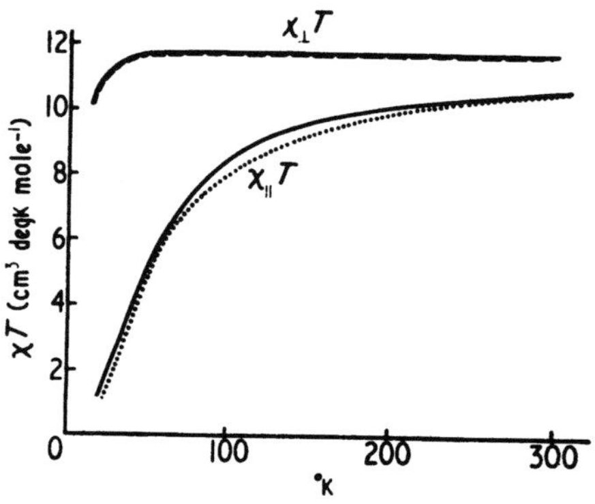

Fig. 2. Magnetic susceptibility of erbium ethyl sulphate parallel and perpendicular to the axis of symmetry as a function of temperature. Broken curves are experimental (10), solid ones are calculated with Erath's (8) crystal field parameters given in Table II.

corrections and perhaps the exceptional case of Ce, real physical significance can be attached to the crystalline field parameters given in Table I. Where different workers obtain somewhat different values it would be interesting to examine in more detail which choice gives optimum agreement with the susceptibility and resonance data.

RARE EARTH CHLORIDES

The chlorides (RCl_3) like the ethyl sulphates have only one atom in the unit cell, and local symmetry of type C_{3h}, so that the crystal potential again has the form given in equation 7. Not quite as much work has been done in evaluating the parameters from combined spectroscopic and magnetic data as in the case of the ethyl sulphates. Also there are no infra-red measurements.

Figure 3, taken from a paper by Burns (11), shows crystalline field parameters for the various ions based largely on the extensive and careful spectroscopic work by Dieke's group at Johns Hopkins. The figure does not exhibit V_6^6, which turns out to be consistently almost exactly $-10V_6^0$. The values in Fig. 3 are based on the optical work of various investigators, and do not include the recent and very careful analysis by Eisenstein (12), who obtained values somewhat, but not greatly, different from those shown in Figure 3. In any event the parameters are seen to vary somewhat irregularly rather than smoothly in going from ion to ion. Explanation of the irregularities would apparently require corrections for configuration interaction of almost prohibitive difficulty. Until recently, susceptibility data on single crystals with which to compare the spectroscopic values has been wanting, but measurements on $CeCl_3$ and $PrCl_3$ have been recently reported by Schneider (13). He does not calculate in detail the curves of susceptibility versus temperature from crystalline field para-

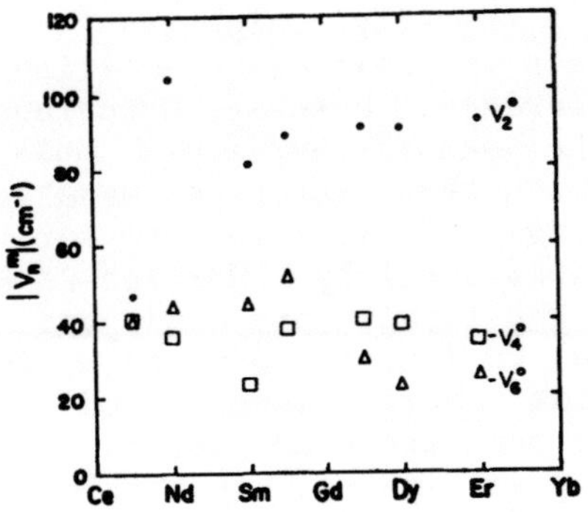

Fig. 3. Crystal field parameters for rare earth chlorides, as given by Burns (11).

meters, but shows that in $PrCl_3$ there is reasonable consistency between the optical resonance and susceptibility data. Reliable parameters for $CeCl_3$ appear to be wanting.

It should be emphasized that in the recent literature on the crystal field parameters in the rare earth chlorides, an attempt is made to fit not merely the Stark components of the lowest J manifold, or even of all the different J manifolds belonging to the lowest L, S state. Instead a fit is sought to all the different Stark levels belonging to all the various Russell-Saunders states of the given configuration (e.g., ^1I, ^3H, ^1G, ^3F, ^1D, ^3P, ^1S for Pr), each of which has its own J manifolds and Stark components thereof. This is an exceedingly ambitious program, as there are only eight disposable parameters (viz., beside the four parameters of Eq. 7, the spin-orbit coupling constant and three Slater F's which may be somewhat different in the solid than in the free ion). Each ion has then a great many levels, not all of which have been found. Even so, Eisenstein in one case tries to fit over 70 levels with these 8 parameters--an exceedingly ambitious undertaking. The fact that he has a standard deviation of a few wave numbers in a totality of levels spanning over 10,000 cm^{-1} is not surprising in view of the neglect of the ever-present but intractible configuration interaction. It is a triumph of crystal field theory that the fit is as good as it is. Because of configuration interaction, the parameters may vary slightly from one J manifold to another, and any small discrepancy with magnetic susceptibility data could doubtless be explained away on the ground that the best parameters for the lowest-J manifold may be different from those that give the best fit in the aggregate for a vast number of upper states.

As compared with the ethyl sulphates, the chlorides have the advantage that the crystal structure is simpler, and it should in principle be possible to compute the constants of the crystalline potential from a point charge model. In the ethyl sulphates there are so many atoms, with such uncertainty in their exact position, that this is not feasible. For the chlorides, however, Hutchings and Ray (14) have made a comparison between the empirical parameters such as are shown in Figure 3 with the theoretical results obtained from a point charge model using the values for the various $\langle r \rangle^n$ calculated for the rare earth 4f shell by a Hartree-Fock self-consistent field procedure (15). The results are disappointing. The empirical ratio of the sixth to fourth order terms is about ten times that given by the point charge model, and the ratio of second to fourth order about one tenth as large. The empirical values of the fourth order terms are of substantially the observed order, but the second and sixth order ones are both an order of magnitude off, being respectively too small and too large. An effort to try and understand the discrepancy has been made by Burns (11), but the results are not too definitive. Burns shows that it is very difficult to estimate the second order terms because of uncertainty in the positions of some of the neighboring ions and that for the other terms there are important but rather elusive polarizability corrections, essentially allowance for the fact that the surrounding ions have finite extent instead of being point charges. There is always the possibility of Jorgensen's (16) suggestion of incipient covalence, but this, if important, should presumably have shown up as transferred hyperfine structure in nuclear resonance experiments.

THE RARE EARTH GARNETS

The garnets have six rare earth ions in the unit cell. In that respect they are as bad as the hydrated sulphates which we have already deprecated. However, the garnets have the merit of being macroscopically cubic, and all six of the rare earth sites are equivalent except as regards the orientation of the local field. Also a great deal of experimental and theoretical work has been done on the garnets because of their technological importance. This importance stems from the fact that it is possible to prepare ferrimagnetic garnets in which certain crystal sites, the so-called (a) and (d) ones, can be filled by Fe^{3+} ions which are highly isotropic and so very useful in certain engineering devices, such as Hogan's gyrator. The rare earth ions occupy almost entirely the (c) sites. We will be concerned in the present paper only with paramagnetic crystals in which the (a) and (d) sites are filled by diamagnetic ions, usually gallium or aluminum, so that the structure is $R_3Ga_5O_{12}$ or $R_3Al_5O_{12}$ rather than $R_3Fe_5O_{12}$. Studying the iron garnets provides a great deal of information about exchange forces, but is outside the province of the present paper.

The local field at the rare earth, or (c) sites, has only rhombic symmetry, so that the explicit form of the potential (2) is

$$\begin{aligned} V = {} & V_2^0 Y_2^0 + V_2^2 (Y_2^2 + Y_2^{-2}) \\ & + V_4^0 Y_4^0 + V_4^2 (Y_4^2 + Y_4^{-2}) + V_4^4 (Y_4^4 + Y_4^{-4}) \\ & + V_6^0 Y_6^0 + V_6^2 (Y_6^2 + Y_6^{-2}) + V_6^4 (Y_6^4 + Y_6^{-4}) + V_6^6 (Y_6^6 + Y_6^{-6}) \end{aligned} \tag{8}$$

Since there are nine different constants it has been impractical, at least to date, to determine all of them from experimental measurements. So, to simplify things, one of the following procedures has been adopted.

1) Assume a field of cubic symmetry. Then there are only two parameters for five of the V's vanish, and there are the further relations $V_4^4 = -5 V_4^0$, $V_6^4 = 21 V_6^0$. The cubic approximation is often fairly good for explaining the magnetic susceptibility, but is powerless to describe the extra components which arise when the cubic degeneracy is lifted by the non-cubic parts of the field, and which are revealed by spectroscopic measurements.

2) Assume a field of tetragonal symmetry such as would result if two opposite faces of a cube are twisted in opposite directions, each with four charges at their respective corners. There is x-ray evidence that the eight ions surrounding the rare earth ion have this kind of distorted octahedral arrangement.

3) Take certain of the coefficients (generally V_4^2, V_6^2, V_6^6 and sometimes also V_2^2) equal to zero, and treat the remaining constants as empirically determined parameters.

4) Treat V_2^0, V_2^2, V_4^0, V_6^0 as empirically determined parameters and use values of the ratios V_4^m/V_4^0, V_6^m/V_6^0 determined by making a point charge calculation. This is obviously only a semi-theoretical procedure, and it may seem pretty ad hoc. The justification is that Hutchings and Ray (14) find that in the chlorides the point charge model gives almost exactly the empirical value of V_6^6/V_6^0 despite its failure to give the right magnitude for V_6^0. Presumably this means that the ratios of harmonics of the same order are less affected by polarizability corrections than are absolute values.

The successes and failures of different investigators with the various methods which we have described are indicated schematically in Table II.

TABLE II

Test of Various Models for the Crystalline Potential in Garnets

Method and Investigator	Ions	Spectroscopic Levels	Reson.g Values	Magnet. Suscept.
1. Lin et al.	Yb	-----	approx.	yes
2. Schieber and VanVleck	Tm	-----		yes
3. Ayant et al.	Er	partly	wrong	yes
4. Koningstein and collaborators	Eu,Nd,Tb Er,Tm,Yb	yes	-----	---
5. Hutchings and Wolf; Pearson et al.	Yb	yes	yes	wrong

The entry "yes" means that reasonably good agreement is obtained and "wrong" indicates that the proposed choice of crystalline field model and parameters yields results not in accord with experiment. The entry "partly" means that only the levels of the lowest J manifold have been analysed and described. In all cases, except Yb, there are so many J manifolds and Stark components that it would be hopeless to try and find them all experimentally or describe them theoretically. Hence "yes" in the spectroscopic column does not imply that every level in the various manifolds has been found and fitted. A dash means that comparison of theory and experiment has not been attempted. The magnetic resonance measurements relate only to the lowest Kramers doublet. In even ions, such as Tb, Ho, Tm, the ground level is a singlet, barring accidental degeneracies, rather than a Kramers doublet, and magnetic resonance measurements involving transitions between Zeeman components cannot be made.

In work not yet published, Ayant and collaborators have shown that the difficulty of incorrect g-values for Er with the tetragonal model II, indicated by the entry "wrong" in Table II, can be rectified by adding to the main tetragonal members a small rhombic potential of the type given in our Eq. (5). On the other hand, it should be mentioned that the agreement with susceptibility data previously reported by Ayant et al. with model II for ErGaG is achieved only by using the empirical energy levels for the ground manifold $J = 15/2$ furnished by Tinkham's infra-red measurements, rather than those calculated directly from their crystalline potential, which seems to work best for the excited manifold $J = 13/2$, only important for optical spectra. The entry "yes" for the susceptibility of Er in Table II is hence somewhat misleading.

The difficulty that the ratio for sixth to fourth order terms is an order of magnitude larger than obtained with a point charge model and Freeman-Watson radii, is just as pronounced as in the case of the chlorides. It is hard to explain away the discrepancy as due to polarizability corrections, since the polarizability of O^{--} ions is low.

For further discussion, and more detail regarding the theoretical and experimental work summarized in Table II, the reader is referred to another paper by the writer (17). To avoid duplication, we do not repeat the details of the comparison here. The most elaborate calculation of parameters is that by Hutchings and Wolf, and extended by Herrmann, Pearson, Wickersheim, and Buchanan. For a while it looked as if it might be completely successful, but it finally turned out that it gives a value about 20% too low for the constant a in the formula $\chi = a + (b/T)$ for the susceptibility of the Yb ion at low temperatures, where the Boltzmann factor for upper states is negligible.

Further study is obviously required, and it is to be hoped that this deficiency can be remedied, and that in the not too far distant future a set of empirical crystal parameters will be found somewhere for the garnets which are almost as successful as those for the chlorides and the ethyl sulphates. It will, however, be a long time, if ever, before one can deduce acceptable values of the parameters for any of these salts from pure theory instead of from experiment.

REFERENCES

1. *K.W.H. Stevens, Proc. Phys. Soc. London 65A, 209-215 (1962): J.M. Baker, B. Bleaney and W. Hayes, Proc. Roy. Soc. 247A, 141-151 (1958); M.T. Hutchings, Solid State Physics (Academic Press, N.Y.) 16, 227-273 (1964).*

2. *W.G. Penney and R. Schlapp, Phys. Rev. 41, 194-207 (1932).*

3. *W.H. Zachariasen, J. Chem. Phys. 3, 197-198 (1935).*

4. *R.J. Elliot and K.W.H. Stevens, Proc. Roy. Soc. 215A, 437-453 (1952); Proc. Roy. Soc. 218A, 553-566 (1952); Proc. Roy. Soc. 219A, 387-404 (1952).*

5. *S. Hufner, Z. Physik 169, 417-426 (1962).*

6. *J.C. Hill, "The Far Infrared Spectra of Rare Earth Sulphates" Ph.D. Dissertation, Yale University, 1966.*

7. *R.R. Judd, Molecular Physics 2, 407-414 (1959).*

8. *E.H. Erath, J. Chem. Phys. 34, 1985-1989 (1961).*

9. *K.H. Hellwege, W. Schembs and B.Schneider, Z. Physik 167, 477-486 (1962).*

10. *A.H. Cooke, R. Lazenby and M.J. Leask, Proc. Phys. Soc. London 85, 767-781 (1965).*

11. *G. Burns, J. Chem. Phys. 42, 377-390 (1965).*

12. *J.C. Eisenstein, J. Chem. Phys. 39, 2134-2140 (1963), J. Chem. Phys. 40, 2044 (1964).*

13. *B. Schneider, Z. Physik 177, 179-189 (1964).*

14. *M.T. Hutchings and D.K. Ray, Proc. Phys. Soc. London 81, 663-676 (1963).*

15. *A.J.Freeman and R.E. Watson, Phys. Rev. 127, 2058-2075 (1962).*

16. *C.K. Jorgenson, R. Pappalardo and H. Schmidtke, J. Chem. Phys. 39, 1422-1430 (1963).*

17. *J.H. Van Vleck, J. Phys. Chem. Solids 27, 1047-1051 (1966).*

DISCUSSION

R.A. WEEKS (Oak Ridge National Laboratory, New York, U.S.A.): *Reasonable agreement is found between observed and calculated g-values of* Yb^{3+} *and* Er^{3+} *doped into* ThO_2 *and* CeO_2 *single crystals.*

THE INTERACTION OF LIGHT ABSORPTION AND EMISSION WITH THE PHONON SPECTRUM OF CRYSTALS AND GLASSES

W.A. Weyl, Visiting Professor in Physical Sciences

The American University in Cairo

Cairo, Egypt, U.A.R.

Certain glass colors, such as the cerium titanium yellow and the carbon amber, as well as deviations from Beer's Law for ionic colors, require a dynamic model of the color centers. The quenching of fluorescence by temperature or by higher concentrations of an activator are attributed to the anharmonicity of thermal vibrations. Weyl and Marboe attribute the effect of temperature upon rate phenomena not only to the increase of the available energy (KT) but also to the increase of the anharmonicity component of the thermal vibrations. The latter can be increased by means other than by raising the temperature. On this basis an explanation is presented for the accelerated gold ruby formation when a glass containing atomic gold is irradiated (S.D.Stookey's photosensitive glasses).

INTRODUCTION

In the year 1949 S.D. Stookey (1) published his invention of photosensitive glasses which seemed to be based upon a very unusual kind of photo-reaction: acceleration of material transport through light absorption. There are strong reasons for assuming that introducing small amounts of gold in any form into a soda lime silica glass causes the formation of "gold vapor" in which gold atoms interact only by their relatively strong van der Waals forces (high ionization potential) with the network of oxygen ions This relatively strong van der Waals interaction makes it possible for a glass to keep in solution more gold than would be present in its vapor at the glass melting temperature. Nevertheless, the absolute concentration of gold that can be introduced is small, less than 1/100 of one percent but it increases with the refractive index of the base glass. Due to the greater London dispersion forces, glasses with a high PbO or BaO content are better solvents for gold than soda lime glasses.

On reheating the molded glass, the gold atoms aggregate and go through a colloidal stage which gives the glass a beautiful red to purple color. This process, called the "striking" of the gold ruby, takes place in the region of 500°C to 700°C depending on the glass composition and its thermal history. Stookey discovered that this process can be accelerated by irradiating the glass or parts of it at relatively low temperatures with intense visible and ultraviolet light. Heating after partial exposure to light causes the irradiated parts to develop the ruby color faster than the non-irradiated parts of the same glass.

THE PROBLEM

Sttokey explained the phenomenon by an electron transfer that causes Au^{+} and Au^{3+} ions to become Au^{0} atoms. For this purpose light sensitive ions such as the couple Ce^{3+} - Ce^{4+} or Sn^{2+} - Sn^{4+} had to be introduced. This reasoning seems to be a logical one because it would bring this photochemical reaction in line with many solarization processes which involve an electron transfer.

There are, however, several reasons why it is improbable that this process is a reaction analogous to solarization, namely:

1. Gold ruby glasses are not particularly sensitive to fluctuations in the furnace atmosphere such as occur in technical furnaces. Glasses containing couples such as Mn^{2+} - Mn^{3+} or Cr^{3+} - Cr^{6+} are sensitive to oxidizing and reducing conditions.

2. Our attempts to introduce the intensely yellow Au^{3+} ions into low melting (200-300°C) nitrate glasses showed this color, at the best, only temporarily. The oxidizing character of these melts and the low temperatures should be favorable to the Au^{3+} state.

3. If a substantial part of the gold would participate in a ruby glass as Au^{+} or Au^{3+} ions it is hard to explain why the gold concentration in a soda lime glass cannot be raised to 1%.

4. Many electron transfer reactions are known to occur in glasses but normally solarization processes are reversed at temperatures above 200°C, some of them revert at ordinary temperature in the dark and faster in infrared light.

This raises a difficult question: if electron transfer is not the process by which gold atoms form and aggregate faster under the influence of light what can be the possible mechanism for accelerated striking? Apparently no observations have been recorded which support the view that light affects the diffusion rate of neutral particles such as gold in gold sols. H. Nordenson (2) made extensive studies of the effect of light on the coagulation of gold in aqueous systems and reviewed the early literature on this subject. Light has no effect upon diffusion rates and upon the coagulation of a sol provided it does not raise its temperature. When my colleague, E. C. Marboe (3), and I discussed diffusion phenomena in glasses we arrived at an explanation for the effect of light upon the diffusion of gold that might be of general interest to all workers in the field of radiation.

EFFECT OF THE ANHARMONICITY OF THERMAL VIBRATIONS ON RATE PHENOMENA

The Weyl-Marboe treatment of reaction rates emphasizes specifically the asymmetry component of thermal vibrations in addition to their amplitudes which determine the value of KT available to the system.

As long as one deals with a crystal of a pure substance no such distinction is justified because the temperature determines the phonon spectrum of a single crystal. All modes of thermal vibrations, as well as their amplitudes, change with its temperature. For a heterogeneous or a polycrystalline material this is no longer true. When J.A. Hedvall (4) shocked the scientific world by announcing the formation of Rinmann's green and of other pigments by heating the mixtures of the oxides in the absence of a liquid phase, it was not only the solid state reaction in itself that was surprising but the extremely low temperatures at which they occur. Rinmann's green, cobalt zincate, begins to form at 500°C. Magnesium oxide, a highly refractory material that requires temperatures of more than 1600°C in order to sinter reacts far below 1000°C with CoO or NiO. We attribute the greater reactivity of these mixtures to the anharmonic vibrations or to the disproportionation of forces initiated by the interaction of two different cations, e.g. Mg^{2+} and Co^{2+} with the same oxygen ion. The difference in their masses and in their ionization potentials causes a distortion of the electron density distribution of the O^{2-} ion that in a crystal of MgO is symmetrically surrounded by Mg^{2+} ions. This establishes a relationship between diffusion and reaction rates and the phonon spectrum of a solid, in particular the anharmonicity components of the thermal spectrum at a given temperature.

We assume that an electron transfer between two constituents of Stookey's photosensitive glasses, e.g., a cerium and a tin ion produces a space charge in the rigid glass that enhances anharmonicity and with it the diffusion rate of gold atoms at temperatures where in the absence of radiation the mobility of Au atoms is negligible. Each ion tends to move into a potential well when the glass melt is cooled. The free energy is lowered by bringing all cations into positions of maximum screening by the proper number of anions and by establishing electroneutrality in the smallest possible volume. The phonon spectrum of such a glass will be modified by any absorption process in which one ion donates and a distant one receives an electron. Such an electron transfer process in the glass does not need to involve the gold atoms directly. This interaction is the reverse process of the modification of absorption and emission spectra of color and fluorescence centers by a change of the phonon spectrum of the host lattice.

EFFECT OF THE ANHARMONICITY OF VIBRATIONS ON COLOR AND FLUORESCENCE (TEMPERATURE QUENCHING)

Color centers absorb a part of the electromagnetic spectrum and change radiant energy into heat. As a rule, smaller light quanta can be absorbed by a color center if the temperature is raised. Colorless ultraviolet absorbing crystals such as ZnO turn yellow on heating in a reversible way. The same has been found for ultraviolet absorbing cerium or lead ions in silicate glasses. Yellow crystals (Na_2CrO_4 or CdS) turn orange and red on heating. We attribute these phenomena to the increase of the anharmonicity of thermal vibrations. Anharmonicity or lowering the symmetry of thermal vibrations can be achieved by structural changes in addition to raising the temperature. CdS is yellow at ordinary and colorless at liquid air temperatures; it becomes orange-red at 300°C. W.O. Milligan (5) caused CdS to change its color permanently toward red at ordinary temperatures by disturbing its lattice through shear stresses. One can produce orange and red pigments from CdS by replacing a part of the Cd^{2+} ions by the heavier Hg^{2+} ions or a part of the S^{2-} ions by the heavier Se^{2-} ions. The crystals $Cd_{1-x}Hg_xS$ and Cd $S_{1-x}Se_x$ are orange to red, depending on the value of x. Pure CdSe is grey. In these examples the change of thermal vibrations is the result of the different masses. Partial replacement of Cd by Hg or S by Se changes other properties of the crystal, in particular it increases the refractive index or the polarizability of the system. That masses are actually involved in a color change could be shown by the synthesis of colored crystals with identical structures that contain atoms of chemically nearly identical elements. Because of lanthanide contraction, some atoms, e.g. Ta and Nb, have the same valency, and nearly the same size and ionization potentials but very different masses.

L. Tcheichvili and W.A. Weyl (6) synthesized pigments containing Ta^{5+} ions and corresponding ones with Nb^{5+} ions. Ni^{2+} ions in their maximum coordination with oxygen are either green (aqueous solution) or yellow (some silicate glasses). Using ThO_2

as a host and compensating the difference in excess charges in one structure by Ta^{5+} and in the other by Nb^{5+} ions, it was found that

the pigment $Th^{4+}_{1-3x}\ Ni^{2+}_{x}\ Ta^{5+}_{2x}\ O^{2-}_{2}$ is green

and the pigment $Th^{4+}_{1-3x}\ Ni^{2+}_{x}\ Nb^{5+}_{2x}\ O^{2-}_{2}$ is yellow.

Analogous pigments could be made by introducing into a suitable host lattice Fe^{3+} or Cr^{3+} ions and compensating the difference in charges by Ta^{5+} and Nb^{5+} respectively. The ratio of their masses (Ta/Nb) being 2.0 suggests that the color center involves the Fe^{3+} or the Cr^{3+} ions and the vibrational spectrum of the host and not just the first coordination of six oxygens.

Similar phenomena are well known to exist in glasses. The blue color of Cu^{2+} ions in a soda lime glass corresponds to aqueous solutions. It can be changed into green either by increasing the Cu^{2+} ion concentration or by raising the temperature or by replacing a part of the Si^{4+} ions by the heavier Ti^{4+} or Pb^{2+} ions. Increasing the concentration of Cu^{2+} ions or introducing other heavy atoms into a soda lime silica glass has the same effect as raising its temperature a few hundred degrees: Its phonon spectrum becomes more complex.

Low concentrations of molybdenum, manganous, or ferric ions (<0.1%) do not materially affect the light absorption of a soda lime glass. Molybdenum in an oxygen network is colorless and Fe^{3+} ions or Mn^{2+} ions in an oxygen network are faint yellow. Replacing a small fraction of the O^{2-} ion by the more polarizable and heavier S^{2-} ion produces beautiful amber and deep reddish brown glasses. Before the origin of this color was understood it was attributed to carbonaceous material that had to be introduced for reducing Na_2SO_4 to form S^{2-} ions, hence these glasses are still called "carbon amber" glass. The heavier sulfur atoms participating in an oxygen net work introduce new modes of thermal vibrations which intensity and modify the relatively weak absorption of the MoO_6, MnO_6 and FeO_6 polyhedra through an asymmetry: $Fe\ O_{6-x}\ S_x$. The value of x has to be small, otherwise the metal sulfides precipitate as a separate phase and black glasses result.

The origin of the "carbon amber" color is the same as that of the cerium titanium amber where the colorless ultraviolet absorbing CeO_6 groups become absorbers for violet and blue radiation if about 5% TiO_2 is substituted for SiO_2. This Ce-Ti amber colored glass was developed by W.C. Taylor (7).

Color centers which absorb light and cannot readily return from their excited state to their normal state are likely to fluoresce. For glasses the mechanism of fluroescence of Mn^{2+} ions was elucidated by S.H. Linwood and W.A. Weyl (8). N.J. Kreidl (9) used

Mn^{2+} ions as fluorescence indicators for studying the effects of composition changes upon the constitution. The Mn^{2+} ion in fourfold coordination emits green light and in sixfold coordination red light.

The ability of MnO_4 and MnO_6 groups to dissipate their absorbed energy in the form of heat depends upon the complexity of the phonon spectrum. Introducing heavy atoms such as Ti or Pb and even Mn itself makes the phonon spectrum more complex and can increase the probability of the excited Mn^{2+} ion returning to a lower energy state without light emission.

CONCENTRATION QUENCHING OF FLUORESCENCE

Thermal quenching involves not only the absolute temperature, or the amplitude of the thermal vibrations but also their anharmonicity. For a given temperature the anharmonicity can be increased by introducing heavier atoms into a host. Even the activator atoms themselves modify the structure and have a quenching effect upon the fluorescence of glasses and crystals.

For a given parent glass containing an optimum concentration of Mn^{2+} ions, the fluorescence intensity decreases when more Mn^{2+} ions are added or when the glass is heated. Depending on the participation of the Mn^{2+} ions in the structure of the glass, MnO_4 groups and MnO_6 groups show different sensitivity to concentration quenching and to temperature quenching.

The phenomenon has its analogue in crystalline phosphors and in solution. For systems in which the activator is an organic molecule, a third kind of quenching can be observed that has been called internal quenching. The strongly fluorescent sodium salt of resorcinephthaleine (uranine, fluoresceine) lowers its fluorescence intensity when some hydrogen atoms are replaced by the heavier atoms chlorine, bromine, and iodine. These substitutions within the molecule affect its vibrational spectrum. It can absorb smaller light quanta and can interact with the solvent so that emission of the absorbed energy becomes unnecessary. The yellow color of the fluoresceine is changed to red by adding bromine (eosine) and to purple (rose bengale) by introducing both Cl and I atoms. This substitution causes the fluorescence to vanish.

EFFECT OF ATOMIC MASSES ON OPTICAL PROPERTIES

A similar shift of light absorption was observed by the author when he replaced the proton in nitrophenol by Li^+, Na^+, K^+, Rb^+ and Cs^+ ions. Nitropenol is yellow but its cesium salt is deep red. The colors of other alkali salts lie between yellow and red. Halochromie is well known for violuric acid. It would be very difficult to explain the colors of the Li salt (orange), K salt (blue) and Cs salt (green) by major structural changes in the violuric acid. In their different vibrations these ions must affect the light absorption by electronic shifts, polarization or dipole induction.

The light absorption of the platinum cyanide complexes is vested in the $(Pt\ (CN)_4)^{2-}$ group. The hydrated magnesium salt can assume a great variety of colors depending upon its water content. For the trihydrates of the alkali salts the colors are

$Na_2Pt(CN)_4 \cdot 3H_2O$ colorless

$NaKPt(CN)_4 \cdot 3H_2O$ orange

$LiKPt(CN)_4 \cdot 3H_2O$ red

$LiRbPt(CN)_4 \cdot 3H_2O$ yellow, green fluorescence

The different colors are the results for the different polarization of the $Pt(CN)_4$ group by different alkalies and for color and fluorescence it is not the average polarization that counts but the maximum degree of polarization that occurs during thermal vibration.

REFERENCES

1. *S.D. Stookey, a) J. Am. Ceram. Soc. 32, 246 (1949).*
   b) Ind. Eng. Chem. 41, 856 (1949).

2. *H. Nordenson, Z. Phys. Chem. 90, 603 (1915).*

3. *W.A. Weyl and E. C. Marboe, The Constitution of Glasses, A Dynamic Interpretation, Vol. II, page 723, Interscience Publishing Company, New York 1964.*

4. *J.A. Hedvall, Ber. Deutsch. Chem. Ges. 45, 2095 (1912).*

5. *W.O. Milligan, J. Phys. Chem. 38, 797 (1934).*

6. *L. Tcheichvili and W.A. Weyl, a) The Glass Industry 43 (1963), January to March.*
   b) Bol. Sociedad Espanola de Ceramica 4, 461 (1965).

7. *W.C. Taylor, U.S. Patent 1, 292, 147 (1919).*

8. *S.H. Linwood and W.A. Weyl, J. Optical Soc. Am. 32, 443 (1942).*

9. *N.J. Kreidl, J. Optical Soc. 35, 249 (1945).*

MICROWAVE EMISSION FROM INDIUM ANTIMONIDE

H. Motz

St. Catherine's College

Oxford, England

Microwave emission was recently observed from Indium Antimonide in the 8 mm. band at nitrogen temperatures. Similar observations have also been reported by Larrabee and Ancker-Johnson.

Other emission phenomena have been reported by various authors. In the paper experimental evidence is reviewed in detail and it is shown that emissions occurring when high DC electric fields are applied have a character different from those at low electric fields. Various theories have been advanced which involve instabilities of steams of electrons and holes moving in opposite directions. These theories are reviewed and it is seen that they cannot fully explain what is observed. We have advanced a further development of instability theory which involves a high density gradient likely to exist near the semiconductor surface. This may explain some of the experiments obtained with very thin samples and relatively low magnetic fields. Thicker samples in high magnetic fields show emission which is fairly independent of magnetic field values which may be due to photo-mixing of recombination radiation.

INTRODUCTION

Microwave emission from indium antimonide was first observed by Larrabee (1) (1964) and more fully reported by Larrabee and Hicinbotham (2). A number of other authors have pursued the subject experimentally and theoretically, but the mechanism of the emission process is by no means clearly understood to date. Much useful information can be found in an earlier review by Ancker-Johnson (3). Several theories have been proposed and it seems likely that some of them are involved, separately under some experimental conditions and jointly in other cases.

All the authors use samples of rectangular cross sections with a length of the order of one or several millimeters. Typical dimensions of the cross section range from 0.016 mm. to 0.5 mm. In some cases thin platelets are used, e.g. with dimensions 3x0.49x0.16 mm. The samples are cut from high purity n-type or p-type InSb of carrier concentration $5x10^{12}cm^{-3}$ - $1.3x10^{14}$ and mobility $3x10^{3}cm^{2}/V$ sec for p type and $1.7x10^{5}cm^{2}/V$ sec for n type material.

Experiments have been carried out with the samples placed in a magnetic field up to several thousand gauss oriented at various angles with respect to the longest dimension. In other experiments the magnetic field is perpendicular to the longest dimension, which we shall call the axis, and rotated around the cross section.

In all cases a D.C. field applied in the axial direction ranging up to several hundred volts/cm (600 V/cm) causes an axial current to flow in the sample. This current tends to heat up the sample, and for this reason voltage pulses of a duration of the order of a microsecond are applied to restrict the heating effects. Most experiments are carried out at 77°K. Typically the sample is placed in a waveguide with the axis across the narrow dimension of the guide as shown in Figure 1 and the D.C. connections are made by indium joints or other good ohmic contacts.

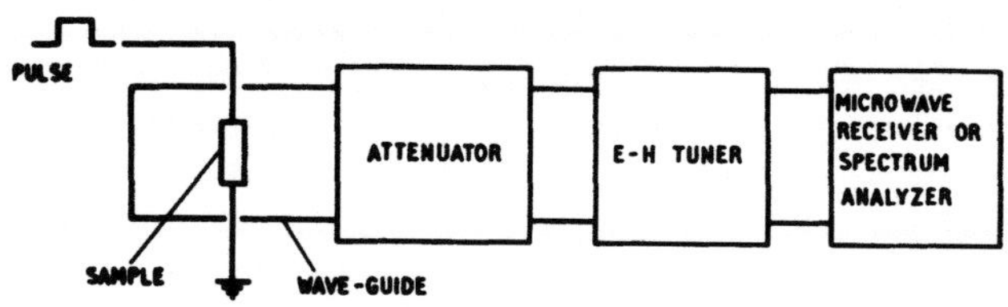

Fig. 1. Measurement set up Block diagram

As far as the author knows, no microwave emissions have been observed in the absence of a magnetic field, although the enhancement of microwave transmission has been observed by Lazarus in our laboratory and the enhancement of reflection of microwaves by a sample has been reported by Köchner (4).

Fig. 2 shows an oscillogram obtained by the former, recording the amplified microwave signal at 75 Gc/sec passed through a waveguide down into the nitrogen dewar, inside which a sample was mounted, and up again through a parallel guide, the coupling of the two guides being provided by an opening with a coaxially mounted sample of rectangular cross section. The lower trace records the D.C. pulse voltage applied (deflections are downward in each case).

Fig. 3 shows an oscillogram obtained by Köchner. The smaller trace is the (amplified) microwave signal after reflection from a sample mounted across a guide and the upper trace records the D.C. pulse (here deflections are upward). The frequency was 35 Gc/sec. Köchner (4b) has also measured the conductivity of n-type single crystals at 35 Gc/sec. He measured the Q-factor of a resonator with a hole in the wall covered by the sample. He found that, within a narrow region (100-225 V/cm) of applied voltages the conductivity depended on the orientation of the D.C. field with respect to the RF field, thus indicating a microwave interaction.

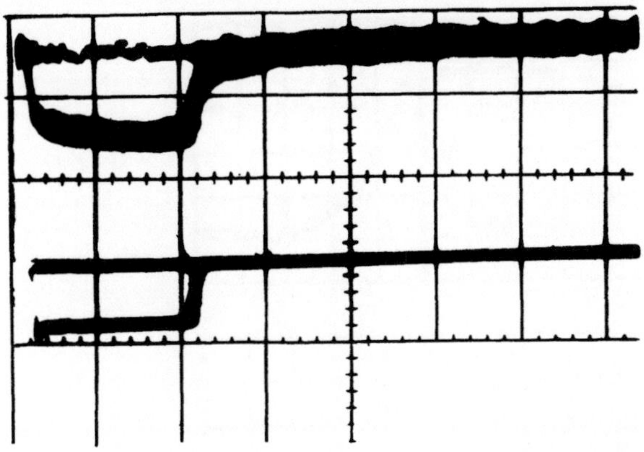

Fig. 2. Transmission 75 Gc/sec. 200 V/cm. 1 μsec/div.

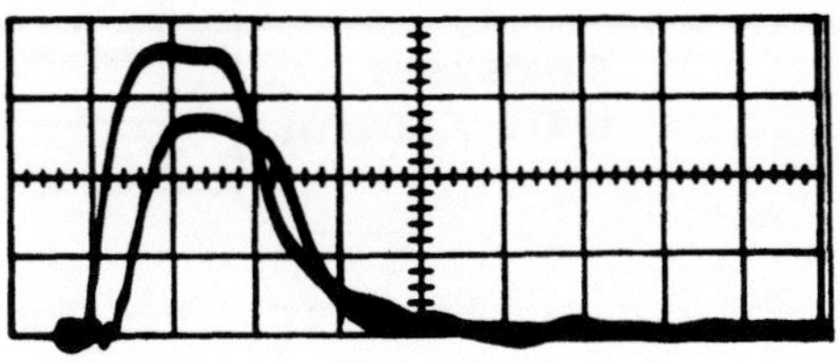

Fig. 3. Reflection, 35 Gc/sec.

(after Köchner).

Turning to experiments with magnetic fields we reproduce in Figure 4 a result of Larrabee. The receiver was tuned to 15 Gc/sec and the magnetic field inclined to the sample axis under 20°. The applied voltage was 183 Volts/cm and the current pulse of the order of $10^4 A/cm^2$. The lower output shows the microwave output with order of magnitude 1 microwatt. Note that the emission appears to persist after the current pulse, a feature which is probably partly due to instrumentation delays, such that the emission peak occurs at the end of the voltage pulse.

In our laboratory, Auchterlonie (5) has carried out extensive investigations of the emission phenomena with n-type and p-type material. Broadly speaking, there is no essential difference in the performance of the two substances. He observed that the emission is delayed with respect to the start of the voltage pulse, beyond instrumental error. Above a certain voltage there is a sudden enhancement of emission. As soon as this appears the voltage across the sample shows irregular high frequency oscillations. This is illustrated in Figure 5. The uppermost trace shows the microwave output at 37.5 Gc/sec. before enhancement and the upper curve of the lower oscillogram shows the enhancement, while the lower trace shows the oscillations of the voltage pulse. For higher applied fields the oscillation amplitude grows considerably.

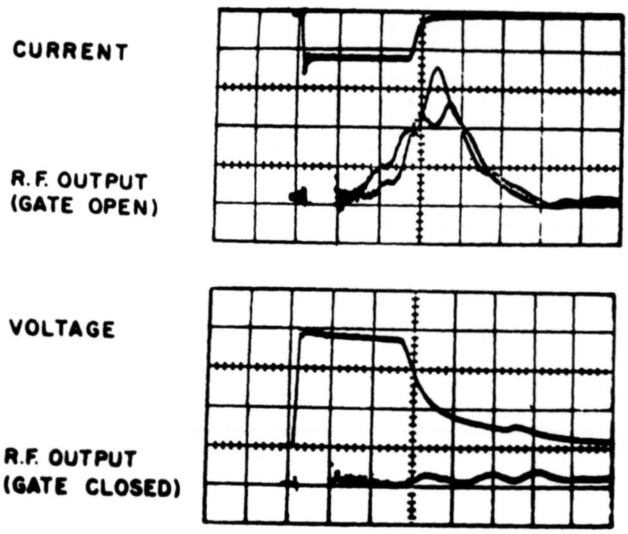

Fig. 4. Magnetic field experiment: frequency = 15 Gc/sec, applied voltage = 183 V/cm, current pulse = 10^4 A/cm^2 (after Larrabee).

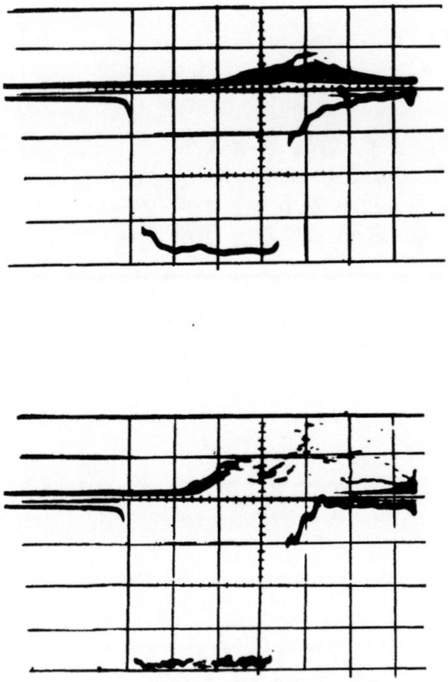

Fig. 5. H = 3.8 kGs
E and H inclined at 300°: 10 pulses

Horizontal scale: 0.1 μs/division
Top trace; vertical scale 1 V/division: microwave signal
Bottom trace: vertical scale 50 V/division: applied E pulse.

The sample size in this experiment was 2.24 x 0.45 x 0.49 mm. and the intrinsic carrier concentration $2x10^{12}$. It should be mentioned here that the character of the emission is very different for samples much thinner in one of the transverse dimensions.

With his samples Auchterlonie studied the emission as a function of the angle between the magnetic field and the applied field axis. Some of his observations are summarized in the polar plots of Figures 6 and 7 for n-type and p-type materials.

It will be seen that the "lobes" of the polar plot are not smooth but show sharp jumps in certain angular regions; one gains the impression that for different angles, different phenomena come into play.

Morisaki and Inuishi (6) carried out experiments with samples not very different from ours but of lesser purity (carrier concentration $1.5x10^{14}$). The magnetic field was perpendicular to the applied field axis and the observations were carried out by means of a microwave receiver with a total gain of 120 db, of bandwidth 5 mc/sec, tuned to a frequency of 9.4 Gc/sec. They summarize their experiment in a contour map shown in Figure 8.

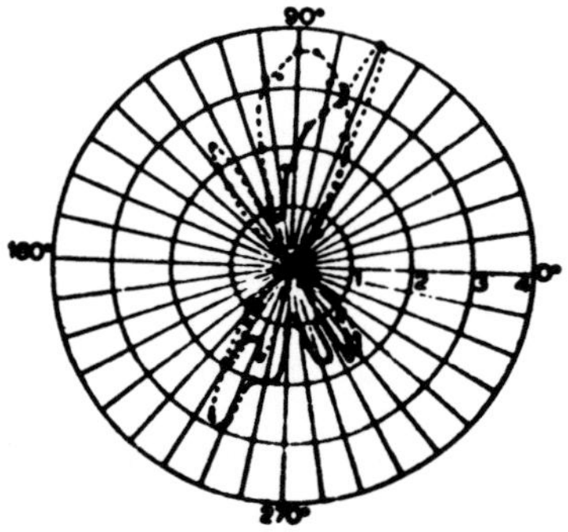

Fig. 6. $H = 3.8$ kGs
frequency = 34.80 Gc/s
—o— $E = 8.04 \times 10^4$ V/m
—●— $E = 8.48 \times 10^4$ V/m

Fig. 7. $H = 3.8$ kGs
$E = 5.05 \times 10^4$ V/m
—o— 34.66 Gc/s
—●— 34.80 Gc/s.

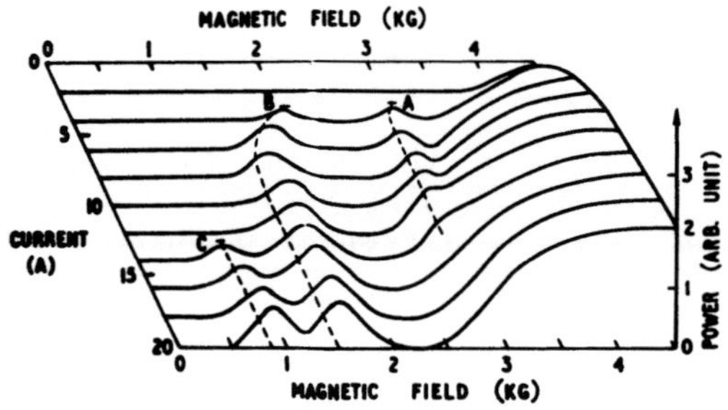

Fig. 8. Microwave power as a function of the applied current and magnetic fields.

The observed emissions fall into two categories. At rather high magnetic fields there is a region which appears above certain thresholds of applied voltage and magnetic fields, which has characteristics similar to those of the emission reported by Larrabee. The second category consists of emissions observed at magnetic fields of 29 kG (A), 1.5 kG (B), and 0.9 kG (C). These emissions (except for a slight change at low values) are not influenced by the current density, and the thresholds are so high that breakdown has certainly happened.

The frequency spectrum of these emissions from InSb is very broad. They have been observed at frequencies ranging from 2 to 35 kMc/sec.

K. Suzuki (7) has observed the frequency spectrum using very thin samples with dimensions 2.5 x 0.4 x 0.05 mm. in two regions: one (region I) from 7 - 26 Gc/sec, and another (region II) from 2-24 Gc/sec, with transverse magnetic field; he found maxima in the frequency spectrum in region I at 7,9,13.5,15.5,17.5,22,24 and 26 Gc/sec, and minima at 8,10.5,14.5,16.5,23.5, and 25.5 Gc/sec, independent of current density and between 3000 and 5400 Gauss independent of magnetic field strength and orientation around the axis. These may be interpreted with fair accuracy as harmonics of a fundamental frequency of 2.2 Gc/sec, and it seems unlikely that they are spurious resonances. Near this fundamental, i.e. in region I, he finds maxima and minima in the spectrum at intervals of 170 and 130 MC/sec; the emitted power was much lower in region I.

With thin samples (3 x 0.49 x 0.016) K. Suzuki (8) obtained emission spikes during the pulse, whose occurrence depended critically on the magnetic field; the magnetic field was oriented perpendicular to the sample axis and the smallest dimension axis. The voltage pulse showed an initial rise followed by a flat portion. The emission rose linearly as a function of the height of the latter. With somewhat thicker samples, 3.17 x 0.305 x 0.05 mm.. the initial peak in the voltage pulse did not appear. In both cases the slope of the current-voltage curve shows sharp changes as the onset of the emission. The power increases with current density for fixed magnetic field, and the threshold current density decreases with magnetic field. There was almost no emission when the magnetic field pointed in the applied field direction, and the angular dependence shows fairly smooth lobes rising to a maximum when the magnetic field makes an angle of 90° with the axis.

There is yet another emission phenomenon which was observed by Buchsbaum, Chynoweth and Feldmann (9). This occurs with n-type samples at low fields with a threshold of 12 V/cm and a magnetic threshold of 1300 gauss. It was detected with a low noise receiver tuned to 4.2 Mc/sec (noise temperature 160°K and bandwidth 30 Mc/sec). It is characteristic of this phenomenon that the highest power is radiated when the magnetic field is parallel to the axis, and almost none when it is perpendicular. The intensity of the emission was measured to be equivalent to that of a black body at 10^5°K. For p-type material the performance is similar, but although the threshold field is higher (45 V/cm.) it is still insufficient to cause breakdown. The dimensions of the sample were 9 x 1 x 1 mm. It was shown that the microwaves were polarized with the E-vector parallel to the sample axis.

Eidson and Kino (10) have also found this type of emission, but with samples 1 x k x 0.25 with tin contacts; they also observed narrow band emissions at low electric field strength depending on the polarity of the applied voltage which suggests that the nature of the contact is important.

At very low fields Musha, Lindvall and Hägglund (11) have, above a threshold of 3 V/cm, observed short emission spikes during a 3 m sec. pulse. This phenomenon too seemed to depend on the nature of the contact and was confined to small magnetic field angles.

THEORETICAL WORK RELATED TO THE EMISSION PHENOMENA

No D.C. Magnetic Field

The search for microwave emissions was stimulated by a theoretical paper by Pines and Schrieffer (12) on two stream instabilities in solid state plasma associated with electron drift under the application of an electric field. They discuss whether double stream instabilities of a nature similar to those known in gaseous plasma can exist in semiconductors or metals. The model then is an electron-hole plasma in which the holes of density n_+ are endowed with an effective mass m_+ and the electrons with density n_- correspondingly with effective mass m_-, while τ_+ and τ_- are their lifetimes against extraneous scattering mechanisms. It is clearly necessary to have a substance with $\omega\tau_\pm > 1$ if one wishes to find emissions at a frequency ω. Several authors, e.g. Jackson, have studied the condition for the starting of double stream instabilities in an electron ion plasma, and they find that the drift velocity v_d of the electrons must exceed 1.32 v_- where v_- is the thermal velocity of the electrons. This is for the case of equal temperatures T_+ and T_- of the charge carriers. When they are very unequal the necessary drift velocity is reduced, and is of the order $(m_-/m_+)^{1/2}$ v_- or smaller.

In InSb the electron mobility is very high (10^5 cm^2/v sec), and the effective masses being $m_- = 0.013\ m_e$, $m_+ = 0.18\ m_e$, the ratio m_+/m_- is large ($\sim$14); the hole mobility is consequently an order of magnitude smaller than the electron mobility, and as a result one gets much less hole heating than electron heating, so that large values of T_-/T_+ are favored.

The dispersion equation is derived in the usual way from the collisionless Boltzmann equation and Poisson's equation. Choosing a coordinate system in which the holes have zero drift velocity they obtain

$$1 + \frac{k_+^2}{k^2} W(z_+) + \frac{k_-^2}{k^2} W(z_-) = 0 \qquad (1)$$

where

$$z_+ = \frac{\omega}{kv_+}, \quad z_- = \left(\frac{\omega - \underline{k}\cdot\underline{v}_d}{k\cdot v_-}\right)$$

$W(z)$ is the well known plasma dispersion function, and the Debye wave numbers $k_\pm$ are defined by

$$k_\pm^2 = 4\pi\, n_\pm\, e^2/e_o\, \kappa T$$

For $T_+ = T_-$ and real ω, Jackson shows that there is an instability: the wave number k is complex, i.e. waves are exponentially growing. For infinitely long wavelength they begin to grow for drift velocities larger than

$$v_d = 0.926\, v_-\left[1 + (m_-/m_+)^{\frac{1}{2}}\right]$$

Physically, the instability arises because, when the phase velocity of the waves is smaller than the drift velocity around which the electron distribution function is centered there are more electrons with higher than with lower energy (the distribution function is rising), and thus there is a population inversion. (It may also be shown that the electrons stay a longer time in the higher energy states).

Pines and Schrieffer discuss the case of unequal temperatures $T_+ << T_-$ and unequal densities ($n_+ \neq n_-$) and they show that the critical drift voltage for the excitation of plasma oscillations can be made much lower. They show that for the best situations in semiconductors T_-/T_+ is between 5 and 10. The electrons would have to have a drift velocity between 0.22 and 0.40 times v_- to excite a growing wave instability. Collisions have the effect of lowering the net growth rate. The maximum growth rate to be obtained is shown to be of the order of $\omega_+/15$ where ω_+ is the hole plasma frequency and it is thus required that $\omega_+\tau_+ \geqslant 15$. We cannot here reproduce the detailed arguments regarding scattering mechanisms and heating processes which leads the authors to the conclusion that it is marginally possible but somewhat doubtful whether the conditions for the occurrence of the two-stream instability can be met in the most promising material, InSb.

In the absence of a D.C. magnetic field emission of waves has not been reported, but it appears likely that Köchner's observation of impedance changes show that a double stream instability can be driven although it will not appear spontaneously.

The fact that Lazarus, in our laboratory, has observed enhancement in reflection leads us to think that an active process is involved in the impedance change. Further work is necessary to confirm this view.

D.C. Magnetic Field

Magnetic field parallel to applied field. We now turn to theories involving magnetic fields. Bok and Nozieres (13) have studied the case of a magnetic field parallel to the drift velocity. A plasma can support transverse waves in this case which are known as whistler modes by workers in the field of atmospheric propagation of radio waves, and as helicon waves by workers in solid state plasma physics. From Boltzmann's equation they obtain, in quasistatic approximation, i.e. when $\omega \ll \omega_c$, the cyclotron frequency $k\, v_d \ll \omega$, the dispersion equation

$$k^2c^2 = \frac{\omega_+^2\,(\omega - k\,v_{d+})}{-\nu\omega_{c+} + i/\tau_+} + \frac{\omega_-^2\,(\omega - k\,v_{d-})}{-\nu\omega_{c-} + i/\tau_-} \tag{2}$$

where ω_{c+} and v_{d+} are the cyclotron frequency and drift velocity of the holes which have a sign opposite to that of the corresponding quantities v_{d-} and ω_{c-} of the electrons, and $\nu = +1$ or -1 correspond to the two circular polarization directions of the waves (for $\omega\nu > o$, E rotates in the same direction as electrons in the magnetic field). The condition for instability which they formulate is that for the solution of (2), Im (ω) should be negative. They obtain

$$\omega_+^2\tau_+\,(1 + \omega_{c-}^2\tau_-^2) + \omega_-^2\tau_-\,(1 + \omega_{c+}^2\tau_+^2)\,k^2c^2 + \nu\,k\omega_+^2\omega_-^2\tau_+\tau_-(\omega_{c-}\tau_- - \omega_{c+}\tau_+)(v_{d+} - v_{d-}) < 0 \tag{2a}$$

which can only be satisfied if

$$k\nu(v_{d+} - v_{d-}) < 0$$

and since v_{d+} and v_{d-} have opposite signs only one polarization of the waves can lead to instability for a given applied electric

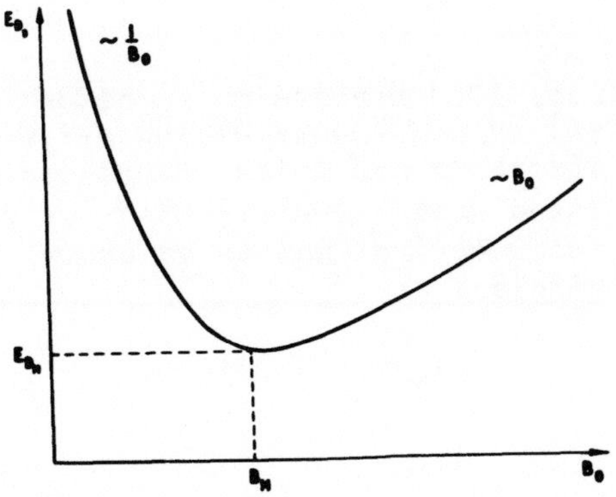

Fig. 9. The threshold d.c. electric field E_{Do} as a function of the longitudinal d.c. magnetic field B_o.

field direction. From (2a), expressing the drift velocities in terms of the mobilities and electric fields, follows the threshold value of the electric field as a function of the magnetic field, which is plotted in Figure 9.

Such minima have not been observed, and in the case when the threshold shows a magnetic field dependence, as in the experiments of K. Suzuki, the maximum emission occurred with the magnetic field perpendicular to the drift axis, and there was none when it was parallel to it. In InSb the highest drift velocities are of the order of 1.5 times the electron thermal velocities. Under such circumstances one would expect strong Landau damping. It is known, on the other hand, that there is no Landau damping for propagation perpendicular to the magnetic field.

Magnetic field parallel to applied electric field. A case of orientation of magnetic field perpendicular to the drift axis was analyzed by Hasegawa (14), who assumes propagation of the waves in the direction of the drift axis. He obtains (from the Lundquist equations without temperature terms) the dispersion equation

$$1 - \left\{\omega_-^2/\left[(\omega-kv_{d-})^2 - \omega_{c-}^2\right]\right\} - \left\{\omega_+^2/\left[(\omega-kv_{d+})^2 - \omega_{c+}^2\right]\right\} = 0 \qquad (3)$$

in quasistatic approximation. Geometrically the zeros of the denominators in equation 3 correspond to two pairs of straight lines in the ω, k plane with four intersections (Fig. 10). The lines may be thought of as representing hybrid modes whose intersections is governed by equation 3

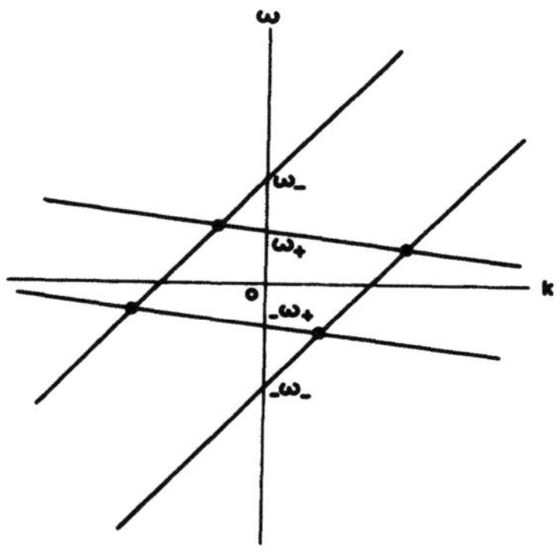

Fig. 10. Dispersion relation of the hybrid waves of the drifting electrons and holes. Couplings of these waves are expected at the intersecting points as shown by circles.

Exploring the interaction region by expanding ω around the points he obtains

$$\delta\omega = \pm\left[\omega_-^2\ \omega_+^2/2(\omega - kv_{d-})(\omega - kv_{d+})\right]^{1/2} \tag{4}$$

and the growth rate of the instability is given by

$$\delta\omega = \pm\left[\omega_-\ \omega_+/\left\{(\omega_-^2 + \omega_{c-}^2)(\omega_+^2 + \omega_{c+}^2)^{1/2}\right\}^{1/2} j\right] \tag{5}$$

So far the theory does not take temperature effects into account. When this is done using the Boltzmann equation and solving it in the manner of Bernstein (15), the minimum frequency ω_m for instability is obtained for $\omega_m - k_m v_{d-} < 0$. As expected from well known kinetic power theorems, the slower wave is responsible for the interaction. The minimum frequency ω_m is given by

$$\omega_m = \frac{2\omega_+^2\tau_+}{\omega_-^2\tau_-}\ \frac{\omega_{c-}}{F_1(k_m v_\pm/\omega_{c-})} + k_m v_{d+} \tag{6}$$

where

$$F_1(x) = \frac{2}{x^2}\, e^{-x^2}\, I_1(x) \tag{7}$$

and I_1 is a modified Bessel function. Assuming $n_+ = n_-$, the ratio $\omega_+^2\tau_+/\omega_-^2\tau_-$ equals the ratio μ_+/μ_- of the mobilities. With the ratio $\mu_-/\mu_+ = 200$ and with $v_{d-}/v_- = 0.45$, $1/F_1$ reaches its smallest value ~ 55. Equation 6 holds only provided that

$$\tau_-\omega_-^2 F_1\ (k_m v_-/\omega_{c-}) \geq 4\omega_{c-} \tag{8}$$

and with the above values this becomes

$$\omega_-^2\tau_-/\omega_c \gtrsim 200 \tag{9}$$

which cannot be satisfied with densities $n_+ = n_- \sim 10^{20}$ and the known electron lattice and hole lattice collision rates $\nu_+=1/\tau_+ = 2 \times 10^{12}$, $\nu_- = 1/\tau_- = 2 \times 10^{11}$/sec. We shall see later that there exist density gradients and that the density reaches 10^{22} at the semiconductor surface. A density gradient is caused by the Suhl effect. It is well known that the magnetic field which deflects oppositely charged counterstreaming holes and electrons into the same direction causes a density gradient leading to high densities near one of the semiconductor surfaces. The space charge appearing at the surface causes the so-called Hall electric fields to appear near that surface.

Before passing on to the theories of K. Suzuki or of McNeill (16) of our laboratory which bring density gradients and Hall voltages into play, we note that the transport equation theory of the type just discussed leads to a dispersion equation which has infinitely many solutions associated with the cyclotron harmonic frequencies and corresponding instabilities. In a gaseous plasma, Etiévant et al. (17) have observed instabilities due to the interaction of many cyclotron harmonic helicon waves and space charge modes of counterstreaming electrons.

McNeill starts from the macroscopic equations

$$\delta n_\pm/\delta t + \nabla . (n_\pm \underline{v_\pm}) = 0 \tag{10}$$

$$n_\pm m_\pm \left[\delta \underline{v_\pm}/\delta t + \underline{v_\pm}.\nabla \underline{v_\pm}\right] \mp n_\pm q\left[\underline{E} + \underline{v_\pm} \times \underline{B}\right] - \nabla\hat{\psi} = 0 \tag{11}$$

where $\hat{\psi}$ is the pressure tensor which he simplifies in accordance with the adiabatic isotropic approximation.

He first obtains a steady state solution which yields the density distribution

$$n_o(x) = N_o \exp - \alpha(d-x),\ \alpha = - \frac{E_x(|y_+| + |y_-|)}{\kappa(T_+ + T_-)} q \tag{12}$$

where d is the thickness of the sample in the direction of the Lorentz force and y the product of mobility and magnetic field. A small perturbation of this distribution causes charge separation which in turn sets up an electric field which induces a plasma flux in the x-direction. However, since there is a density gradient in this direction, the divergence of this flux is not zero and it turns out that the gradient is further increased. Instabilities arise when the increase in the perturbation due to this Hall flux exceeds the losses due to recombination and diffusion.

The waves are considered to propagate in the z-direction, the direction of the applied field. Accordingly

$$n' = An_o \exp i(\omega t - kz), \quad E' = B \exp i(\omega t - kz) \tag{13}$$

Using a formalism developed by the author the dispersion equation of the waves is found from Maxwell's equations in quasistatic approximation by inserting in these equations a current density

$$j = \hat{T}\,\underline{E} \tag{14}$$

where the conductivity tensor is given by

$$\hat{T} = \sum \frac{\omega_p^2 \, \epsilon_o}{i\omega \, \omega_{Dp}} \left\{ 1 - \frac{\omega_{cp}}{\omega_{Dp}^2} + \frac{\gamma \, \kappa \, T_p}{m_p \, \omega_{Dp}^2} \left(\frac{\alpha k \omega}{\omega_{Dp}} - k^2 \right) \right\}^{-1}$$

$$\begin{vmatrix} \omega_{Dp}\left(1 - \dfrac{k^2 \gamma \kappa T_p}{m_p \, \omega_{Dp}^2}\right), & 0 & , \dfrac{-i\omega \, \omega_{cp}}{\omega_{Dp}} \\ 0 & , \omega_{Dp}\left(1 - \dfrac{k^2 \gamma \kappa T_p}{m_p \, \omega_{Dp}^2}\right), & 0 \\ \dfrac{i\omega}{\omega_{Dp}} \left(\omega_{cp} - \dfrac{k\alpha\gamma\kappa T_p}{m_p \, \omega_{Dp}}\right), & 0 & , \dfrac{\omega^2}{\omega_{Dp}} \end{vmatrix} \tag{15}$$

in the absence of collisions, and the sum is taken over the charges of both species where in accordance with our previous notation p = sign q is - for electrons and + for holes. $\omega_{D\pm} = \omega - kv_{d\pm}$

The dispersion equation is given by

$$\sum \frac{\omega_p^2}{\omega_{Dp}^2 - \omega_{cp}^2 + \dfrac{\gamma\kappa T_p}{m_p} \left(\dfrac{\alpha k \omega_{cp}}{\omega_{Dp}} - k^2 \right)} = 1 \tag{16}$$

which may be written by means of the dimensionless parameters

$$w = \frac{\omega}{\omega_{c-}},\quad z = \frac{kv_{d-}}{\omega_{c-}},\quad R = \frac{m_-}{m_+},\quad T = \frac{|v_{d+}|}{|v_{d-}|},\quad S = \frac{\omega_-^2}{\omega_{c-}^2},\quad X = \frac{\gamma\kappa T}{mv_{d-}^2},\quad Y = \frac{-\alpha v_{d-}}{\omega_{c-}} \tag{17}$$

in the form

$$\frac{S}{(w-z)^2 - 1 - Xz^2 + XYz/(w-z)} + \frac{RS}{(w+Tz)^2 - R^2 - RXz^2 - RXYz/(\omega+Tz)} = 1 \tag{18}$$

In the case of the cold plasma without density gradient there will be interactions near the intersections of the four hybrid modes

$$(w - z)^2 = 1 + S \ ,\quad (w + Tz)^2 = RS + R^2 \tag{19}$$

When the carrier temperature is finite but the density gradient is neglected the dispersion relation represents two hyperbolas with their center at the origin; the diagram resembles Fig. 10, but the lines are now curved. Above a certain carrier temperature the intersections disappear, i.e. interaction ceases and the instabilities predicted by the two-stream analysis of cold plasma theory will be eliminated by diffusion.

In the presence of a density gradient, space charge waves travelling in synchronism with the drifting carriers can interact with the hybrid modes. Solutions have been obtained using parameters appropriate for indium antimonide, and the form of the interaction between the hole space charge wave and the hole hybrid mode has been determined. Computer results are presented in Figures 11 and 12 which differ only in the plasma frequency. They show that the interaction which occurs for the lower density of Figure 11 somewhat above the whole cyclotron disappears in the denser plasma.

The mechanism may be responsible for the emissions observed by Morisaki and Inuishi. The peaks seen in Figure 8 correspond roughly to the hole cyclotron frequency and its harmonics but only if the mass 0.6 m_c is used. There is some evidence for the existence of such heavy holes. Under similar conditions K. Suzuki found that for a thinner sample the threshold density for emission was lower than for a thicker one, and the intensity of the radiation was much greater in the former case. This may be explained by pointing out that according to our theory the plasma with higher density produced by the Suhl effect at the surface of the thick-

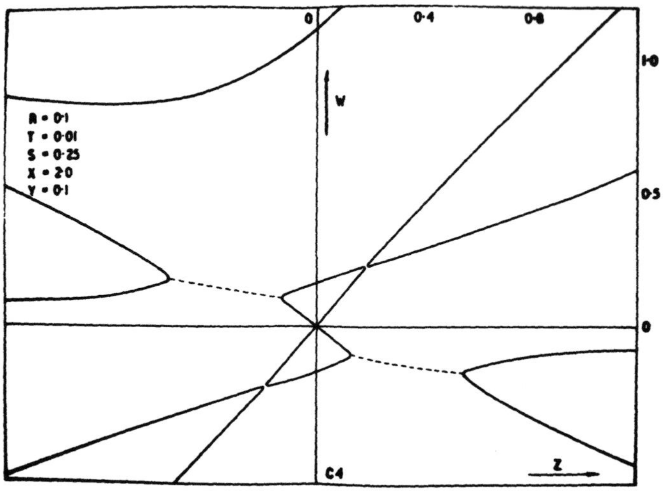

Fig. 11. Computor results for the case of indium antimonide (S = 0.25).

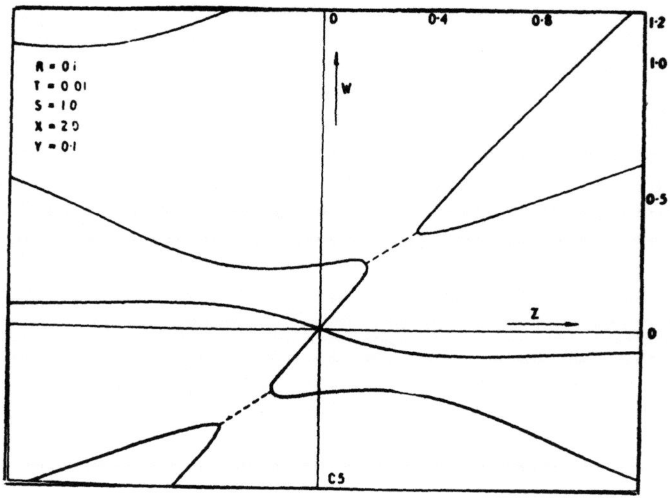

Fig. 12. Computor results for the case of indium antimonide (S = 1.0).

er plasma can absorb some of the radiation generated at greater depth. The variation of the output power with orientation of the magnetic field in a plane perpendicular to the drift axis can be similarly explained. The power is largest when the magnetic field is parallel to the surface of the thin plate and decreases rapidly as the angle is changed. One can see that the sample resembles a thick sample more and more as the field component in the thin direction increases.

A theory predicting features similar to those sketched above has been offered by T. Suzuki (18). It is rather different, however in that he assumes the propagation vector k to be parallel to the direction of the Hall voltage and the density gradient, i.e. normal to the surface of the flat platelet. We believe that this is rather the direction of propagation of the low frequency oscillations which, according to K. Suzuki, modulate the high frequency ones with about 150 Mc/sec, and which presumably correspond to the oscillations of the pulse top after a critical field has been applied, as observed by Auchterlonie and others.

In McNeill's theory it is assumed that the threshold for breakdown has been exceeded by the applied field and that, due to heating, the resistance has increased and the current has decreased until the generation rate of charge carriers is balanced by the recombination rate. On this assumption he has no source term in his continuity equation.

Toda(19), on the other hand, has based a model of an oscillation mechanism on dependence of the generation rate γ on the field strength, which he writes in the form

$$\gamma = aE^m \tag{20}$$

The recombination is field-independent. There is thus an A.C. term representing a source in the continuity equation. It is now assumed that, since it takes time for electrons to accelerate, there is a delay of the current phase with respect to the field phase, so that the A.C. source term is given by

$$\gamma' n_o = n_o m \gamma E'/E_o(1 + i\omega\tau) \qquad \omega\tau < 0 \tag{21}$$

and τ represents the delay. Toda then shows that the capacity between the end contacts, which can be deliberately increased by an external shunt capacity, together with this delay, can lead to a phase difference between the generation rate and a density perturbation greater than 90°, so that positive feedback results. He was able to demonstrate experimentally the existence of (damped) low frequency oscillations depending on the parameters as predicted by this theory. He also, in another paper, proposes a mechanism for microwave generation which applies in the case of a transverse magnetic field. For strong magnetic fields, the Hall electric field causes breakdown with the generation rate following a power law (20). This field, together with the magnetic field, causes a current change in the direction of the originally applied field. For certain frequencies, when the resulting phase relations result in positive feedback, instability occurs.

It will be seen that this mechanism may well exist at the verge of breakdown, when the intrinsic carriers predominate, so that space charge is created by the current which in turn sets up the Hall field. If the plasma densities created by breakdown far exceed the intrinsic density the dielectric behavior will have to be examined carefully.

Perhaps this mechanism is responsible for some of the oscillations near threshold.

The emissions in the high field region of Morisaki and Inuishi's contour map of Figure 8 occur in such a broad band and are so insensitive to field changes that a theory recently proposed by Steel (20) commends itself. He postulates that the emission is due to photoconductive mixing of infrared radiation components with frequency difference equal to the frequency of the observed microwaves. The infrared radiation results from recombination radiation, which is generated by a mechanism analogous to superradiance in a laser.

The population inversion which gives rise to the superradiance comes about by an increase in carrier density owing to breakdown. If the electron density is high enough, the quasi-Fermi level of the electrons will be situated above the bottom of the conduction band. Steel assumes that the quasi-Fermi level of the holes is situated near the top of the valence band owing to the much greater density of states available in that band. There thus exists a population inversion for recombination across the gap. The width of the spectrum is equal to the height of the quasi-Fermi level of the electrons above the gap.

If, now, different frequency components of the infrared spectrum to which these transitions correspond are to be mixed, the radiation must be re-absorbed and lead to photo emission. Photoemissive mixing of laser lines has been observed e.g. by Di Domenico, Pantell, Svelto and Weaver (21).

However, if the semiconductor is uniformly filled with holes and electrons, no strong reabsorption will take place, because the quantum energy available is only of the same order as that of the band gap. Here is where the magnetic field comes into the picture. By creating a density gradient due to the Suhl effect, it makes the quasi-Fermi level higher in the regions of greater density, and the radiation can be absorbed in the lower density region.

The superradiance produces radiation in a single pass. From this one can estimate the coherence time of the radiation as

$$\tau_\kappa = \frac{L\sqrt{e_L}}{c} \tag{22}$$

where L is the length of the crystal, and e_L the relative dielectric constant. For L = 1 cm, $\tau_\kappa = 10^{-10}$ sec. Steele, Larrabee and Hicinbotham have measured a coherence time of this order. Our own experiments indicate a much longer time for much shorter crystals. A voltage threshold is predicted by this theory as well as by McNeill's theory. Breakdown has to occur to raise the Fermi level above the band gap. Above a certain magnetic field sufficient to create a density gradient the output is expected to increase, because the photo-mixing process becomes more efficient, and then to saturate. It is difficult to see how the theory could account for the discontinuous change of emission with orientation of the magnetic field shown in Auchterlonie's polar plot of Figs. 6 and 7. This behavior could be more easily explained by a plasma instability.

The dispersion diagram of Figs. 11 and 12 shows a convective instability in the infrared. It is interesting to speculate whether this could, for high magnetic fields, provide an amplification mechanism for the superradiance. It may also be possible for plasma instabilities to amplify the mixer output.

Turning to the low field emissions, we do not propose to discuss theoretical interpretations in this paper. Attention is drawn to a theory of Musha and Bers (22) which explains these effects by the stimulated emission of phonons by the electrons drifting in a longitudinal magnetic field. The quantum mechanical treatment considers two processes: Transitions within a Landau level which are an acoustic analogue of the Cherenkov effect, and transitions between Landau levels coupled to the phonons. The emitted phonons must somehow couple to the wave guide field at the semiconductor surface.

ACKNOWLEDGMENTS

The preparation of this paper has been assisted by the Air Force Office of Scientific Research, under Grant AF 65-37, with the European Office of Aerospace Research (OAR), United States Air

Force. It is a pleasure to thank Mr. P. R. McNeill, Mr. L. Auchterlonie and Dr. M. J. Lazarus for the help they have given to the author.

REFERENCES

1. *R.D. Larrabee, Bull. Amer. Phys. Soc. II, 9 (1964) 258.*

2. *R. D. Larrabee and W. A. Hicinbotham, Symposium on Plasma Effects in Solids, Paris 1964 (Dunod 1965).*

3. *Betsy Ancker-Johnson, Plasmas in Semiconductors and Semimetals, Semiconductors and Semimetals, Vol. I Academic Press, N.Y.*

4a. *W. Köchner, Untersuchung der Verstärkungsmöglichkeit von Millimeterwellen in Halbleiterplasmen, Institut für Hochfrequenztechnik, Technische Hochschule, Wien.*

4b. *W. Köchner, Proc. IEEE 53 No. 9 (1965) 1234-1235.*

5. *L. Auchterlonie, Electr. Letters I No. 9 (1965).*

6. *H. Morisaki and Y. Inuishi, Japan J. Appl. Phys. 5 (1966) 343-344.*

7. *K. Suzuki, special issue on Semiconductor Bulk Effects and Transit Time Devices, Jan. 1966.*

8. *K. Suzuki, Japan J. Appl. Phys. 4 No. 1 (1965) 42-52.*

9. *S.J. Buchsbaum, A.G. Chynoweth and W.L. Feldmann, Appl. Phys. Letters 6 (1965) 67.*

10. *J. C. Eidson and G. S. Kino, Appl. Phys. Letters 8 (1966) 183-185.*

11. *T. Musha, F. Lindvall and J. Hägglund, Appl. Phys. Letters 8 (1966) 157-159.*

12. *D. Pines and J.R. Schrieffer, Phys. Rev. 124 (1961) 1387-1400.*

13. *J.Bok and P.Nozieres, J. Phys. Chem. Solids 24 (1963) 709-714.*

14. *A. Hasegawa, J. Appl. Phys. 36 (1965) 3590-3595.*

15. *I.B. Bernstein, Phys. Rev. 109 (1950) 10.*

16. *P.R. McNeill, Instabilities in a Drifting Semiconductor Plasma*

Proc. IEEE Letter accepted for publication.

17. *J. Olivain, C. Etiévant and M. Perulli, J. de Physique 26 (1965) 505-507.*

18. *T. Suzuki, IEEE Transaction on Electron Devices, Jan. 1965 202-203.*

19. *M. Toda, J. Appl. Phys. 37 (1966) 32-36, J. Appl. Phys. 37 (1966) 37-39.*

20. *M.C. Steele, RCA Review, June 1966 263-271.*

21. *R. H. Pantell, M. Didomenico, O. Svelto and J. N. Weaver, Quantum Electronics III, Paris Dunod and Columbia Press (1964) 1811-1818.*

22. *T. Mucha and A. Bers, Quart. Progr. Rpt. No. 79 MIT 1966.*

DISCUSSIONS

G. PAKE (Washington University, Missouri, U.S.A.): *What is the nature of the microwave frequency spectrum?*

H. MOTZ: *The spectrum is very broad. Each sample emits radiation in a band ranging from about 6 mm. to 3 cm. or more. The spectrum appears to have a periodic variation of intensity with frequency.*

RECENT PHOTON ECHO EXPERIMENTS

N.A. Kurnit and S.R. Hartmann

Columbia University,

New York, U.S.A.

A sample irradiated by resonant light pulses at $t = 0$ and $t=\tau$ can exhibit a macroscopic oscillating dipole moment at $t=2\tau$, resulting in the emission of a burst of coherent spontaneous radiation. Experiments are described which have been performed to study the effect of applied magnetic fields on the relaxation of these "photon echoes" in ruby. In the limit of magnetic fields large compared to the effective field ($\approx$ 1 kg) of the Cr ions at neighboring Al sites, one can expect that the Al nuclei will be "locked" out of the influence of the Cr ions and consequently will not degrade the photon echo as the field is tilted away from the optic axis. This conjecture is examined, and experiments bearing on its validity are presented.

INTRODUCTION

Since the initial detection of photon echoes in ruby (1), a number of experiments have been performed to study echo relaxation caused by phonon-induced transitions in the R_1 level (2) and by spin-spin interactions with the neighboring Al nuclei (3). The latter experiments have now been extended by the application of magnetic fields greater than four kilogauss, and several new phenomena have been observed which will be discussed in this paper. These results are supplemented by a study of the relaxation of recently observed stimulated photon echoes (4).

The photon echo is the optical frequency analog of the spin echo (5), and it can be explained in the same way. In our experiments two intense pulses from a Q-switched liquid N_2- cooled ruby laser are made incident on a liquid He-cooled ruby sample with the result that a third pulse is generated by the ruby sample delayed from the second excitation pulse by the time between the excitation pulses. The photon-echo mechanism is quite straightforward as one need only conisder an ensemble of two-level systems interacting with a classical electromagnetic field (2,3). The atoms are initially in their ground state represented by a wave function

$$\psi_o > = \psi_- >_j \quad , \tag{1}$$

where the subscript j refers to the jth atom.

The first excitation pulse is adjusted to put each atom in a linear superposition of the ground and excited states:

$$\psi_{90}(t) >_j = \frac{1}{\sqrt{2}} \psi_+ >_j \; e^{-iE_{+j}t/\hbar} + \frac{1}{\sqrt{2}} \psi_- >_j \; e^{-iE_{-j}t/\hbar} \tag{2}$$

where $\psi_{90}(t)$ is the wave function at a time t after this 90° excitation pulse. $\psi_+ >$ is the wave function of the excited atom, and E_{-j} and E_{+j} are the energies of the ground and excited states respectively. The atom in the state described by Eq.(2) now exhibits an electric dipole moment since there is a matrix element of the electric dipole moment operator between the $\psi_+ >$ and $\psi_- >$ wave functions. Since all the atoms have been excited in phase, their moments sum to yield a macroscopic oscillating electric dipole moment. This dipole moment dephases quickly, however, since crystalline field strains produce a distribution in the energy separations $(E_{+j} - E_{-j})$ of the excited atoms. If a second pulse is applied at a time τ with twice the electric field vector as the first but of the same duration, the functions $\psi_+ >$ and $\psi_- >$ are interchanged to yield

$$\psi_{180}(t) >_j = \frac{1}{\sqrt{2}} \psi_+ >_j \; e^{-iE_{+j}(t-\tau)/\hbar} \; e^{-iE_{-j}\tau/\hbar}$$
$$- \frac{1}{\sqrt{2}} \psi_- >_j \; e^{-iE_{-j}(t-\tau)/\hbar} \; e^{-iE_{+j}\tau/\hbar} \quad , \tag{3}$$

which results in an oscillating electric dipole moment given by

$$<\vec{\rho}\,(t) >_j = - \mathrm{Re}\,\{<+|P|-> e^{i(E_{+j} - E_{-j})(t-2\tau)/\hbar}\} , \tag{4}$$

where $<+|P|->$ is the matrix element of the dipole moment operator between the ground and excited states. The echo is then produced at $t = 2\tau$ since at this time all the dipole moments will be in phase independent of their energy separation $(E_{+j} - E_{-j})$.

The application of three unequally spaced pulses leads to the stimulated echo. Here again the echo formation proceeds just as it does for spin echoes and can be analyzed in the same way (5,6). The stimulated echo is formed at a time after the third excitation pulse equal to the separation between the first two pulses and can be most simply thought of as being formed by a 90° pulse followed by a 180° pulse which is made up of two temporally separated 90° pulses. Relaxation of the stimulated echo during the time between the last two excitation pulses depends only on the net shift in the energy separation ($E_{+j} - E_{-j}$) during this interval and not on its detailed time development. One therefore expects the stimulated echo to be much longer lived than the ordinary photon echo. Of course, in ruby the ground and excited states are multiplets, and one must also consider the possibility of transitions within their levels which may relax the stimulated echo.

RESULTS AND DISCUSSION

The apparatus used to generate the stimulated echoes is the same as that used for ordinary photon-echo generation except for the addition of a second delay line and is shown schematically in Figure 1. A 200-kW 10-nsec pulse from the ruby laser is split into three parts which are directed onto the ruby sample 1) directly, 2) through the left delay line, and 3) through the right delay line. The crystal used in these experiments was 0.05% by weight Cr_2O_3, and its optic axis was oriented perpendicular to the direction of propagation of the light. The Kerr-cell shutter placed before the photomultiplier prevents it from saturating as does the aperture stop which blocks off the excitation pulses and allows the echo through by making use of their spatial separation properties which are similar to those of ordinary photon-echoes (3). Fig. 2a shows a typical oscilloscope trace of the stimulated echo.

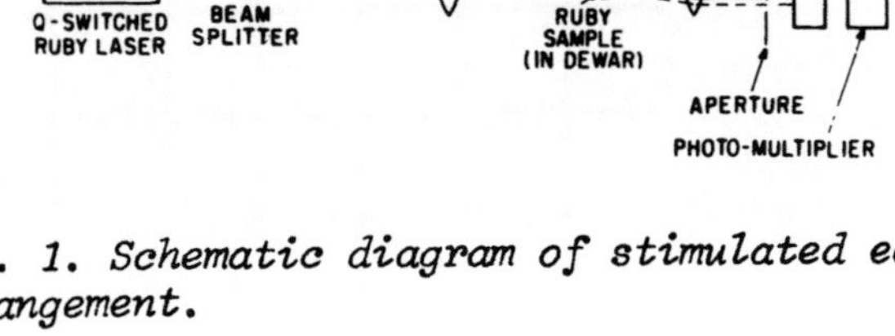

Fig. 1. Schematic diagram of stimulated echo arrangement.

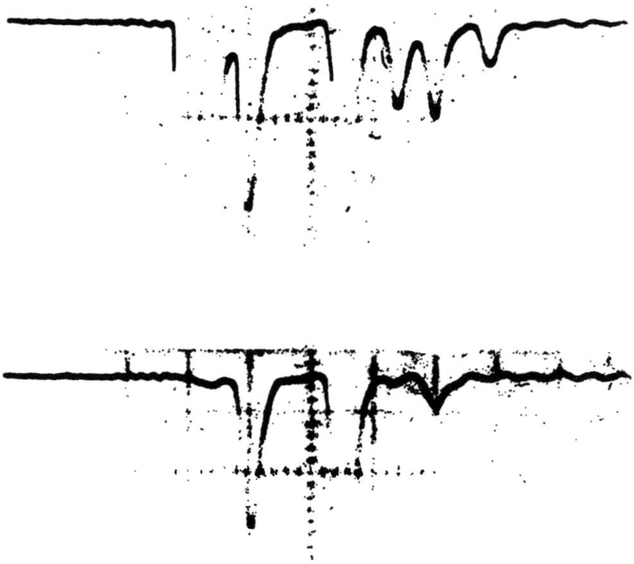

Fig. 2 a) Observation of a stimulated photon echo. Kerr cell is opened after third excitation pulse to permit observation of, respectively, stimulated echo, echo due to pulses (2) and (3), and echo due to pulses (1) and (3). Time scale is 50nsec/ division. First pulse is blocked in bottom trace.

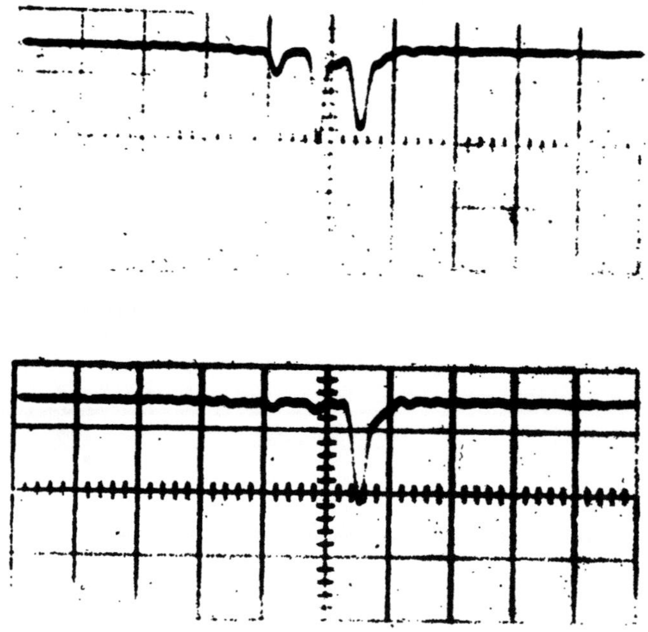

Fig. 2b) Observation of photon echo in zero applied field (top trace) and in field of 500 G (bottom trace, gain reduced by ten) 34nsec pulse separation.

The initial photon-echo experiments demonstrated the necessity for applying a magnetic field in order to observe photon echoes (1) and showed that the effect of the magnetic field is to inhibit a relaxation mechanism (3). It has now been possible to reduce the saturation effects of closely spaced excitation pulses sufficiently to enable the observation of photon echoes in zero magnetic field. A typical result is shown in Figure 2b. The values of the zero-field and high-field echo amplitudes allow one to estimate a relaxation time of the order of 20 nsec in zero magnetic field.

A series of experiments was carried out to measure the amplitude of the photon echo as a function of the applied magnetic field, and the results of three experiments are shown in Figure 3. With the magnetic field H_z applied along the optic axis, the experimental data are well approximated for τ of the order of $\leqslant$100nsec by a curve of the form

$$S = S_0 \; e^{-\alpha(\tau)\frac{H_L^2}{H_L^2 + H_z^2}} \tag{5}$$

where the field $H_L \approx 15$ G is comparable to the ESR line width (7). Superimposed upon these curves are two sharp, pronounced dips in echo amplitude at 2 and 4 kG associated with level crossings in the ground state (8). Earlier experiments (3) have shown that as the magnetic field is tilted away from the optic axis, the echo amplitude falls off nearly exponentially with angle. When the magnetic field is increased to above 1 kG, however, it is now found that a transverse component of the magnetic field is no longer as effective in destroying the echo. This is illustrated by the third curve in Fig. 3 in which the magnetic field is misaligned by 2° from the optic axis. The rise in echo amplitude at $\approx$ 2 kG is believed to be associated with the applied magnetic field at the Al sites becoming larger than the effective field due to the Cr^{3+} ions at these sites (7).

The initial argument (3) used to explain the effect of magnetic fields reasoned that since the g value of the Cr^{3+} electron in the ground state was isotropic and in the excited state was anisotropic, then one would expect that the orientation of the Cr^{3+} magnetic moment would be nonadiabatically changed on being excited by the excitation pulses, resulting in a sudden change of hyperfine field direction at the Al sites. If one postulates that the Al nuclei are effectively polarized in this hyperfine field, then the reorientation of the field will cause a forced precession of the Al moments about the new field resulting in a modulation of the Zeeman splitting at the Cr^{3+} sites which will relax the echo. Although these arguments are suspect, they do suggest that if one

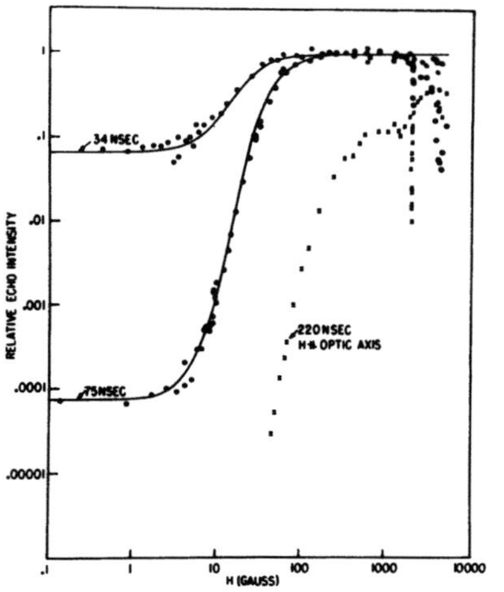

Fig. 3. Amplitude of photon echo as a function of applied magnetic field. Field is parallel to optic axis for 34 and 75nsec pulse separations and is misaligned by 2° from optic axis for 220 nsec sepation.

were to apply a magnetic field large compared to the hyperfine field of the Cr^{3+} at the Al sites, then the reorientation of the Cr^{3+} moments would not play an important role, and the echo would not be relaxed on their reorientation. This is what has now been observed experimentally. However, since one cannot expect any appreciable Al polarizations at the fields and temperatures used, we propose to suggest an alternative explanation of the magnetic field relaxation results.

A proper treatment of the photon-echo relaxation involves a density matrix calculation that takes into account the multiplicity of the levels associated with the ground and excited states. A general calculation that neglects interactions of Al nuclei shows that the application of a magnetic field of any magnitude and orientation does not relax the echo. This problem is solvable in closed form and shows that after a small transient effect which will average out if the field is inhomogeneous, one always obtains an unrelaxed echo. The inclusion of Al interactions to provide time-dependent relaxation fields at the Cr^{3+} sites complicates the problem, but two important features can be displayed. When the applied magnetic field becomes larger than the internal local fields at the Cr^{3+} sites, one must eliminate interactions in which the Cr^{3+} electron flips in the applied field as they do not conserve energy. When the applied field is larger than the hyperfine fields at the Al nuclei, then those interactions involving Al spin flips which do not conserve energy in this field must also be eliminated. The successive elimination of certain spin-spin interactions at higher fields reduces the effectiveness of the echo dephasing mechanism, and consequently one observes larger echo amplitudes as the magnetic field is increased, particularly at fields corresponding to: 1) the local field at the Cr^{3+} sites and 2) the local field at the Al sites.

Preliminary results showing the relaxation of the stimulated echo are given in Fig. 4. These results are given to indicate the expected relative independence of the echo amplitude on the pulse separation of the last two excitation pulses.

An additional result that we wish to report is that of being able to produce photon echoes with excitation pulses polarized parallel to the optic axis of the ruby. The photon echoes so produced are also polarized parallel to the optic axis, and it is found that these echoes are enhanced by the application of a small magnetic field along the optic axis in essentially the same way as the echoes produced in the standard way. These new echoes are two orders of magnitude smaller than the regular echoes, due in part to their smaller dipole matrix element. An important feature of these results was that the amplitude of the new "parallel" echoes as a function of the angle between the magnetic field and the optic axis was essentially the same as that observed with the ordinary perpendicular echoes.

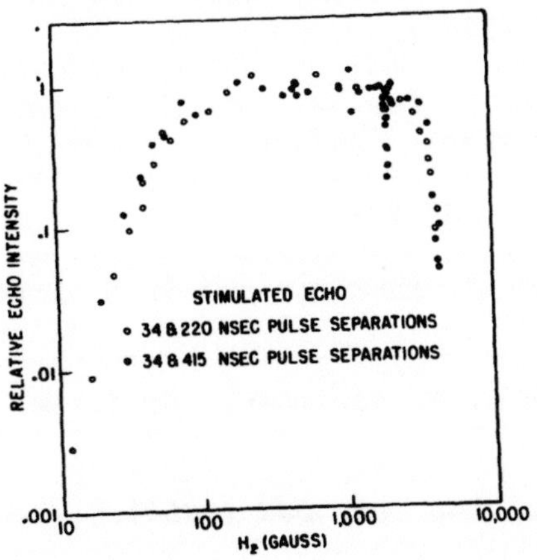

Fig. 4. Amplitude of stimulated echo as a function of magnetic field applied along optic axis, first two pulses separated by 34 nsec, first and third by 220 and 415 nsec.

CONCLUSION

The experimental results have indicated that the relaxation of the photon echo is inhibited by the presence of a magnetic field and that there are two critical fields at which the character of the relaxation changes. The critical fields are the $\approx$ 12-G local field at the Cr^{3+} sites and the $\approx$ 1-kG field at the neighboring Al sites. Below 12 G the echo is relaxed with a time constant of the order of 20 nsec but is observable if sufficient care is taken experimentally. As the field is increased above $\approx$ 12 G and then $\approx$1 kG, the effectiveness of the relaxation decreases since conservation of energy serves to eliminate certain relaxation interactions. The stimulated echo is observed to be relatively stable as expected.

ACKNOWLEDGMENT

This work was supported in part by the Army Research Office-Durham under Contract DA-31-124-ARO-D-341, and in part by the Joint Services Electronics Program under Contract DA-28-043 AMC-00099(E).

REFERENCES

1. *N. A. Kurnit, I. D. Abella, and S. R. Hartmann, Phys. Rev. Letters 13, 567 (1964).*

2. *N. A. Kurnit, I. D. Abella, and S. R. Hartmann, Physics of Quantum Electronics - Conference Proceedings, eds., P. L. Kelley, B. Lax, and P. E. Tannenwald (McGraw-Hill Book Co., N. Y., 1966), pp. 267-279.*

3. *I. D. Abella, N. A. Kurnit, and S. R. Hartmann, Phys. Rev. 141, 391 (1966).*

4. *N. A. Kurnit and S. R. Hartmann, Bull. Am. Phys. Soc. II, 11, 112 (1966).*

5. *E. L. Hahn, Phys. Rev. 80, 580 (1950).*

6. *A. K. Saha and T. P. Das, Theory and Applications of Nuclear Induction (Saha Institute of Nuclear Physics, Calcutta, India, 1957).*

7. *N. Laurence, E. C. McIrvine, and J. Lambe, J. Phys. Chem. Solids 23, 515 (1962); A. A. Manenkov and V. B. Fedorov, J. Exptl. Theoret. Phys. 38, 1042 (1960) (Soviet Physics JETP 11, 751 (1960)).*

8. *E. O. Schulz-Dubois, Bell System Tech. J. 38, 271 (1959).*

DISCUSSION

H. MOTZ (St. Catherine's College, Oxford, ENGLAND): *How did you obtain Q switching in ruby at 77°K? We spent a year learning just how to do this.*

N.A. KURNIT: *It certainly is difficult, but we were satisfied to obtain pulses of 100 to 200 KW intensity rather than the much higher intensity pulses obtainable at room temperature.*

MIX
Papier aus verantwortungsvollen Quellen
Paper from responsible sources
FSC® C105338

If you have any concerns about our products,
you can contact us on
ProductSafety@springernature.com

In case Publisher is established outside the EU,
the EU authorized representative is:
Springer Nature Customer Service Center GmbH
Europaplatz 3, 69115 Heidelberg, Germany

Printed by Libri Plureos GmbH
in Hamburg, Germany